HANDBOOK OF WATER AND WASTEWATER TREATMENT TECHNOLOGIES

Nicholas P. Cheremisinoff, Ph.D.
N&P Limited

Boston Oxford Auckland Johannesburg Melbourne New Delhi

A member of the Reed Elsevier group

Recognizing the importance of preserving what has been written, Butterworth–Heinemann prints its books on acid-free paper whenever possible.

ISBN: 0-7506-7498-9

The publisher offers special discounts on bulk orders of this book.
For information, please contact:

Manager of Special Sales
Butterworth–Heinemann
225 Wildwood Avenue
Woburn, MA 01801-2041
Tel: 781-904-2500
Fax: 781-904-2620

For information on all Butterworth-Heinemann publications available, contact our World Wide Web home page at: http://www.bh.com

10 9 8 7 6 5 4 3 2 1

Printed in the United States of America

CONTENTS

Preface

This volume covers the technologies that are applied to the treatment and purification of water. Those who are generally familiar with this field will immediately embrace the subject as a treatise on solid-liquid separations. However, the subject is much broader, in that the technologies discussed are not just restricted to pollution control hardware that rely only upon physical methods of treating and purifying wastewaters. The book attempts to provide as wide a coverage as possible those technologies applicable to both water (e.g., drinking water) and wastewater (i.e., industrial and municipal) sources. The methods and technologies discussed are a combination of physical, chemical and thermal techniques.

There are twelve chapters. The first of these provides an orientation of terms and concepts, along with reasons why water treatment practices are needed. This chapter also sets the stage for the balance of the book by providing an organizational structure to the subjects discussed. The second chapter covers the A-B-Cs of filtration theory and practices, which is one of the fundamental unit operations addressed in several chapters of the book. Chapter 3 begins to discuss the chemistry of wastewater and focuses in on the use of chemical additives that assist in physical separation processes for suspended solids. Chapters 4 through 7 cover technology-specific filtration practices. There is a wide range of hardware options covered in these three chapters, with applications to both municipal and industrial sides of the equation. Chapter 8 covers the subjects of sedimentation, clarification flotation, and coalescence, and gets us back into some of the chemistry issues that are important achieving high quality water. Chapter 9 covers membrane separation technologies which are applied to the purification of drinking water. Chapter 10 covers two very important water purification technologies that have found applications not only in drinking water supply and beverage industry applications, but in groundwater remediation applications. These technologies are ion exchange and carbon adsorption. Chapter 11 covers chemical and non-chemical water sterilization technologies, which are critical to providing high quality drinking water. The last chapter focuses on the solid waste of wastewater treatment - sludge. This chapter looks not only at physico-chemical and thermal methods of sludge dewatering, but we explore what can be done with these wastes and their impact on the overall costs that are associated with a water treatment plant operation. Sludge, like water, can be conditioned and sterilized, thereby transforming it from a costly waste, requiring disposal, to a useful byproduct that can enter into secondary markets. Particular emphasis is given to pollution prevention technologies that are not only more environmentally friendly than conventional waste disposal practices, but more cost effective.

What I have attempted to bring to this volume is some of my own philosophy in dealing with water treatment projects. As such, each chapter tries to embrace the individual subject area from a first-principles standpoint, and then explore case-specific approaches. Tackling problems in this field from a generalized approach oftentimes enables us to borrow solutions and approaches to water treatment from a larger arsenal of information. And a part of this arsenal is the worldwide Web. This is not only a platform for advertising and selling equipment, but there is a wealth of information available to help address various technical aspects of water treatment. You will find key Web sites cited throughout the book, which are useful to equipment selection and sizing, as well as for troubleshooting treatment plant operational problems.

Most chapters include a section of recommended resources that I have relied upon in my own consulting practice over the years, and believe you will also. In addition, you will find a section titled ***Questions for Thinking and Discussing*** in eleven of the twelve chapters. These chapter sections will get you thinking about the individual subject areas discussed, and challenge you into applying some of the calculation methods and methodologies reviewed. Although my intent was not to create a college textbook, there is value in using this volume with engineering students, either as a supplemental text or a primary text on water treatment technologies. If used as such, instructors will need to gauge the level of understanding of students before specifying the book for a course, as well as integrate the sequence and degree of coverage provided in this volume, for admittedly, for such a broad and complex subject, it is impossible to provide uniform coverage of all areas in a single volume. My own experience in teaching shows that the subject matter, at the level of presentation in this volume is best suited to students with at least 3 years of engineering education under their belts.

Another feature that is incorporated into each chapter is the use of sidebar discussions. These highlight boxes contain information and facts about each subject area that help to emphasize important points to remember, plus can assist plant managers in training technical staff, especially operators on the specific technologies relied upon in their operations. Finally, there is a ***Glossary*** of several hundred terms at the end of the book. This will prove useful to you not only when reading through the chapters, but as a general resource reference.

In some cases equipment suppliers and tradenames are noted, however these citations should not be considered an endorsement of products or services. They are cited strictly for illustrative purposes. Also recognize, that neither I, nor the publisher guarantee any designs emanating from the use of resources or discussions presented herein. Final designs must be based upon strict adherence to local engineering codes, and federal safety and environmental compliance standards.

A heartfelt thanks is extended to Butterworth-Heinemann Publishers for their fine production of this volume, and in sharing my vision for this series, and to various companies cited throughout the book that contributed materials and their time

Nicholas P. Cheremisinoff, Ph.D.
Washington, D.C.

In Memory

This volume is dedicated to the memory of Paul Nicholas Cheremisinoff, P.E., who fathered a generation of pollution control and prevention specialists at New Jersey Institute of Technology.

About the Author

Nicholas P. Cheremisinoff is a private consultant to industry, lending institutions, and donor agencies, specializing in pollution prevention and environmental management. He has more than twenty years experience in applied research, manufacturing and international business development, and has worked extensively throughout Russia, Eastern Europe, Korea, Latin America, and the United States. Dr. Cheremisinoff has contributed extensively to the industrial press, having authored, co-authored or edited more than 100 technical reference books, and several hundred articles, including Butterworth-Heinemann's ***Green Profits: The Manager's Handbook for ISO 14001 and Pollution Prevention***. He received his B.S., M.S. and Ph.D. degrees in chemical engineering from Clarkson College of Technology. He can be reached by email at ncheremisi@aol.com.

Foreword

This volume constitutes the beginning of what Butterworth-Heinemann Publishers and I hope to provide to environmental and pollution control engineers/managers, namely an authoritative and extensive reference series covering control equipment and technologies. As a chemical engineer and a consultant, I not only had the great fortune of having a father, who was famous in the field of pollution control, but the opportunity to work in consulting practice with him on a broad spectrum of environmental problems within industry. We oftentimes talked and planned on writing an authoritative volume on the hardware and technologies available to solve pollution problems in the belief that, although there are many great works in the technical literature, the levels of presentations of this important subject vary dramatically and the information is fragmented. With my father's untimely death in 1994, and my commitment to a multi-year assignment, dealing with environmental responsible care and the development of national environmental policies in Ukraine and Russia, as part of contracts commitments to the U.S. Agency for International Development and the European Union, the original volume we intended was never written. Only now, having the opportunity to try and bring this work forward, I recognize that no single volume can do adequate justice to the subject area.

Also, there is the misconception among a younger generation of engineers that pollution control can be displaced by pollution prevention practices, and hence recent times have de-emphasized the need for engineering innovative pollution controls. I am a strong proponent of pollution prevention, and indeed have developed an international consulting practice around it. However, we should recognize that oftentimes pollution prevention relies upon essentially the identical technologies that are applied to so-called "end-of-pipe" treatment. It is the manner in which these technologies are applied, along with best management practices, which enable pollution prevention to be practiced. As such, pollution prevention does not replace the need for pollution controls, nor does it replace entire processes aimed at cleaning or preventing pollutants from entering the environment. What it does do is channel our efforts into applying traditional end-of-pipe treatment technologies in such manners that costly practices for the disposal of pollutants are avoided, and savings from energy efficiency and materials be achieved.

The volume represents the initial fulfillment of a series, and is aimed at assisting process engineers, plant managers, environmental consultants, water treatment plant operators, and students. Subsequent volumes are intended to cover air pollution controls, and solid waste management and minimization.

This volume is a departure from the style of technical writing that I and many of my colleagues have done in the past. What I have attempted is to discuss the subject, rather than to try and teach or summarize the technologies, the hardware, and selection criteria for different equipment. It's a subject to discuss and explore, rather than to present in a dry, strictly technical fashion. Water treatment is not only a very important subject, but it is extremely interesting. Its importance is simply one of environmental protection and public safety, because after all, water is one of the basic natural elements we rely upon for survival. Even if we are dealing with non-potable water supplies, the impact of poor quality water to process operations can be devastating in terms of achieving acceptable process efficiencies in heat exchange applications, in minimizing the maintenance requirements for heat exchange and other equipment, in the quality of certain products that rely on water as a part of their composition and processing, and ultimately upon the economics of a process operation. It's a fascinating subject, because the technology is both rapidly changing, and cost-effective, energy-saving solutions to water treatment require innovative solutions.

Chapter 1

AN OVERVIEW OF WATER AND WASTE-WATER TREATMENT

INTRODUCTION

We may organize water treatment technologies into three general areas: Physical Methods, Chemical Methods, and Energy Intensive Methods. Physical methods of wastewater treatment represent a body of technologies that we refer largely to as solid-liquid separations techniques, of which filtration plays a dominant role. Filtration technology can be broken into two general categories conventional and non-conventional. This technology is an integral component of drinking water and wastewater treatment applications. It is, however, but one unit process within a modern water treatment plant scheme, whereby there are a multitude of equipment and technology options to select from depending upon the ultimate goals of treatment. To understand the role of filtration, it is important to make distinctions not only with the other technologies employed in the cleaning and purification of industrial and municipal waters, but also with the objectives of different unit processes.

Chemical methods of treatment rely upon the chemical interactions of the contaminants we wish to remove from water, and the application of chemicals that either aid in the separation of contaminants from water, or assist in the destruction or neutralization of harmful effects associated with contaminants. Chemical treatment methods are applied both as stand-alone technologies, and as an integral part of the treatment process with physical methods.

Among the energy intensive technologies, thermal methods have a dual role in water treatment applications. They can be applied as a means of sterilization, thus providing high quality drinking water, and/or these technologies can be applied to the processing of the solid wastes or sludge, generated from water treatment applications. In the latter cases, thermal methods can be applied in essentially the same manner as they are applied to conditioning water, namely to sterilize sludge contaminated with organic contaminants, and/or these technologies can be applied to volume reduction. Volume reduction is a key step in water treatment operations, because ultimately there is a tradeoff between polluted water and hazardous solid waste.

Energy intensive technologies include electrochemical techniques, which by and large are applied to drinking water applications. They represent both sterilization and conditioning of water to achieve a palatable quality.

All three of these technology groups can be combined in water treatment, or they may be used in select combinations depending upon the objectives of water treatment. Among each of the general technology classes, there is a range of both hardware and individual technologies that one may select from. The selection of not only the proper unit process and hardware from within each technology group, but the optimum combinations of hardware and unit processes from the four groups depends upon such factors as:

1. How clean the final water effluent from our plant must be;
2. The quantities and nature of the influent water we need to treat;
3. The physical and chemical properties of the pollutants we need to remove or render neutral in the effluent water;
4. The physical, chemical and thermodynamic properties of the solid wastes generated from treating water; and
5. The cost of treating water, including the cost of treating, processing and finding a home for the solid wastes.

To understand this better, let us step back and start from a very fundamental viewpoint. All processes are comprised of a number of unit processes, which are in turn made up of unit operations. Unit processes are distinct stages of a manufacturing operation. They each focus on one stage in a series of stages, successfully bringing a product to its final form. In this regard, a wastewater treatment plant, whether industrial, a municipal wastewater treatment facility, or a drinking water purification plant, is no different than, say, a synthetic rubber manufacturing plant or an oil refinery. In the case of a rubber producing plant, various unit processes are applied to making intermediate forms of the product, which ultimately is in a final form of a rubber bale, that is sold to the consumer. The individual unit processes in this case are comprised of: (1) a catalyst reparation stage - a pre-preparation stage for monomers and catalyst additives; (2) polymerization - where an intermediate stage of the product is synthesized in the form of a latex or polymer suspended as a dilute solution in a hydrocarbon diluent; (3) followed by finishing - where the rubber is dried, residual diluent is removed and recovered, and the rubber is dried and compressed into a bale and packaged for sale. Each of these unit process operations are in turn comprised of individual unit operations, whereby a particular technology or group pf technologies are applied, which, in turn, define a piece of equipment that is used along the production line. Drinking water and wastewater treatment plants are essentially no different. There are individual unit processes that comprise each of these types of plants that are applied in a succession of operations, with each stage aimed at improving the quality of the water as established by a set of product-performance criteria. The criteria focuses on the quality of the final water, which in the case of drinking water is established based upon legal criteria (e.g., the Safe Drinking Water Act, *SDWA*), and if non-potable or process plant water, may be operational criteria (e.g., non-brackish waters to prevent scaling of heat exchange equipment).

The number and complexity of unit processes and in turn unit operations comprising a water purification or wastewater treatment facility are functions of the legal and operational requirements of the treated water, the nature and degree of contamination of the incoming water (raw water to the plant), and the quantities of water to be processed. This means then, that water treatment facilities from a design and operational standpoints vary, but they do rely on overlapping and even identical unit processes.

If we start with the first technology group, then filtration should be thought of as both a unit process and a unit operation within a water treatment facility. As a separate unit process, its objective is quite clear: namely, to remove suspended solids. When we combine this technology with chemical methods and apply sedimentation and clarification (other physical separation methods), we can extend the technology to removing dissolved particulate matter as well. The particulate matter may be biological, microbial or chemical in nature. As such, the operation stands alone within its own block within the overall manufacturing train of the plant. Examples of this would be the roughening and polishing stages of water treatment. In turn, we may select or specify specific pieces of filtration equipment for these unit processes.

The above gives us somewhat of an idea of the potential complexity of choosing the optimum group of technologies and hardware needed in treating water. To develop a cost-effective design, we need to understand not only what each of the unit processes are, but obtain a working knowledge of the operating basis and ranges for the individual hardware. That, indeed, is the objective of this book; namely, to take a close look at the equipment options available to us in each technology group, but not individually. Rather, to achieve an integrated and well thought out design, we need to understand how unit processes and unit operations compliment each other in the overall design.

This first chapter is for orientation purposes. Its objectives are to provide an overview of water treatment and purification roles and technologies, and to introduce terminology that will assist you in understanding the relation of the various technologies to the overall schemes employed in waster treatment applications. Recommended resources that you can refer to for more in-depth information are included at the end of each chapter. The organization of these resources are generally provided by subject matter. Also, you will find a section for the student at the end of each chapter that provides a list of ***Questions for Thinking and Discussing***. These will assist in reinforcing some of the principles and concepts presented in each chapter, if the book is used as a primary or supplement textbook.

We should recognize that the technology options for water treatment are great, and quite often the challenge lies with the selection of the most cost-effective combinations of unit processes and operations. In this regard, cost-factors are examined where appropriate in our discussions within later chapters.

WHAT WE MEAN BY WATER PURIFICATION

When we refer to water purification, it makes little sense to discuss the subject without first identifying the contaminants that we wish to remove from water. Also, the source of the water is of importance. Our discussion at this point focuses on drinking water. Groundwater sources are of a particular concern, because there are many communities throughout the U.S. that rely on this form. The following are some of the major contaminants that are of concern in water purification applications, as applied to drinking water sources, derived from groundwater.

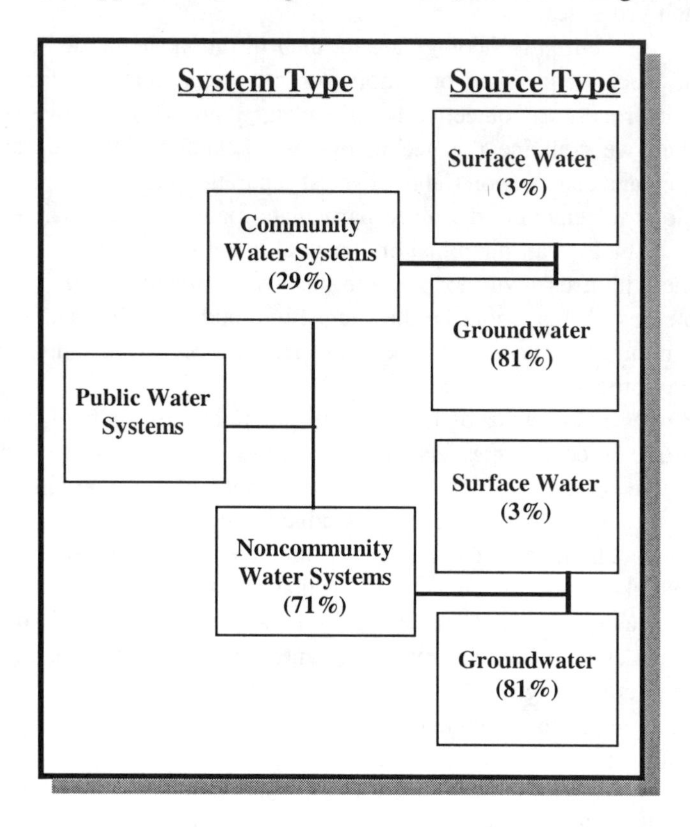

Heavy Metals - Heavy metals represent problems in terms of groundwater pollution. The best way to identify their presence is by a lab test of the water or by contacting county health departments. There are concerns of chronic exposure to low levels of heavy metals in drinking water.

Turbidity - Turbidity refers to suspended solids, i.e. muddy water, is very turbid. Turbidity is undesirable for three reasons:

- aesthetic considerations,
- solids may contain heavy metals, pathogens or other contaminants,

- turbidity decreases the effectiveness of water treatment techniques by shielding pathogens from chemical or thermal damage, or in the case of UV (ultra violet) treatment, absorbing the UV light itself.

Organic Compounds - Water can be contaminated by a number of organic compounds, such as chloroform, gasoline, pesticides, and herbicides from a variety of industrial and agricultural operations or applications. These contaminants must be identified in a lab test. It is unlikely groundwater will suddenly become contaminated, unless a quantity of chemicals is allowed to enter a well or penetrating the aquifer. One exception is when the aquifer is located in limestone. Not only will water flow faster through limestone, but the rock is prone to forming vertical channels or sinkholes that will rapidly allow contamination from surface water. Surface water may show great variations in chemical contamination levels due to differences in rainfall, seasonal crop cultivation, and industrial effluent levels. Also, some hydrocarbons (the chlorinated hydrocarbons in particular) form a type of contaminant that is especially troublesome. These are a group of chemicals known as dense nonaqueous phase liquids, or DNAPLs. These include chemicals used in dry cleaning, wood preservation, asphalt operations, machining, and in the production and repair of automobiles, aviation equipment, munitions, and electrical equipment. These substances are heavier than water and they sink quickly into the ground. This makes spills of DNAPLs more difficult to handle than spills of petroleum products. As with petroleum products, the problems are caused by groundwater dissolving some of the compounds in these volatile substances. These compounds can then move with the groundwater flow. Except in large cities, drinking water is rarely tested for these contaminants. Disposal of chemicals that have low water solubility and a density greater than water result in the formation of distinct areas of pure residual contamination in soils and groundwater. These chemicals are typically solvents and are collectively referred to as Dense Non-Aqueous Phase Liquids (DNAPLs). Because of their relatively high density, they tend to move downward through soils and groundwater, leaving small amounts along the migratory pathway, until they reach an impermeable layer where they collect in discrete pools. Once the DNAPLs have reached an aquitard they tend to move laterally under the influence of gravity and to slowly dissolve into the groundwater, providing a long-term source for low level contamination of groundwater. Because of their movement patterns DNAPL contamination is difficult to detect, characterize and remediate.

Pathogens - These include protozoa, bacteria, and viruses. Protozoa cysts are the largest pathogens in drinking water, and are responsible for many of the waterborne disease cases in the U.S. Protozoa cysts range is size from 2 to 15 μm (*a micron is one millionth of a meter*), but can squeeze through smaller openings. In order to insure cyst filtration, filters with a absolute pore size of 1μm or less should be used. The two most common protozoa pathogens are *Giardia lamblia* (Giardia) and *Cryptosporidium* (Crypto). Both organisms have caused numerous deaths in recent years in the U.S. and Canada, the deaths occurring in the young and elderly, and the sick and immune compromised. Many deaths were a result of more than one of these conditions. Neither disease is likely to be fatal to a healthy adult, even if untreated. For example in Milwaukee in April of 1993, of 400,000 who were diagnosed with Crypto, only 54 deaths were linked to the outbreak, 84% of whom were AIDS patients. Outside of the U.S. and other developed countries, protozoa are responsible for many cases of amoebic dysentery, but so far this has not been a problem in the U.S., due to the application of more advanced wastewater treatment technologies. This could change during a survival situation. Tests have found Giardia and/or Crypto in up to 5% of vertical wells and 26% of springs in the U.S.

Bacteria are smaller than protozoa and are responsible for many diseases, such as typhoid fever, cholera, diarrhea, and dysentery. Pathogenic bacteria range in size from 0.2 to 0.6 μm, and a 0.2 μm filter is necessary to prevent transmission. Contamination of water supplies by bacteria is blamed for the cholera epidemics, which devastate undeveloped countries from time to time. Even in the U.S., ***E. coli*** is frequently found to contaminated water supplies. Fortunately, E. coli is relatively harmless as pathogens go, and the problem isn't so much with E. coli found, but the fear that other bacteria may have contaminated the water as well. Never the less, dehydration from diarrhea caused by E. coli has resulted in fatalities.

One of hundreds of strains of the bacterium *Escherichia coli*, *E. coli* O157:H7 is an emerging cause of food borne and waterborne illness. Although most strains of *E. coli* are harmless and live in the intestines of healthy humans and animals, this strain produces a powerful toxin and can cause severe illness. *E. coli* O157:H7 was first recognized as a cause of illness during an outbreak in 1982 traced to contaminated hamburgers. Since then, most infections are believed to have come from eating undercooked ground beef. However, some have been waterborne. The presence of *E. coli* in water is a strong indication of recent sewage or animal waste contamination. Sewage may contain many types of disease-causing organisms. Since *E. coli* comes from human and animal wastes, it most often enters drinking water sources via rainfalls, snow melts, or other types of precipitation, *E. coli* may be washed into creeks, rivers, streams, lakes, or groundwater. When these waters are used as sources of drinking water and the water is not treated or inadequately treated, *E. coli* may end up in drinking water. *E. coli* O157:H7 is one of hundreds of strains of the bacterium *E. coli*. Although most strains are harmless and live in the intestines of healthy humans and animals, this strain produces a powerful toxin and can cause severe illness. Infection often causes severe bloody diarrhea and abdominal cramps; sometimes the infection causes non-bloody diarrhea. Frequently, no fever is present. It should be noted that these symptoms are common to a variety of diseases, and may be caused by sources other than contaminated drinking water. In some people, particularly children under 5 years of age and the elderly, the infection can also cause a complication, called hemolytic uremic syndrome, in which the red blood cells are destroyed and the kidneys fail. About 2%-7% of infections lead to this complication. In the U.S. hemolytic uremic syndrome is the principal cause of acute kidney failure in children, and most cases of hemolytic uremic syndrome are caused by *E. coli* O157:H7. Hemolytic uremic syndrome is a life-threatening condition usually treated in an intensive care unit. Blood transfusions and kidney dialysis are often required. With intensive care, the death rate for hemolytic uremic syndrome is 3%-5%. Symptoms usually appear within 2 to 4 days, but can take up to 8 days. Most people recover without antibiotics or other specific treatment in 5-10 days. There is no evidence that antibiotics improve the course of disease, and it is thought that treatment with some antibiotics may precipitate kidney complications. Antidiarrheal agents, such as loperamide (Imodium), should also be avoided. The most common methods of treating water contaminated with E. coli is by using chlorine, ultra-violet light, or ozone, all of which act to kill or inactivate *E. coli*. Systems, using surface water sources, are required to disinfect to ensure that all bacterial contamination is inactivated, such as *E. coli*. Systems using ground water sources are not required to disinfect, although many of them do. According to EPA regulations, a system that operates at least 60 days per year, and serves 25 people or more or has 15 or more service connections, is regulated as a public water system under the Safe Drinking Water Act (SDWA). If a system is not a public water system as defined by EPA's regulations, it is not regulated under the SDWA, although it may be regulated by state or local authorities. Under the SDWA, EPA requires public water systems to monitor for coliform bacteria. Systems analyze first for total coliform, because this test is faster to produce results. Any time that a sample is positive for total coliform, the same sample must be analyzed for either fecal coliform or *E. coli*. Both are indicators of contamination with animal waste or human sewage. The largest public water systems (serving millions of people) must take at least 480 samples per month. Smaller systems must take at least five samples a month, unless the state has conducted a sanitary survey - a survey in which a state inspector examines system components and ensures they will protect public health - at the system within the last five years.

Viruses are the 2nd most problematic pathogen, behind protozoa. As with protozoa, most waterborne viral diseases don't present a lethal hazard to a healthy adult. Waterborne pathogenic viruses range in size from 0.020-0.030 μm, and are too small to be filtered out by a mechanical filter. All waterborne enteric viruses affecting humans occur solely in humans, thus animal waste doesn't present much of a viral threat. At the present viruses don't present a major hazard to people drinking surface water in the U.S., but this could change in a survival situation as the level of human sanitation is reduced. Viruses do tend to show up even in remote areas, so a case can be made for eliminating them now.

THE DRINKING WATER STANDARDS

When the objective of water treatment is to provide drinking water, then we need to select technologies that are not only the best available, but those that will meet local and national quality standards. The primary goals of a water treatment plant

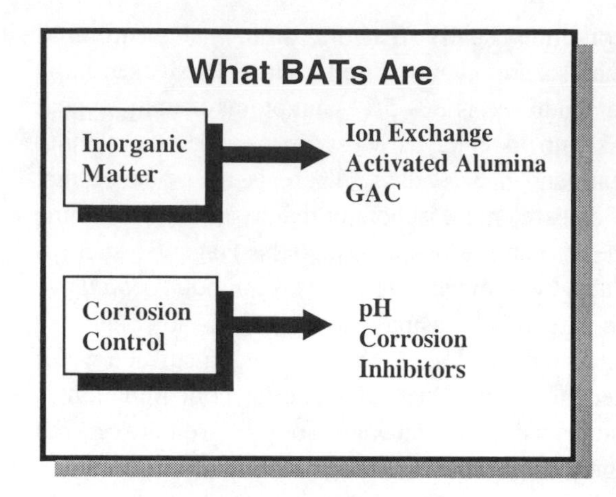

for over a century have remained practically the same: namely to produce water that is biologically and chemically safe, is appealing to the consumer, and is noncorrosive and nonscaling. Today, plant design has become very complex from discovery of seemingly innumerable chemical substances, the multiplying of regulations, and trying to satisfy more discriminating palates. In addition to the basics, designers must now keep in mind all manner of legal mandates, as well as public concerns and en-vironmental considerations, to provide an initial prospective of water works engineering planning, design, and operation.

The growth of community water supply systems in the United States started in the early 1800s. By 1860, over 400, and by the turn of the century over 3000 major water systems had been built to serve major cities and towns. Many older plants were equipped with slow sand filters. In the mid 1890s, the Louisville Water Company introduced the technologies of coagulation with rapid sand filtration.

The first application of chlorine in potable water was introduced in the 1830s for taste and odor control, at that time diseases were thought to be spread by odors. It was not until the 1890s and the advent of the germ theory of disease that the importance of disinfection in potable water was understood. Chlorination was first introduced on a practical scale in 1908 and then became a common practice.

Federal authority to establish standards for drinking water systems originated with the enactment by Congress in 1883 of the Interstate Quarantine Act, which authorized the Director of the United States Public Health Services (USPHS) to establish and enforce regulations to prevent the introduction, transmission, or spread of communicable diseases.

Today resource limitations have caused the United States Environmental Protection Agency (USEPA) to reassess schedules for new rules. A 1987 USEPA survey indicated there were approximately 202,000 public water systems in the United States. About 29 percent of these were community water systems, which serve approximately 90 percent of the population. Of the 58,908 community systems that serve about 226 million people, 51,552 were classified as "small" or "very small." Each of these systems at an average serves a population of fewer than 3300 people. The total population served by these systems is approximately 25 million people. These figures provide us with a magnitude of scale in meeting drinking water demands in the United States. Compliance with drinking water standards is not uniform. Small systems are the most frequent violators of federal regulations. Microbiological violations account for the vast majority of cases, with failure to monitor and report. Among others, violations exceeding SDWA maximum contaminant levels (MCLs) are quite common. Bringing small water systems into compliance requires applicable technologies, operator ability, financial resources, and institutional arrangements. The 1986 SDWA amendments authorized USEPA to set the best available technology (BAT) that can be incorporated in the design for the purposes of complying with the National Primary Drinking Water Regulations (NPDWR). Current BAT to maintain standards are as follows:

For turbidity, color and microbiological control in surface water treatment: filtration. Common variations of filtration are conventional, direct, slow sand, diatomaceous earth, and membranes.

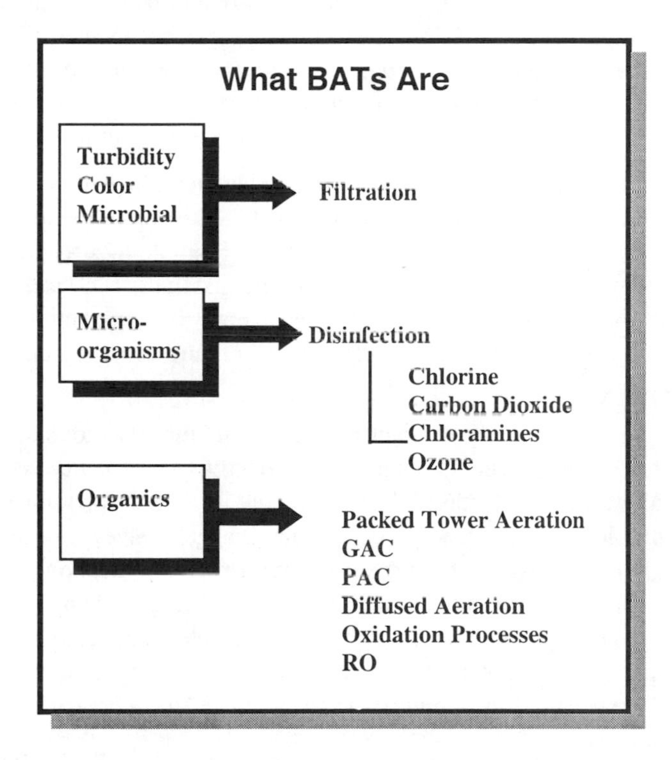

For inactivation of microorganisms: disinfection. Typical disinfectants are chlorine, chlorine dioxide, chloramines, and ozone.

For organic contaminant removal from surface water: packed-tower aeration, granular activated carbon (GAC), powdered activated carbon (PAC), diffused aeration, advanced oxidation processes, and reverse osmosis (RO).

For inorganic contaminants removal: membranes, ion exchange, activated alumina, and GAC.

For corrosion control: typically, pH adjustment or corrosion inhibitors. The implications of the 1986 amendments to the SDWA and new regulations have resulted in rapid development and introduction of new technologies and equipment for water treatment and monitoring over the last two decades. Biological processes in particular have proven effective in removing biodegradable organic carbon that may sustain the regrowth of potentially harmful microorganisms in the distribution system, effective taste and odor control, and reduction in chlorine demand and DBP formation potential. Both biologically-active sand or carbon filters provide cost effective treatment of micro-contaminants than do physicochernical processes in many cases. Pertinent to the subject matter cover in this volume, *membrane technology* has been applied in drinking water treatment, partly because of affordable membranes and demand to removal of many contaminants. *Microfiltration, ultrafiltration, nanofiltration* and others have become common names in the water industry. Membrane technology is experimented with for the removal of microbes, such as *Giardia and Cryptosporidium* and for selective removal of nitrate. In other instances, membrane technology is applied for removal of DBP precursors, VOCs, and others.

Other treatment technologies that have potential for full-scale adoption are photochemical oxidation using ozone and UV radiation or hydrogen peroxide for destruction of refractory organic compounds. One example of a technology that was developed outside North America and later emerged in the U.S. is the Haberer process. This process combines contact flocculation, filtration, and powdered activated carbon adsorption to meet a wide range of requirements for surface water and groundwater purification.

Utilities are seeking not only to improve treatment, but also to monitor their supplies for microbiological contaminants more effectively. *Electro-optical sensors* are used to allow early detection of algal blooms in a reservoir and allow for diagnosis of problems and guidance in operational changes. *Gene probe* technology was first developed in response to the need for improved identification of microbes in the field of clinical microbiology. Attempts are now being made by radiolabeled and nonradioactive gene-probe assays with traditional detection methods for enteric viruses and protozoan parasites, such as *Giardia and Cryptosporidium.* This technique has the potential for monitoring water supplies for increasingly complex groups of microbes.

In spite of the multitudinous regulations and standards that an existing public water system must comply with, the principles of conventional water treatment process have not changed significantly over half a century. Whether a filter contains sand, anthracite, or both, slow or rapid rate, constant or declining rate, filtration is still filtration, sedimentation is still sedimentation, and disinfection is still disinfection. What has changed, however, are many tools that we now have in our engineering arsenal. For example, , a supervisory control and data acquisition (SCADA) system can provide operators and managers with accurate process control variables and operation and maintenance records. In addition to being able to look at the various options on the computer screen, engineers can conduct pilot plant studies of the multiple variables inherent in water treatment plant design. Likewise, operators and managers can utilize an ongoing pilot plant facility to optimize chemical feed and develop important information needed for future expansion and upgrading.

Technology and ultimately equipment selection depends on the standards set by the regulations. Drinking water standards are regulations that EPA sets to control the level of contaminants in the nation's drinking water. These standards are part of the Safe Drinking Water Act's "multiple barrier" approach to drinking water protection, which includes assessing and protecting drinking water sources; protecting wells and collection systems; making sure water is treated by qualified operators; ensuring the integrity of distribution systems; and making information available to the public on the quality of their drinking water. With the involvement of EPA, states, tribes, drinking water utilities, communities and citizens, these multiple barriers ensure that tap water in the U.S. and territories is safe to drink. In most cases, EPA delegates responsibility for implementing drinking water standards to states and tribes. There are two categories of drinking water standards:

- A National Primary Drinking Water Regulation (NPDWR or primary standard) is a legally-enforceable standard that applies to public water systems. Primary standards protect drinking water quality by limiting the levels of specific contaminants that can adversely affect public health and are known or anticipated to occur in water. They take the form of Maximum Contaminant Levels (MCL) or Treatment Techniques (TT).
- A National Secondary Drinking Water Regulation (NSDWR or secondary standard) is a non-enforceable guideline regarding contaminants that may cause cosmetic effects (such as skin or tooth discoloration) or aesthetic effects (such as taste, odor, or color) in drinking water. EPA recommends secondary standards to water systems but does not require systems to comply. However, states may choose to adopt them as enforceable standards. This information focuses on national primary standards.

Drinking water standards apply to public water systems (PWSs), which provide water for human consumption through at least 15 service connections, or regularly serve at least 25 individuals. Public water systems include municipal water companies, homeowner associations, schools, businesses, campgrounds and shopping malls. EPA considers input from many individuals and groups throughout the rule-making process. One of the formal means by which EPA solicits the assistance of its stakeholders is the National Drinking Water Advisory Council (NDWAC). The 15-member committee was created by the Safe Drinking Water Act. It is comprised of five members of the general public, five representatives of state and local agencies concerned with water hygiene and public water supply, and five representations of private organizations and groups demonstrating an active interest in water hygiene and public water supply, including two members who are associated with small rural public water systems.

NDWAC advises EPA's Administrator on all of the agency's activities relating to drinking water. In addition to the NDWAC, representatives from water utilities, environmental groups, public interest groups, states, tribes and the general public are encouraged to take an active role in shaping the regulations, by participating in public meetings and commenting on proposed rules. Special meetings are also held to obtain input from minority and low-income communities, as well as representatives of small businesses.

The 1996 Amendments to Safe Drinking Water Act require EPA to go through several steps to determine, first, whether setting a standard is appropriate for a particular contaminant, and if so, what the standard should be. Peer-reviewed science and data support an intensive technological evaluation, which includes many factors: occurrence in the environment; human exposure and risks of adverse health effects in the general population and sensitive subpopulations; analytical methods of detection; technical feasibility; and impacts of regulation on water systems, the economy and public health. Considering public input throughout the process, EPA must (1) identify drinking water problems; (2) establish priorities; and (3) set standards.

EPA must first make determinations about which contaminants to regulate. These determinations are based on health risks and the likelihood that the contaminant occurs in public water systems at levels of concern. The National Drinking Water Contaminant Candidate List (CCL), published March 2, 1998, lists contaminants that (1) are not already regulated under SDWA; (2) may have adverse health effects; (3) are known or anticipated to occur in public water systems; and (4) may require regulations under SDWA. Contaminants on the CCL are divided into priorities for regulation, health research and occurrence data collection.

In August 2001, EPA selected five contaminants from the regulatory priorities on the CCL and determined whether to regulate them. To support these decisions, the Agency determined that regulating the contaminants presents a meaningful opportunity to reduce health risk. If the EPA determines regulations are necessary, the Agency must propose them by August 2003, and finalize them by February 2005. In addition, the Agency will also select up to 30 unregulated contaminants from the CCL for monitoring by public water systems serving at least 100,000 people. Currently, most of the unregulated contaminants with potential of occurring in drinking water are pesticides and microbes. Every five years, EPA will repeat the cycle of revising the CCL, making regulatory determinations for five contaminants and identifying up to 30 contaminants for unregulated monitoring. In addition, every six years, EPA will re-evaluate existing regulations to determine if modifications are necessary. Beginning in August 1999, a new National Contaminant Occurrence Database was developed to store data on regulated and unregulated chemical, radiological, microbial and physical contaminants, and other such contaminants likely to occur in finished, raw and source waters of public water systems.

After reviewing health effects studies, EPA sets a **Maximum Contaminant Level Goal (MCLG),** the maximum level of a contaminant in drinking water at which no known or anticipated adverse effect on the health of persons would occur, and which allows an adequate margin of safety. MCLGs are non-enforceable public health goals. Since MCLGs consider only public health and not the limits of detection and treatment technology, sometimes they are set at a level which water systems cannot meet. When determining an MCLG, EPA considers the risk to sensitive subpopulations (infants, children, the elderly, and those with compromised immune systems) of experiencing a variety of adverse health effects.

SOME IMPORTANT DEFINITIONS

Maximum Contaminant Level (MCL) - *The highest level of a contaminant that is allowed in drinking water. MCLs are set as close to MCLGs as feasible using the best available treatment technology and taking cost into consideration. MCLs are enforceable standards.*
Maximum Contaminant Level Goal (MCLG) - *The level of a contaminant in drinking water below which there is no known or expected risk to health. MCLGs allow for a margin of safety and are non-enforceable public health goals.*
Maximum Residual Disinfectant Level (MRDL) - *The highest level of a disinfectant allowed in drinking water. There is convincing evidence that addition of a disinfectant is necessary for control of microbial contaminants.*
Maximum Residual Disinfectant Level Goal (MRDLG) - *The level of a drinking water disinfectant below which there is no known or expected risk to health. MRDLGs do not reflect the benefits of the use of disinfectants to control microbial contaminants.*
Treatment Technique - *A required process intended to reduce the level of a contaminant in drinking water.*

Non-Carcinogens (excluding microbial contaminants): For chemicals that can cause adverse non-cancer health effects, the MCLG is based on the reference dose. A **reference dose** (RFD) is an estimate of the amount of a chemical that a person can be exposed to on a daily basis that is not anticipated to cause adverse health effects over a person's lifetime. In RFD calculations, sensitive subgroups are included, and uncertainty may span an order of magnitude. The RFD is multiplied by typical adult body weight (70 kg) and divided by daily water consumption (2 liters) to provide a Drinking Water Equivalent Level (DWEL). Note that the DWEL is multiplied by a percentage of the total daily exposure contributed by drinking water to determine the MCLG. This empirical factor is usually 20 percent, but can be a higher value.

Chemical Contaminants (Carcinogens): If there is evidence that a chemical may cause cancer, and there is no dose below which the chemical is considered safe, the MCLG is set at zero. If a chemical is carcinogenic and a safe dose can be deter mined, the MCLG is set at a level above zero that is safe.

Microbial Contaminants: For microbial contaminants that may present public health risk, the MCLG is set at zero because ingesting one protozoa, virus, or bacterium may cause adverse health effects. EPA is conducting studies to determine whether there is a safe level above zero for some microbial contaminants. So far, however, this has not been established.

Once the MCLG is determined, EPA sets an enforceable standard. In most cases, the standard is a **Maximum Contaminant Level (MCL)**, the maximum permissible level of a contaminant in water which is delivered to any user of a public water system. The MCL is set as close to the MCLG as feasible, which the Safe Drinking Water Act defines as the level that may be achieved with the use of the best available technology, treatment techniques, and other means which EPA finds are available(after examination for efficiency under field conditions and not solely under laboratory conditions) are available, taking cost into consideration. When there is no reliable method that is economically and technically feasible to measure a contaminant at particularly low concentrations, a **Treatment Technique (TT)** is set rather than an MCL. A treatment technique (TT) is an enforceable procedure or level of technological performance which public water systems must follow to ensure control of a contaminant. Examples of Treatment Technique rules are the Surface Water Treatment Rule (disinfection and filtration) and the Lead and Copper Rule (optimized corrosion control). After determining a MCL or TT based on affordable technology for large systems, EPA must complete an economic analysis to determine whether the benefits of that standard justify the costs. If not, EPA may adjust the MCL for a particular class or group of systems to a level that "maximizes health risk reduction benefits at a cost that is justified by the benefits."

WHAT THE CURRENT DRINKING WATER STANDARDS ARE

The following matrices provide you with a summary of the NPDWRs or primary standards. You should visit the EPA Web site (www.epa.gov) and become familiar with the various documents that are publically available. You will not only find these regulations there, but detailed information that explains the reasoning behind each MCLG. You will also find the entire legislation on this site and can become familiar with all of the subtleties of this piece of complex environmental legislation. Tables 1 through 5 are derived from *EPA Web site - www.epa.gov/safewater*.

Table 1. NPDW Regulations for Microorganisms.

Microorganisms	MCLG[1] (mg/L)[2]	MCL or TT[1] (mg/L)[2]	Potential Health Effects from Ingestion of Water	Sources of Contaminant in Drinking Water
Cryptosporidium	0	as of 01/01/02: TT [3]	Gastrointestinal illness (e.g., diarrhea, vomiting, cramps)	Human and animal fecal waste
Giardia lamblia	0	TT[3]	Gastrointestinal illness (e.g., diarrhea, vomiting, cramps)	Human and animal fecal waste
Heterotrophic plate count	n/a	TT[3]	HPC has no health effects, but can indicate how effective treatment is at controlling microorganisms.	HPC measures a range of bacteria that are naturally present in the environment
Legionella	0	TT[3]	Legionnaire's Disease, commonly known as pneumonia	Found naturally in water; multiplies in heating systems
Total Coliforms (including fecal coliform and *E. Coli*)	0	5.0%[4]	Used as an indicator that other potentially harmful bacteria may be present[5]	Coliforms are naturally present in the environment; fecal coliforms and E. coli come from human and animal fecal waste.
Turbidity	n/a	TT[3]	Turbidity, a measure of water cloudiness, is used to indicate water quality and filtration effectiveness (e.g., whether disease-causing organisms are present). Higher turbidity is associated with higher levels of microorganisms such as viruses, parasites and some bacteria. These organisms can cause symptoms such as nausea, cramps, diarrhea, and associated headaches.	Soil runoff
Viruses (enteric)	0	TT[3]	Gastrointestinal illness (e.g., diarrhea/vomiting)	Human and animal fecal waste

Table 2. NPDW Regulations for Disinfectants and Disinfection Byproducts.

Disinfectants & Disinfection Byproducts	MCLG[1] (mg/L)[2]	MCL or TT[1] (mg/L)[2]	Potential Health Effects from Ingestion of Water	Sources of Contaminant in Drinking Water
Bromate	as of 01/01/02: zero	as of 01/01/02: 0.010	Increased risk of cancer	Byproduct of drinking water disinfection
Chloramines (as Cl_2)	as of 01/01/02: MRDLG = 4[1]	as of 01/01/02: MRDL = 4.0[1]	Eye/nose irritation; stomach discomfort, anemia	Water additive used to control microbes
Chlorine (as Cl_2)	as of 01/01/02: MRDLG = 4[1]	as of 01/01/02: MRDL = 4.0[1]	Eye/nose irritation; stomach discomfort	Water additive used to control microbes
Chlorine dioxide (as ClO_2)	as of 01/01/02: MRDLG =0.8[1]	as of 01/01/02: MRDL=0 .8[1]	Anemia; infants & young children: nervous system effects	Water additive used to control microbes
Chlorite	as of 01/01/02: 0.8	as of 01/01/02: 1.0	Anemia; infants & young children: nervous system effects	Byproduct of drinking water disinfection
Haloacetic acids (HAA5)	as of 01/01/02: n/a[6]	as of 01/01/02: 0.060	Increased risk of cancer	Byproduct of drinking water disinfection
Total Trihalomethanes (TTHMs)	none[7] ---------- as of 01/01/02: n/a[6]	0.10 ---------- as of 01/01/02: 0.080	Liver, kidney or central nervous system problems; increased risk of cancer	Byproduct of drinking water disinfection

Table 3. NPDW Regulations for Inorganic Chemicals.

Inorganic Chemicals	MCLG[1] (mg/L)[2]	MCL or TT[1] (mg/L)[2]	Potential Health Effects from Ingestion of Water	Sources of Contaminant in Drinking Water
Antimony	0.006	0.006	Increase in blood cholesterol; decrease in blood glucose	Discharge from petroleum refineries; fire retardants;

Inorganic Chemicals	MCLG[1] (mg/L)[2]	MCL or TT[1] (mg/L)[2]	Potential Health Effects from Ingestion of Water	Sources of Contaminant in Drinking Water
				ceramics; electronics; solder
Arsenic	none[7]	0.05	Skin damage; circulatory system problems; increased risk of cancer	Erosion of natural deposits; runoff from glass & electronics production wastes
Asbestos fiber > 10 micrometer	7 million fibers per liter	7 MFL	Increased risk of developing benign intestinal polyps	Decay of asbestos cement in water mains; erosion of natural deposits
Barium	2	2	Increase in blood pressure	Discharge of drilling wastes; discharge from metal refineries; erosion of natural deposits
Beryllium	0.004	0.004	Intestinal lesions	Discharge from metal refineries and coal-burning factories; discharge from electrical, aerospace, and defense industries
Cadmium	0.005	0.005	Kidney damage	Corrosion of galvanized pipes; erosion of natural deposits; discharge from metal refineries; runoff from waste batteries and paints
Chromium (total)	0.1	0.1	Some people who use water containing chromium well in excess of the MCL over many years could experience allergic dermatitis	Discharge from steel and pulp mills; erosion of natural deposits
Copper	1.3	TT[8] Action Level = 1.3	Short term exposure: Gastrointestinal distress. Long term exposure: Liver or kidney damage.	Corrosion of household plumbing systems; erosion of natural deposits
Cyanide (as free cyanide)	0.2	0.2	Nerve damage or thyroid problems	Discharge from steel/metal factories; discharge from plastic and fertilizer factories

Inorganic Chemicals	MCLG[1] (mg/L)[2]	MCL or TT[1] (mg/L)[2]	Potential Health Effects from Ingestion of Water	Sources of Contaminant in Drinking Water
Fluoride	4.0	4.0	Bone disease (pain and tenderness of the bones); Children may get mottled teeth.	Water additive which promotes strong teeth; erosion of natural deposits; discharge from fertilizer and aluminum factories
Lead	zero	TT[8]; Action Level = 0.015	Infants and children: Delays in physical or mental development Adults: Kidney problems; high blood pressure	Corrosion of household plumbing systems; erosion of natural deposits
Mercury (inorganic)	0.002	0.002	Kidney damage	Erosion of natural deposits; discharge from refineries and factories; runoff from landfills and cropland
Nitrate (measured as Nitrogen)	10	10	"Blue baby syndrome" in infants under six months - life threatening without immediate medical attention. Symptoms: Infant looks blue and has shortness of breath.	Runoff from fertilizer use; leaching from septic tanks, sewage; erosion of natural deposits
Nitrite (measured as Nitrogen)	1	1	"Blue baby syndrome" in infants under six months - life threatening without immediate medical attention. Symptoms: Infant looks blue and has shortness of breath.	Runoff from fertilizer use; leaching from septic tanks, sewage; erosion of natural deposits
Selenium	0.05	0.05	Hair or fingernail loss; numbness in fingers or toes; circulatory problems	Discharge from petroleum refineries; erosion of natural deposits; discharge from mines
Thallium	0.0005	0.002	Hair loss; changes in blood; kidney, intestine, or liver problems	Leaching from ore-processing sites; discharge from electronics, glass, and pharmaceutical plants.

Table 4. NPDW Regulations for Organic Chemicals.

Organic Chemicals	MCLG[1] (mg/L)[2]	MCL or TT[1] (mg/L)[2]	Potential Health Effects from Ingestion of Water	Sources of Contaminant in Drinking Water
Acrylamide	zero	TT[9]	Nervous system or blood problems; increased risk of cancer	Added to water during sewage/wastewater treatment
Alachlor	zero	0.002	Eye, liver, kidney or spleen problems; anemia; increased risk of cancer	Runoff from herbicide used on row crops
Atrazine	0.003	0.003	Cardiovascular system problems; reproductive difficulties	Runoff from herbicide used on row crops
Benzene	zero	0.005	Anemia; decrease in blood platelets; increased risk of cancer	Discharge from factories; leaching from gas storage tanks and landfills
Benzo(a)pyrene (PAHs)	zero	0.0002	Reproductive difficulties; increased risk of cancer	Leaching from linings of water storage tanks and distribution lines
Carbofuran	0.04	0.04	Problems with blood or nervous system; reproductive difficulties.	Leaching of soil fumigant used on rice and alfalfa
Carbon tetrachloride	zero	0.005	Liver problems; increased risk of cancer	Discharge from chemical plants and other industrial activities
Chlordane	zero	0.002	Liver or nervous system problems; increased risk of cancer	Residue of banned termiticide
Chlorobenzene	0.1	0.1	Liver or kidney problems	Discharge from chemical and agricultural chemical factories
2,4-D	0.07	0.07	Kidney, liver, or adrenal gland problems	Runoff from herbicide used on row crops

Organic Chemicals	MCLG[1] (mg/L)[2]	MCL or TT[1] (mg/L)[2]	Potential Health Effects from Ingestion of Water	Sources of Contaminant in Drinking Water
Dalapon	0.2	0.2	Minor kidney changes	Runoff from herbicide used on rights of way
1,2-Dibromo-3-chloropropane (DBCP)	zero	0.0002	Reproductive difficulties; increased risk of cancer	Runoff/leaching from soil fumigant used on soybeans, cotton, pineapples, and orchards
o-Dichlorobenzene	0.6	0.6	Liver, kidney, or circulatory system problems	Discharge from industrial chemical factories
p-Dichlorobenzene	0.075	0.075	Anemia; liver, kidney or spleen damage; changes in blood	Discharge from industrial chemical factories
1,2-Dichloroethane	zero	0.005	Increased risk of cancer	Discharge from industrial chemical factories
1,1-Dichloroethylene	0.007	0.007	Liver problems	Discharge from industrial chemical factories
cis-1,2-Dichloroethylene	0.07	0.07	Liver problems	Discharge from industrial chemical factories
trans-1,2-Dichloroethylene	0.1	0.1	Liver problems	Discharge from industrial chemical factories
Dichloromethane	zero	0.005	Liver problems; increased risk of cancer	Discharge from pharmaceutical and chemical factories
1,2-Dichloropropane	zero	0.005	Increased risk of cancer	Discharge from industrial chemical factories
Di(2-ethylhexyl) adipate	0.4	0.4	General toxic effects or reproductive difficulties	Leaching from PVC plumbing systems; discharge from chemical factories
Di(2-ethylhexyl) phthalate	zero	0.006	Liver problems; increased risk of cancer	Discharge from rubber and chemical factories

Organic Chemicals	MCLG[1] (mg/L)[2]	MCL or TT[1] (mg/L)[2]	Potential Health Effects from Ingestion of Water	Sources of Contaminant in Drinking Water
Dinoseb	0.007	0.007	Reproductive difficulties	Runoff from herbicide used on soybeans and vegetables
Dioxin (2,3,7,8-TCDD)	zero	0.00000003	Reproductive difficulties; increased risk of cancer	Emissions from waste incineration and other combustion; discharge from chemical factories
Diquat	0.02	0.02	Cataracts	Runoff from herbicide use
Endothall	0.1	0.1	Stomach and intestinal problems	Runoff from herbicide use
Endrin	0.002	0.002	Nervous system effects	Residue of banned insecticide
Epichlorohydrin	zero	TT[9]	Stomach problems; reproductive difficulties; increased risk of cancer	Discharge from industrial chemical factories; added to water during treatment process
Ethylbenzene	0.7	0.7	Liver or kidney problems	Discharge from petroleum refineries
Ethelyne dibromide	zero	0.00005	Stomach problems; reproductive difficulties; increased risk of cancer	Discharge from petroleum refineries
Glyphosate	0.7	0.7	Kidney problems; reproductive difficulties	Runoff from herbicide use
Heptachlor	zero	0.0004	Liver damage; risk of cancer	Residue of banned termiticide
Heptachlor epoxide	zero	0.0002	Liver damage; risk of cancer	Breakdown of hepatachlor
Hexachlorobenzene	zero	0.001	Liver or kidney problems; reproductive difficulties; risk of cancer	Discharge from metal refineries and agricultural chemical factories

Organic Chemicals	MCLG[1] (mg/L)[2]	MCL or TT[1] (mg/L)[2]	Potential Health Effects from Ingestion of Water	Sources of Contaminant in Drinking Water
Hexachlorocyclopen tadiene	0.05	0.05	Kidney or stomach problems	Discharge from chemical factories
Lindane	0.0002	0.0002	Liver or kidney problems	Runoff/leaching from insecticide used on catttle, lumber, gardens
Methoxychlor	0.04	0.04	Reproductive difficulties	Runoff/leaching from insecticide used on fruits, vegetables, alfalfa, livestock
Oxamyl (Vydate)	0.2	0.2	Slight nervous system effects	Runoff/leaching from insecticide used on apples, potatoes, and tomatoes
Polychlorinated biphenyls (PCBs)	zero	0.0005	Skin changes; thymus gland problems; immune deficiencies; reproductive or nervous system difficulties; increased risk of cancer	Runoff from landfils; discharge of waste chemicals
Pentachlorophenol	zero	0.001	Liver or kidney problems; increased risk of cancer	Discharge from wood preserving factories
Picloram	0.5	0.5	Liver problems	Herbicide runoff
Simazine	0.004	0.004	Problems with blood	Herbicide runoff
Styrene	0.1	0.1	Liver, kidney, and circulatory problems	Discharge from rubber and plastic factories; leaching from landfills
Tetrachloroethylene	zero	0.005	Liver problems; increased risk of cancer	Discharge from factories and dry cleaners
Toluene	1	1	Nervous system, kidney, or liver problems	Discharge from petroleum factories

Organic Chemicals	MCLG[1] (mg/L)[2]	MCL or TT[1] (mg/L)[2]	Potential Health Effects from Ingestion of Water	Sources of Contaminant in Drinking Water
Toxaphene	zero	0.003	Kidney, liver, or thyroid problems; increased risk of cancer	Runoff/leaching from insecticide used on cotton and cattle
2,4,5-TP (Silvex)	0.05	0.05	Liver problems	Residue of banned herbicide
1,2,4-Trichlorobenzene	0.07	0.07	Changes in adrenal glands	Discharge from textile finishing factories
1,1,1-Trichloroethane	0.20	0.2	Liver, nervous system, or circulatory problems	Discharge from metal degreasing sites and other factories
1,1,2-Trichloroethane	0.003	0.005	Liver, kidney, or immune system problems	Discharge from industrial chemical factories
Trichloroethylene	zero	0.005	Liver problems; increased risk of cancer	Discharge from petroleum refineries
Vinyl chloride	zero	0.002	Increased risk of cancer	Leaching from PVC pipes; discharge from plastic factories
Xylenes (total)	10	10	Nervous system damage	Discharges from petroleum and chemical plants

Table 5. NPDW Regulations for Radionuclides.

Radionuclides	MCLG[1] (mg/L)[2]	MCL or TT[1] (mg/L)[2]	Potential Health Effects from Ingestion of Water	Sources of Contaminant in Drinking Water
Alpha particles	none[7] ---------- as of 12/08/03: zero	15 picocuries per Liter (pCi/L)	Increased risk of cancer	Erosion of natural deposits

Radionuclides	MCLG[1] (mg/L)[2]	MCL or TT[1] (mg/L)[2]	Potential Health Effects from Ingestion of Water	Sources of Contaminant in Drinking Water
Beta particles and photon emitters	none[7] ---------- as of 12/08/03: zero	4 millirems per year	Increased risk of cancer	Decay of natural and man-made deposits
Radium 226 and Radium 228 (combined)	none[7] ---------- as of 12/08/03: zero	5 pCi/L	Increased risk of cancer	Erosion of natural deposits
Uranium	as of 12/08/03: zero	as of 12/08/03: 30 ug/L	Increased risk of cancer, kidney toxicity	Erosion of natural deposits

The following footnotes apply to the above tables.

[1] Definitions: Refer to the discussion box on page 12.

[2] Units are in milligrams per liter (mg/L) unless otherwise noted. Milligrams per liter are equivalent to parts per million.

[3] EPA's surface water treatment rules require systems using surface water or ground water under the direct influence of surface water to (1) disinfect their water, and (2) filter their water or meet criteria for avoiding filtration so that the following contaminants are controlled at the following levels:

- *Cryptosporidium:* (as of January 1, 2002) 99% removal/inactivation
- *Giardia lamblia:* 99.9% removal/inactivation
- Viruses: 99.99% removal/inactivation
- *Legionella:* No limit, but EPA believes that if *Giardia* and viruses are removed/inactivated, *Legionella* will also be controlled.
- Turbidity: At no time can turbidity (cloudiness of water) go above 5 nephelolometric turbidity units (NTU); systems that filter must ensure that the turbidity go no higher than 1 NTU (0.5 NTU for conventional or direct filtration) in at least 95% of the daily samples in any month. As of January 1, 2002, turbidity may never exceed 1 NTU, and must not exceed 0.3 NTU in 95% of daily samples in any month
- HPC: No more than 500 bacterial colonies per milliliter.

[4] No more than 5.0% samples total coliform-positive in a month. (For water systems that collect fewer than 40 routine samples per month, no more than one sample can be total coliform-positive). Every sample that has total coliforms must be analyzed for fecal coliforms. There may not be any fecal coliforms or *E. coli*.
[5] Fecal coliform and *E. coli* are bacteria whose presence indicates that the water may be contaminated with human or animal wastes. Disease-causing microbes (pathogens) in these wastes can cause diarrhea, cramps, nausea, headaches, or other symptoms. These pathogens may pose a special health risk for infants, young children, and people with severely compromised immune systems.
[6] Although there is no collective MCLG for this contaminant group, there are individual MCLGs for some of the individual contaminants:

- Trihalomethanes: bromodichloromethane (zero); bromoform (zero); dibromochloromethane (0.06 mg/L). Chloroform is regulated with this group but has no MCLG.
- Haloacetic acids: dichloroacetic acid (zero); trichloroacetic acid (0.3 mg/L). Monochloroacetic acid, bromoacetic acid, and dibromoacetic acid are regulated with this group but have no MCLGs.

[7] MCLGs were not established before the 1986 Amendments to the Safe Drinking Water Act. Therefore, there is no MCLG for this contaminant.

[8] Lead and copper are regulated by a Treatment Technique that requires systems to control the corrosiveness of their water. If more than 10% of tap water samples exceed the action level, water systems must take additional steps. For copper, the action level is 1.3 mg/L, and for lead is 0.015 mg/L.
[9] Each water system must certify, in writing, to the state (using third party or manufacturer's certification) that when acrylamide and epichlorohydrin are used in drinking water systems, the combination (or product) of dose and monomer level does not exceed the levels specified, as follows:

- Acrylamide = 0.05% dosed at 1 mg/L (or equivalent)
- Epichlorohydrin = 0.01% dosed at 20 mg/L (or equivalent)

NATIONAL SECONDARY DRINKING WATER REGULATIONS

National Secondary Drinking Water Regulations (NSDWRs or secondary standards) are non-enforceable guidelines regulating contaminants that may cause cosmetic effects (such as skin or tooth discoloration) or aesthetic effects (such as taste, odor, or color) in drinking water. EPA recommends secondary standards to water systems but does not require systems to comply. However, states may choose to adopt them as enforceable standards. The following table summarizes the secondary standards.

Table 6. Summary of National Secondary Drinking Water Regulations.

Contaminant	Secondary Standard
Aluminum	0.05 to 0.2 mg/L
Chloride	250 mg/L
Color	15 (color units)
Copper	1.0 mg/L
Corrosivity	noncorrosive
Fluoride	2.0 mg/L
Foaming Agents	0.5 mg/L
Iron	0.3 mg/L
Manganese	0.05 mg/L
Odor	3 threshold odor number
pH	6.5-8.5
Silver	0.10 mg/L
Sulfate	250 mg/L
Total Dissolved Solids	500 mg/L
Zinc	5 mg/L

THE CLEAN WATER ACT

Drinking water standards are not the only regulations we need to comply with in the U.S. Today's Clean Water Act has its origins from the late 1940s. The original 1948 statute (Chapter 758; PL 845), the Water Pollution Control Act, authorized the Surgeon General of the Public Health Service, in cooperation with other federal, state, and local entities, to prepare comprehensive programs for eliminating or reducing the pollution of interstate waters and tributaries and improving the sanitary condition of surface and underground waters. Since 1948, the original statute has been amended extensively to authorize additional water quality programs, standards and procedures to govern allowable discharges, and funding for construction grants or general programs. Amendments in other years provided for continued authority to conduct program activities or administrative changes to related activities. This legislation was originally enacted as the Federal Water Pollution Control Act of 1972, and was amended in 1977 and renamed the Clean Water Act. It was reauthorized in 1991.

The Clean Water Act strives to restore and maintain the chemical, physical, and biological integrity of the nation's water. The act sets up a system of water quality standards, discharge limitations, and permits. If a project may result in the

placement of material into waters, a Corps of Engineers' Dredge and Fill Permit (Section 404) may be required. The permit also pertains to activities in wetlands and riparian areas. Certain Federal projects may be exempt from the requirements of Section 404, if the conditions set forth in section 404(r) are met. Before either a National Pollutant Discharge Elimination System (NPDES) (Section 402) or Section 404 permit can be issued, the applicant must obtain a Section 401 certification. This declaration states that any discharge complies with all applicable effluent limitations and water quality standards.
If the water quality of a water body is potentially affected by a proposed action (e.g., construction of a wastewater treatment plant), a NPDES permit may be required. The Environmental Protection Agency is responsible for this program; however, in most cases, has turned this responsibility over to the states as long as the individual state program is acceptable.
Section 319, Nonpoint Source Management Programs, was added to the Clean Water Act by PL 100-4 to have the states establish nonpoint source management plans designed. Section 319 (k) requires each Federal department and agency to allow states to review individual development projects and assistance applications and accommodate, in accordance with Executive Order 12372, the concerns of the state regarding the consistency of these applications or projects with the state nonpoint source pollution management program.

Congress enacted the most recent major amendments to the Clean Water Act in 1987 (P.L. 100-4). Since then, the EPA, states, and others have been working to implement the many program changes and additions mandated in the law. At issue today, as it has been for some time, is what progress EPA and the states are making. In general, many states and environmental groups fault EPA for delays in issuing guidance and assistance needed to carry out the provisions of the law. EPA and others are critical of states, in turn, for not reaching beyond conventional knowledge and institutional approaches to address their water quality problems. Environmental groups have been criticized for insufficient recognition of EPA's and states' need for flexibility to implement the Act. Finally, Congress has been criticized for not providing adequate funding and resources to meet EPA and state needs.
Three issues have predominated recently in connection with implementation of the law. The first involves implementation of requirements under current law for states to develop total maximum daily loads (TMDLs) to restore pollution-impaired waters. The second issue involves the nonpoint pollution management provisions added in 1987. States are developing management programs describing methods that will be used to reduce nonpoint pollution, which may be responsible for as much as 50% of the nation's remaining water quality problems. Most observers agree that implementation of nonpoint source control measures is significantly hindered by lack of resources, including federal assistance. EPA adopted program guidance intended to give states more flexibility and to speed up progress in nonpoint source control. The third issue is funding to construct municipal wastewater treatment plants under the State Revolving Fund provisions of the 1987

amendments. Budgetary constraints on federal aid for wastewater treatment and large remaining funding needs are a continuing concern.
Since 1993, EPA has begun a number of agency-wide and program-specific reforms focusing on flexibility and "common sense" approaches to regulation. Many of these will affect implementation of water quality programs. In February 1998, the Clinton Administration released a multi-agency Clean Water Action Plan intended to build on the environmental successes of the Act and address the nation's remaining water quality challenges. But a new administration in the White House may alter this. Reauthorization of the Act was on the agenda of the 104th Congress, when the House passed H.R. 961, but no Clean Water Act amendments were enacted. Comprehensive legislation was not introduced in the 105th Congress, and no major House or Senate committee activity occurred. Similarly, no major activity occurred in the 106th Congress, although there was action on individual program areas within the Act. Recent attention has focused on new EPA rules for the Act's TMDL program. Congressional oversight of the TMDL program is likely in the 107th Congress (see CRS Issue Brief IB10069, *Clean Water Act Issues in the 107th Congress*.)

For our discussions, it is important to recognize that the Federal Water Pollution Control Act, or Clean Water Act, is the principal law governing pollution in the nation's streams, lakes, and estuaries. Originally enacted in 1948, it was totally revised by amendments in 1972 (P.L. 92-500) that gave the Act its current form and spelled out ambitious programs for water quality improvements that are now being put in place by industries and cities. Congress made certain fine-tuning amendments in 1977 (P.L. 95-217) and 1981 (P.L. 97-117). The Act consists of two major parts: regulatory provisions that impose progressively more stringent requirements on industries and cities in order to meet the statutory goal of zero discharge of pollutants, and provisions that authorize federal financial assistance for municipal wastewater treatment construction. Industries were to meet pollution control limits first by use of Best Practicable Technology (BPTs) and later by improved Best Available Technology (BAT). Cities were to achieve secondary treatment of municipal wastewater (roughly 85% removal of conventional wastes), or better if needed to meet water quality standards. Both major parts are supported by research activities authorized in the law, plus permit and penalty provisions for enforcement. These programs are administered by the EPA, while state and local governments have the principal day-to-day responsibility for implementing the law. The most recent amendments, enacted in February 1987, are the Water Quality Act of 1987 (P.L. 100-4). These amendments culminated 6 years of congressional efforts to extend and revise the Act and are the most comprehensive amendments to it since 1972. They recognize that, despite much progress to date, significant water quality problems persist. Among its many provisions, the 1987 legislation:

- established a comprehensive program for controlling toxic pollutant discharges, beyond that already provided in the Act, to respond to so-called "toxic hot spots;"

- added a program requiring states to develop and implement programs to control nonpoint sources of pollution, or rainfall runoff from farm and urban areas, plus construction, forestry, and mining sites;
- authorized a total of $18 billion in aid for wastewater treatment assistance under a combination of the Act's traditional construction grants program through FY1990 and, as a transition to full state funding responsibility, a new program of grants to capitalize State Revolving Funds, from FY1989-1994;
- authorized or modified a number of programs to address water pollution problems in diverse geographic areas such as coastal estuaries, the Great Lakes, and the Chesapeake Bay; and
- revised many of the Act's regulatory, permit, and enforcement programs.

Section 303(d) of the Clean Water Act requires states to identify pollutant-impaired water segments and develop "total maximum daily loads" (TMDLs) that set the maximum amount of pollution that a water body can receive without violating water quality standards. If a state fails to do so, EPA is required to develop a priority list for the state and make its own TMDL determination. Most states have lacked the resources to do TMDL analyses, which involve complex assessment of point and nonpoint sources and mathematical modeling, and EPA has both been reluctant to override states and has also lacked resources to do the analyses. Thus, there has been little implementation of the provision that Congress enacted in 1972. In recent years, national and local environmental groups have filed more than 40 lawsuits in 38 states against EPA and states for failure to fulfill requirements of the Act. Of the suits tried or settled to date, 19 have resulted in court orders requiring expeditious development of TMDLs. EPA and state officials have been concerned about diverting resources from other high-priority water quality activities in order to meet the courts' orders. In October 1996, EPA created an advisory committee to solicit advice on the TMDL problem. Recommendations from the advisory committee, received in June 1998, formed the basis of program changes that EPA proposed in August 1999. The 1999 proposal set forth criteria for states, territories, and authorized Indian tribes to identify impaired waters and establish all TMDLs within 15 years. It would require more comprehensive assessments of waterways, detailed cleanup plans, and timetables for implementation.

The 1987 amendments added a new Section 319 to the Act, under which states were required to develop and implement programs to control nonpoint sources of pollution, or rainfall runoff from farm and urban areas, as well as construction, forestry, and mining sites. Previously, the Act had largely focused on controlling point sources, while helping states and localities to plan for management of diverse nonpoint sources. Yet, as industrial and municipal sources have abated pollution, uncontrolled nonpoint sources have become a relatively larger portion of remaining water quality problems -- perhaps contributing as much as 50% of the nation's water pollution. Table 7 provides a summary of the major sections of the Clean Water Act.

Table 7. Major U.S. Code Sections of the Clean Water Act[1] (codified generally as 33 U.S.C. 1251-1387)

33 U.S.C.	Section Title	Clean Water Act (as amended)
Subchapter I -	Research and Related Programs	
1251	Congressional declaration of goals and policy	sec. 101
1252	Comprehensive programs for water pollution control	sec. 102
1253	Interstate cooperation and uniform laws	sec. 103
1254	Research, investigations, training and information	sec. 104
1255	Grants for research and development	sec. 105
1256	Grants for pollution control programs	sec. 106
1257	Mine water pollution demonstrations	sec. 107
1258	Pollution control in the Great Lakes	sec. 108
1259	Training grants and contracts	sec. 109
1260	Applications for training grants and contracts, allocations	sec. 110
1261	Scholarships	sec. 111
1262	Definitions and authorization	sec. 112
1263	Alaska village demonstration project	sec. 113
1265	In-place toxic pollutants	sec. 115
1266	Hudson River reclamation demonstration project	sec. 116
1267	Chesapeake Bay	sec. 117
1268	Great Lakes	sec. 118
1269	Long Island Sound	sec. 119
1270	Lake Champlain management conference	sec. 120
Subchapter II -	Grants for Construction of Treatment Works	
1281	Congressional declaration of purpose	sec. 201

33 U.S.C.	Section Title	Clean Water Act (as amended)
1282	Federal share	sec. 202
1283	Plans, specifications, estimates, and payments	sec. 203
1284	Limitations and conditions	sec. 204
1285	Allotment of grant funds	sec. 205
1286	Reimbursement and advanced construction	sec. 206
1287	Authorization of appropriations	sec. 207
1288	Areawide waste treatment management	sec. 208
1289	Basin planning	sec. 209
1290	Annual survey	sec. 210
1291	Sewage collection system	sec. 211
1292	Definitions	sec. 212
1293	Loan guarantees	sec. 213
1294	Wastewater recycling and reuse information and education	sec. 214
1295	Requirements for American materials	sec. 215
1296	Determination of priority	sec. 216
1297	Guidelines for cost effective analysis	sec. 217
1298	Cost effectiveness	sec. 218
1299	State certification of projects	sec. 219
Subchapter III -	Standards and Enforcement	
1311	Effluent Limitations	sec. 301
1312	Water quality-related effluent limitations	sec. 302
1313	Water quality standards and implementation plans	sec. 303
1314	Information and guidelines	sec. 304
1315	State reports on water quality	sec. 305
1316	National standards of performance	sec. 306
1317	Toxic and pretreatment effluent standards	sec. 307
1318	Records and reports, inspections	sec. 308
1319	Enforcement	sec. 309

33 U.S.C.	Section Title	Clean Water Act (as amended)
1320	International pollution abatement	sec. 310
1321	Oil and hazardous substance liability	sec. 311
1322	Marine sanitation devices	sec. 312
1323	Federal facility pollution control	sec. 313
1324	Clean lakes	sec. 314
1325	National study commission	sec. 315
1326	Thermal discharges	sec. 316
1327	Omitted (alternative financing)	sec. 317
1328	Aquaculture	sec. 318
1329	Nonpoint source management program	sec. 319
1330	National estuary study	sec. 320
Subchapter IV -	Permits and Licenses	
1341	Certification	sec. 401
1342	National pollutant discharge elimination system	sec. 402
1343	Ocean discharge criteria	sec. 403
1344	Permits for dredge and fill materials	sec. 404
1345	Disposal or use of sewage sludge	sec. 405
Subchapter V -	General Provisions	
1361	Administration	sec. 501
1362	Definitions	sec. 502
1363	Water pollution control advisory board	sec. 503
1364	Emergency powers	sec. 504
1365	Citizen suits	sec. 505
1366	Appearance	sec. 506
1367	Employee protection	sec. 507
1368	Federal procurement	sec. 508
1369	Administrative procedure and judicial review	sec. 509
1370	State authority	sec. 510

33 U.S.C.	Section Title	Clean Water Act (as amended)
1371	Authority under other laws and regulations	sec. 511
1372	Labor standards	sec. 513
1373	Public health agency coordination	sec. 514
1374	Effluent standards and water quality information advisory committee	sec. 515
1375	Reports to Congress	sec. 516
1376	Authorization of appropriations	sec. 517
1377	Indian tribes sec.	sec. 518
Subchapter VI -	State Water Pollution Control Revolving Funds	
1381	Grants to states for establishment of revolving funds	sec. 601
1382	Capitalization grant agreements	sec. 602
1383	Water pollution control revolving loan funds	sec. 603
1384	Allotment of funds	sec. 604
1385	Corrective actions	sec. 605
1386	Audits, reports, fiscal controls, intended use plan	sec. 606
1387	Authorization of appropriations	sec. 607

Footnote: *This table shows only the major code sections. For more detail and to determine when a section was added, the reader should consult the official printed version of the U.S. Code.*

INTRODUCING THE PHYSICAL TREATMENT METHODS

The following technologies are among the most commonly used physical methods of purifying water:

Heat Treatment - Boiling is one way to purify water of all pathogens. Most experts feel that if the water reaches a rolling boil it is safe. A few still hold out for maintaining the boiling for some length of time, commonly 5 or 10 minutes, plus an extra minute for every 1000 feet of elevation. One reason for the long period of boiling is to inactivate bacterial spores (which can survive boiling), but these spore are unlikely to be waterborne pathogens. Water can also be treated at below boiling temperatures, if contact time is increased. Commercial units are available for residential use, which treat 500 gals of water per day at an estimated cost of \$1/1000 gallons for the energy. The process is similar to milk pasteurization, and holds the water at 161° F for 15 seconds. Heat exchangers recover most of the energy used to warm the water. Solar pasteurizers have also been built that can heat three gallons of water to 65° C and hold the temperature for an hour. A higher temperature could be reached, if the device was rotated east to west during the day to follow the sunlight. Regardless of the method, heat treatment does not leave any form of residual to keep the water free of pathogens in storage.

Reverse Osmosis - Reverse osmosis forces water, under pressure, through a membrane that is impermeable to most contaminants. The membrane is somewhat better at rejecting salts than it is at rejecting non-ionized weak acids and bases and smaller organic molecules (molecular weight below 200). In the latter category are undissociated weak organic acids, amines, phenols, chlorinated hydrocarbons, some pesticides and low molecular weight alcohols. Larger organic molecules and all pathogens are rejected. Of course, it is possible to have a imperfection in the membrane that could allow molecules or whole pathogens to pass through. Using reverse osmosis to desalinate seawater requires considerable pressure (1000 psi) to operate. Reverse osmosis filters are available that will use normal municipal or private water pressure to remove contaminates from water. The water produced by reverse osmosis, like distilled water, will be close to pure H_2O. Therefore mineral intake may need to be increased to compensate for the normal mineral content of water in much of the world.

Distillation - Distillation is the evaporation and condensation of water to purify water. Distillation has two disadvantages: 1) A large energy input is required and 2) If simple distillation is used, chemical contaminants with boiling points below water will be condensed along with the water. Distillation is most commonly used to remove dissolved minerals and salts from water. The simplest form of a distillation for use in the home is a solar still. A solar still uses solar radiation to evaporate water below the boiling point, and the cooler ambient air to condense the vapor. The water can be extracted from the soil, vegetation piled in the still, or contaminated water (such as radiator fluid or salt water) can be added to the still. While per still output is low, they are an important technique if water is in short supply. Other forms of distillation require a concentrated heat source to boil water which is then condensed. Simple stills use a coiling coil to return this heat to the environment. Efficient distillations plants use a vapor compression cycle where the water is boiled off at atmospheric pressure, the steam is compressed, and the condenser condenses the steam above the boiling point of the water in the boiler, returning the heat of fusion to the boiling water. The hot condensed water is run through a second heat exchanger, which heats up the water feeding into the boiler. These plants normally use an internal combustion engine to run the compressor. Waste heat from the engine, including the exhaust, is used to start the process and make up any heat loss.

Microfilters - Microfilters are small-scale filters designed to remove cysts, suspended solids, protozoa, and, in some cases, bacteria from water. Most filters use a ceramic or fiber element that can be cleaned to restore performance as the units are used. Most units and almost all made for camping use a hand pump to force the water through the filter. Others use gravity, either by placing the water to be filtered above the filter (e.g. the Katadyn drip filter), or by placing the filter in the water, and running a siphon hose to a collection vessel located below the filter (e.g. Katadyn siphon filter). Microfilters are the only method, other than boiling, to remove Cryptosporidia. Microfilters do not remove viruses, which many experts do not consider to be a problem in North America. Despite this, the Katadyn microfilter has seen considerable use around the world by NATO-member militaries, WHO, UNHCR, and other aid organizations. Microfilters share a problem with charcoal filter in having bacteria grow on the filter medium. Some handle this by impregnating the filter element with silver, such as the Katadyn, others advise against storage of a filter element after it has been used. Many microfilters may include silt prefilters, activated charcoal stages, or an iodine resin. Most filters come with a stainless steel prefilter, but other purchased or improvised filters can be added to reduce the loading on the main filter element. Allowing time for solids to settle, and/or prefiltering will also extend filter life. Iodine matrix filters will kill viruses that will pass through the filter, and if a charcoal stage is used it will remove much of the iodine from the water. Charcoal filters will also remove other dissolved natural or manmade contaminates. Both the iodine and the charcoal stages do not indicate when they reach their useful life, which is much shorter than the filter element.

Slow Sand Filter - Slow sand filters pass water slowly through a bed of sand. Pathogens and turbidity are removed by natural die-off, biological action, and filtering. Typically the filter will consist of a layer of sand, then a gravel layer in which the drain pipe is embedded. The gravel doesn't touch the walls of the filter, so that water can't run quickly down the wall of the filter and into the gravel. Building the walls with a rough surface also helps. A typical loading rate for the filter is 0.2 meters/hour day (the same as 0.2 m^3/m^2 of surface area). The filter can be cleaned several times before the sand has to be replaced. Slow sand filters should only be used for continuous water treatment. If a continuous supply of raw water can't be insured (say, using a holding tank), then another method should be chosen. It is also important for the water to have as low turbidity (suspended solids) as possible. Turbidity can be reduced by changing the method of collection (for example, building an infiltration gallery, rather than taking water directly from a creek), allowing time for the material to settle out (using a raw water tank), prefiltering or flocculation (adding a chemical, such as alum to cause the suspended material to floc together.) The SSF filter itself is a large box. The walls should be as rough as possible to reduce the tendency for water to run down the walls of the filter, bypassing the sand. The bottom layer of the filter is a gravel bed, in which a slotted pipe is placed to drain off the filtered water. The slots or the gravel should be no closer than 20 cm to the walls, again, to prevent the water from bypassing the sand. The sand for a SSF needs to be clean and uniform, and of the correct size. The sand can be cleaned in clean running water , even if it is in a creek. The ideal specs on sand are effective size (sieve size through which 10% of the sand passes) between 0.15 and 0.35 mm, uniformity coefficient (ratio of sieve sizes through which 60% pass and through which 10% pass) of less than 3; maximum size of 3 mm, and minimum size of 0.1 mm. The sand is added to a SSF to a minimum depth of 0.6 meters. Additional thickness will allow more cleanings before the sand must be replaced. 0.3 to 0.5 meters of extra sand will allow the filter to work for 3-4 years. An improved design uses a *geotextile* layer on top of the sand to reduce the frequency of cleaning. The outlet of a SSF must be above the sand level, and below the water level. The water must be maintained at a constant level to insure an even flow rate throughout the filter. The flow rate can be increased by lowering the outlet pipe, or increasing the water level. While the SSF will begin to work at once, optimum treatment for pathogens will take a week or more. During this time the water should be chlorinated, if at all possible (iodine can be substituted). After the filter has stabilized, the water should be safe to drink, but chlorinating of the output is still a good idea, particularly to prevent recontamination. As the flow rate slows down the filter will have to be cleaned by draining and removing the top few inches of sand. If a geotextile filter is used, only the top ½" may have to be removed. As the filter is refilled, it will take a few days for the biological processes to reestablish themselves.

Activated Charcoal Filter - Activated charcoal filters water through adsorption; chemicals and some heavy metals are attracted to the surface of the charcoal, and are attached to it. Charcoal filters will filter some pathogens, though they will quickly use up the filter adsorptive ability, and can even contribute to contamination, as the charcoal provides an excellent breeding ground for bacteria and algae. Some charcoal filters are available impregnated with silver to prevent this, though current research concludes that the bacteria growing on the filter are harmless, even if the water wasn't disinfected before contacting the filter. Activated charcoal can be used in conjunction with chemical treatment. The chemical (iodine or chlorine) will kill the pathogens, while the carbon filter will remove the treatment chemicals. In this case, as the filter reaches its capacity, a distinctive chlorine or iodine taste will be noted. The more activated charcoal in a filter, the longer it will last. The bed of carbon must be deep enough for adequate contact with the water. Production designs use granulated activated charcoal (effective size or 0.6 to 0.9 mm for maximum flow rate). Home or field models can also use a compressed carbon block or powered activated charcoal (effective size 0.01) to increase contact area. Powered charcoal can also be mixed with water and filtered out later. As far as life of the filter is concerned, carbon block filters will last the longest for a given size, simply due to their greater mass of carbon. A source of pressure is usually needed with carbon block filters to achieve a reasonable flow rate.

INTRODUCING CHEMICAL TREATMENT

CHLORINE

Chlorine is familiar to most people as it is used to treat virtually all municipal water systems in the United States. Chlorine has a number of problems when used for field treatment of water. When chlorine reacts with organic material, it attaches itself to nitrogen containing compounds (ammonium ions and amino acids), leaving less free chlorine to continue disinfection. Carcinogenic trihalomethanes are also produced, though this is only a problem with long-term exposure. Trihalomethanes can also be filtered out with a charcoal filter, though it is more efficient to use the same filter to remove organics before the water is chlorinated. Unless free chlorine is measured, disinfection can not be guaranteed with moderate doses of chlorine. One solution is superchlorination, the addition of far more chlorine than is needed. This must again be filtered through activated charcoal to remove the large amounts of chlorine, or hydrogen peroxide can be added to drive the chlorine off. Either way there is no residual chlorine left to prevent recontamination. This isn't a problem, if the water is to be used at once.
Chlorine is sensitive to both the pH and temperature of the treated water. Temperature slows the reaction for any chemical treatment, but chlorine treatment is particularly susceptible to variations in the pH as at lower pHs, hypochlorous acid is formed, while at higher pHs, it will tend to dissociate into hydrogen and chlorite ions, which are less effective as a disinfectant. As a result, chlorine effectiveness drops off when the pH is greater than 8.Ordinary household bleach (such as Clorox) in the U.S. contains 5.25% sodium hypochlorite (NaOCL) and can be used to purify water if it contains no other active ingredients, scents, or colorings. Some small treatment plants in Africa produce their own sodium hypochlorite on site from the electrolysis of brine. Power demands range from 1.7 to 4 kWh per lb. of NaOCL. 2 to 3.5 lbs. of salt are needed for each pound of NaOCL. These units are fairly simple and are made in both the U.S. and the U.K. Another system, designed for China, where the suitable raw materials were mined or manufactured locally, used a reaction between salt, manganese dioxide, and sulfuric acid to produce chlorine gas. The gas was then allowed to react with slaked lime to produce a bleaching powder that could then be used to treat water. A heat source is required to speed the reaction up. Bleaching Powder (or Chlorinated Lime) is sometimes used at the industrial scale. Bleaching powder is 33-37% chlorine when produced, but losses its chlorine rapidly, particularly when exposed to air, light or moisture.
Calcium Hypochlorite, also known as High Test Hypochlorite (HTH) is supplied in crystal form; it is nearly 70% available chlorine. One product, the Sanitizer (formally the Sierra Water Purifier) uses these crystals to superchlorinate the water to insure pathogens were killed off, then hydrogen peroxide is added to drive off

the residual chlorine. This is the most effective method of field chlorine treatment. The U.S. military and most aid agencies also use HTH to treat their water, though a test kit, rather than superchlorination, is used to insure enough chlorine is added. This is preferable for large-scale systems, as the residual chlorine will prevent recontamination. Usually bulk water treatment plants first dilute to HTH to make a 1% working solution at the rate of 14g HTH per liter of water. While testing to determine exact chlorine needs are preferable, the solution can be used at the dose rate of 8 drops/gallon, or for larger quantities, 1 part of 1% solution to 10,000 parts clear water. Either of these doses will result in 1 PPM chlorine and may need to be increased, if the water wasn't already filtered by other means.

IODINE

Iodine's use as a water purification method emerged after World War 2, when the U.S. military was looking for a replacement for Halazone tablets. Iodine was found to be in many ways superior to chlorine for use in treating small batches of water. Iodine is less sensitive to the pH and organic content of water, and is effective in lower doses. Some individuals are allergic to iodine, and there is some question about long term use of iodine. The safety of long-term exposure to low levels of iodine was proven when inmates of three Florida prisons were given water disinfected with 0.5 to 1.0 ppm iodine for 15 years. No effects on the health or thyroid function of previously healthy inmates was observed. Of 101 infants born to prisoners drinking the water for 122- 270 days, none showed detectable thyroid enlargement. However, 4 individuals with preexisting cases of hyperthyroidism became more symptomatic, while consuming the water. Nevertheless, experts are reluctant to recommend iodine for long term use. Average American iodine intake is estimated at 0.24 to 0.74 mg/day, higher than the RDA of 0.4 mg/day. Due to a recent National Academy of Science recommendation that iodine consumption be reduced to the RDA, the EPA discourages the use of iodized salt in areas where iodine is used to treat drinking water. Iodine is normally used in doses of 8 PPM to treat clear water for a 10 minute contact time. The effectiveness of this dose has been shown in numerous studies. Cloudy water needs twice as much iodine or twice as much contact time. In cold water (Below 41° F or 5° C) the dose or time must also be doubled. In any case doubling the treatment time will allow the use of half as much iodine. These doses are calculated to remove all pathogens (other than cryptosporida) from the water. Of these, giardia cysts are the hardest to kill, and are what requires the high level of iodine. If the cysts are filtered out with a microfilter (any model will do since the cysts are 6 μm), only 0.5 ppm is needed to treat the resulting water.

Water treated with iodine can have any objectionable taste removed by treating the water with vitamin C (ascorbic acid), but it must be added after the water has stood for the correct treatment time. Flavored beverages containing vitamin C will accomplish the same thing. Sodium thiosulfate can also be used to combine with free iodine, and either of these chemicals will also help remove the taste of chlorine as well. Usually elemental iodine can't be tasted below 1 ppm, and below 2 ppm the taste isn't objectionable. Iodine ions have an even higher taste threshold of 5 ppm. Note that removing the iodine taste does not reduce the dose of iodine ingested by the body.

SILVER

Silver has been suggested by some for water treatment and may still be available outside the U.S. Its use is currently out of favor due to the EPA's establishment of a 50 ppb MCL (Maximum Contaminate Level) limit on silver in drinking water. This limit is set to avoid *argyrosis*, a cosmetic blue/gray staining of the skin, eyes, and mucous membranes. As the disease requires a net accumulation of 1 g of silver in the body, one expert calculated that you could drink water treated at 50 ppb for 27 years before accumulating 1 g. Silver has only be proven to be effective against bacteria and protozoan cysts, though it is quite likely also effective against viruses. Silver can be used in the form of a silver salt, commonly silver nitrate, a colloidal suspension, or a bed of metallic silver. Electrolysis can also be used to add metallic silver to a solution. Some evidence has suggested that silver deposited on carbon block filters can kill pathogens without adding as much silver to the water.

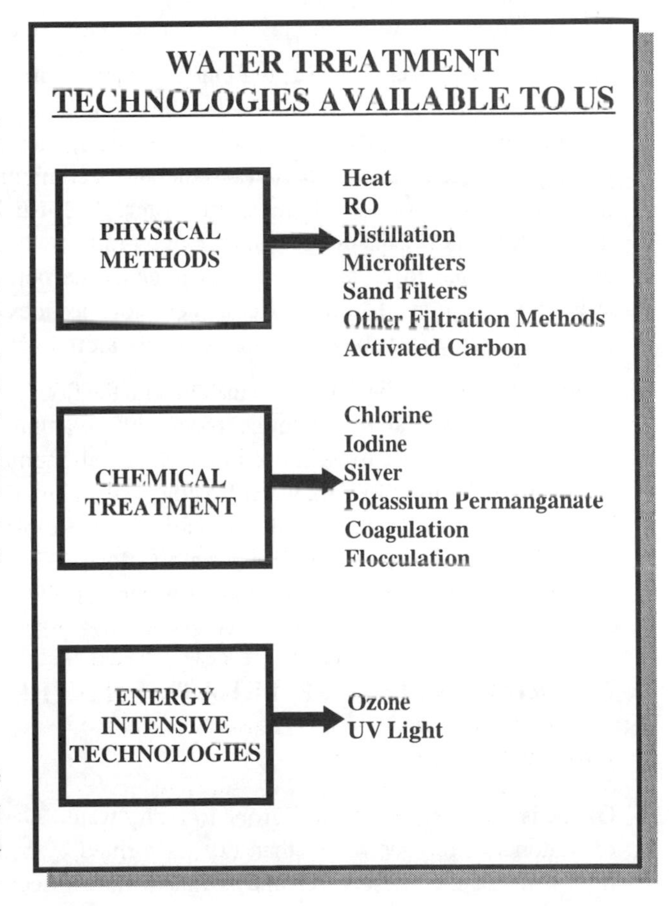

POTASSIUM PERMANGANATE

Potassium permanganate is no longer commonly used in the developed world to kill pathogens.

It is much weaker than the other alternatives cited, more expensive, and leaves a objectionable pink or brown color. Still, some underdeveloped countries rely on it, especially in home-use applications. If it must be used, 1 gram per liter would probably be sufficient against bacteria and viruses (no data is available on it effectiveness against protozoan cysts). Hydrogen Peroxide can be used to purify water if nothing else is available. Studies have shown of 99 percent inactivation of poliovirus in 6 hr with 0.3 percent hydrogen peroxide and a 99% in-activation of rhinovirus with a 1.5% solution in 24 minutes. Hydrogen Peroxide is more effective against bacteria, though Fe^{+2} or Cu^{+2} needs to be present as a catalyst to get a reasonable concentration-time product.

COAGULATION/FLOCCULATION AGENTS

While flocculation doesn't kill pathogens, it will reduce their levels along with removing particles that could shield the pathogens from chemical or thermal destruction, and organic matter that could tie up chlorine added for purification. 60-98% of coliform bacteria, 65-99% of viruses, and 60-90% of giardia will be removed from the water, along with organic matter and heavy metals.

Some of the advantages of coagulation/flocculation can be obtained by allowing the particles to settle out of the water with time (sedimentation), but it will take a while for them to do so. Adding coagulation chemicals, such as alum, will increase the rate at which the suspended particles settle out by combining many smaller particles into larger floc, which will settle out faster. The usual dose for alum is 10-30 mg/liter of water. This dose must be rapidly mixed with the water, then the water must be agitated for 5 minutes to encourage the particles to form flocs. After this at least 30 minutes of settling time is need for the flocs to fall to the bottom, and them the clear water above the flocs may be poured off.

Most of the flocculation agent is removed with the floc, nevertheless, some question the safety of using alum due to the toxicity of the aluminum in it. There is little to no scientific evidence to back this up. Virtually all municipal plants in the US dose the water with alum. In bulk water treatment, the alum dose can be varied until the idea dose is found. The needed dose varies with the pH of the water and the size of the particles. Increase turbidity makes the flocs easier to produce not harder, due to the increased number of collisions between particles.

ENERGY INTENSIVE TREATMENT TECHNOLOGIES

OZONE

Ozone is used extensively in Europe to purify water. Ozone, a molecule composed of 3 atoms of oxygen rather than two, is formed by exposing air or oxygen to a high voltage electric arc. Ozone is much more effective as a disinfectant than chlorine, but no residual levels of disinfectant exist after ozone turns back into O_2. (One source quotes a half life of only 120 minutes in distilled water at 20 °C). Ozone is expected to see increased use in the U.S. as a way to avoid the production and formation of trihalomethanes, and while ozone does break down organic molecules, sometimes this can be a disadvantage as ozone treatment can produce higher levels of smaller molecules that provide an energy source for microorganisms. If no residual disinfectant is present (as would happen if ozone were used as the only treatment method), these microorganisms will cause the water quality to deteriorate in storage. Ozone also changes the surface charges of dissolved organics and colloidially suspended particles. This causes microflocculation of the dissolved organics and coagulation of the colloidal particles.

UV LIGHT

Ultraviolet light has been known to kill pathogens for a long time. A low pressure mercury bulb emits between 30 to 90 % of its energy at a wave length of 253.7 nm, right in the middle of the UV band. If water is exposed to enough light, pathogens will be killed. The problem is that some pathogens are hundreds of times less sensitive to UV light than others. The least sensitive pathogens to UV are protozoan cysts. Several studies show that Giardia will not be destroyed by many commercial UV treatment units. Fortunately, these are the easiest pathogens to filter out with a mechanical filter. The efficiency of UV treatment is very dependent on the turbidity of the water. The more opaque the water is, the less light that will be transmitted through it.

The treatment units must be run at the designed flow rate to insure sufficient exposure, as well as insure turbulent flow rather than plug flow. Another problem with UV treatment is that the damage done to the pathogens with UV light can be reversed if the water is exposed to visible light (specifically 330-500 nm) through a process known as photoreactivation. UV treatment, like ozone or mechanical filtering, leaves no residual component in the water to insure its continued disinfection. Any purchased UV filter should be checked to insure it at least complies with the 1966 HEW standard of 16 mW-s/cm^2 with a maximum water depth of 7.5 cm. ANSI/NSF require 38 mW-s/cm^2 for primary water treatment systems. This level was chosen to give better than 3 log (99.9%) inactivation of Bacillus subtillis. This level is of little use against Giardia, and of no use against Crypto.

The US EPA explored UV light for small scale water treatment plants and found it compared unfavorably with chlorine due to 1) higher costs, 2) lower reliability, and 3) lack of a residual disinfectant.

WATER TREATMENT IN GENERAL

Water must have eye appeal and taste appeal before we will drink it with much relish. Instinctively we draw back from the idea of drinking dirty, smelly water. Actually far more important to our well-being is whether or not a water is safe to drink. If it holds disease bacteria, regardless of its clarity and sparkle, we should avoid it. Let's consider these two highly important aspects of water: potability and palatability.

Regardless of any other factors, water piped into the home must be potable. To be potable, it should be completely free of disease organisms. Water is the breeding ground for an almost unbelievably large variety of organisms. Water does not produce these organisms. It merely is an ideal medium in which they can grow. These organisms gain entry into water through a variety of sources. They enter from natural causes, surface drainage and sewage. Many of the organisms in water are harmless. In fact, they are extremely beneficial to man. Others have a wide nuisance value and still others are the source of disease. In general, we are primarily concerned here with organisms which are potential disease-producers. These are of five types: bacteria, protozoa, worms, viruses, fungi. The presence of certain organisms of these various types can lead to such infectious diseases as typhoid fever, dysentery, cholera, jaundice, hepatitis, undulant fever and tularaemia.

There are other diseases as well, which spread through drinking unsafe water. Tremendous strides have been made in the control of these diseases within recent years. Much of the credit must go to sanitary engineers for their careful, consistent control of public water supplies. Biologically, there are two major classifications for our purposes. We can classify water organisms either as members of the plant or animal kingdoms.The following ways are the natural ways, in which water is purified: Bacteria and algae consume organic waste; Micro-organisms devour bacteria and algae; Oxidation renders organic matter harmless; Ultra-violet rays of sun have germicidal effects.

Under the broad heading of plant forms, we can classify the following:

Algae - These organisms are found throughout the world. They constitute the chief group of aquatic plants both in sea and fresh water. Algae range in size from microscopic organisms to giant seaweeds several hundred feet in length. They contain chlorophyll and other pigments which give them a variety of colors. They manufacture their food by photosynthesis. Algae thrive well in stagnant surface waters especially during the warm weather. Algae gives water fishy, grassy and other even more objectionable odors. While algae-laden waters are repulsive to man, animals will drink them and the presence of blue-green algae has been known to cause the death of cattle drinking this water.

Diatoms - Diatoms belong to the algae family. Some exist as single cells; others are found as groups or colonies. More than 15,000 forms of diatoms are known to exist. Diatoms have silica-impregnated cell walls. At time they release essential oils which give water a fishy taste.

Fungi - Fungi are another large group of plant forms. Like the algae, fungi have many varieties included among these are molds and bacteria. Fungi are not able to manufacture their own food. They exist by feeding on living things or on dead organic matter. Depending on their individual characteristics, they are usually colorless but may vary in this respect.

Molds - One important category of fungi is molds. This group of fungi feeds entirely on organic matter. They decompose carbohydrates such as sugars, starches and fats as well as proteins and other substances. They thrive ideally in water that has a temperature range of approximately 80 degrees to 100 °F. The presence of molds is generally a strong indicator of heavy pollution of water.

Bacteria - Bacteria are another important class of fungi. Again numerous smaller groupings are possible. Among the higher organisms in this group are the iron, manganese and sulphur bacteria. These higher bacteria gain their energy from the oxidation of simple organic substances. Lower forms of bacteria can be grouped as those that are helpful and those that are harmful to man. Those harmful to man are mainly the disease-producing organisms. Helpful organisms hasten the process of decomposing organic matter and by feeding on waste material; they aid in the purifying of water. All bacteria are sensitive to the temperature and pH of water. Some bacteria can tolerate acid water. But for the most part, they thrive best in waters that have a pH between 6.5 to 7.5, that is essentially neutral waters. As to temperature, most pathogenic or disease bacteria thrive best in water of body temperature. Beyond this no hard and fast statements can be made. Some bacteria are more resistant to heat than are others. Some are more sensitive to cold. At low temperatures, for example, some bacteria may become dormant for long periods of time but will still continue to exist. Interestingly enough, the waste products of their own growth can hamper bacteria and may even prove toxic to them.

Animal forms like plant life thrive in water providing conditions are right. Among the higher forms of animal life found in water are fish, amphibians (turtles and frogs), mollusks (snails and shellfish) and anthropoids (lobsters, crabs, water insects, water mites and others). Concern here is with those lower forms of animal life in water. Again, some are helpful to man as scavengers; others are injurious as possible sources of infection. Principal ones of concern are as follows:

Worms - There are three types of worms found in water. For the most part, they dwell in the bed of the material at the bottom of lakes and streams. There they do important work as scavengers. The rotifiers are the only organisms in this category at or near the surface. They live primarily in stagnant fresh water. The eggs and larvae of various intestinal worms found in man and warm-blooded animals pollute the water at times. They do not generally cause widespread infection for several reasons. They are relatively few in number and are so large they can be filtered out of water with comparative ease.

Protozoa - Another basic classification in the animal kingdom is that group of microscopic animals known as protozoa. These one-celled organisms live mainly in water either at or near the surface or at great depths in the oceans. Many live as parasites in the bodies of man and animals. Sometimes, drinking water becomes infested with certain protozoa which are not disease-producing. When present, they give the water a fishy taste and odor. Some protozoa are aerobic, that is, they exist only where free oxygen is available. Some exist where no free oxygen is available. Others can either be aerobic or anaerobic.
Nematodes - Nematodes belong to the worm family. They have long, cylindrical bodies which have no internal segments. Interestingly enough, those nematodes which are found in the bodies of men and warm-blooded animals are large enough to be visible to the naked eye; those living in fresh water or the soil are microscopic. Nematodes can be a problem in drinking water because they impart objectionable tastes and odors to water. They are also under suspicion of being carriers of the type of disease-bearing bacteria found in the intestines of warm-blooded animals though studies show that possibility is somewhat remote. Nematodes are apt to be found in municipal waters derived from surface supplies.
Viruses - As yet not too well understood is that group of parasitic forms known as viruses. Too small to be seen under a microscope, viruses are capable of causing disease in both plants and animals. Viruses can pass through porcelain filters that are capable of screening out bacteria. At least one virus that produces infectious hepatitis is water born. Drinking water contaminated with this virus is hazardous.

As the reader can see from even this brief summary, there is a tremendous variety of living organisms in water. To understand and classify the countless varieties requires an immense amount of knowledge and time. These organisms, whether plant or animal forms, are pathogenic or disease-producing; they make water unsafe to drink. For obvious reasons, even where there is just a possibility that water contains pathogenic organisms, it must be considered contaminated.

While there is a large and varied number of pathogens, no single contaminated water supply is apt to contain more than a few of these countless varieties. On one hand this is fortunate but, at the same time it makes detection of pathogens extremely difficult in terms of a routine water analysis. Since both speed and accuracy are essential, laboratory scientists need a sure way to expedite detection of pathogens. They have a dependable answer in a group of readily identified organisms that indicate possible contamination. These indicator organisms are the coliform bacteria.
Study has proved that these coliform bacteria indicate the presence of human or animal wastes in water. Coliform bacteria naturally exist in the intestines of humans and certain animals. Thus, the presence of these bacteria in water is accepted proof that the water has been contaminated by human or animal wastes. Although such water may contain no pathogens, an infected person, animal or a carrier of disease, could add pathogens at any moment! Thus, immediate corrective action must be taken. The presence of coliform bacteria shows water is contaminated by human

wastes and is potentially contaminated with pathogens. In short, these bacteria become a measure of guilt by association. The other side is this: the mere absence of coliform bacteria does not assure there are no pathogens. However, this is considered unlikely. Just how can water be tested for the presence of coliform bacteria? These organisms cause the fermentation of lactose (the crystalline sugar compound in milk). When water containing coliform bacteria is placed in a lactose culture, it will cause fermentation resulting in the form of gas. This confirms the suspicions. The Maximum Acceptable Concentration (MAC) for coliforms in drinking water is zero organisms detectable per 100 mL. Recognizing the danger, what can be done to provide adequate protection against contamination? When a water supply becomes contaminated, correct the problem at once. This means going beyond treatment alone - important as this may be. It is a basic rule of water sanitation to get to the source of the problem and eliminate it. If a well, for example, becomes badly contaminated, it is necessary to trace the contamination to its source and, if possible, remedy the situation. It may even be necessary to seek out a new source of supply.

WATER DISINFECTION ONE MORE TIME

Treatment of a water supply is a safety factor, not a corrective measure. There are a number of ways of purifying water. In evaluating the methods of treatment available, the following points regarding water disinfectants should be considered:

a. A disinfectant should be able to destroy all types of pathogens and in whatever number present in water.

b. A disinfectant should destroy the pathogens within the time available for disinfection.

c. A disinfectant should function properly regardless of any fluctuations in the composition or condition of the water.

d. A disinfectant should not cause the water to become toxic or unpalatable.

e. A disinfectant should function within the temperature range of the water.

f. A disinfectant should be safe and easy to handle.

g. A disinfectant should be such that it is easy to determine its concentration in the water.

h. A disinfectant should provide residual protection against recontamination.

Techniques such as filtration may remove infectious organisms from water. They are, however, no substitute for disinfection. The following are the general methods used for disinfecting water:

Boiling - This involves bringing the water to its boiling point in a container over heat. The water must be maintained at this temperature 15 to 20 minutes. This will disinfect the water. Boiling water is an effective method of treatment because no important waterborne diseases are caused by heat-resisting organisms.
Ultraviolet Light - The use of ultraviolet light is an attempt to imitate nature. As you recall, sunlight destroys some bacteria in the natural purification of water. Exposing water to ultraviolet light destroys pathogens. To assure thorough treatment, the water must be free of turbidity and color. Otherwise, some bacteria will be protected from the germ-killing ultraviolet rays. Since ultraviolet light adds nothing to the water, there is little possibility of its creating taste or odor problems. On the other hand, ultraviolet light treatment has no residual effect. Further, it must be closely checked to assure that sufficient ultraviolet energy is reaching the point of application at all times.

Use of Chemical Disinfectants - The most common method of treating water for contamination is to use one of various chemical agents available. Among these are chlorine, bromine, iodine, potassium permanganate, copper and silver ions, alkalis, acids and ozone. Bromine is an oxidizing agent that has been used quite successfully in the disinfecting of swimming pool waters. It is rated as a good germicidal agent. Bromine is easy to feed into water and is not hazardous to store. It apparently does not cause eye irritation among swimmers nor are its odors troublesome.
One of the most widely used disinfecting agents to ensure safe drinking water is chlorine. Chlorine in cylinders is used extensively by municipalities in purification work. However, in this form chlorine gas (Cl_2) is far too dangerous for any home purpose. For use in the home, chlorine is readily available as sodium hypochlorite (household bleach) which can be used both for laundering or disinfecting purposes. This product contains a 5.25% solution of sodium hypochlorite which is equivalent to 5% available chlorine. Chlorine is also available as calcium hypochlorite which is sold in the form of dry granules. In this form, it is usually 70% available chlorine. When calcium hypochlorite is used, this chlorinated lime should be mixed thoroughly and allowed to settle, pumping only the clear solution. For a variety of reasons not the least of which is convenience, chlorine in the liquid form (sodium hypochlorite) is more popular for household use. Chlorine is normally fed into water with the aid of a chemical feed pump. The first chlorine fed into the water is likely to be consumed in the oxidation of any iron, manganese or hydrogen sulphide that may be present. Some of the chlorine is also neutralized by organic matter normally present in any supply, including bacteria, if present. When the "chlorine demand" due to these materials has been satisfied, what's left over -the chlorine that has not been consumed - remains as "chlorine residual". The rate of feed is normally adjusted with a chemical feed pump to provide a chlorine residual of 0.5 - 1.0 ppm after 20 minutes of contact time. This is enough to kill coliform bacteria but may or may not kill any viruses or cysts which may be present. Such a chlorine residual not only serves to overcome intermittent trace contamination from coliform bacteria but, also provides for minor variations in the chlorine

demand of the water. The pathogens causing such diseases as typhoid fever, cholera and dysentery succumb most easily to chlorine treatment. The cyst-like protozoa causing dysentery are most resistant to chlorine. As yet, little is known about viruses, but some authorities place them at neither extreme in resistance to chlorination.

There are three basic terms used in the chlorination process: chlorine demand, chlorine dosage and chlorine residual. Chlorine demand is the amount of chlorine which will reduced or consumed in the process of oxidizing impurities in the water. Chlorine dosage is the amount of chlorine fed into the water. Chlorine residual is the amount of chlorine still remaining in water after oxidation takes place. For example, if a water has 2.0 ppm chlorine demand and is fed into the water in a chlorine dosage of 5.0 ppm, the chlorine residual would be 3.0 ppm.

For emergency purposes, iodine may be used for treatment of drinking water. Much work at present is being done to test the effect of iodine in destroying viruses which are now considered among the pathogens most resistant to treatment. Tests show that 20 minutes exposure to 8.0 ppm of iodine is adequate to render a potable water. As usual, the residual required varies inversely with contact time. Lower residuals require longer contact time while higher residuals require shorter contact time. While such test results are encouraging, not enough is yet known about the physiological effects of iodine-treated water on the human system. For this reason, its use must be considered only on an emergency basis.

Silver in various forms has been used to destroy pathogens. It can be added to the water as a liquid or through electrolytic decomposition of metallic silver. It has also been fed into water through an absorption process from silver-coated filters. Various household systems have been designed to yield water with a predetermined silver concentration. However, fluctuations in the flow rate often result in wide variations in the amount of silver in the water. In minute concentrations, silver can be highly destructive in wiping out disease-bearing bacteria. While long contact time is essential, silver possesses residual effect that can last for days. Silver does not produce offensive tastes or odors when used in water treatment. Further, organic matter does not interfere with its power to kill bacteria as in the case with free chlorine. Its high cost and the need for long periods of exposure have hindered its widespread acceptance.

Copper ions are used quite frequently to destroy algae in surface waters but these ions are relatively ineffective in killing bacteria.

Disease-bearing organisms are strongly affected by the pH of a water. They will not survive when water is either highly acidic or highly alkaline. Thus, treatment which sharply reduces or increases pH in relation to the normal range of 6.5 to 7.5 can be an effective means of destroying organisms.

There are numerous other agents which have proved to be successful in destroying pathogens. Many of these must still be subjected to prolonged testing with regard to their physiological effect on man. Among these are certain surfactants and chlorine dioxide. There are several types of surfactants which aid in destroying pathogens. The cationic detergents readily kill pathogens. Anionic detergents are only weakly effective in destroying pathogens. Surfactants have not been seriously considered for treating drinking water because of their objectionable flavor and possible toxic effects. Chlorine dioxide has unusually good germ killing power. Up to the present time, no valid tests for its use have been developed because of the lack of means for determining low residual concentrations of this agent. It's such a strong oxidizing agent, a larger residual of chlorine dioxide would probably be needed than is the case with chlorine. At present, chlorination in one form or another is regarded as the most effective disinfectant available for all general purposes. It has full acceptance of health authorities. Still there are certain factors which affect its ability to disinfect waters. These should always be kept in mind. They are:

a. "Free" chlorine residuals are more effective than "combined" or "chloramine" residuals. Disinfection regardless of the type of chlorine becomes more effective with increased residuals. Chloramine is the compound formed by feeding both chlorine and ammonia to the water. This treatment has been used for controlling bacteria growth in long pipelines and in other appliances where its slower oxidizing action is of particular benefit.

b. A pH of 6.0 to 7.0 makes water a far more effective medium for chlorine as a disinfecting agent than do higher pH values of around 9.0 to 10.0.

c. The effectiveness of chlorine residuals increase with higher temperatures within the normal water temperature range.

d. The effectiveness of disinfection increases with the amount of contact time available.

e. All types of organisms do not react in the same way under various conditions to chlorination.

f. An increase in the chlorine demand of a water increases the amount of chlorine necessary to provide a satisfactory chlorine residual.

In order to ensure the destruction of pathogens, the process of chlorination must achieve certain control of at least one factor and, preferably two, to compensate for fluctuations that occur. For this reason, some authorities on the subject stress the fact that the type and concentration of the chlorine residual must be controlled to ensure adequate disinfection. Only this way, they claim, can chlorination adequately take into account variations in temperature, pH, chlorine demand and types of organisms in the water. While possible to increase minimum contact times, it is difficult to do so. Five to ten minutes is normally all the time available with the type of pressure systems normally used for small water supplies. Many experts feel that satisfactory chlorine residual alone can provide adequate control for disinfection. In their opinion, superchlorination-dechlorination does the best job. Briefly, what is this technique and how does it operate?

The success of superchlorination-dechlorination system depends on putting enough chlorine in the water to provide a residual of 3.0 to 5.0 ppm. This is considerably greater than chlorine residual of 0.1 to 0.5 ppm usually found in municipal water supplies when drawn from the tap. A superchlorination-dechlorination systems consists of two basic units. A chlorinator feeds chlorine into raw water. This chlorine feed is stepped up to provide the needed residual. A dechlorinator unit then removes the excess chlorine from the water before it reaches the household taps. The chlorinator should be installed so that it feeds the chlorine into the water before it reaches the pressure tank. A general purpose chemical feed pump will do the job. The size and the placement of the dechlorinator unit depends on the type of treatment necessary. This will usually be an activated carbon filter. If pathogen kill is all that is required, a small dechlorinator can be installed at the kitchen sink. This unit then serves to remove chlorine from water used for drinking and cooking. The advantage in dechlorinating only a part of the water is obvious. A smaller filter unit does the job and since only a small portion of the total water is filtered under such conditions, the unit lasts longer before either servicing or replacement is necessary. Essentially dechlorination is not needed to ensure a safe drinking water. Once the water is chlorinated, the health hazard is gone. The chlorine residual is removed merely to make the water palatable. If the problem is compounded due to the presence or iron and/or manganese, all the water should be filtered. Under such conditions, a larger central filter is necessary and should be placed on the main line after the pressure tank. The prime advantage of the superchlorination-dechlorination process is that it saturates water with enough chlorine to kill bacteria. Simple chlorination sometimes fails of its objective because homeowners may set the chlorine feed rate too low in order to avoid giving their water a chlorine taste.

Sodium Dichloroisocyanurate - Sodium Dichloroisocyanurate can sterilize drinking water, swimming pool, tableware and air, or be used for fighting against infectious diseases as routine disinfection, preventive tableware and environmental sterilization in different places, or act as disinfectant in raising silkworm, livestock, poultry and fish. It can also be used to prevent wool from shrinkage, bleach the textile and clean the industrial circulating water. The product has high efficiency and constant performance with no harm to human beings. It enjoys goods reputation both at home and abroad. Table 8 summarizes some of this chemical's properties.

Table 8. Properties of Sodium Dichloroisocyanurate.

UN No.	2465
CAS No.	2839-78-9
Formula	$C_3O_3N_3Cl_2Na$
Physicochemical Properties	White crystalline powder, granular, or tablets
Specifications	Powder or Granular

Available Chlorine	56% min.	60% min.
pH Value	5.5-7.0	
Packing	25 or 50 kg plastic drums	
Qty/20' FCL (MT)	20	

Trichloroisocyanuric Acid - With strong bleaching and disinfection effects, Trichloroisocyanuric Acid is widely used as high effective disinfectant for civil sanitation, animal husbandry and plant protection as bleaching agent of cotton, gunny, chemical fabrics, or as shrink-proof agent for woolens, battery materials, organic synthesis industry and dry-bleaching agent of clothes. Effervescent Tablets (250, 500, 650 or 1000 mg) of TCCA are available for household use. Table 9 provides some general properties.

Table 9. Properties of Trichloroisocyanuric Acid.

UN No.	2468			
CAS No.	87-90-1			
Formula	$C_3O_3N_3Cl_3$			
Physicochemical Properties	White crystal powder, granular or tablets, with stimulant smell of Hypochloric Acid, slightly soluble in water, easily soluble in Acetone.			
Specifications	Powder	Granular	Tablet (200g)	Tablet (20g)
Available Chlorine	90% min.			
Moisture	0.5% max.			
pH Value (1% W. S.)	2.7-2.9			
Packing	25 or 50 kg plastic drums			
Qty/20' FCL (MT)	20			

Isocyanuric Acid - Cyanuric Acid is widely used for the stabilization of available chlorine swimming pool water treatment. CYA is also the starting compound for the synthesis of many organic derivatives. Table 10 provides some general properties.

Table 10. Properties of Isocyanuric Acid.

CAS No.	108-80-5	
Formula	$C_3H_3N_3O_3$	
Physicochemical Properties	White crystalline solid powder or granular, non- toxic and odorless	
Specifications	Powder	Granular
Cyanuric Acid	98.5% min.	98% min.
Moisture	0.4% max.	0.5% max.
Particle	0.3 mm max. 90% through	0.6-2 mm 90% through
pH Value (1% Water Solution)	4.0 - 4.6	4.0 - 4.6
Melting Point (Centigrade)	330 min.	330 min.
Fe	25 PPM max.	25 PPM max.
Packing	woven bags	fiber drums
Qty/20' FCL (MT)	20	15.89

Discussions thus far have focused on pathogens and methods of destroying them in the process of making water potable - safe to drink. This is highly important but it's not the whole story; for water must be palatable as well as potable. The obvious question to ask is – What makes a water palatable?

To be palatable, a water should be free of detectable tastes and odors. Immediately, we come to a stumbling block. What constitutes a detectable taste or odor? Undoubtedly when you have traveled around the country, you have tasted waters which must have had unpleasant tastes or odors. Natives in the area may be surprised to note you reaction. for after drinking the water for many years, they find nothing peculiar to either the taste or odor of the water. Then, there are those waters which have tastes and odors so obnoxious (hydrogen sulphide water, for example), even the long time inhabitant can't stomach them. Turbidity, sediment and color play important roles in determining whether a water is a delight to drink.

Various odors and tastes may be present in water. They can be traced to many conditions. Unfortunately, the causes of bad taste and odor problems in water are so many, it is impossible to suggest a single treatment that would be universally effective in controlling these problems. Tastes are generally classified in four groups - sour, salt, sweet and bitter.

Odors possess many classifications. There are 20 of them commonly used, all possessing rather picturesque names. In fact the names, in many cases, are far more pleasant than the odors themselves. To name a few of them - nasturtium, cucumber, geranium, fishy, pigpen, earthy, grassy and musty. Authorities further classify these odors in terms of their intensity from very faint, faint, distinct and decided to very strong. Now your taste buds and olfactory organs are not necessarily of the same acuteness as your neighbors. So there may be some disagreement on the subject. Generally you or your neighbor should not be made aware of any tastes or odors in water if there is to be pleasure in drinking it. If you are conscious of a distinct odor, without specifically seeking for such, the water is in need of treatment. In many cases, it is difficult to detect what constitutes a taste or an odor. The reason is obvious. Both the taste buds and olfactory organs work so effectively as a team, it is hard to realize where one leaves off and the other begins. To illustrate: hydrogen sulphide gives water an "awful" taste yet actually it is this gas's unpleasant odor that we detect rather than an unpleasant taste. Unfortunately, there is little in the way of standard measuring equipment for rating tastes and odors. Tastes and odors in water can be traced to at least five factors. They are:

1) decaying organic matter

2) living organism

3) iron, manganese and the metallic product or corrosion

4) industrial waste pollution from substances such as phenol

5) chlorination

6) high mineral concentrations

In general, odors can be traced to living organisms, organic matter and gases in water. Likewise, tastes can be traced generally to the high total minerals in water. There are some tastes due to various algae and industrial wastes. Some tastes and odors, especially those due to organic substances, can be removed from water simply by passing it through an activated carbon filter. Other tastes and odors may respond to oxidizing agents such as chlorine and potassium permanganate. Where these problems are due to industrial wastes and certain other substances, some of the above types of treatment may completely fail. In some cases, for example, chlorination may actually intensify a taste or odor problem. Potassium permanganate has been found to be extremely effective in removing many musty, fishy, grassy and moldy odors. Two factors make this compound valuable - it is a strong oxidizing agent and it does not form obnoxious compounds with organic matter. However, a filter must be used to remove manganese dioxide formed when permanganate is reduced.

Turbidity and suspended matter are not synonymous terms although most of us use the terms more or less interchangeably. Correctly speaking, suspended mat- ter is that material which can be removed from water through filtration or the coagulation process. Turbidity is a measure of the amount of light absorbed by water because of the suspended matter in the water. There is also some danger of confusion regarding turbidity and color. Turbidity is the lack of clarity or brilliance in a water. Water may have a great deal of color - it may even be dark brown and still be clear without suspended matter. The current method of choice for turbidity measurement in Canada is the nephelometric method; the unit of turbidity measured using this method is the nephelometric unit (NTU). Turbidity in excess of 5 NTU becomes apparent and may be objected to by a majority of consumers. Therefore an Aesthetic Objective (AO) of < = 5 NTU has been set for water at the point of consumption. The suspended particles clouding the water may be due to such inorganic substances as clay, rock flour, silt, calcium carbonate, silica, iron, manganese, sulphur or industrial wastes. Again the clouding may be due to a single foreign substance in water, chances are it is probably due to a mixture of several or many substances. These particles may range in size from fine colloidal materials to course grains of sand that remain in suspension only as long as the water is agitated. Those particles which quickly sink to the bottom are usually called, "sediment". There are no hard and fast rules for classifying such impurities.

If you take water from a swiftly flowing river or stream, you generally find that it contains a considerable amount of sediment. In contrast, you find that water taken from a lake or pond is usually much clearer. In these more quiet, non-flowing waters, there is greater opportunity for settling action. Thus all but very fine particles sink to the bottom. Least apt to contain sediment are wells and springs. Sediment is generally strained from these water as they percolate through sand, gravel and rock formations. Turbidity varies tremendously even within these various groupings. Some rivers and streams have water that appears crystal clear with just trace amounts of turbidity in them especially at points near their sources. These same moving waters may contain upwards of 30,000 ppm of turbidity at other points in their course to the oceans. In fact, turbidity in amounts well over 60,000 ppm have been registered. Again there are significant fluctuations in the amount of turbidity in a river at different times in a year. Heavy rainfalls, strong winds and convection currents can greatly increase the turbid state of both lakes and rivers. Warm weather and increases in the temperature can also add to the problem. For with warm weather, micro-organisms and aquatic plants renew their activity in the water. As they grow and later decay, these plant and animal forms substantially add to the turbid state of a water. Also, they frequently cause a heightening of taste, odor and color problems.

Mechanical filtration will remove all forms of turbidity. Of course, the smaller the turbid particles, the finer the filter openings must be in order to strain them out. Under some circumstances, the openings may have to be so small that they cause an excessive pressure drop as the water creeps through the filter and the unit may be impractical. In many cases, filters containing specially graded and sized gravel and sand are effective in screening out turbid particles. With such units, a periodic backwashing to remove the filtered material is all the maintenance necessary. As discussed in later chapters, the use of filter aids is necessary in treating many water sources. A filter aid is a chemical that is added onto the top of the filter bed immediately after backwashing. The filter aid traps fine dirt particles producing a more a sparkling clear water and keeps dirt from penetrating the filter bed, insuring better bed cleansing during backwashing. In some cases, cartridge filters are effective.

Municipal and industrial systems frequently make use of the coagulation process to aid in the removal of turbidity. In this economical process, a coagulating agent such as aluminum sulphate is fed into the water. After rapid mixing, the coagulating agent forms a "floc" generally in the form of a gelatinous precipitate. This floc gives the appearance of a soft, gentle snowfall. A settling period is then needed to allow the floc to fall gently through the water. As the floc forms and settles, it tends to collect or entrap the turbid particles and form them into larger particles which sink to the bottom. On large installations, huge settling basins provide the necessary time and space for the process. After the settling period, the water flows through a filter to remove the last traces of the coagulant and any remaining turbid particles.

An additional water quality parameter of importance is color. Ordinarily we think of water as being blue in color. When artists paint bodies of water, they generally color them blue or blue-green. While water does reflect blue-green light, noticeable in great depths, it should appear colorless as used in the home. Ideally, water from the tap is not blue or blue-green. If such is the case, there are certain foreign substances in the water. Among these substances: Infinitely small microscopic particles add color to water. Colloidal suspensions and non-colloidal organic acids as well as neutral salts also affect the color of water; The color in water is primarily of vegetable origin and is extracted from leaves and aquatic plants; Naturally, water draining from swamps has the most intense coloring. The bleaching action of sunlight plus the aging of water gradually dissipates this color, however. All surface waters possess some degree of color. Like some shallow wells, springs and an occasional deep well can contain noticeable coloring. In general, water from deep wells is practically colorless. An arbitrary standard scale has been developed for measuring color intensity in water samples. When a water is rated as having a color of five units, it means: The color of this water is equal in intensity to the color of distilled water containing 5 milligrams of platinum as potassium chloroplatinate per liter. Highly colored water is objectionable for most process work in the industrial field because excessive color causes stains. While color is not a factor of great concern in relation to household applications, excessive color lacks appeal from an aesthetic standpoint in a potable water. Further, it can cause staining. The Aesthetic Objective (AO) for color in drinking water is $<=$ 15 true color units. The provision of treated water at or below the AO will encourage rapid notification by consumers should problems leading to the formation of color arise in the distribution system. In general, color is reduced or removed from water through the use of coagulation, settling and filtration techniques. Aluminum sulphate is the most widely used coagulant for this purpose. Superchlorination, activated carbon filters and potassium permanganate have been used with varying degrees of success in removing color. Table 11 summarizes water treatment methods currently used.

Table 11. Wastewater Treatment Methods.

Objective of Treatment	Method or Technology
ODOR	
Rotten Egg Smell	a) Manganese green-sand filter up to 6 ppm H_2S with pH not lower than 6.7
	b) Over 6 ppm H_2S constant chlorination by filtration / dechlorination
	c) Open aeration followed by oxidizing-catalyst filter
	The water should be tested at the source for H_2S determination as the gas escapes rapidly.
Petroleum	Locate and eliminate seepage. Activated carbon will adsorb oil and gasoline (most hydrocarbons) on a short term basis. Air-strip with (40:1 air/water ratio) followed by 2 ft^3 carbon units in series
Aromatic, Fishy, Earthy, or Woody Smell	a) Activated carbon type filter, or
	b) Cartridge-activated carbon filter for drinking and cooking
Sharp Metallic Smell	a) Water softener can remove 0.5 ppm or iron (Fe) for every grain/gal. of hardness up to 10 ppm at minimum pH of 6.7 (unaerated water)
	b) Over 10 ppm Fe: chlorination with sufficient retention tank time for full oxidation followed by filtration and dechlorination
	c) pressure aeration plus filtration for up to 20 ppm Fe
APPEARANCE	
Rust	a) up to 10 ppm iron removed by manganese greensand filter if pH is 6.7 or higher; or
	b) Manganese-treated pumicite catalyst filter if pH is 6.8 or higher and oxygen is 15% of total iron content
	c) Downflow water softened with good backwash, up to 10 ppm, use calcite filter followed be downflow water softener
Black Staining	a) Manganese greensand or manganese zeolite-type catalyst-filter to limit of 6 ppm or 15 ppm, respectively (combined Fe and Mn), with pH not lower than 6.7 value
	b) Process used for iron removal usually will handle manganese
	c) Manganese punicite* medium catalyst-filter with ultra-filtration-type membrane element

Objective of Treatment	Method or Technology
	b) For whole-house system, remove by absorption via special macroporous Type 1 anion exchange resin regenerated with NaC1. up to 3 ppm
	c) Above 3 ppm, constant chlorination with full retention time, followed by filtration and/or dechlorination
Gelatinous Slime	a) Destroy iron bacteria with a solution of hydrochloric acid, then constant chlorination, followed by activated carbon filtration or calcite filter.
	b) Potassium permanganate chemical feed followed by MnZ/anthracite filter
Hydrocarbon Sheen	[Same as Petroleum]
Murky	a) For mud, clay, and sediment - use a calcite or pumicite filter, up to 50 ppm b) For sand, grit, or clay - use a hydrocyclone, sand trap, and/or install new well screen
TASTE	
Salinity	a) There is no commercial residential treatment for sodium over 1,800 ppm
	b) Deionize drinking water only with disposable mixed bed-anion/cation resin; or
	c) Reverse osmosis for drinking and cooking water only; or
	d) Home distillation system for drinking water.
Medicine	Single faucet activated carbon filter or whole-house tank-type activated absorption filter
Chemical Tastes (Other)	Pesticides-herbicides: Activated carbon filter will absorb limited amount. Must continue to monitor the product water closely

SOME GENERAL COMMENTS

So there we have it – a broad overview of a complex subject that spans both technical and legal arenas. Much of the discussions have focused on drinking water, but from this point forward we will depart from the subject and only address this in passing. Recognize that there are a large number of technologies that are applied to treating water. The combination of technologies needed for a water treatment application depend on what we are ultimately trying to achieve in terms of final water quality.

Although the term pollution control has fallen out of favor today and what has become fashionable is *Pollution Prevention*, the fact remains that what we are doing is removing unwanted contaminants from water, whether it be to meet drinking water purposes, or to meet a discharge standard to a local (nonpotable) water body. The contaminants may be caused by man, or they simply exist from nature. Either way, we are applying technologies aimed at removing these constituents, and ultimately these concentrated forms of pollutants require disposal. In this regard, physical methods alone are quite limited, because they represent a non-destructive form of treatment. Their objective is both to remove suspended contaminants and to concentrate them within the limitations of the technology or hardware. From that point on, further concentration is required in order reconstitute the collected contaminants in a form that can be readily handled for ultimate disposal and or destruction. This is known as dewatering. But as noted above, water often contains much more than just suspended matter.

For newcomers to this subject, there is a section of general questions for thinking and discussing among your colleagues. These will help reinforce some of the general concepts and principles covered in this first chapter, and help you to prepare for the more technical discussions that follow.

LIST OF ABBREVIATIONS USED IN THIS CHAPTER

BATs best available technologies
CCL contaminant candidate list
CWA Clean Water Act
DNAPLs dense nonaqueous phase liquids
DWEL drinking water equivalent level
GAC granular activated carbon
MCLG maximum contaminant level goal
MCLs maximum contaminant levels
NDWAC National Drinking Water Advisory Council
NPDWR National Primary Drinking Water Regulations
NSDWRs National Secondary Drinking Water Regulations
PAC powdered activated carbon
RFD reference dose
RO reverse osmosis
SCADA supervisory control and data acquisition
SDWA Safe Drinking Water Act

TT	treatment techniques
USEPA	United States Environmental Protection Agency
UV	ultra violet

RECOMMENDED RESOURCES FOR THE READER

You can obtain in-depth information on drinking water plant design issues from the following references.

1. Baker, M. *N. The Questfor Pure Water,* 2d ed., American Water Works Association, Inc., New York, 1981.
2. White, C. G. *The Handbook of Chlorination and Alternative Disinfectants,* 3d ed., Van Nostrand Reinhold, New York, 1992.
3. Pedden, T. M. *Drinking Water and Ice Supplies and Their Relations to Health and Disease,* G. P. Putnam's Sons, The Knickerbocker Press, New York, 1981
4. AWWA. *Water Quality and Treatment,* 4th ed., McGraw-Hill Book Co., New York, 1990.
5. Pontius, F. W. "SDWA - A Look Back," *Jour. AWWA,* vol. 85, no. 2, pp. 22-24 & 94, February 1993.
6. Pontius, F. W. and Robinson, J. A. "The Current Regulatory Agenda: An Update," *Jour AWWA,* vol. 86, no. 2, pp. 54-63, February 1994.
7. Pontius, F. W. "An Update of the Federal Drinking Water Regs," *Jour. AWWA,* vol. 90, no. 3, pp. 48-58, March 1998.
8. USEPA. *The National Public Water System Program,* FY 1988 Compliance Report, Office of Drinking Water, Cincinnati, OH, March 1990.
9. Goodrich, J. A., Adams, J. Q ., Lykins, B. W., and Clark, R. M. "Safe Drinking Water from Small Systems: Treatment Options," *Jour AWWA,* vol. 84, no. 5, pp. 49-55, May 1992.
10. *USEPA. Technologies for Upgrading Existing or Designing New Drinking Water Treatment Facilities,* EPA/625/4- 89/023, Office of Drinking Water, Cincinnati, OH, March 1990.
11. Le Chevallier, M. K., Becker, W. C., Schorr, P., and Lee, R. G. "Evaluating the Performance of Biologically Active Rapid Filters," *Jour. AWWA,* vol. 84, no. 4, pp. 136-146, April 1992.
12. Manem, J. A. and Rittmann, B. E. "Removing Trace-Level Organic

Pollutants in a Biological Filter," *Jour. AWWA,* vol. 84, no. 4, pp. 152-157, April 1992.

13. Adam, S. S., Jacangelo, J. G., and Laine, J. M. "Low Pressure Membranes: Assessing Integrity," *Jour. AWWA,* vol. 87, no. 3, pp. 62-75, March 1995.

14. McCleaf, P. R. and Schroeder, E. D. "Denitrification Using a Membrane Immobilized Biofilm;" *Jour AWWA,* vol. 87, no. 3, pp. 77-86, March 1995.

15. Allgerier, S. C. and Summers, R. C. "Evaluating NF for DBP Control with RBSMT," *Jour AWWA,* vol. 87, no. 3, pp. 87-99, March 1995.

16. Castro, K. and Zander, A. K. "Membrane Air-Stripping Effects of Pretreatments;'Jour AWWA, vol. 87, no. 3, pp. 50-61, March 1995.

17. Glaze, W. H. and Kang, J. W. "Advanced Oxidation Process for Treating Ground Water Contaminated with TCE and PCE: Laboratory Studies:'Jour AWWA, vol. 80, no. 5, pp. 57-63, May 1988.

18. Glaze, W. H., Kang, J. W., and Aieta, M. "Ozone-Hydrogen Peroxide Systems for Control of Organics in Municipal Water Supplies:' *Proceedings of the Second International Conference in the Role of Ozone on Water and Wastewater Treatment,* TekTran International Ltd., Kitchener, Ontario, Canada, pp. 233-244, 1987.

19. Haberar, K. and Schmidth, S. N. "The Haberar Process: Combining Contact Flocculation, Filtration, and PAC Adsorption:' Jour AWWA, vol. 83, no. 9, pp. 82-89, September 1991.

20. Stukenberg, J. R. and Hesby, J. C. "Pilot Testing the Haberar Process in the United States;' Jour AWWA, vol. 83, no. 9, pp. 90-96, September 1991.

21. White, B. N., Kiefer, D. A., Morrow, J. H., and Stolarik, G. F. "Remote Biological Monitoring in an Open Finished-Water Reservoir," *Jour AWWA,* vol. 83, no. 9, pp. 107-112, September 1991.

22. Richardson, K J., Stewart, M. H., and Wolfe, R. L. "Application of Gene Probe Technology to the Water Industry," *Jour AWWA,* vol. 83, no. 90, pp. 71-81, September 1992.

23. James M. Montgomery, Inc. *Water Treatment Principles and Design,* John Wiley & Sons, New York, 1985.

You can obtain in-depth information on the Clean Water Act from the following references.

24. Dubrowski, Fran. *Crossing the Finish Line.* THE ENVIRONMENTAL FORUM. July/August 1997. pp.28-37.

25. Goplerud, C. Peter III. *Water Pollution Law: Milestones from the Past and Anticipation of the Future.* NATURAL RESOURCES&ENVIRONMENT. Fall 1995. pp.7-12.
26. Loeb, Penny. *Very Troubled Waters.* U.S. NEWS & WORLD REPORT, vol.125, no. 12. Sept.28, 1998. pp.39, 41-42.
27. Schneider, Paul. *Clear Progress, 25 Years of the Clean Water Act.* AUDUBON. September/October 1997. pp.36-47,106-107.
28. U.S. Environmental Protection Agency. Office of Water. *Environmental Indicators of Water Quality in the United States.* EPA 841-R-96-002. June 1996. 30 p.
29. *The Quality of Our Nation's Water: 1996. Executive Summary of the National Water Quality Inventory: 1996 Report to Congress.* EPA 841-S-97-001. April 1998. 197p.
30. U.S. General Accounting Office. *Water Quality, A Catalog of Related Federal Programs.* GAO/RCED-96-173. 64p.

QUESTIONS FOR THINKING AND DISCUSSING

1. What are the water treatment technologies available to us and can you rank them in accordance with the likely cost of treatment (from highest to lowest)? *Hint, take into consideration the use of energy.*
2. Of the water treatment technologies discussed in this chapter, which ones could be applied economically to small scale applications like home use?
3. What treatment technologies would be best in destroying E. coli?
4. Define the terms MCLG and TT.
5. Explain the process by which EPA established drinking water quality discharge standards.
6. List the categories of pollutants that the NPDW Regulations address.
7. Pick any three chemicals that you are aware are covered by the NPDW Regulations and discuss the potential health effects that could occur from ingesting water contaminated by them.
8. What are the two most important provisions of the Clean Water Act?
9. Briefly explain how a sand filter operates.
10. Develop a simplified diagram that explains the various pathways by which water can become contaminated.
11. What water quality characteristic would reduce the efficiency of UV in destroying pathogens contained in water?
12. List the BATs for treating inorganic matter in water.

13. List the BATs for treating organic matter in water.

14. Explain what BPTs and BATs are. What are the differences between these two terms?

15. Explain the terms coagulation and flocculation. How do you envision these operations are practiced?

16. What are coliforms and how are they introduced inot water supplies?

17. Make a list of heavy metals that can contaminate water sources? What are some of the health effects associated with ingesting water contaminated by heavy metals?

18. Briefly explain what nanofiltration is.

19. What is Reverse Osmosis and the general principle of operation?

20. What is the role of NSDW Regulations? Are these enforceable?

21. What reasons would a water treatment operation employ microfilters?

22. Briefly explain the principle of adsorbtion and how it is used in water treatment.

23. What parameters impact on the effectiveness of chlorine as a water disinfectant?

24. List several alternative chemicals that compete with chlorine as a disinfectant.

25. Can ozone be used as the only means of water disinfection in a treatment facility? If not, why not, and what other technologies do you think would work in combination with ozone treatment?

26. What is photoreactivation?

27. List the various types of plant forms that can enter into and contaminate drinking water.

28. Make a list of the animal forma that we are most concerned with in contaminating drinking water sources.

29. What makes a water palatable?

30. List several water quality parameters that are important to track and control.

31. Develop a list of the chemicals that are part of the wastewater streams in the plant that you are employed at. Establish whether or not the operation is in compliance with discharge standards, and whether all permits to discharge are current.

Chapter 2

WHAT FILTRATION IS ALL ABOUT

INTRODUCTION

Filtration is a fundamental unit operation that, within the context of this volume, separates suspended particle matter from water. Although industrial applications of this operation vary significantly, all filtration equipment operate by passing the solution or suspension through a porous membrane or medium, upon which the solid particles are retained on the medium's surface or within the pores of the medium, while the fluid, referred to as the filtrate, passes through.

In a very general sense, the operation is performed for one or both of the following reasons. It can be used for the recovery of valuable products (either the suspended solids or the fluid), or it may be applied to purify the liquid stream, thereby improving product quality, or both. Examples of various processes that rely on filtration include adsorption, chromatography, operations involving the flow of suspensions through packed columns, ion exchange, and various reactor engineering applications. In petroleum engineering, filtration principles are applied to the displacement of oil with gas (i.e., liquid-liquid separations), in the separation of water and miscible solvents (including solutions of surface-active agents), and in reservoir flow applications. In hydrology, interest is in the movement of trace pollutants in water systems, the purification of water for drinking and irrigation, and to prevent saltwater encroachment into freshwater reservoirs. In soil physics, applications are in the movement of water, nutrients and pollutants into plants. In biophysics, the subject of flow through a porous media touches upon life processes such as the flow of fluids in the lungs and the kidney. Although there are numerous industry-specific applications of filtration, waster treatment has historically and continues to be the largest general application of this unit operation.

The objective of this chapter is to provide an overview of filtration terminology and basic engineering principles, as well as calculation methods that describe the filtration process in a generalized way. The basis equations describing the generalized process of filtration have been around for nearly 100 years, and with few refinements, continue to be applied to modern design practices.

At the end of this chapter you will find four annexes. The first is a list of important terms and their definitions. These terms will help to orient you to discussions in later chapters. Next you will find a ***Nomenclature*** section that defines all the mathematical terms used in this chapter. Following this is a list of suggested references along with a critique and general synopsis of the references. This will tell you what each of the references covers and the level of coverage, along with the price of the textbook. Finally, there is a section on ***Questions for Thinking and Discussing*** to challenge you. Remember to refer to the ***Glossary*** at the end of the book if you run across any terms that are unfamiliar to you.

TERMINOLOGY AND GOVERNING EQUATIONS

There are essentially four important physical parameters that characterize a filter media and are used as a basis for relating the characteristics of the material to the system flow dynamics. These are porosity, permeability, tortuosity and connectivity.

We may begin by describing any porous medium as a solid matter containing many holes or pores, which collectively constitute an array of tortuous passages. Refer to Figure 1 for an example. The number of holes or pores is sufficiently great that a volume average is needed to estimate pertinent properties. Pores that occupy a definite fraction of the bulk volume constitute a complex network of voids. The manner in which holes or pores are embedded, the extent of their interconnection, and their location, size and shape characterize the porous medium.

The term *porosity* refers to the fraction of the medium that contains the voids. When a fluid is passed over the medium, the fraction of the medium (i.e., the pores) that contributes to the flow is referred to as the *effective porosity* of the media. In a general sense, porous media are classified as either unconsolidated and consolidated and/or as ordered and random. Examples of unconsolidated media are sand, glass beads, catalyst pellets, column packing materials, soil, gravel and packing such as charcoal.

Examples of consolidated media are most of the naturally occurring rocks, such as sandstones and limestones.

Modeling the pore size in terms of a probability distribution function enables a mathematical description of the pore characteristics. The narrower the pore size distribution, the more likely the absoluteness of retention. The particle-size distribution represented by the rectangular block is the more securely retained, by sieve capture, the narrower the pore-size distribution.

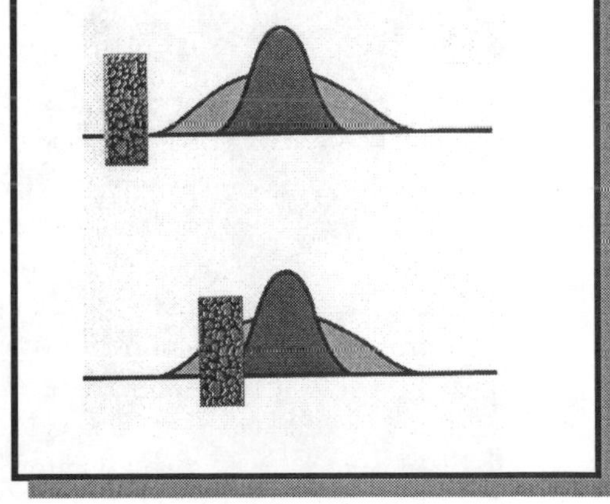

Materials such as concrete, cement, bricks, paper and cloth are manmade consolidated media.

Ordered media are regular packings of various types of materials, such as spheres, column packings and wood. Random media have no particular correlating factor. Porous media can be further categorized in terms of geometrical or structural properties as they relate to the matrix that affects flow and in terms of the flow properties that describe the matrix from the standpoint of the contained fluid. Geometrical or structural properties are best represented by average properties, from which these average structural properties are related to flow properties.

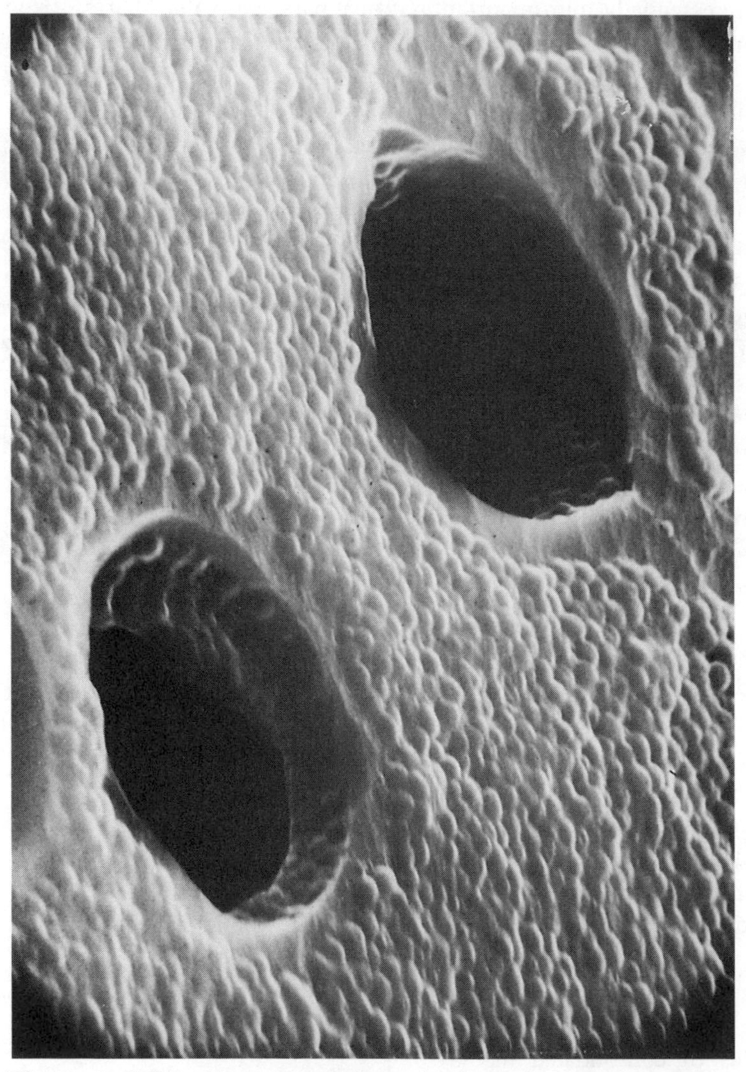

Figure 1. SEM of pores on surface of activated carbon particle.

PORE STRUCTURE

A microscopic description characterizes the structure of the pores. The objective of a pore-structure analysis is to provide a description that relates to the macroscopic or bulk flow properties. The major bulk properties that need to be correlated with pore description or characterization are the four basic parameters: porosity, permeability, tortuosity and connectivity. In studying different samples of the same medium, it becomes apparent that the number of pore sizes, shapes, orientations and interconnections are enormous. Due to this complexity, pore-structure description is most often a statistical distribution of apparent pore sizes. This distribution is apparent because to convert measurements to pore sizes one must resort to models that provide *average* or *model* pore sizes. A common approach to defining a characteristic pore size distribution is to model the porous medium as a bundle of straight cylindrical or rectangular capillaries (refer to Figure 2). The diameters of the model capillaries are defined on the basis of a convenient distribution function.
Pore structure for unconsolidated media is inferred from a particle size distribution, the geometry of the particles and the packing arrangement of particles. The theory of packing is reasonably well established for symmetrical geometries such as spheres and cylinders. Information on particle size, geometry and packing theory allows us to develop relationships between pore size distributions and particle size distributions. A macroscopic description is based on average or bulk properties at sizes much larger than a single pore. In characterizing a porous medium macroscopically, one must deal with the scale of description. The scale used depends on the manner and size in which we wish to model the porous medium. A simplified approach is to assume the medium to be ideal; meaning homogeneous, uniform and isotropic.

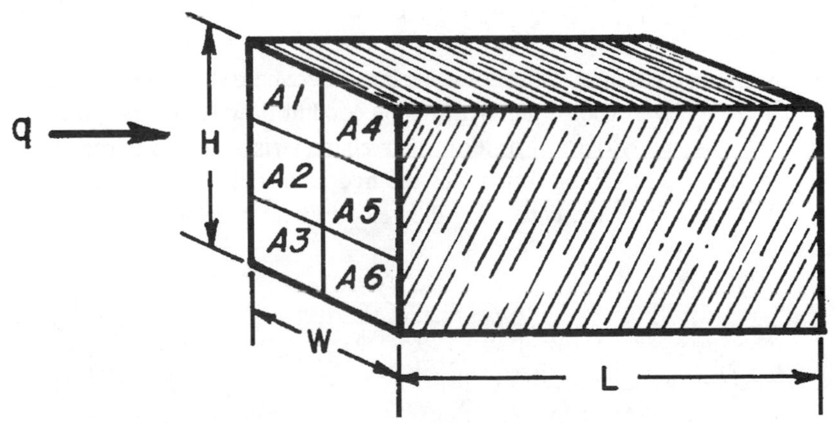

Figure 2. Simplified capillary view of flow through pores.

PERMEABILITY AND DARCY'S LAW

The term *reservoir description* is applied to characterizing a homogeneous system as opposed to a heterogeneous one. A reservoir description defines the reservoir at a level where a property changes sufficiently so that more than a single average must be used to model the flow. In this sense, a reservoir composed of a section of coarse gravel and a section of fine sand, where these two materials are separated and have significantly different permeabilities, is heterogeneous in nature.
Defining dimensions, locating areas and establishing average properties of the gravel and sand constitutes a reservoir description, and is a satisfactory approach for reservoir-level type problems.
The governing flow equation describing flow through as porous medium is known as Darcy's law, which is a relationship between the volumetric flow rate of a fluid flowing linearly through a porous medium and the energy loss of the fluid in motion.

Darcy's law is expressed as:

$$Q = \frac{KA(h_1 - h_2)}{\Delta h} \tag{1}$$

where

$$\Delta h = \Delta z + \frac{\Delta p}{\rho} + constant \tag{2}$$

The parameter, K, is a proportionality constant that is known as the *hydraulic conductivity*.
Darcy's law is considered valid for creeping flow where the Reynolds number is less than one. The Reynolds number in open conduit flow is the ratio of inertial to viscous forces and is defined in terms of a characteristic length perpendicular to flow for the system. Using four times the hydraulic radius to replace the length perpendicular to flow and correcting the velocity with porosity yields a Reynolds number in the form:

$$Re = \frac{D_p v_\infty \rho}{\mu(1 - \phi)} \tag{3}$$

The hydraulic conductivity K depends on the properties of the fluid and on the pore structure of the medium. It is temperature-dependent, since the properties of the fluid (density and viscosity) are temperature-dependent. Hydraulic conductivity can be written more specifically in terms of the intrinsic permeability and the properties of the fluid.

$$K = \frac{k\rho g}{\mu} \tag{4}$$

where k is the intrinsic permeability of the porous medium and is a function only of the pore structure. The intrinsic permeability is not temperature-dependent.

In its differential form, Darcy's equation is:

$$\frac{Q}{A} = q = -\frac{k}{\mu}\frac{dp}{dx} \tag{5}$$

The minus sign results from the definition of Δp, which is equal to $p_2 - p_1$, a negative quantity. The term q is known as the *seepage velocity* and is equivalent to the velocity of approach v_∞, which is also used in the definition of the Reynolds number.

Permeability is normally determined using linear flow in the incompressible or compressible form, depending on whether a liquid or gas is used as the flowing fluid. The volumetric flowrate Q (or Q_m) is determined at several pressure drops. Q (or Q_m) is plotted versus the average pressure p_m. The slope of this line will yield the fluid conductivity K or, if the fluid density and viscosity are known, it provides the intrinsic permeability k. For gases, the fluid conductivity depends on pressure, so that

$$\tilde{K} = K\left(1 + \frac{b}{p}\right) \tag{6}$$

where b depends on the fluid and the porous medium. Under such circumstances a straight line results (as with a liquid), but it does not pass through the origin; instead it has a slope of bK and intercept K. The explanation for this phenomenon is that gases do not always stick to the walls of the porous medium. This slippage shows up as an apparent dependence of the permeability on pressure.

EFFECTS OF HETEROGENEITY, NONUNIFORMITY AND ANISOTROPY ON PERMEABILITY

Heterogeneity, nonuniformity and anisotropy are terms which are defined in the volume-average sense. They may be defined at the level of Darcy's law in terms of permeability. Permeability, however, is more sensitive to conductance, mixing and capillary pressure than to porosity.

Heterogeneity, nonuniformity and anisotropy are defined as follows. On a macroscopic basis, they imply averaging over elemental volumes of radius ϵ about a point in the media, where ϵ is sufficiently large that Darcy's law can be applied for appropriate Reynolds numbers. In other words, volumes are large relative to that of a single pore. Further, ϵ is the minimum radius that satisfies such a condition. If ϵ is too large, certain nonidealities may be obscured by burying their effects far within the elemental volume.

Heterogeneity, nonuniformity and anisotropy are based on the probability density distribution of permeability of random macroscopic elemental volumes selected from the medium, where the permeability is expressed by the one-dimensional form of Darcy's law.

Permeability is the conductance of the medium and has direct relevance to Darcy's law. Permeability is related to the pore size distribution, since the distribution of the sizes of entrances, exits and lengths of the pore walls constitutes the primary resistance to flow. This parameter reflects the conductance of a given pore structure.

Permeability and porosity are related to each other; if the porosity is zero the permeability is zero. Although a correlation between these two parameters may exist, permeability cannot be predicted from porosity alone, since additional parameters that contain more information about the pore structure are needed. These additional parameters are tortuosity and connectivity.

Permeability is a volume-averaged property for a finite but small volume of a medium. Anisotropy in natural or manmade packed media may result from particle (or grain) orientation, bedding of different sizes of particles or layering of media of different permeability. A dilemma arises when considering whether to treat a directional effect as anisotropy or as an oriented heterogeneity.

In an oriented porous medium, the resistance to flow differs depending on the direction. Thus, if there is a pressure gradient between two points and a particular fluid particle is followed, unless the pressure gradient is parallel to oriented flow paths, the fluid particle will not travel from the original point to the point which one would expect. Instead, the particle will drift.

TORTUOSITY

Tortuosity is defined as the relative average length of a flow path (i.e., the average length of the flow paths to the length of the medium). It is a macroscopic measure of both the sinuosity of the flow path and the variation in pore size along the flow path. Both porosity and tortuosity correlate with permeability, but neither can be used alone to predict permeability. Tortuosity and connectivity are difficult to relate to the nonuniformity and anisotropy of a medium. Attempts to predict permeability from a pore structure model require information on tortuosity and connectivity.

CONNECTIVITY

Connectivity is a term that describes the arrangement and number of pore connections. For monosize pores, connectivity is the average number of pores per junction. The term represents a macroscopic measure of the number of pores at a junction. Connectivity correlates with permeability, but cannot be used alone to predict permeability except in certain limiting cases. Difficulties in conceptual simplifications result from replacing the real porous medium with macroscopic parameters that are averages and that relate to some idealized model of the medium. Tortuosity and connectivity are different features of the pore structure and are useful to interpret macroscopic flow properties, such as permeability, capillary pressure and dispersion.

THE KOZENY EQUATION

Porous media is typically characterized as an ensemble of channels of various cross sections of the same length. The Navier-Stokes equations for all channels passing a cross section normal to the flow can be solved to give:

$$S^2 = \frac{c\phi}{k} \tag{7}$$

Where parameter c is known as the Kozeny constant, which is interpreted as a shape factor that is assigned different values depending on the configuration of the capillary (as a point of reference, c = 0.5 for a circular capillary). S is the specific surface area of the channels. For other than circular capillaries, a shape factor is included:

$$r^2 = \frac{ck}{\phi} \tag{8}$$

The specific surface for cylindrical pores is:

$$S_A = \frac{n2\pi rL}{n\pi r^2 L} = \frac{2}{r} \tag{9}$$

and

$$S_A^2 = \frac{2\phi}{8^{1/2}k} \tag{10}$$

Replacing $2/8^{1/2}$ with shape parameter c and S_A with a specific surface, the Kozeny equation is obtained.

$$S = \phi S_A \tag{11}$$

Tortuosity τ is basically a correction factor applied to the Kozeny equation to account for the fact that in a real medium the pores are not straight (i.e., the length of the most probable flow path is longer than the overall length of the porous medium):

$$S^2 = \frac{c\phi^3}{\tau k} \tag{12}$$

To determine the average porosity of a homogeneous but nonuniform medium, the correct mean of the distribution of porosity must be evaluated. The porosities of natural and artificial media usually are normally distributed. The average porosity of a heterogeneous nonuniform medium is the volume-weighted average of the number average:

$$\langle\langle\phi\rangle\rangle = \frac{\sum_{i=1}^{m} V_i \langle\phi_i\rangle}{\sum_{i=1}^{m} V_i} \tag{13}$$

The average nonuniform permeability is spatially dependent. For a homogeneous but nonuniform medium, the average permeability is the correct mean (first moment) of the permeability distribution function. Permeability for a nonuniform medium is usually skewed. Most data for nonuniform permeability show permeability to be distributed log-normally. The correct average for a homogeneous, nonuniform permeability, assuming it is distributed log-normally, is the geometric mean, defined as:

$$\langle k\rangle = \left[\prod_{i=1}^{n} k_i\right]^{1/n} \tag{14}$$

For flow in heterogeneous media, the average permeability depends on the arrangement and geometry of the nonuniform elements, each of which has a different average permeability. To explain this, consider the flow into the face of a rectangular element with overall dimensions of height H, width W and length L. Within that rectangular system, consider a series of smaller, parallel rectangular conduits, such that the cross-sectional area of each flow element is A_1, A_2, A_3, etc.

Since flow is through parallel elements of different constant area, Darcy's law for each element, assuming the overall length of each element is equal, is:

$$Q_1 = \frac{A_1 \langle k_1 \rangle \Delta p}{\mu L}$$
$$Q_2 = \frac{A_2 \langle k_2 \rangle \Delta p}{\mu L} \tag{15}$$

The flowrate through the entire system of elements is $Q=Q_1+Q_2+\ldots$ Combining these expressions we obtain:

$$A \langle\langle k \rangle\rangle_p = A_1 \langle k_1 \rangle + A_2 \langle k_2 \rangle + \ldots \tag{16a}$$

or

$$\langle\langle k \rangle\rangle_p = \frac{A_1 \langle k_1 \rangle + A_2 \langle k_2 \rangle + \ldots}{A} \tag{16b}$$

This means that the average permeability for this heterogeneous medium is the area-weighted average of the average permeability of each of the elements. If the permeability of each element is log-normally distributed, these are the geometric means.

Reservoirs and soils are usually composed of heterogeneities that are nonuniform layers, so that only the thickness of the layers varies. This means that $\langle\langle k_p \rangle\rangle$ simplifies to:

$$\langle\langle k \rangle\rangle_{ph} = \frac{h_1 \langle k_1 \rangle + h_2 \langle k_2 \rangle + \ldots}{h} \tag{17}$$

If all the layers have the same thickness, then

$$\langle\langle k \rangle\rangle_p = \frac{\sum_{i=1}^{h} k_i}{n} \tag{18}$$

where n is the number of layers.

From an industrial viewpoint, the objective of the unit operation of filtration is the separation of suspended solid particles from a process fluid stream which is accomplished by passing the suspension through a porous medium that is referred to as a filter medium. In forcing the fluid through the voids of the filter medium, fluid alone flows, but the solid particles are retained on the surface and in the medium's pores. The fluid discharging from the medium is called the filtrate. The operation may be performed with either incompressible fluids (liquids) or slightly to highly compressible fluids (gases). The physical mechanisms controlling filtration, although similar, vary with the degree of fluid compressibility. Although there are marked similarities in the particle capture mechanisms between the two fluid types, design methodologies for filtration equipment vary markedly. This reference volume concentrates only on process liquid handling (i.e., incompressible fluid flow and processing).

WHAT IS THE DIFFERENCE BETWEEN INCOMPRESSIBLE AND COMPRESSIBLE FLUIDS?

The dependency of liquid volume on pressure may be expressed in terms of the coefficient of compressibility. The coefficient is constant over a wide range of pressures for a particular material, but is different for each substance and for the solid and liquid states of the same material. For liquids, volume decreases linearly with pressure. For gases volume is observed to be inversely proportional to pressure/. If water in its liquid state is subjected to a pressure change from 1 to 2 atm, then less than a 10^{-3} % reduction in volume occurs (the compressibility coefficient is very small). However, when the same pressure differential is applied to water vapor, a volume reduction in excess of 2 occurs.

FILTRATION DYNAMICS

When a suspension of solids passes through a porous media, the solid particles are collected on the feed side of the plate while the filtrate is forced through the media and carried away on the leeward side. A filter medium is, by nature, inhomogeneous, with pores nonuniform in size, irregular in geometry and unevenly distributed over the surface. Since flow through the medium takes place through the pores only, the micro-rate of liquid flow may result in large differences over the filter surface. This implies that the top layers of the generated filter cake are inhomogeneous and, furthermore, are established based on the structure and properties of the filter medium. Since the number of pore passages in the cake is large in comparison to the number in the filter medium, the cake's primary structure depends strongly on the structure of the initial layers. As a result, the cake and filter medium influence each other. Pores with passages extending all the way through the filter medium are capable of capturing solid particles that are smaller than the narrowest cross section of the passage. This is generally attributed to the phenomenon of particle bridging or, in some cases, physical adsorption. Take a close look at Figure 3 to see examples of particle bridging. Depending on the particular filtration technique, different filter media can be employed. Examples of common media are sand, diatomite, coal, cotton or wool fabrics, metallic wire cloth, porous plates of quartz, chamotte, sintered glass, metal powder, and powdered ebonite. The average pore size and configuration (including tortuosity and connectivity) are established from the size and form of individual elements from which the medium is manufactured. On the average, pore sizes are greater for larger medium elements. In addition, pore configuration tends to be more uniform with more uniform medium elements. he fabrication method of the filter medium also affects average pore size and form. For example, pore characteristics are altered when fibrous media are first pressed together.

adsorption - is the grouping together of molecules on the surface of a solid or liquid; such "groupings" are the result of attractive forces between molecules. Activated carbons are highly porous; they contain mazes of interconnecting channels. An imbalance of molecular forces in the walls attracts many substances; these are physically held (adsorbed) by the carbon surfaces. After much use, the carbon may be re-generated and used again.

Pore characteristics also depend on the properties of fibers in woven fabrics, as well as on the exact methods of sintering glass and metal powders. Some filter media, such as cloths (especially fibrous layers), undergo considerable compression when subjected to typical pressures employed in industrial filtration operations. Other filter media, such as ceramic, sintered plates of glass and metal powders, are stable under the same operating conditions. In addition, pore characteristics are greatly influenced by the separation process occurring within the pore passages, as this leads to a decrease in effective pore size and consequently an increase in flow resistance. This results from particle penetration into the pores of the filter medium. The separation of solid particles from a liquid via filtration is a complicated process. For practical reasons filter medium openings are designed to be larger than the average size of the particles to be filtered. The filter medium chosen should be capable of retaining solids by adsorption. Furthermore, interparticle cohesive forces should be large enough to induce particle flocculation around the pore openings.

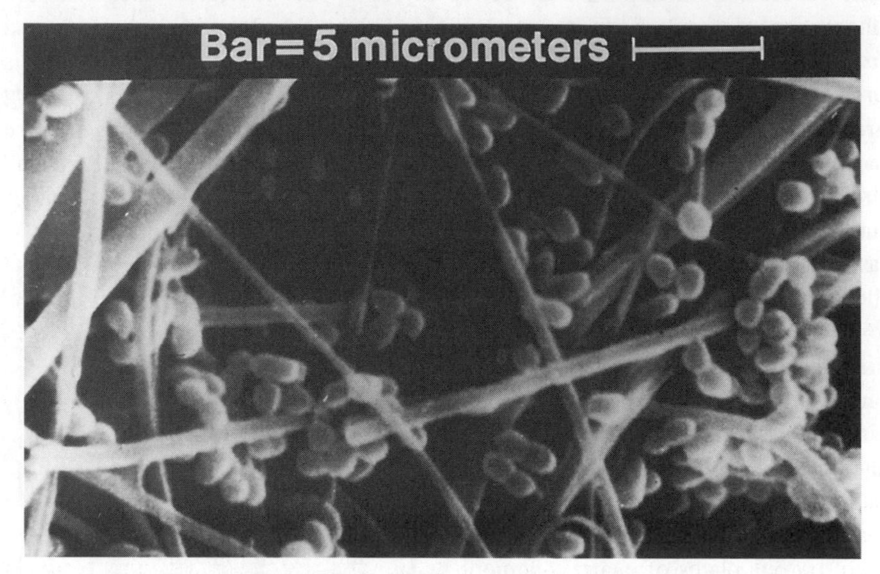

Figure 3. SEM of fibrous depth filter.

PROCESS CLASSIFICATION

There are two major types of filtration: "cake" and "filter-medium" filtration. In the former, solid particulates generate a cake on the surface of the filter medium. In filter-medium filtration (also referred to as clarification), solid particulates become entrapped within the complex pore structure of the filter medium. The filter medium for the latter case consists of cartridges or granular media. Among the most common examples of granular materials are sand or anthracite coal.

When specifying filtration equipment for an intended application one must first account for the parameters governing the application and then select the filtration equipment best suited for the job. There are two important parameters that must be considered, namely the method to be used for forcing liquid through the medium, and the material that will constitute the filter medium.

When the resistance opposing fluid flow is small, gravity force effects fluid transport through a porous filter medium. Such a device is simply called a gravity filter.

When gravity is insufficient to induce flow, the pressure of the atmosphere is allowed to act on one side of the filtering medium, while a negative or suction pressure is applied on the discharge side. This type of filtering device is referred to as a vacuum filter. The application of vacuum filters is typically limited to 15 psi pressure, although there are applications where this value can be exceeded. (*Note*

- *filtration is often used in combination with clarification*).
If still greater force is required, a positive pressure in excess of atmospheric can be applied to the suspension by a pump. This motive force may be in the form of compressed air introduced in a montejus, or the suspension may be directly forced through a pump acting against the filter medium (as in the case of a filter press), or centrifugal force may be used to drive the suspension through a filter medium as is done in screen centrifuges.
In all of these cases, the process of filtration may be characterized as a hydrodynamic process in which the fluid's volumetric rate is directly proportional to the existing pressure gradient across the filter medium, and inversely proportional to the flow resistance imposed by the connectivity, tortuosity and size of the medium's pores, and generated filter cake. The pressure gradient constitutes the driving force responsible for the flow of the suspension.
Regardless of how the pressure gradient is generated, the driving force increases proportionally. However, in most cases, the rate of filtration increases more slowly than the rate at which the pressure gradient rises. The reason for this is that as the gradient rises, the pores of filter medium and cake are compressed and consequently the resistance to flow increases. For highly compressible cakes, both driving force and resistance increase nearly proportionally and any rise in the pressure drop has a minor effect on the filtration rate.

> ***clarification*** - clarifiers are designed to efficiently remove undissolved substances from wastewater; removal is dependent upon density differences and is often enhanced by chemical means. Clarifiers are tank-like structures that may be either circular or rectangular in shape. When wastewaters enter these treatment areas denser undissolved substances settle out, others rise to the surface. A scraper (rake) moves across the bottom of the clarifier; settled matter (sludge) is moved to a collection area. A skimmer moves across the water's surface collecting floating material.

DYNAMICS OF CAKE FORMATION

Filtration operations are capable of handling suspensions of varying characteristics ranging from granular, incompressible, free-filtering materials to slime-like compositions, as well as finely divided colloidal suspensions in which the cakes are incompressible. These latter materials tend to contaminate or foul the filter medium. The interaction between the particles in suspension and the filter medium determines to a large extent the specific mechanisms responsible for filtration.
In practice, cake filtration is used more often than filter-medium filtration. Upon achieving a certain thickness, the cake must be removed from the medium. This can be accomplished by the use of various mechanical devices or by reversing the flow of filtrate back through the medium (hence, the name *backflushing*).
To prevent the formation of muddy filtrate at the beginning of the subsequent

filtration cycle, a thin layer of residual particles is sometimes deposited onto the filter medium. For the same reason, the filtration cycle is initiated with a low, but gradually increasing pressure gradient at an approximately constant flowrate. The process is then operated at a constant pressure gradient while experiencing a gradual decrease in process rate.

The structure of the cake formed and, consequently, its resistance to liquid flow depends on the properties of the solid particles and the liquid phase suspension, as well as on the conditions of filtration. Cake structure is first established by hydrodynamic factors (cake porosity, mean particle size, size distribution, and particle specific surface area and sphericity). It is also strongly influenced by some factors that can conditionally be denoted as physicochemical. These factors are:

1. the rate of coagulation or peptization of solid particles,
2. the presence of tar and colloidal impurities clogging the pores,
3. the influence of electrokinetic potentials at the interphase in the presence of ions, which decreases the effective pore cross section, and
4. the presence of solvate shells on the solid particles (this action is manifested at particle contact during cake formation).

Due to the combining effects of hydrodynamic and physicochemical factors, the study of cake structure and resistance is extremely complex, and any mathematical description based on theoretical considerations is at best only descriptive.

The influence of physicochemical factors is closely related to surface phenomena at the solid-liquid boundary. It is especially manifested by the presence of small particles in the suspension. Large particle sizes result in an increase in the relative influence of hydrodynamic factors, while smaller sizes contribute to a more dramatic influence from physicochemical factors. No reliable methods exist to predict when the influence of physicochemical factors may be neglected. However, as a general rule, for rough evaluations their influence may be assumed to be most pronounced in the particle size range of 15-20 μm.

In specifying and designing filtration equipment, attention must be given to those methods that minimize high cake resistance. This resistance is responsible for losses in filtration capacity, which in turn impact on operating time and removal efficiency. One option for achieving a required filtration capacity is the use of a large number of filter modules. Increasing the physical size of equipment is feasible only within certain limitations as dictated by design considerations, allowable operating conditions, and economic constraints.

A more flexible option from an operational viewpoint is the implementation of process-oriented enhancements that intensify particle separation. This can be achieved by two different methods. In the first method, the suspension to be separated is pretreated to obtain a cake with minimal resistance. This involves the addition of filter aids, flocculants or electrolytes to the suspension. In the second method, the period during which suspensions are formed provides the opportunity to alter suspension properties or conditions that are more favorable to low-resistance cakes. For example, employing pure initial substances or performing a prefiltration operation under milder conditions tends to minimize the formation of tar and colloids. Similar results may be achieved through temperature control, by limiting the duration of certain operations immediately before filtering such as crystallization, or by controlling the rates and sequence of adding reagents.

FILTRATION CONDITIONS

Two significant operating parameters influence the process of filtration: the pressure differential across the filtering plate, and the temperature of the suspension. Most cakes may be considered compressible and, in general, their rate of compressibility increases with decreasing particle size. The temperature of the suspension influences the liquid-phase viscosity, which subsequently affects the ability of the filtrate to flow through the pores of the cake and the filter medium. In addition, the filtration process can be affected by particle inhomogeneity and the ability of the particles to undergo deformation when subjected to pressure and settling characteristics due to the influence of gravity. Particle size inhomogeneity influences the geometry of the cake structure not only at the moment of its formation, but also during the filtration process. During filtration, small particles retained on the outer layers of the cake are often entrained by the liquid flow and transported to layers closer to the filter medium, or even into the pores themselves. This results in an increase in the resistances across the filter medium and the cake that is formed. Particles that undergo deformation when subjected to transient or high pressures are usually responsible for the phenomenon known as *pore clogging*. The addition of coagulating and peptizing agents can greatly improve filterability. These are additives which can drastically alter the cake properties and, subsequently lower flow resistance and ultimately increase the filtration rate and the efficiency of separation. Filter aids may be used to prevent the penetration of fine particles into the pores of a filter plate when processing low concentration suspensions. Filter aids build up a porous, permeable, rigid lattice structure that retains solid particles on the filter medium surface, while permitting liquid to pass through. They are often employed as precoats with the primary aim of protecting the filter medium. They may also be mixed with a suspension of diatomaceous silica type earth (>90% silica content). Cellulose and asbestos fiber pulps were typically employed for many years as well.

WASHING AND DEWATERING

When contaminated, polluted, or valuable suspension liquors are present, it becomes necessary to wash the filter cake to effect clean separation of solids from the mother liquor or to recover the mother liquor from the solids. The operation known as *dewatering* involves forcing a clean fluid through the cake to recover residual liquid retained in the pores, directly after filtering or washing. Dewatering is a complex process on a microscale, because it involves the hydrodynamics of two-phase flow. Although washing and dewatering are performed on a cake with an initially well defined pore structure, the flows become greatly distorted and complex due to changing cake characteristics. The cake structure undergoes compression and disintegration during both operations, thus resulting in a dramatic alteration of the pore structure.

WASTEWATER TREATMENT APPLICATIONS

Subsequent chapters address the application of filtration techniques to wastewater treatment in some detail. For now, only some general comments and terminology are introduced as part of this introductory chapter.

In a very general sense, there are two types of wastewater flows - municipal and industrial. Although municipal wastewaters vary in composition, there are ranges of properties that enable filtration equipment to be readily selected and specified. This is not always the situation when treating industrial wastewater streams. The compositions and properties of industrial wastewaters vary significantly, and even within specific industry sectors, these flows can be dramatically different. This is important to realize because although filtration is a physical process, it depends upon and is integrally a part chemical treatment processes such as preconditioning, buffering and filter aid conditioning. These chemical treatment methods must be properly specified along with the filtration equipment itself in order to ensure that a properly designed filtration system is being applied.

Filtration equipment selection can be complex not only because of the wide variations in suspension properties, but also because of the sensitivities of suspension and cake properties to different process conditions and to the variety of filtering equipment available. Generalities in selection criteria are, therefore, few; however, there are some guidelines applicable to certain classes of filtration applications. One example is the choice of a filter whose flow orientation is in the same direction as gravity when handling polydispersed suspensions. Such an arrangement is more favorable than an upflow design, since larger particles will tend to settle first on the filter medium, thus preventing pores from clogging within the medium structure.

A further recommendation, depending on the application, is not to increase the pressure difference for the purpose of increasing the filtration rate. The cake may, for example, be highly compressible; thus, increased pressure would result in significant increases in the specific cake resistance. We may generalize the selection process to the extent of applying three rules to all filtration problems:

1. The objectives of a filtration operation should be defined;
2. Physical and/or chemical pretreatment options should be evaluated for the intended application based on their availability, cost, ease of implementation and ability to provide optimum filterability; and
3. Final filtration equipment selection should be based on the ability to meet all objectives of the application within economic constraints.

In applying these general criteria, one should focus on the intended application. In wastewater treatment applications, filtration can be applied at various stages. It can be applied as a pretreatment method, in which case the objective is often to remove coarse, gritty materials from the waste-stream. This is a preconditioning step for waste waters which will undergo further chemical and physical treatment downstream.

Filtration may also serve as the preparatory step for the operation following it. The latter stages may be drying or incineration of solids, concentration or direct use of the filtrate. Filtration equipment must be selected on the basis of their ability to deliver the best feed material to the next step. Dry, thin, porous, flaky cakes are best suited for drying where grinding operations are not employed. In such cases, the cake will not ball up, and quick drying can be achieved. A clear, concentrated filtrate often aids downstream treatment, whereby the filter can be operated to increase the efficiency of the downstream equipment without affecting its own efficiency.

Filtration may also be applied as a part of the final stages of treatment in the process. This is most commonly referred to as a polishing operation. Indeed, filtration may be applied both as pretreatment and polishing stages, and even as an intermediate stage in the wastewater treatment process. Filtration equipment selection depends upon the specific operation that the equipment must perform.

Proper pH control can result in clarification that might otherwise not be feasible, since an increase in alkalinity or acidity may change soft, slimy solids into firm, free-filtering ones. In some cases precoats are employed, not because of the danger of filter cloth clogging, but to allow the use of a coarser filter medium, such as metallic cloth.

EQUIPMENT SELECTION METHODOLOGY

Equipment selection is seldom based on rigorous equations or elaborate mathematical models. Where equations are used, they function as a directional guide in evaluating data or process arrangements. Projected results are derived most reliably from actual plant operational data and experience where duplication is desired; from standards set up where there are few variations from plant to plant, so that results can be anticipated with an acceptable degree of confidence (as in municipal water filtration); or from pilot or laboratory tests of the actual material to be handled. Pilot plant runs are typically designed for short durations and to closely duplicate actual operations.

Proper selection of equipment may be based on experiments performed in the manufacturer's laboratory, although this is not always feasible. Sometimes the material to be handled cannot readily be shipped; its physical or chemical conditions change during the time lag between shipping and testing, or special conditions must be maintained during filtration that cannot be readily duplicated, such as refrigeration, solvent washing and inert gas use. A filter manufacturer's laboratory has the advantage of having numerous types of filters and apparatus available with experienced filtration engineers to evaluate results during and after test runs.

The use of pilot-plant filter assemblies is both common and a classical approach to design methodology development. These combine the filter with pumps, receivers, mixers, etc., in a single compact unit and may be rented at a nominal fee from filter manufacturers, who supply operating instructions and sometimes an operator. Preliminary tests are often run at the filter manufacturer's laboratory. Rough tests indicate what filter type to try in the pilot plant.

Comparative calculations of specific capacities of different filters or their specific filter areas should be made as part of the evaluation. Such calculations may be performed on the basis of experimental data obtained without using basic filtration equations. In designing a new filtration unit after equipment selection, calculations should be made to determine the specific capacity or specific filtration area. Basic filtration equations may be used for this purpose, with preliminary experimental constants evaluated. These constants contain information on the specific cake resistance and the resistance of the filter medium.

The basic equations of filtration cannot always be used without introducing corresponding corrections. This arises from the fact that these equations describe the filtration process partially for ideal conditions when the influence of distorting factors is eliminated. Among these factors are the instability of the cake resistance during operation and the variable resistance of the filter medium, as well as the settling characteristics of solids. In these relationships, it is necessary to use statistically averaged values of both resistances and to introduce corrections to account for particle settling and other factors. In selecting filtration methods and evaluating constants in the process equations, the principles of similarity modeling are relied on heavily.

Within the subject of filtration, a distinction is made between micro- and macromodeling. The first one is related to modeling cake formation. The cake is assumed to have a well defined structure, in which the hydrodynamic and physicochemical processes take place. Macromodeling presents few difficulties, because the models are process-oriented (i.e., they are specific to the particular operation or specific equipment). If distorting side effects are not important, the filtration process may be designed according to existing empirical correlations. In practice, filtration, washing and dewatering often deviate substantially from theory. This occurs because of the distorting influences of filter features and the unaccounted for properties of the suspension and cake.

Existing statistical methods permit prediction of macroscopic results of the processes without complete description of the microscopic phenomena. They are helpful in establishing the hydrodynamic relations of liquid flow through porous bodies, the evaluation of filtration quality with pore clogging, description of particle distributions and in obtaining geometrical parameters of random layers of solid particles.

KEY WORDS

The following are key words you will run across in reading over the next several chapters.

air flotation - This treatment usually follows oil-water separation. Wastewater is pressurized to 3-5 times normal atmospheric pressure in the presence of air to produce a saturated air-water solution. When this solution is released to normal pressure in the flotation unit, tiny air bubbles form throughout the liquid; the same effect is observed when a bottle of pop is opened. As the air bubbles form they become attached to tiny oil droplets and to suspended particles; a froth of bubbles and attached wastes rises to the surface and is skimmed off.

biological treatment - Microorganisms in aquatic systems feed on dissolved/suspended organic matter; their digestion processes decompose organic wastes. Oxygen is consumed during these decomposition processes thus decreasing the supply of dissolved oxygen in the water. If the rate of decomposition is excessive, the resultant oxygen depletion produces stresses on aquatic organisms. Biological (secondary) treatment systems address problems associated with biochemical oxygen demand. Break-down of wastes is transferred from natural waterways to lagoons and/or vessels where conditions can be controlled and therefore decomposition occurs efficiently. Biological treatment is simply a concentrated, controlled, application of a natural process.

carbon adsorption--Carbon adsorption is a technology that has been used widely in the drinking water treatment industry, and that is being used with increasing frequency in the wastewater and hazardous waste industry. The process takes advantage of the highly adsorptive properties of specially prepared carbon known as activated carbon. The porous structure of the carbon provides a large internal surface area onto which organic molecules may become attached. Many organic substances, including chlorinated solvents, PCBs, PAHs, pesticides, and others, may be removed from solution using carbon adsorption. Carbon adsorption is achieved by passing water residues through one or more columns containing granular activated carbon operated in parallel or in series. Carbon columns may be operated in either an upflow (expanded bed) or a downflow (fixed bed) mode. In theory, spent carbon may be regenerated. In practice, however, spent carbon must frequently be discarded, especially if high concentrations of PCBs are present.

coagulating agents - Wastes that are removed by this process are classed as suspended or colloidal. Colloids consist of small particles that are constantly moving; gravity does not cause them to settle out. Coagulating agents, eg. alum and ferric chloride, reduce the effects of electrical charges which keep the particles of waste separate from each other. The particles then join together to form masses called flocs. These flocs then rise to the surface or settle to the bottom. Flocculating agents are frequently used to bond flocs together - this speeds the rate at which they rise to the surface or settle to the bottom.

equalization - Equalization systems contain large reservoirs together with piping and treatment processes. These systems minimize fluctuations in wastewater flows and thus give stability, ensuring that wastewater treatment is carried out under the best possible conditions.

flocculating agents - Flocculating agents are routinely used in municipal and industrial wastewater treatment in conjunction with clarifiers. There are many proprietary surfactant-type polymers designed for this purpose, although inorganic chemicals such as ferric chloride may also be used.

ion exchange--Ion exchange is a process in which ions held by electrostatic forces of charged functional groups on the surface of a solid are exchanged for ions of similar charge in a solution in which the solids are immersed. The "solids" are specific resins (usually in the form of beads) that have an affinity for metallic ions. The most common configuration is the fixed bed system, in which the wastewater flows through resin contained in a column. Ion exchange resins are either highly selective for specific metal contaminants or non-specific for a wide variety of metals.

leachate - The term "leachate" refers specifically to water that has flowed through the sediment, such as pore water, or precipitation that has infiltrated sediments in a CDF or landfill. The volume of leachate is generally much smaller than that of effluent, but the concentration of dissolved contaminants is typically higher. The flow rate of effluents and leachates is highly dependent on their source. The effluent from a CDF during filling operations from a hydraulic dredge can be quite substantial--hundreds or even thousands of liters per minute. The duration of such discharges, however, is limited to the duration of dredging, which is typically on the order of weeks or months. Sidestreams from pretreatment or treatment operations are technology-dependent, but generally will produce smaller flows over a longer period of time (months to years). Once the remediation project is completed, the need for effluent treatment is limited to storm water (runoff), which could remain a long-term source if water comes into contact with contaminated sediments. Leachate is generated over very long time periods, and therefore a permanent leachate collection and treatment system is a common requirement at municipal and industrial landfills.

metals removal technologies - Metal contaminants are primarily associated with suspended particulates in most water residues from sediment remedial alternatives. Suspended solids removal technologies should therefore be sufficient to address metals removal needs for the majority of applications. Removal of dissolved metals from water residues can be conducted using ion exchange or precipitation. These technologies have been widely used for industrial wastewater treatment.

neutralization - Blending acidic and basic wastes is a function of most equalization systems. This action (neutralization) is essential before wastes are directed to biological treatment processes where microorganisms feed on organic substances. Extreme changes in pH often kill microorganisms.

oil separation--Some sediments contain very high concentrations of oil and grease. In most cases, the oil and grease will remain attached to the sediment particulates and be captured by suspended solids removal technologies. In some cases, oil and grease is released from sediment particles, forming a slick, a suspension of discrete particles, or an emulsion in the water residue. In such cases, the oil and grease must be captured or removed prior to treatment processes such as ion exchange, carbon adsorption, and filtration, because oily compounds will foul the surfaces of exchange resins and filters. Oil booms and skimmers are routinely used in CDFs to capture oil and floating debris. Coalescing plate separators employ a medium that provides a surface for the aggregation of small, emulsified oil droplets, which can then be removed by gravity separation. Emulsified oils are much more difficult to separate from water. Chemical de-emulsifying agents, heat, and/or acids are generally effective for breaking emulsions. Once the emulsion is broken, the oil is amenable to treatment processes.

oxidation - Oxidation is used to partially or completely degrade organic compounds. Complete oxidation of organic compounds can theoretically reduce complex molecules to carbon dioxide and water. Halogenated organic compounds will produce minor amounts of mineral acids (e.g., hydrochloric acid). However, oxidation is often not complete, resulting in the formation of simpler "daughter" compounds that are usually much less toxic or persistent than the original contaminants. Two forms of oxidation that might be applicable to water residues from sediment remedial alternatives are chemical oxidation and UV-assisted oxidation. Chemical oxidants suitable for treating wastewater include oxygen, ozone (O_3), hydrogen peroxide (H_2O_2), potassium permanganate, chlorine (or hypochlorites), and chlorine dioxide. The oxidizing power of hydrogen peroxide and ozone can be significantly enhanced through the use of UV light. This technology is effective for treating a wide variety of organic compounds, including PCBs and PAHs. Common oxidizing agents, in addition to oxygen, are chlorine, ozone, hydrogen peroxide and potassium permanganate. These substances oxidize wastes to make them more biodegradable and/or more readily removed by adsorption. Oxidation can be enhanced through control of pH and also through using catalysts.

precipitation - Precipitation is a chemical process in which soluble chemicals are removed from solution by the addition of a reagent with which they react to form a (solid) precipitate. This precipitate can then be removed by standard flocculation, sedimentation, and/or filtration processes. Most heavy metals can be precipitated from water as hydroxides with the addition of a caustic (e.g., sodium hydroxide or lime). Alternatively, sodium sulfide or ferric sulfide may be added to precipitate metals as sulfides. The sulfide process is effective for certain metals, such as mercury, which do not precipitate as hydroxides. Precipitation processes produce a sludge that may have to be managed as hazardous waste due to the presence of concentrated heavy metals. Disposal costs for these sludges may therefore be significant. Lime and caustic soda are common sources of hydroxide (OH-) ions.

OH- ions combine with ions of some metals to form insoluble metal hydroxides (precipitation). Precipitated metals settle out and thus are removed from the water; adsorption, using activated carbon, improves this separation process. Iron is one of many metals which is commonly removed in this way.

residue management - Residues are materials, products, or waste streams generated by components of a sediment remedial alternative. Residues may be water, wastewater, solids, oil fractions, or air and gas emissions. The management of these residues may involve treatment, containment, or discharge to the environment. The types of residues anticipated from most sediment remedial alternatives and management options for them are discussed in various sections of this book. Some sediment treatment technologies may generate unique residues, requiring special management considerations. At a minimum, the inert solid particles that were present in the original, untreated sediment, will still be present following the application of any treatment technology.

roughing filters - Roughing filters are used primarily as pretreatment for filter systems that may not be able to tolerate high turbidity or suspended solids in the source water. Many designs are upflow gravel filters. The upflow feature allows for maximum removal efficiency coupled with simple maintenance. By opening the downwash valves, flow through the filter is reversed at a high rate flushing filtered particles out. This system allows for years of use of the filter without gravel replacement. Typical applications include Remove sand and silt (settleable solids); Remove 50-80% turbidity (cloudiness); Pre-treatment for slow sand filters.

sedimentation--Sedimentation is the basic form of primary treatment employed at most municipal and industrial wastewater treatment facilities. There are a number of process options available to enhance gravity settling of suspended particles, including chemical flocculants, CDFs, sedimentation basins, and clarifiers. Of these, gravity settling in CDFs has been used most extensively with contaminated sediments. CDFs have long served the dual role of a settling basin and storage or disposal facility for dredged sediments. Gravity settling in CDFs, with proper design and operation, can take a hydraulically dredged slurry (typically having 10-15 percent solids by weight) and produce an effluent with 1-2 g/L suspended solids. Many CDFs on the Great Lakes produce effluents with suspended solids less than 1 g/L (e.g., 100 mg/L) by gravity settling alone. At most CDFs, a hydraulically dredged slurry is discharged into the CDF at one end and effluent is released over a fixed or adjustable overflow weir at the opposite end, as shown in Settling times of several days are commonly achieved at larger CDFs. Improved settling efficiencies can be achieved by dividing the CDF into two or more cells or through operational controls to increase the detention time and prevent short-circuiting. As the CDF becomes filled, and detention times shorten, dredging production rates may have to be reduced or mechanical dredging used instead of hydraulic dredging to provide suitable settling efficiencies. Sedimentation basins or clarifiers are typically open, concrete or steel tanks with some type of solids collection system that operates on the bottom. Inclined plates may be incorporated into the tanks to improve solids capture for a given flow rate and reduce the size of the clarifier. Rectangular and circular clarifiers are commonly used in municipal and industrial wastewater treatment, but have only been used on a limited basis in applications with contaminated sediments.

solid residues - Solid residues include the bulk of sediment solids following treatment as well as smaller fractions of solids separated from the sediments or produced by the treatment processes. For most remedial alternatives involving a properly designed and thorough treatment system, the treated solids will not require additional treatment and can be disposed using the technologies discussed in Chapter 8. Exceptions to this may include solid residues with special physical properties or concentrations of contaminants requiring special handling. Some treatment technologies produce small volumes of sludges. Other solid residues include debris and oversized materials separated during dredging or pretreatment, sludges from water or wastewater treatment systems, spent media from granular filters or carbon adsorption systems, and particulates collected from air pollution control systems.

stripping - As steam rises through the column it removes (strips) contaminants from wastewater that is moving in the opposite direction. Hydrogen sulphide and ammonia are two contaminants that are stripped from refinery wastes. Methods that are used to recover contaminants from the flow of steam (as it exits from the stripper) include condensation/vacuum recovery systems, biox and incineration.

suspended solids removal - The removal of suspended matter is generally the most important process in the treatment of effluents and leachates from sediment remedial alternatives because most of the contaminants in water residues are associated with the solid particles. An effective solids removal system can significantly reduce contaminant concentrations, leaving behind only those contaminants that are dissolved or associated with colloidal material. Solids removal is a frequently required pretreatment for processes that remove dissolved contaminants (e.g., ion exchange, carbon adsorption). The primary technology types for suspended solids removal are sedimentation and filtration.

wastewater reduction - Reductions in wastewater production almost always result in decreased amounts of wastes that enter the river. Typical reductions include: replacing once-through cooling water (OTCW) with water that has been recirculated through cooling towers; keeping clean stormwater separate from wastewater that requires treatment; reusing treated wastewater eg. in process units; in cooling towers; as feed water for boilers.

water residues - Water is likely to be the most important residue for consideration at most sediment remediation projects simply because of the volumes generated. The removal and transport technologies selected will have a profound effect on how much water residue is generated through the treatment process. For example, if the sediments are dredged hydraulically and transported by pipeline, a large area will probably be needed for gravity settling. In contrast, if the sediments were removed with a mechanical dredge and transported by truck, there would be much less "free water" to handle. Some pretreatment and treatment processes may require the addition of even more water. For final disposal of sediments and solids residues, most of this water must be removed. Depending on how the sediments are handled, treated, and disposed, the volume of water that must ultimately be managed can be less than one-half of the volume of sediments (in place) dredged, or greater than five times this volume. Water residues from a sediment remedial alternative are commonly referred to as effluent or leachate. The term "effluent" may be applied to a wide variety of water residues, including: Discharges from an active CDF; Surface runoff from a landfill or CDF; Sidestreams from a dewatering process (e.g., filtrate from a filter press or centrate from a centrifuge); Wastewater or condensate from a pretreatment or treatment process.

water residue treatment - Technologies for treating wastewater from municipal and industrial sources are well established and well documented (Weber 1972; Metcalf & Eddy, Inc. 1979; Corbitt 1990). Averett et al. (in prep.) evaluated the applicability of these technologies to effluent and leachate from sediment remedial alternatives on the basis of cost, effectiveness, implementability, and availability. Effluent/leachate treatment technologies may be categorized according to the type(s) of contaminants that are removed. This chapter discusses technologies that remove the following contaminant categories: Suspended solids; Metals; Organic compounds. While there is some degree of overlap between the processes, these categories reflect the primary areas of treatment.

NOMENCLATURE

A = area (m^2)

b = parameter in slip flow expression for K (sec^2-m/kg)

c = shape factor, known as Kozeny constant

D = diameter (m)

D_p = particle diameter (m)

g = acceleration due to gravity (m/sec^2)

h = hydraulic head (m)

k = intrinsic permeability (k^2)

K = hydraulic conductivity (m/sec)

L = characteristic macroscopic length (m)

n = number of pore layers

p = pressure (kg/sec^2-m)

q = seepage velocity (m/sec)

Q = volumetric flowrate (m^3/sec)

Q_m = volumetric flowrate at average pressure p_m (m^3/sec)

r = radius

Re = Reynolds number

S = specific surface (m^2)

V = volume (m^3)

v_∞ = velocity of approach (m/sec)

x = coordinate (m)

z = coordinate (in direction of gravity) (m)

Greek Symbol

μ = viscosity (kg/m-sec)

ρ = density (kg/m^3)

τ = tortuosity

ϕ = porosity

RECOMMENDED RESOURCES FOR THE READER

The following references have been reviewed and are recommended for your further reading. Some of these references will greatly enhance your understanding of the subject matter covered in subsequent chapters.

REVERSE OSMOSIS A PRACTICAL GUIDE FOR INDUSTRIAL USERS
First Printing 1995, ISBN 0-927188-03-1; By Wes Byrne
461 pages, hard cover, $90
Reverse osmosis (RO) is a relatively new technology whose applications are rapidly growing. Several books have been written emphasizing the theory of membrane technology. This book, however, emphasizes the application of reverse osmosis, specifically in industrial markets. It explains how RO systems are designed and used and offers some sample computer programs to facilitate this process. It defines pretreatment requirements based upon the characteristics of the application. The book is therefore of value to design and process engineers working for RO

equipment manufacturers and engineering companies. The book is written as a training guide. It explains how to monitor and maintain RO systems. This knowledge is critical for the daily evaluation of RO system performance. The book goes on to cover various aspects of membrane cleaning and sanitization. It also discusses most of the common RO system problems that should be investigated when troubleshooting an RO concern. These topics make the book an excellent tool for technicians who maintain RO systems. Some of the more common and some of the more interesting applications of reverse osmosis are discussed in detail. These include two-pass RO for high-purity water production, pharmaceutical water treatment, seawater desalinization, application of RO for juice concentration, plating metal recycling, the treatment of secondary sewage effluent, and the final filtration of deionized (DI) water. For anyone who works with industrial RO systems, the book will provide practical insights into reverse osmosis technology and will serve as an excellent reference.

PRACTICAL PRINCIPLES OF ION-EXCHANGE WATER TREATMENT
Second Edition 1995, ISBN 0-927188-00-7; By Dean L. Owens
210 pages, hard cover, $55
The Practical Principles of Ion Exchange Water Treatment develops the fundamentals of understanding the ion exchange water treatment process for those people who work with or are concerned with the operation of systems employing this technology. Included are the basics of the mechanics and equipment of ion-exchange water treatment, and the effect of these on the operation of the systems. Operation cycles of backwash, chemical regeneration, and rinses are detailed with explanations of each step. The in-depth explanations of what, how, and why of this water treatment technology should provide the reader of this text with a basis for improving the operation and performance of installed systems. Troubleshooting and problem solving is approached in a new manner by outlining problem sources, with cross-references to basic explanations of the causes and suggestions as to where to look for solutions. With very few established courses on this subject in schools or in industry, this volume has served as a useful basic training text on ion exchange. Over 3,400 copies of the first edition were sold. The second edition contains an updated troubleshooting section, as well as sections on waste treatment and resin sampling and testing.

BASIC PRINCIPLES OF WATER TREATMENT
1st Printing 1996; ISBN 0-927188-05-8 By Cliff Morelli.
Hardcover, 8272 pages. $60.
Virtually all industries utilize water in their manufacturing process, either as a heat-transfer medium, as in the steam-electric generating industry or in the employment of cooling towers; or as a product as in the case of beverages, foods and pharmaceutical preparations. Water is also employed as a working solvent, as in the case of manufacturing microelectronics, where the final product yield is highly dependent on the purity of the water. In addition, all manufacturing processes that employ water in their process, must also deal with a waste stream. Water treatment

is of almost universal importance, yet, most engineers and chemists responsible for the water treatment facility have not had academic training in industrial water treatment. This book presents an elementary overview of the water treatment process, from "why" to "how." This book is designed for the person who is juct entering this field and needs an overview of the basic principles. For industry veterans, this is an excellent book to give to a new person entering their department.

COAGULANTS AND FLOCCULANTS--Theory and Practice
1st Printing 1995; ISBN 0-927188-04-X By Dr. Yong H. Kim.
Softcover, 96 pages. Price: $23.
The purpose of this book is to provide engineers and plant operators who are involved in solid-liquid separation with a fundamental understanding of coagulation and flocculation. Applications include water and wastewater treatment, pulp and paper industries, mineral processing, and enhanced oil recovery. The readers of the book will obtain an appropriate knowledge of coagulants and polymeric flocculants, their chemistry, properties, and utilization. As a result, they should be able to optimize the performance of pertinent processes. They will especially find useful information about the dissolution of polymeric flocculants-a previously neglected topic, but one which has serious effects on the subsequent flocculation process. Considering the wide -spread use of polymeric flocculants in the field, there are very limited sources readily available to many interested readers. This book will serve as a useful text for them to be well equipped on the subject.

FUNDAMENTALS OF FLUID FILTRATION-- A TECHNICAL PRIMER
2nd edition 1998; ISBN 0-927188-01-5; By Peter R. Johnston
Softcover, 136 pages . $29.
This book is written as the book one wants to read when first starting the study of filtration. This book is intended for the person who has some technical background outside of filtration, and who now wants to "get his feet wet" in this subject This book is also for the "old filtration hand" who, while he has some experience in some special field of filtration, may now want to step back and see a larger view. And it is intended as a vehicle for sorting out and explaining words and phrases that appear in technical and advertising literature. The incentive to write this book comes from the author's experience, as a charter member of ASTM's Committee F21 on Filtration, as chair of the subcommittee on liquid filtration (for 15 years), and as an instigator of the 1986 symposium on filtration sponsored by ASTM.

QUESTIONS FOR THINKING AND DISCUSSING

1. It is recognized that pore blockage in a filter medium occurs because: (1) Pores become blocked by the lodging of a single particle in the pore passage, (2) Gradual blockage can occur due to the accumulation of many

particles in pore passages; and (3) Blockage may occur during intermittent filtration practices. Consider a 1 m^2 surface of filter medium containing N_p number of pores. The average pore radius and length are r_p and ℓ_p, respectively. Assuming laminar flow, develop an expression to calculate the volume of filtrate through a single pore per unit of time. (Hint - Are you familiar with the *Hagen-Poiseuille expression*?).

2. Continuing with question 1, develop an expression for the initial filtration rate per unit area of filtration.

3. Consider 1 m^3 of very dilute suspension containing n number of suspended particles. Develop an expression describing the rate of filtration through the pores, taking into consideration the number of blocked pores.

4. Explain how the term *reservoir description* is applied to characterizing a homogeneous floe system.

5. Go to a standard handbook like Perry's Chemical Engineer's Handbook and obtain several hydraulic conductivities (say for sand, crushed stone, gravel, diatomaceous earth, other). Assume a constant head of fluid over a bed of each material, and apply Darcy's law to calculate flowrates and compare the results. Try several hydraulic head calculations and plot the results. Which of the materials studied shows the highest hydraulic resistance?

6. What parameters influence permeability? Can you list them in terms of first-order, second-order and lower order effects?

7. Explain the term *connectivity* and its relevance to the filtration process.

8. Here's some library or Internet work for you. Compile a table of values on the porosity of common materials (e.g., soils, clay, glass beads, crushed stone, charcoal, other materials). Or if you are really ambitious, apply the equations provided in this chapter along with physical properties data obtained from your search and estimate the porosities.

9. What does the term *backflushing mean*? How do you think this os performed on a filtering machine.?

Chapter 3

CHEMICAL ADDITIVES THAT ENHANCE FILTRATION

INTRODUCTION

Filtration generally requires the use of various chemicals known as filter aids. These chemicals include those used in the process of coagulation, and to accelerate and enhance the particle sedimentation process (called *flocculants*), as well as to provide a precoat filter to enhance the suspended particle removal efficiency. Both an understanding of the physical and chemical forces involved in the filtration process are important to proper selection and operation of filters in waster treatment applications. This chapter covers widely used chemical additives employed in water filtration operations. After reviewing some of the properties of these chemical additives, our attention will focus on the actual mechanisms and roles that these chemicals play in the filtration process. An understanding of the role that chemicals play in the process is important to equipment selection and sizing, but also, this will help us to understand the operating regimes and limitation of filtration hardware. Remember to refer to the ***Glossary*** at the end of the book if you run across any terms that are unfamiliar to you.

ALUMINUM BASED CHEMICAL ADDITIVE COMPOUNDS

Aluminum based chemicals have been used for many years in wastewater treatment applications for suspended solids removal. These include dry and liquid alum, with sodium aluminate used in activated sludge plants for phosphorus removal.
The commercial *dry alum* most often used in wastewater treatment is known as *filter alum*, and has the approximate chemical formula $Al_2(SO_4)_3 \cdot 14H_2O$ and a molecular weight of about 600. Alum is white to cream in color and a 1 percent solution has a pH of about 3.5. The commercially available grades of alum and their corresponding bulk densities and angles of repose are given in Table 1.

Each of these grades has a minimum aluminum content of 27 percent, expressed as $A1_2 0_3$, and maximum $Fe_2 0_3$ and soluble contents of 0.75 percent and 0.5 percent, respectively.
Since dry alum is only partially hydrated, it has hygroscopic tendencies. However, it is relatively stable when stored under normal temperature and humidity conditions encountered.

Table 1. General Properties of Commercial Grades of Alum

Grade	Angle of repose	Bulk density (lbs. /cubic feet)
Lump	-	62 to 68
Ground	43	60 to 71
Rice	38	57 to 71
Powdered	65	38 to 45

Some typical values for the solubility of commercial dry alum at various temperatures are reported in Table 2.

Table 2. Solubility Data on Alum

Temperature (°F)	Solubility (lbs./gal)
32	6.03
50	6.56
68	7.28
86	8.45
104	10.16

Dry alum is not corrosive unless it absorbs moisture from the air, such as during prolonged exposure to humid atmospheres. Therefore, precautions should be taken to ensure that the storage space is free of moisture.
Alum is shipped in 100 lb bags, drums, or in bulk (minimum of 40,000 lb) by truck or rail. Bag shipments may be ordered on wood pallets if desired.
Ground and rice alum are the grades most commonly used by utilities because of their superior flow characteristics. These grades have less tendency to lump or arch in storage and therefore provide more consistent feeding qualities. Hopper agitation is seldom required with these grades, and in fact may be detrimental to feeding because of the possibility of packing the bin.

Alum dust is present in the ground grade and will cause minor irritation of the eyes and nose on breathing. A respirator may be worn for protection against alum dust. Gloves may be work to protect the hands. Because of minor irritation in handling and the possibility of alum dust causing rusting of adjacent machinery, dust removal equipment is desirable. Alum dust should be thoroughly flushed from the eyes immediately and washed from the skin with water.

Bulk alum can be stored in mild steel or concrete bins with dust collector vents located in, above, or adjacent to the equipment room. Recommended storage capacity is about 30 days. Dry alum in bulk form can be transferred or metered by means of screw conveyors, pneumatic conveyors, or bucket elevators made of mild steel. Pneumatic conveyor elbows should have a reinforced backing as the alum can contain abrasive impurities.

Bags and drums of alum should be stored in a dry location to avoid caking. Bag or drum-loaded hoppers should have a nominal Storage capacity for eight hours at the nominal maximum feed rate so that personnel are not required to charge the hopper more than once per shift. Converging hopper sections should have a minimum slope of 60 degrees to prevent arching.

Bulk storage hoppers should terminate at a bin gate so that the feeding equipment may be isolated for servicing. The bin gate should be followed by a flexible connection, and a transition hopper chute or hopper which acts as a conditioning chamber over the feeder.

A typical feed system includes all of the components required for the proper preparation of the chemical solution. Capacities and assemblies should be selected to fulfill individual system requirements. Three basic types of chemical feed equipment are used: volumetric, belt gravimetric, and loss-in-weight gravimetric.

Volumetric feeders are usually used where initial low cost and usually lower capacities are the basis of selection. Volumetric feeder mechanisms are usually exposed to the corrosive dissolving chamber vapors which can cause corrosion of discharge areas. Manufacturers usually control this problem by use of an electric heater to keep the feeder housing dry or by using plastic components in the exposed areas.

Volumetric dry feeders in general use are of the screw type. Screw-feed mechanisms allow even withdrawal across the bottom of the feeder hopper to prevent hopper dead zones. Some screw designs are based on a variable-pitch configuration with the pitch expanding unevenly to the discharge point. Other screw designs are based on constant-pitch type expanding evenly to the discharge point. This type of screw design is known as the constant-pitch-reciprocating type. This type has each half of the screw turned in opposite directions so that the turning and reciprocating motion alternately fills one half of the screw while the other half of the screw is discharging. The variable-pitch screw has one point of discharge, while the constant-pitch-reciprocating screw has two points of discharge, one at each end of the screw. The accuracy of volumetric feeders is influenced by the character of the material being fed and ranges between $\pm$ 1 percent for free-flowing materials

and ± 7 percent for cohesive materials. This accuracy is volumetric and should not be related to accuracy by weight (gravimetric).
Where the greatest accuracy and the most economical use of chemicals is desired, the loss-in-weight-type feeder should be selected. This feeder is limited to the low and intermediate feed rates up to a maximum rate of approximately 4,000 lb/hr. The loss-in-weight-type feeder consists of a material hopper and feeding mechanism mounted on enclosed scales. The feed-rate controller retracts the scale poise weight to deliver the dry chemical at the desired rate. The feeding mechanism must feed at this rate to maintain the balance of the scale. Any unbalance of the scale beam causes a corrective change in the output of the feeding mechanism. Continuous comparison of actual hopper weight with set hopper weight prevents cumulative errors.
Belt-type gravimetric feeders span the capacity ranges of volumetric and loss-in-weight feeders and can usually be sized for all applications encountered in wastewater treatment applications. Initial expense is greater than for the volumetric feeder and slightly less than for the loss-in-weight feeder. Belt-type gravimetric feeders consist of a basic belt feeder incorporating a weighing and control system. Feed rates can be varied by changing either the weight per foot of belt, or the belt speed, or both. Controllers in general use are mechanical, pneumatic, electric, and mechanical-vibrating. Accuracy specified for belt-type gravimetric feeders should be within ± 1 percent of set rate. Materials of construction of feed equipment normally include mild steel hoppers, stainless steel mechanism components, and rubber-surfaced feed belts.
Because alum solution is corrosive, dissolving or solution chambers should be constructed of type 316 stainless steel, fiberglass reinforced plastic (FRP), or plastics. Dissolvers should be sized for preparation of the desired solution strength. The solution strength usually recommended is 0.5 lb of alum to 1 gal. of water, or a 6 percent solution. The dissolving chamber is designed for a minimum detention time of 5 minutes at the maximum feed rate. Because excessive dilution may be detrimental to coagulation, eductors, or float valves that would ordinarily be used ahead of centrifugal pumps are not recommended. Dissolvers should be equipped with water meters and mechanical mixers so that the water-to-alum ratio may be properly established and controlled.

FRP, plastics (polyvinyl chloride, polyethylene, polypropylene, and other similar materials), and rubber are general use and are recommended for alum solutions. Care must be taken to provide adequate support for these piping systems, with close attention given to spans between supports so that objectionable deflection will not be experienced. The alum solution should be injected into a zone of rapid mixing or turbulent flow.
Solution flow by gravity to the point of discharge is desirable. When gravity flow is not possible, transfer components should be selected that require little or no dilution. When metering pumps or proportioning weir tanks are used, return of

excess flow to a holding tank should be considered. Metering pumps are discussed further in the section on liquid alum.

Standard instrument control and pacing signals are generally acceptable for common feeder system operation. Volumetric and gravimetric feeders are usually adaptable to operation from any standard instrument signals.

When solution must be pumped, consideration should be given to use of holding tanks between the dry feed system and feed pumps, and the solution water supply should be controlled to prevent excessive dilution. The dry feeders may be started and stopped by tank level probes. Variable-control metering pumps can then transfer the alum stock solution to the point of application without further dilution.

Means should be provided for calibration of the chemical feeders. Volumetric feeders may be mounted on platform scales. Belt feeders should include a sample chute and box to catch samples for checking actual delivery with set delivery.

Gravimetric feeders are usually furnished with totalizers only. Remote instrumentation is frequently used with gravimetric equipment, but seldom used with volumetric equipment.

Liquid alum is shipped in rubber-lined or stainless steel insulated tank cars or trucks. Alum shipped during the winter is heated prior to shipment so that crystallization will not occur during transit. Liquid alum is shipped at a solution strength of about 8.3 percent as $A1_2O_3$ or about 49 percent as $A1_2(SO_4)_3 \cdot 14H_2O$. The latter solution weighs about 11 lb/gal at 60°F and contains about 5.4 lb dry alum (17 percent Al_2O_3) per gal of liquid. This solution will begin to crystallize at 30° F and freezes at about 18° F.

Bulk unloading facilities usually must be provided at the treatment plant. Rail cars are constructed for top unloading and therefore require an air supply system and flexible connectors to pneumatically displace the alum from the car. U.S. Department of Transportation regulations concerning chemical tank car unloading should be observed. Tank truck unloading is usually accomplished by gravity or by a truck mounted pump.

Established practice in the treatment field has been to dilute liquid alum prior to application. However, recent studies have shown that feeding undiluted liquid alum results in better coagulation and settling. This is reportedly due to prevention of hydrolysis of the alum.

No particular industrial hazards are encountered in handling liquid alum. However, a face shield and gloves should be worn around leaking equipment. The eyes or skin should be flushed and washed upon contact with liquid alum. Liquid alum becomes very sick upon evaporation and therefore spillage should be avoided.

Storage tanks may be open if indoors but must be closed and vented if outdoors. Outdoor tanks should also be heated, if necessary, to keep the temperature above 450F to prevent crystallization. Storage tanks should be constructed of type 316 stainless steel, FRP, steel lined with rubber, polyvinyl chloride, or lead. Liquid alum can be stored indefinitely without deterioration.

Storage tanks should be sized according to maximum feed rate, shipping time required, and quantity of shipment. Tanks should generally be sized for 1.5 times the quantity of shipments. A ten-day to two-week supply should be provided to allow for unforeseen shipping delays.
$Al_2(SO_4)_3$ in aqueous solution is most commonly known as aluminum sulfate solution. The CAS (Chemical Abstract Service) Index is "Sulfuric Acid, Aluminum Salt (3:2)". Its CAS number is 10043-01-3. Aluminum sulfate, solution, technical grade is a clear, white to slightly yellow brown liquid.
Reactions between alum and the normal constituents of wastewaters are influenced by many factors; hence, it is impossible to predict accurately the amount of alum that will react with a given amount of alkalinity, lime, or soda ash which may have been added to the wastewater. Theoretical reactions can be written which will serve as a general guide, but in general the optimum dosage in each case must be determined by laboratory jar tests.
The simplest case is the reaction of Al^{3+} with OH^- ions made available by the ionization of water or by the alkalinity of the water.

Solution of alum in water produces:

$$A1_2(SO_4)^3 \rightleftharpoons 2A1^{3+} + 3(SO_4)^{2-}$$

Hydroxyl ions become available from ionization of water:

$$H_2O \rightleftharpoons H^+ + OH^-$$

The aluminum ions ($A1^{3+}$) then react:

$$2\ A1^{3+} + 6OH^- \rightleftharpoons 2\ Al(OH)_3$$

Consumption of hydroxyl ions will result in a decrease in the alkalinity. Where the alkalinity of the wastewater is inadequate for the alum dosage, the pH must be increased by the addition of hydrated lime, soda ash, or caustic soda.
When amounts of alkali when added to wastewater they will maintain the alkalinity of the water unchanged when 1 mg/1 of alum is added. For example, if no alkalinity is added, 1 mg/1 of alum will reduce the alkalinity of 0.50 mg/1 as $CaC0_3$ but alkalinity can be

The reactions of alum with the common alkaline reagents are...

$Al_2(SO_4)_3 + 3\ Ca(HCO_3)_3 \rightarrow 2\ Al(OH)_3 \downarrow + 3\ CaSO_4 + 6\ CO_2 \uparrow$

$Al_2(SO_4)_3 + 3\ Na2CO_3 + 3\ H_2O \rightarrow 2\ Al(OH)_3 \downarrow + 3\ CO_2 \uparrow$

$Al_2(SO_4)_3 + 3\ Ca(OH)_2 \rightarrow 2\ Al(OH)_3 \downarrow + 3\ CaSO_4$

maintained unchanged if 0.39 mg/l of hydrated lime is added. This lowering of natural alkalinity is desirable in many cases to attain the pH range for optimum coagulation. For each mg/1 of alum dosage, the sulfate (SO_4) content of the water will be increased approximately 0.49 mg/1 and theCO_2 content of the water will be increased approximately 0.44 mg/1.

IRON-BASED COMPOUNDS

Iron compounds have pH coagulation ranges and floc characteristics similar to aluminum sulfate. The cost of iron compounds may often be less than the cost of alum. However, the iron compounds are generally corrosive and often present difficulties in dissolving, and their use may result in high soluble iron concentrations in process effluents. Among the most commonly used iron compounds used in wastewater treatment applications are ferric chloride, ferrous chloride, ferric sulfate, ferrous sulfate

Liquid ferric chloride is a corrosive, dark brown oily-appearing solution having a weight as shipped and stored of 11.2 to 12.4 lb/gal (35 percent to 45 percent $FeCl_3$). The ferric chloride content of these solutions, as $FeCl_3$, is 3.95 to 5.58 lb/gal. Shipping concentrations vary from summer to winter due to the relatively high crystallization temperature of the more concentrated solutions. The pH of a 1 percent solution is about 2.0.

The molecular weight of ferric chloride is 162.22. Viscosities of ferric chloride solutions at various temperatures are can be found in reference 13.

Ferric chloride solutions are corrosive to many common materials and cause stains which are difficult to remove. Areas which are subject to staining should be protected with resistant paint or rubber mats.

Normal precautions should be employed when cleaning ferric chloride handling equipment. Workers should wear rubber gloves, rubber apron, and goggles or a face shield. If ferric chloride comes in contact with the eyes or skin, flush with copious quantities of running water and call a physician. If ferric chloride is ingested, induce vomiting and call a physician.

Ferric chloride solution can be stored as shipped. Storage tanks should have a free vent or vacuum relief valve. Tanks may be constructed of FRP, rubber-lined steel, or plastic-lined steel. Resin-impregnated carbon or graphite are also suitable materials for storage containers.

It may be necessary in most instances to house liquid ferric chloride tanks in heated areas or provide tank heaters or insulation to prevent crystallization. Ferric chloride can be stored for long periods of time without deterioration. The total storage capacity should be 1.5 times the largest anticipated shipment, and should provide at least a ten-day to two-week supply of the chemical at the design average dosage. It may not be desirable to dilute the ferric chloride solution from its shipping concentration to a weaker feed solution because of possible hydrolysis. Ferric chloride solutions may be transferred from underground storage to day tanks with impervious graphite or rubber-lined self-priming centrifugal pumps having Teflon rotary and stationary seals. Because of the tendency for liquid ferric chloride to stain or deposit, glass-tube rotameters; should not be used for metering this solution. Rotodip feeders and diaphragm metering pumps are often used for ferric chloride, and should be constructed of materials such as rubber-lined steel and plastics.

Materials for piping and transporting ferric chloride should be rubber or Saran-lined steel, hard rubber, FRP, or plastics. Valving should consist of rubber or resin-lined diaphragm valves. Saran-lined valves with Teflon diaphragms, rubber-sleeved pinch-type valves, or plastic ball valves. Gasket material for large openings such as manholes in storage tanks should be soft rubber; all other gaskets should be graphite-impregnated blue asbestos, Teflon, or vinyl. System pacing and control requirements are similar to those discussed previously for liquid alum.

Ferrous chloride, $FeCl_2$, as a liquid is available in the form of waste pickle liquor from steel processing. The liquor weighs between 9.9 and 10.4 lb/gal and contains 20 percent to 25 percent $FeCl_2$ or about 10 percent available Fe^{2+}. A 22 percent solution of $FeCl_2$ will crystallize at a temperature of - 4 °F. The molecular weight of $FeCl_2$ is 126.76. Free acid in waste pickle liquor can vary from 1 percent to 10 percent and usually averages about 1.5 percent to 2.0 percent. Ferrous chloride is slightly less corrosive than ferric chloride.

Waste pickle liquor is available in 4,000 gal truckload lots and a variety of carload lots. In most instances the availability of waste pickle liquor will depend on the proximity to steel processing plants.

Since ferrous chloride or waste pickle liquor may not be available on a continuous basis, storage and feeding equipment should be suitable for handling ferric chloride. Therefore, the ferric chloride section should be referred to for storage and handling details.

Ferric sulfate is marketed as dry, partially-hydrated granules with the formula $Fe_2(SO_4)_3 \cdot X\ H_2O$, where X is approximately 7. Typical properties of a commercial product are given in Table 3.

Table 3. Typical Properties of Ferric Sulfate

Property	Value	Units
Molecular Weight	526	
Bulk Density	56-60	lb/cu ft
Water Soluble Iron Expressed as Fe	21.5	percent
Water Soluble Fe^{+3}	19.5	percent
Water Soluble Fe^{+2}	2.0	percent
Insolubles Total	4.0	percent

Property	Value	Units
Free Acid	2.5	percent
Moisture @ 105° C	2.0	percent

Reactions involving ferric sulfate are...

$Fe_2(SO4)_3 + 3Ca(HCO_3)_2 \rightarrow 2\ Fe(OH)_3 \downarrow + 3\ CaSO_4 + 6\ CO_2 \uparrow$

$Fe_2(SO4)_3 + 3Na2CO_3 + 3\ H2O \rightarrow 2Fe(OH)_3 \downarrow + 3\ Na_2SO_4 + 3CO_2 \uparrow$

$Fe_2(SO4)_3 + 3Ca(OH)_2 \rightarrow 2\ Fe(OH)_3 \downarrow + 3CaSO_4$

Ferric sulfate is shipped in car and truck load lots of 50 lb and 100 lb moisture-proof paper bags and 200 lb and 400 lb fiber drums.

General precautions should be observed when handling ferric sulfate, such as wearing goggles and dust masks, and areas of the body that come in contact with the dust or vapor should be washed promptly.

Aeration of ferric sulfate should be held to a minimum because of the hygroscopic nature of the material, particularly in damp atmospheres. Mixing of ferric sulfate and quicklime in conveying and dust vent systems should be avoided as caking and excessive heating can result. The presence of ferric sulfate and lime in combination has been known to destroy cloth bags in pneumatic unloading devices. Because ferric sulfate in the presence of moisture will stain, precautions similar to those discussed for ferric chloride should be observed.

Ferric sulfate is usually stored in the dry state either in the shipping bags or in bulk in concrete or steel bins. Bulk storage bins should be as fight as possible to avoid moisture absorption, but dust collector vents are permissible and desirable. Hoppers on bulk storage bins should have a minimum slope of 36°; however, a greater angle is preferred.

Bins may be located inside or outside and the material transferred by bucket elevator, screw, or air conveyors. Ferric sulfate stored in bins usually absorbs some moisture and forms a thin protective crust which retards further absorption until the crust is broken.

Feed solutions are usually made up at a water to chemical ratio of 2:1 to 8:1 (on a weight basis) with the usual ratio being 4:1 with a 20-minute detention time. Care must be taken not to dilute ferric sulfate solutions to less than 1 percent to prevent hydrolysis and deposition of ferric hydroxide. Ferric sulfate is actively corrosive in solution, and dissolving and transporting equipment should be fabricated of type 316 stainless steel, rubber, plastics, ceramics, or lead,

Dry feeding requirements are similar to those for dry alum except that belt-type feeders are rarely used because of their open type of construction. Closed construction, as found in the volumetric and loss-in-weight-type feeders, generally exposes a minimum of operating components to the vapor, and thereby minimizes maintenance. A water jet vapor remover should be provided at the dissolver to protect both the machinery and operator,

Chemical reactions involving ferric sulfate and ferrous hydroxide are...

$FeSO_4$ + Ca(HCO3)2 → Fe(OH)2 ↓ + Ca SO4 + 2CO2 ↑

$FeSO_4 + Ca(OH)_2 \rightarrow Fe(OH)_2 \downarrow + Ca\ SO_4$

$4\ Fe(OH)_2 + O_2 + 2H_2 \rightarrow 4\ Fe(OH)_3 \downarrow$

Ferrous sulfate or ***copperas*** is a by-product of pickling steel and is produced as granules, crystals, powder, and lumps. The most common commercial form of ferrous sulfate is $FeSO_4 \cdot 7H_2O$, with a molecular weight of 278, and containing 55 percent to 58 percent $FeSO_4$ and 20 percent to 21 percent Fe. The product has a bulk density of 62 to 66 lb/cu ft. When dissolved, ferrous sulfate is acidic. The composition of ferrous sulfate may be quite variable and should be established by consulting the nearest manufacturers.

Ferrous sulfate is also available in a wet state in bulk form from some plants. This form is likely to be difficult to handle and the manufacturer should be consulted for specific information and instructions.

Dry ferrous sulfate cakes at storage temperatures above 68° F, is efflorescent in dry air, and oxidizes and hydrates further in moist air. General precautions similar to those for ferric sulfate, with respect to dust and handling acidic solutions, should be observed when working with ferrous sulfate. Mixing quicklime and ferrous sulfate produces high temperatures and the possibility of fire.

The optimum chemical-to-water ratio for continuous dissolving is 0.5 lb/ gal. of 6 percent with a detention time of 5 minutes in the dissolver. Mechanical agitation should be provided in the dissolver to assure complete solution. Lead, rubber, iron, plastics, and type 304 stainless steel can be used as construction materials for handling solutions of ferrous sulfate.

Ferric sulfate and ferric chloride react with the alkalinity of wastewater or with the added alkaline materials such as lime or soda ash. The reactions may be written to show precipitation of ferric hydroxide, although in practice, as with alum, the reactions are more complicated than this.

Ferrous hydroxide is rather soluble and oxidation to the more insoluble ferric hydroxide is necessary if high iron residuals in effluents are to be avoided. Flocculation with ferrous iron is improved by addition of lime or caustic soda at a rate of 1 to 2 mg/mg Fe to serve as a floc-conditioning agent. Polymers are also generally required to produce a clear effluent.

LIME

Lime is among a family of chemicals which are alkaline in nature and contain principally calcium, oxygen and, in some cases, magnesium. In this grouping are included quicklime, dolomitic lime, hydrated lime, dolomitic hydrated lime, limestone, and dolomite. The most commonly used additives are quicklime and hydrated lime, but the dolomitic counterparts of these chemicals (i.e., the high-magnesium forms) are also widely used in wastewater treatment and are generally similar in physical requirements.

Quicklime, CaO, has a density range of approximately 55 to 75 lb/cu ft, and a molecular weight of 56.08. A slurry for feeding, called milk of lime, can be prepared with up to 45 percent solids. Lime is only slightly soluble, and both lime dust and slurries are caustic in nature. A saturated solution of lime has a pH of about 12.4.

The CaO content of commercially available quicklime can vary quite widely over an approximate range of 70 percent to 96 percent. Content below 88 percent is generally considered below standard in the municipal use field. Purchase contracts are often based on 90 percent CaO content with provisions for payment of a bonus for each 1 percent over and a penalty for each 1 percent under the standard. A CaO content less than 75 percent probably should be rejected because of excessive grit and difficulties in slaking.

Pebble quicklime, all passing a 34-in. screen and not more than 5 percent passing a No. 100 screen, is normally specified because of easier handling and less dust. Hopper agitation is generally not required with the pebble form. Published slaker capacity ratings require "soft or normally burned" limes which provide fast slaking and temperature rise, but poorer grades of limes may also be satisfactorily slaked by selection of the appropriate slaker retention time and capacity.

Storage of bagged lime should be in a dry place, and preferably elevated on pallets to avoid absorption of moisture. System capacities often make the use of bagged quicklime impractical. Maximum storage period is about 60 days. Bulk lime is stored in airtight concrete or steel bins having a 55-degree to 60-degree slope on the bin outlet. Bulk lime can be conveyed by conventional bucket elevators and screw, belt, apron, drag-chain, and bulk conveyors of mild steel construction. Pneumatic conveyors subject the lime to air slaking and particle sizes may be reduced by attrition. Dust collectors should be provided on manually and pneumatically-filled bins.

Quicklime feeders are usually limited to the belt or loss-in-weight gravimetric types because of the wide variation of the bulk density. Feed equipment should have an adjustable feed range of at least 20:1 to match the operating range of the associated slaker. The feeders should have an over-under feed rate alarm to immediately warn of operation beyond set limits of control. The feeder drive should be instrumented to be interrupted in the event of excessive temperature in the slaker compartment.

Lime slakers for wastewater treatment should be of the continuous type, and the major components should include one or more slaking compartments, a dilution compartment, a grit separation compartment, and a continuous grit remover. Commercial designs vary in regard to the combination of water to lime, slaking temperature, and slaking time in obtaining the "milk of lime" suspensions. The *paste-type* slaker admits water as required to maintain a desired mixing viscosity. This viscosity therefore sets the operating retention time of the slaker. The paste slaker usually operates with a low water-to-lime ratio (approximately 2:1 by weight), elevated temperature, and 5-minute slaking time at maximum capacity. The *detention-type* slaker admits water to maintain a desired ratio with the lime, and therefore the lime feed rate sets the retention time of the slaker. The detention slaker operates with a wide range of water-to-lime ratios (2.5:1 and 6:1), moderate temperature, and a 1 0-minute slaking time at maximum capacity. A water-to-lime ratio of from 3.5:1 to 4:1 is most often used. The operating temperature in lime slakers is a function of the water-to-lime ratio, lime quality, heat transfer, and water temperature. Lime slaking evolves heat in hydrating the CaO to $Ca(OH)_2$ and therefore vapor removers are required for feeder protection.

Lime slurry should be transported by gravity in open channels wherever possible. Piping channels and accessories may be rubber, iron, steel, concrete, and plastics. Glass tubing, such as that in rotameters, will cloud rapidly and therefore should not be used. Any abrupt directional changes in piping should include plugged tees or crosses to allow rodding out of deposits. Long sweep elbows should be provided to allow the piping to be cleaned by the use of a cleaning "pig." Daily cleaning is desirable.

Milk-of-lime transfer pumps should be of the open impeller centrifugal type. Pumps having an iron body and impeller with bronze trim are suitable for this purpose. Rubber-lined pumps with rubber-covered impellers are also frequently used. Makeup tanks are usually provided ahead of centrifugal pumps to ensure a flooded suction at all times. *Plating out* of lime is minimized by the use of soft water in the makeup tank and slurry recirculation. Turbine pumps and eductors should be avoided in transferring milk of lime because of scaling problems.

Lime slaker water proportioning is integrally controlled or paced from the feeder. Therefore, the feeder-slaker system will follow pacing controls applied to the feeder only. As discussed previously, gravimetric feeders are adaptable to receive most standard instrumentation pacing signals. Systems can be instrumented to allow remote pacing with telemetering of temperature and feed rate to a central panel for control purposes.

The lime feeding system may be controlled by an instrumentation system integrating both plant flow and pH of the wastewater after lime addition. However, it should be recognized that pH probes require daily maintenance in this application to monitor the pH accurately. Deposits tend to build up on the probe and necessitate frequent maintenance. The low pH lime treatment systems (pH 9.5 to 10.0) can be more readily adapted to this method of control than high-lime treatment systems (pH 11.0 or greater) because less maintenance of the pH equipment is required. In a close-loop pH-flow control system, milk of lime is prepared on a batch basis and transferred to a holding tank with variable output feeders set by the flow and pH meters to proportion the feed rate.

Hydrated lime, $Ca(OH)_2$, is usually a white powder (200 to 400 mesh); has a bulk density of 20 to 50 lb/cu ft, contains 82 percent to 98 percent $Ca(OH)_2$, is slightly hydroscopic, tends to flood the feeder, and will arch in storage bins if packed. The modular weight is 74.08. The dust and slurry of hydrated lime are caustic in nature. The pH of hydrated lime solution is the same as that given for quicklime.

Hydrated lime is slaked lime and needs only enough water added to form milk of lime. Wetting or dissolving chambers are usually designed to provide 5-minutes detention with a ratio of 0.5 lb/gal of water or 6 percent slurry at the maximum feed rate. Hydrated lime is usually used where maximum feed rates do not exceed 250 lb/hr., i.e., in smaller plants. Hydrated lime and milk of lime will irritate the eyes, nose, and respiratory system and will dry the skin. Affected areas should be washed with water.

Information given for quicklime also applies to hydrated lime except that bin agitation must be provided. Bulk bin outlets should be provided with nonflooding rotary feeders. Hopper slopes vary from 60 degrees to 66 degrees.

Volumetric or gravimetric feeders may be used, but volumetric feeders are usually selected only for installations where comparatively low feed rates are required. Dilution does not appear to be important, therefore, control of the amount of water used in the feeding operation is not considered necessary. Inexpensive hydraulic jet agitation may be furnished in the wetting chamber of the feeder as an alternative to mechanical agitation. The jets should be sized for the available water supply pressure to obtain proper mixing.

Controls as listed for dry alum apply to hydrated lime. Hydraulic jets should operate continuously and only shut off when the feeder is taken out of service. Control of the feed rate with pH as well as pacing with the plant flow may be used with hydrated lime as well as quicklime.

Lime is somewhat different from the hydrolyzing coagulants. When added to wastewater it increases pH and reacts with the carbonate alkalinity to precipitate calcium carbonate. If sufficient lime is added to reach a high pH, approximately 10.5, magnesium hydroxide is also precipitated. This latter precipitation enhances clarification due to the flocculant nature of the $Mg(OH)_2$. Excess calcium ions at high pH levels may be precipitated by the addition of soda ash. The preceding reactions are shown as follows:

$$Ca(OH)2 + Ca(HCO3)_2 \rightarrow 2\ CaCO_3\downarrow + 2H_2O$$

$$2\ Ca(OH)_2 + Mg(HCO_3)_2 \rightarrow 2\ CaCO_3\downarrow + Mg(OH)_2\downarrow + 2H_2O$$

$$Ca(OH)_2 + Na2CO_3 \rightarrow CaCO_3\downarrow + 2\ NaOH$$

Reduction of the resulting high pH levels may be accomplished in one or two stages. The first stage of the two-stage method results in the precipitation of calcium carbonate through the addition of carbon dioxide according to the following reaction:

$$Ca(OH)_2 + CO_2 \rightarrow CaCO_3 \downarrow + H_2O$$

Single-stage pH reduction is generally accomplished by the addition of carbon dioxide, although acids have been employed, This reaction, which also represents the second stage of the two-stage method, is as follows:

$$Ca(OH)_2 + 2CO_2 \rightarrow Ca(HCO_3)_2$$

As noted for the other chemicals, the preceding reactions are merely approximations to the more complex interactions which actually occur in waste waters. The lime demand of a given wastewater is a function of the buffer capacity or alkalinity of the wastewater.

SODA ASH

Soda ash, Na_2CO_3, is available in two forms. Light soda ash has a bulk density range of 35 to 50 lb/cu ft and a working density of 41 lb/cu ft. Dense soda ash has a density range of 60 to 76 lb/cu ft and a working density of 63 lb/cu ft. The pH of a 1 percent solution of soda ash as 11.2. It is used for pH control and in lime treatment. The molecular weight of soda ash is 106. Commercial purity ranges from 98 percent to greater than 99 percent Na_2CO_3. Soda ash by itself is not particularly corrosive, but in the presence of lime and water caustic soda is formed which is quite corrosive.

Dense soda ash is generally used in municipal applications because of superior handling characteristics. It has little dust, good flow characteristics, and will not arch in the bin or flood and feeder. It is relatively hard to dissolve and ample dissolver capacity must be provided. Normal practice calls for 0.5 lb of dense soda ash per gallon of water or a 6 percent solution retained for 20 minutes in the dissolver.

The dust and solution are irritating to the eyes, nose, lungs, and skin and therefore general precautions should be observed and the affected areas should be washed promptly with water.

Soda ash is usually stored in steel bins and where pneumatic-filling equipment is used, bins should be provided with dust collectors. Bulk and bagged soda ash tend to absorb atmospheric CO_2 and water, forming the less active sodium bicarbonate ($NaHCO_3$). Material recommended for unloading facilities is steel.

Feed equipment as described for dry alum is suitable for soda ash. Dissolving of soda ash may be hastened by the use of warm dissolving water. Mechanical or hydraulic jet mixing should be provided in the dissolver. Materials of construction for piping and accessories should be iron, steel, rubber, and plastics.

LIQUID CAUSTIC SODA

Anhydrous caustic soda (NaOH) is available but its use is generally not considered practical in water and wastewater treatment applications. Consequently, only liquid caustic soda is discussed here. Liquid caustic soda is generally shipped at two concentrations, 50 percent and 73 percent NaOH. The densities of the solutions as shipped are 12.76 lb/gal for the 50 percent solution and 14.18 lb/gal for the 73 percent solution. These solutions contain 6.38 lb/gal NaOH and 10.34 lb/gal NaOH, respectively. The crystallization temperature is 53° F for the 50 percent solution and 165° F for the 73 percent solution. The molecular weight of NaOH is 40. The pH of a 1 percent solution of caustic soda is 12.9.

Truckload lots of 1,000 to 4,000 gallons are available in the 50 percent concentration only. Both shipping concentrations can be obtained in 8,000, 10,000 and 16,000 gal carload lots. Tank cars can be unloaded through the dome eduction pipe using air pressure or through the bottom valve by gravity or by using air pressure or a pump. Trucks are usually unloaded by gravity or with air pressure or a truck-mounted pump.

Liquid caustic soda is received in bulk shipments, transferred to storage, and diluted as necessary for feeding to the points of application. Caustic soda is poisonous and is dangerous to handle. U.S. Department of Transportation Regulations for "White Label" materials must be observed. However, if handled properly caustic soda poses no particular industrial hazard. To avoid accidental spills, all pumps, valves, and lines should be checked regularly for leaks. Workers should be thoroughly instructed in the precautions related to the handling of caustic soda. The eyes should be protected by goggles at all times when exposure to mist or splashing is possible. Other parts of the body should be protected as necessary to prevent alkali bums. Areas exposed to caustic soda should be washed with copious amounts of water for 15 minutes to 2 hours. A physician should be called when exposure is severe. Caustic soda taken internally should be diluted with water or milk and then neutralized with dilute vinegar or fruit juice. Vomiting may occur spontaneously but should not be induced except on the advice of a physician.

Liquid caustic soda may be stored at the 50 percent concentration. However, at this solution strength, it crystallizes at 53° F. Therefore, storage tanks must be located indoors or provided with heating and suitable insulation if outdoors. Because of its relatively high crystallization temperature, liquid caustic soda is often diluted to a concentration of about 20 percent NaOH for storage. A 20 percent solution of NaOH has a crystallization temperature of about - 20° F. Recommendations for dilution of both 73 percent and 50 percent solutions should be obtained from the manufacturer because special considerations are necessary.

Storage tanks for liquid caustic soda should be provided with an air vent for gravity flow. The storage capacity should be equal to 1.5 times the largest expected delivery, with an allowance for dilution water, if used, or two-weeks supply at the anticipated feed rate, whichever is greater. Tanks for storing 50 percent solution at a temperature between 75° F and 140° F may be constructed of mild steel. Storage temperatures above 140° F require more elaborate systems.

FILTER AIDS

Filter aids as well as flocculants are employed to improve the filtration characteristics of hard-to-filter suspensions. A filter aid is a finely divided solid material, consisting of hard, strong particles that are, en masse, incompressible. The most common filter aids are applied as an admix to the suspension. These include diatomaceous earth, expanded perlite, Solkafloc, fly ash, or carbon. Filter aids build up a porous, permeable, and rigid lattice structure that retains solid particles and allows the liquid to pass through. These materials are applied in small quantities in clarification or in cases where compressible solids have the potential to foul the filter medium.

FILTER AIDS *are fine, chemically inert powders applied in both process and waste microfiltrations to maintain high flowrates while giving brilliant clarity. For difficult separations this long-established technology is the economical way to produce high quality fluids and manageable solid residues. Examples of filter aids are:*

DIATOMITE - *Manufactured from either marine or fresh water deposits.*

PERLITE - *Low density, low crystalline silica grades suit a wide range of process, water and wastewater applications.*

CELLULOSE - *As fibrous precoat aids or where special chemical compatibilities are required cellulosebased additives achieves separations that would otherwise be difficult or impossible.*

Filter aids may be applied in one of two ways. The first method involves the use of a precoat filter aid, which can be applied as a thin layer over the filter before the suspension is pumped to the apparatus. A precoat prevents fine suspension particles from becoming so entangled in the filter medium that its resistance becomes exces-sive. In addition it facilitates the removal of filter cake at the end of the filtration cycle. The second application method involves incorporation of a certain amount of the material with the suspension before introducing it to the filter. The addition of filter aids increases the porosity of the sludge, decreases its compressibility, and reduces the resistance of the cake. In some cases the filter aid displays an adsorption action, which results in particle separation of sizes down to 0.1 μ. The adsorption ability of certain filter aids, such as bleached earth and activated charcoals, is manifest by a decoloring of the suspension's liquid phase. This practice is widely used for treating fats and oils. The properties of these additives are determined by the characteristics of their individual components. For any filter aid, size distribution and the optimal dosage are of great importance. Too low a dosage results in poor clarity; too great a dosage will result in the formation of very thick cakes. In general, a good filter aid should form a cake having high porosity (typically 0.85 to 0.9), low surface area, and good particle-size distribution. An acceptable filter aid should have a much lower filtration resistance than the material with which it is being mixed. It should reduce the filtration resistance by 67 percent to 75 percent with the addition of no more than 25 percent by weight of filter aid as a fraction of the total solids. The addition of only a small amount of filter aid (e.g., 5 percent of the sludge solids) can actually cause an increase in the filtration resistance. When the amount of filter aid is so small that the particles do not interact, they form a coherent structure, and resistance may be affected adversely.

Filter aids are evaluated in terms of the rate of filtration and clarity of filtrate. Finely dispersed filter aids are capable of producing clear filtrate; however, they contribute significantly to the specific resistance of the medium. As such, applications must be made in small doses. Filter aids comprised of coarse particles contribute considerably less specific resistance; consequently, a high filtration rate can be achieved with their use. Their disadvantage is that a muddy filtrate is produced.

The optimum filter aid should have maximum pore size and ensure a prespecified filtrate clarity. Desirable properties characteristics for the optimum filter aid include:

1. The additive should provide a thin layer of solids having high porosity (0.85 to 0.90) over the filter medium's external surface. Suspension particles will ideally form a layered cake over the filter aid cake layer. The high porosity of the filter aid layer will ensure a high filtration rate. Porosity is not determined by pore size alone. High porosity is still possible with small size pores.
2. Filter aids should have low specific surface, since hydraulic resistance results from frictional losses incurred as liquid flows past particle surfaces. Specific surface is inversely proportional to particle size. The rate of particle dispersity and the subsequent difference in specific surface determines the deviations in filter aid quality from one material to another. For example, most of the diatomite species have approximately the same porosity; however, the coarser materials experience a smaller hydraulic resistance and have much less specific surface than the finer particle sizes.
3. Filter aids should have a narrow fractional composition. Fine particles increase the hydraulic resistance of the filter aid, whereas coarse particles exhibit poor separation. Desired particle-size distributions are normally prepared by air classification, in which the finer size fractions are removed.
4. In applications where the filter aid layer is to be formed on open-weave synthetic fabric or wire screens, wider size distributions may have to be

prepared during operation. Filter aids should have the flexibility to be doped with amounts of coarser sizes. This provides rapid particle bridging and settling of the filter aid layer. For example, diatomite having an average particle size of 8 μ may be readily applied to a screen with a mesh size of 175 μ by simply adding a small quantity of filter aid with sizes that are on the same order but less in size than the mesh openings. Particle sizes typically around 100 μ will readily form bridges over the screen openings and prevent the loss of filter aid in this example.

5. The filter aid should be chemically inert to the liquid phase of the suspension and not decompose or disintegrate in it.

The ability of an admix to be retained on the filter medium depends on both the suspension's concentration and the filtration rate during this initial precoat stage. The same relationships for porosity and the specific resistance of the cake as functions of suspension concentration and filtration rate apply equally to filter aid applications.

Filter aids are added in amounts needed for a suspension to acquire desirable filtering properties and to prepare a homogeneous suspension before the actual filtration process begins. Essentially, filter aids increase the concentration of solids in the feed suspension. This promotes particle bridging and creates a rigid lattice structure for the cake. In addition, they decrease the flow deformation tendency. Irregular or angular-shaped particles tend to have better bridging characteristics than spherical particles. Generally, the weight of filter aid added to the suspension should equal the particle weight in suspension. Typical filter aid additions are in the range of 0.01 percent to 4 percent by weight of suspension; however, the exact amount should be determined from experiments. Excess amounts of filter aid will decrease the filter rate. Operations based on the addition of admixes to their suspensions may be described by the general equations of filtration with cake formation. A plot of filtration time versus filtrate volume on rectangular coordinates results in a nearly parabolic curve passing through the origin. The same plot on logarithmic coordinates, assuming that the medium resistance may be neglected, results in a straight line. This convenient linear relationship allows results obtained from short-time filtration tests to be extrapolated to long-term operating performance (i.e., for several hours of operation). This reduces the need to make frequent, lengthy tests and saves time in the filter selection process.

In precoating, the prime objective is to prevent the filter medium from fouling. The volume of initial precoat normally applied should be 25 to 50 times greater than that necessary to fill the filter and connecting lines. This amounts to about 5-10 lb/100 ft^2 of filter area, which typically results in a 1/16-in. to 1/8-in. precoat layer over the outer surface of the filter medium. An exception to this rule is in the precoating of continuous rotary drum filters where a 2-in. to 4-in. cake is deposited before filtration. The recommended application method is to mix the precoat material with clear liquor (which may consist of a portion of the filtrate). This mixture should be recycled until all the precoat has been deposited onto the filter medium. The unfiltered liquor follows through immediately without draining off excess filter aid liquor. This operation continues until a predetermined head loss develops, when the filter is shut down for cleaning and a new cycle.

In precoating, regardless of whether the objective is to prevent filter medium clogging or to hold back fines from passing through the medium to contaminate the filtrate, the mechanical function of the precoat is to behave as the actual filter medium. Since it is composed of incompressible, irregularly shaped particles, a high-porosity layer is formed within itself, unless it is impregnated during operation with foreign compressible materials. Ideally, a uniform layer of precoat should be formed on the surface of the filter medium. However, a nonuniform layer of precoat often occurs due to uneven medium resistance or fluctuations in the feed rate of filter aid suspension. Cracks can form on the precoat layer that will allow suspension particles to penetrate into the medium. To prevent cracking, the filter aid may be applied as a compact layer. On a rotating drum filter, for example, this may be accomplished by applying a low concentration of filter aid (2 percent to 4 percent) at the maximum drum rpm. In other filter systems, maintaining a lowpressure difference during the initial stages of precoating and then gradually increasing it with increasing layer thickness until the start of filtration will help to minimize cake cracking. Also, with some filter aids (such as diatomite or perlite), the addition of small amounts of fibrous material will produce a more compact precoat cake. At low-suspension concentrations (typically 0.01 percent), filter aids serve as a medium under conditions of gradual pore blocking. In this case the amount of precoat is 10 to 25 N/m^2 of the medium and its thickness is typically 3 to 10 mm. In such cases, the filter aid chosen should have sufficient pore size to allow suspension particle penetration and retention within the precoat layer. Commonly used filter aids include diatomite, perlite,

☞Filter presses used to dewater sludge in industrial wastewater treatments usually suffer severely sub-optimal performance as fine particles blind their cloths after a brief period in service. Even sophisticated synthetic fabrics selected for their cake release properties are subject to blinding, and cake adhesion especially by metal hydroxide sludges or flocked oil and grease. As a result, high labor costs for manual press cleaning are incurred. In some cases the filter press is bypassed, or sludge with high moisture content is trucked away to a special treatment facility, representing a heightened environmental danger during transport and, again, excessive cost.

☞A precoat filteraid applied to the filter septum with clean water before the introduction of the sludge protects the cloth from blinding, and permits long filter cycles. Because the filter chambers can now be completely filled, dewatering performance is greatly enhanced. With the precoat acting as a release agent, the well-dried cakes fall readily with minimal manual cleaning of the press.

☞In some cases, the optimization of an existing press by precoating avoids the necessity of buying more filter capacity as a waste treatment plant expands.

Diatomaceous earth, widely-known and long-used as a filteraid in process and waste filtrations, has a high microcrystalline silica content. As well as being a respiratory hazard in the workplace, the silica is being scrutinized in some jurisdictions as a potentially hazardous dust in landfills in which spent filter cakes are deposited.

cellulose, sawdust, charcoal and flysah, as well as an abundant of commercial additives. The most important filter aids from a volume standpoint are the ***diatomaceous silica type*** (90 percent or better silica). These are manufactured from the siliceous fossil remains of tiny marine plants known as diatoms. Diatomaceous filter aids are available in various grades. This is possible because the natural product can be modified by calcining and processing, and because filter aids in different size ranges and size distributions have different properties. The filter aids may be classified on the basis of cake permeability to water and water flow rate. Finer grades are the slower-filtering products; however, they provide better clarification than do faster-filtering grades. Thus a fast-filtering aid may not provide the required clarification. However, by changing the physical character of the impurities (e.g., by proper coagulation), the same clarity may be obtained by using the fast-filtering grades. Calcinated diatomaceous additives are characterized by their high retention ability with relatively low hydraulic resistance. Calcining dramatically affects the physical and chemical properties of diatomite, making it heat resistant and practically insoluble in strong acids. Further information is given in the literature. Diatomaceous earth is a natural occurring siliceous sedimentary mineral compound from microscopic skeletal remains of unicellular algae-like plants called diatoms. These plants have been part of the earth's ecology since prehistoric times. Diatoms are basic to the oceanic cycle, and the food for minute animal life which in turn becomes the food for higher forms of marine life. As living plants, diatoms weave microscopic shells from the silica they extract from the water, then as they die, deposits are formed and then fossilized in what are now dried lake and ocean beds. The material is then mined, ground and screened to various grades, for the countless uses in today's products and processes, from toothpaste to cigars, plastics to paprika, filter media in swimming pools to home fish tanks, as well as insect and parasite control in animals and grains. It is a natural (not calcined or flux calcined) compound with many elements which include:

Silicon Dioxide, SiO_2 = 83.7 %
Aluminum Oxide, $A1_2O_3$ = 5.6 %

Iron Oxide, Fe_2O_3 = 2.3 %

Calcium Oxide, CaO = 0.4 %

Magnesium Oxide, MgO = 0.3 %

Other Oxides = 1.9 %

Ignition Loss at 1000 = 5.3 %

Semi quantitive spectrographic analysis of other elements:

Copper:	2ppm
Strontium:	100ppm
Titanium:	1800ppm
Manganese:	200ppm
Sodium:	2000ppm
Vanadium:	500ppm
Boron:	50ppm
Zirconium:	200ppm

Diatomaceous earth's unique combination of physical properties include:

High Porosity: Up to eighty-five percent of the volume of diatomaceous earth is made up of tiny interconnected pores and volds. It is quite literally more air than diatom.

High Absorption: Diatomaceous earth can generally absorb up to 1 times, its own weight in liquid and still exhibit the properties of dry powder.

Particle Structure/High Surface Area: Diatom particles are characterized by their very irregular shapes, generally spiny structures and pitted surface area. They average only 5 to 20 microns in diameter, yet have a surface area several times greater than any other mineral with the same particle size. Diatomaceous earth increases bulk without adding very much weight. These features, it is believed, are what make it an ideal mineral for internal parasite control in animals: It is approved by the USDA up to 2% by weight of total ration for use as an inert carrier or anti-caking agent in animal feed. It is not necessary to use this percent of product on a continual basis. It may be varied to suit individual purposes.

Grain Storage: A rate of seven pounds per ton of grain in barley, buckwheat, corn, wheat, oats, rice, rye, sorghum and mixtures of these grains. It is most effective when grain is treated directly after harvest by coating the outside surface of the gain. This can be done by applying the powder at the elevator or auger when grain is being moved into storage. When used at proper rates, diatomaceous earth has been effective against ants, aphids, bollworm, salt marsh caterpillar, cockroach, cornworm, earwig, house fly, fruit fly, lead perforator, leaf hopper, lygus bug, mite, pink boll weevil, red spider mite, slugs, snail, termites, Japanese beetle (grub stage) and many other insects. Diatomaceous earth is a natural grade diatomite. It requires no warning label on the bag or container. However, the continual breathing of any dust should he absolutely avoided.

Perlite and Solka-floc® are finely divided powders manufactured from a volcanic mineral and from wood pulp respectively, which have filtration properties very similar to those of diatomite. Like diatomite, they are inert to a wide range of process liquids. Like diatomite, they are available in a range of particle-size distributions to give the desired clarity and flowrate in different applications. On a cost-of-use basis, they are as economical as, or more economical than, diatomite.

Although less known than diatomite, these products have been in wide use for many years so that there exists a sound body of applications knowledge upon which to base grade selection, dosage, and procedures. Perlite and Solka-floc® have the same availability in bagged, semi-bulk, or bulk formats as diatomite.

Perlite is glass-like volcanic rock, called volcanic glass, consisting of small particles with cracks that retain 2 percent to 4 percent water and gas. Natural perlite is transformed to a filter aid by heating it to its melting temperature (about 1000° C), where it acquires plastic properties and expands due to the emission of steam and gas. Under these conditions its volume increases by a factor of 20. Beads of the material containing a large number of cells are formed. The processed material is then crushed and classified to provide different grades. The porosity of perlite is 0.85 to 0.9 and its volumetric weight is 500 to 1,000 N/m^3. Compared to diatomite, perlite has a smaller specific weight and compatible filter applications typically require 30 percent less additive. Perlite is used for filtering glucose solutions, sugar, pharmaceutical substances, natural oils, petroleum products, industrial waters, and beverages. The principal advantage of perlite over diatomite is its relative purity. There is a danger that diatomite may foul filtering liquids with dissolved salts and colloidal clays. Perlite is not a trade name but a generic term for a naturally occurring siliceous rock. The distinguishing feature that sets perlite apart from other volcanic glasses is that when heated to a suitable point in its softening range, it expands from four to twenty times its original volume. This expansion is due to the presence of two to six percent combined water in the crude perlite rock. Then quickly heated to above 1600° F (871° C), the crude rock pops in a manner similar to popcorn as the combined water vaporizes and creates countless tiny bubbles that account for the amazing light weight and other exceptional physical properties of expanded perlite. The expansion process also creates one of perlite's most distinguishing characteristics: its white color. While the crude rock may range from transparent light gray to glossy black, the color of expanded perlite ranges from snowy white to grayish white. Expanded perlite can be manufactured to weigh as little as 2 pounds per cubic foot (32 kg/m^3) making it adaptable for numerous applications. Since perlite is a form of natural glass, it is classified as chemically inert and has a pH of approximately 7. Refer to Tables 4 and 5 for some general properties data.

Table 4. Elemental Analysis of Perlite.

Component	Weight Percent
Silicon	33.8
Aluminum	7.2
Potassium	3.5
Sodium	3.4
Iron	0.6
Calcium	0.6
Magnesium	0.2
Trace	0.2
Oxygen (by difference)	47.5
Net Total	97.0
Bound water	3.0
Total	100.0

Table 5.Physical Properties of Perlite.

Property	Characteristic
Color	White
Refractive index	1.5
Free moisture, maximum	0.5 %
pH of water slurry	6.5 - 8.0
Specific gravity	2.2 - 2.4
Bulk density, normal	2-15 lb/ft^3
Mesh Sizes (normal)	4 - 40 and finer mesh

Property	Characteristic
Softening Point	1600-2000° F
Fusion Point	2300-2450° F
Specific Heat	0.2BTU/lb-°F
Thermal Conductivity	0.27 - 0.41 BTU.in/h.ft^2 °F

Clay - The use of clay based flocculating agent(s) in conjunction with a strong metal precipitator has proven successful in many wastewater treatment applications where the objectives are aimed at metals removal. Clay based flocculants cleans the wastewater and in some cases replaces multistage conventional treatment system and saves the traditional operational difficulties of treatment with several chemicals such as metal hydroxide precipitation, coagulant, flocculants and other methods. Commercial clay-based flocculants usually consist of bentonite and other proprietary ingredients. These products are in a granulated form to minimize any dust exposure. Betonite mixtures are often used for removing many wastewater constituents including, but not limited to heavy metals, oil and grease, pigments, and phosphates. It encapsulates the metals and other wastewater constituents in a clay barrier, producing an easily disposable waste sludge and treated water. Industrial wastewater from rinses and cleaning operation can be mixed with the flocculant at an average of <1% (by weight) dosage rate in a mixing tank. Even though the process of flocculation takes few minutes, a rapid multi-stage complex chemistry starts working. These hidden reaction stages, simulate the

Note that filter aid selection must be based on planned laboratory tests. Guidelines for selection may only be applied in the broadest sense, since there is almost an infinite number of combinations of filter media, filter aids, and suspensions that will produce varying degrees of separation. The hydrodynamics of any filtration process are highly complex; filtration is essentially a multiphase system in which interaction takes place between solids from the suspension, filter aid, and filter medium, and a liquid phase. Experiments are mandatory in most operations not only in proper filter aid selection but in defining the method of application. Some general guidelines can be applied to such studies: the filter aid must have the minimum hydraulic resistance and provide the desired rate of separation; an insufficient amount of filter aid leads to a reduction in filtrate quality -- excess amounts result in losses is filtration rate; and it is necessary to account for the method of application and characteristics of filter aids.

phenomena of attraction, coagulation, precipitation, and separation of metals, oil, grease, and pigment due to a strong affinity of muti-layer positively charged of the bentonite crystal structure and other blended additives in each package. Depending on the constituent of the effluent, blended specialty products are formulated such that in addition to metal removal, it has added –value chemistries that have strong tendencies to remove chlorinated solvents as well as oil & grease –all in one chemical package system. After addition of the flocculant, the fully reacted mass is a bonded and complex formula that strongly encapsulates wastewater contaminants. These bonds are generally categorized as weak Van der Waals, as well as strong electrostatic forces. The clay has also tendency to entrap and agglomerate the surrounding suspended solids very efficiently. During this stage, some Pozzolanic reactions also occur, in which, a cementatious particles settles down to bottom of the reaction vessel. It is very interesting that the entire micro-encapsulation process occurs in few minutes as long as the granulated clays are fed into the system, leaving clear solution for reuse, recycle, or discharge. The solidified and flocculated waste sludge is often classified as non-leachable and in many occasions, depending the waste stream, it may pass the TCLP and STLC tests. If the test passes, it confirms that contaminant is surrounded by a barrier of clay particles and is not accessible to external leaching solutions or processes.

Cellulose fiber is applied to cover metallic cloths. The fibers form a highly compressed cake with good permeability for liquids, but a smaller retention ability for solid particles than that of diatomite or perlite. The use of cellulose is recommended only in cakes where its specific properties are required. These properties include a lack of ashes and good resistance to alkalies. The cost of cellulose is higher than those of diatomite and Perlite.

Sawdust may be employed in cases where the suspension particles consist of a valuable product that may be roasted. For example, titanium dioxide is manufactured by calcining a mixture of sawdust and metal titanium acid. The mixture is obtained as a filter cake after separating the corresponding suspension with a layer of filter aid.

Charcoal is not only employed in activated form for decoloring and adsorbing dissolved admixtures but also in its unactivated form as a filter aid. It can be used in suspensions consisting of aggressive liquids (e.g., strong acids and alkalies). As with sawdust, it can be used to separate solids that may be roasted. On combustion, the charcoal leaves a residue of roughly 2 percent ash. Particles of charcoal are porous and form cakes of high density but that have a lesser retention ability than does diatomite.

Fly ash has a number of industrial filtering applications but primarily is applied to dewatering sewage sludge. The precoat is built up to 2-in. thick from a 60 percent solid slurry. On untreated sludges, filtration rates of 25 lb/ft^2-hr are obtainable. This rate can be doubled with treated sludges. The sludge is reduced from a liquid to a semidry state. Fly ash may also be used as a precoat in the treatment of papermill sludge.

Polymeric flocculants are high molecular weight organic chains with ionic or other functional groups incorporated at intervals along the chains. Because these compounds have characteristics of both polymers and electrolytes, they are frequently called *polyelectrolytes*. They may be of natural or synthetic origin. All synthetic polyelectrolytes can be classified on the basis of the type of charge on the polymer chain. Thus, polymers possessing negative charges are called *anionic* while those carrying positive charges are *cationic*. Certain compounds carry no electrical charge and are called *nonionic* polyelectrolytes. Because of the great variety of monomers available as starting material and the additional variety that can be obtained by varying the molecular weight, charge density, and ionizable groups, it is not surprising that a great assortment of polyelectrolytes are available to the wastewater plant operator. Extensive use of any specific polymer as a flocculant is of necessity determined by the size, density, and ionic charge of the colloids to be coagulated. As other factors need to be considered (such as the coagulants used, pH of the system, techniques and equipment for dissolution of the polyelectrolyte, and so on), it is mandatory that extensive jar testing be performed to determine the specific polymer that will perform its function most efficiently. These results should be verified by plant-scale testing. Types of polymers vary widely in characteristics. Manufacturers should be consulted for properties, availability, and cost of the polymer being considered. Dry polymer and water must be blended and mixed to obtain a recommended solution for efficient action. Solution concentrations vary from fractions of a percent up. Preparation of the stock solution involves wetting of the dry material and usually an aging period prior to application. Solutions can be very viscous, and close attention should be paid to piping size and length and pump selections. Metered solution is usually diluted just prior to injection to the process to obtain better dispersion at the point of application. Two types of systems are frequently combined to feed polymers. The solution preparation system includes a manual or automatic blending system with the polymer dispensed by hand or by a dry feeder to a wetting jet and then to a mixing-aging tank at a controlled ratio. The aged polymer is transported to a holding tank where metering pumps or rotodip feeders dispense the polymer to the process. It is generally advisable to keep the holding or storage time of polymer solutions to a minimum, one to three days or less, to prevent deterioration of the product. Selection must be made after determination of the polymer; however, type 316 stainless steel or plastics are generally used. The solution preparation system may be an automatic batching system that fills the holding tank with aged polymer as required by level probes.

Ultra High Molecular Weight Flocculants - Water soluble polymers with an average molecular weight of $> 10^6$ g/mol are generally considered to be high molecular weight flocculants. In recent years there has been a trend for flocculants to be manufactured with ever increasing molecular weight. This new generation of flocculants are referred to as 'ultra high molecular weight'. The use of flocculants with ultra high molecular weights can lead to stronger flocs compared to lower molecular weight alternatives. The floc strength is of particular importance where high degrees of mixing energy prevail at the flocculant dosing point of a given dewatering application. Typically, the process operator has no control over this mixing shear, and consequently the choice of flocculant is paramount for ensuring that effective flocculation and subsequent solid/liquid separation occurs. A further advantage of high molecular weight flocculants is their propensity for dose efficient performance. The nature of high molecular weight flocculants, with long chain lengths, increase the likelihood of effective inter-particle polymer bridges compared to lower molecular weight, shorter-chain-length analogues. This dose efficient performance is desirable as a means of minimizing the cost of chemical pre-treatments that aid solid/liquid separation unit operations.

Sewage Sludge Pre-treatment Flocculants - Sewage sludge must be dewatered to facilitate economic disposal. Dewatered sludge decreases the cost of transportation to landfill, or if the sludge is to be incinerated the removal of water combustion. Ultra high molecular weight cationic flocculants are increasingly used for pre-treatment of sewage sludge prior to dewatering because they are particularly effective in the thickening of surplus activated sludges and centrifuge applications. Centrifuges exert a very large shearing force on flocs as they enter the centrifuge bowl. The resilience of flocs formed with ultra high molecular weight flocculants, readily explain the usefulness of such flocculants for centrifuge applications. It can be observed that increasing molecular weight has improved the dose-efficiency of the flocculant, making the product more cost-effective.

Red Mud Flocculation - In the Bayer process, crushed bauxite is digested in concentrated sodium hydroxide in order to dissolve and extract aluminium. An insoluble residue known as red mud is produced, which is composed primarily of iron oxides, quartz, sodium aluminosilicates, calcium carbonate, calcium aluminate and titanium dioxide. To recover the aluminium, the solid red mud must be removed from the liquor. This is typically achieved by a series of sedimentation tanks referred to as the 'wash train'. The sediment from the primary sedimentation is washed in order to recover entrained aluminium rich liquor. This washing process is repeated several times. The use of ultra high molecular weight anionic flocculants promote a desirable sedimentation rate at an economic dose, and can lead to a more compact sediment than that obtained using lower molecular weight flocculants. The more compact the sediment, the greater the liquor recovery, and therefore the higher the aluminium recovery.

Drainage and Retention Aids - Paper formation is critical for both aesthetic and practical reasons i.e. visual appearance and paper strength, respectively. Increased mixing, which improves the distribution of fibre and filler within the paper has been highlighted as a means of achieving a superior formation. Drainage and retention aids are added to induce the rapid and efficient release of water and to capture and retain the filler within the paper. Over the last decade, the throughput rate on paper machines has increased significantly, while the length of the paper machine, specifically the drainage area, has decreased. Due to the requirements and changes described above, the use of highly soluble, ultra high molecular weight flocculants as drainage and retention aids has increased. A more robust, shear stable floc is produced which withstands the increased mixing intensities and dewatering forces and leads to a better drainage and retention performance compared to lower molecular weight analogues.

Crosslinked Flocculants - The accepted theory of flocculation dictates the need for a water-soluble polymer, which is as linear as possible. However, controlled levels of polymer crosslinking or branching can provide unexpected benefits to enhance solid/liquid separation. The mechanism by which crosslinked flocculants operate is not well documented. It is possible that the crosslinked polymer lies on the surface of a solid particle and, by virtue of its structure, a proportion of the polymer charge is available to interact with adjacent particles, i.e. the crosslinked polymer cannot fully adsorb all its charge to one particle. One could perceive the charge remains available even after a transient bond has been broken. Thus, flocculation and re-flocculation mechanisms may explain the unique dewatering characteristics engendered by crosslinked flocculants. In recent times, the trend has been for manufacturers to develop commercially available flocculants that contain higher levels of crosslinking.

Sewage Sludge Pre-treatment Flocculants - Highly crosslinked cationic flocculants have gained popular use in the pre-treatment of sewage sludge prior to dewatering by gravity belt thickening and centrifugation. In the laboratory the maximum filtrate volume (after a drainage time of 5 seconds) increases with the degree of flocculant crosslinking. This rapid rate of filtration observed in the laboratory translates into increased throughput rates when highly crosslinked flocculants are used on full-scale plants. As the degree of flocculant crosslinking increases, so does the flocculant dose requirement to achieve the desirable enhanced performance. A similar increase in dose requirement is necessary in order to gain performance benefits of highly crosslinked flocculants, when used to pretreat sludge prior to centrifugation. Advantages of highly crosslinked flocculants for centrifugation are increased throughput rates, increased percentage solid content of centrifuge cakes and cleaner centrates.

Encapsulated Flocculants - A recent innovation in commercially available flocculants is a counter ionic system, where one charged moiety is encapsulated and suspended in the counter charged product. This facilitates a mechanism of delayed release of the encapsulated product.

Coal Slurries - Two component treatments prove to be effective on a commercial scale for coal slurry flocculation. The use of encapsulated flocculant suspended in a counter charged flocculant provides the robustness of traditional dual component systems, but with additional performance advantages, which include reduction in filter cake moisture content and an increased throughput rate. Figure 5 illustrates the typical filter cake moisture content obtained by a conventional treatment system compared to the encapsulated treatment system using coal tailings as the substrate.

Microparticulate Systems - Although dual combination treatment systems comprising microparticles and flocculants have been used in the paper industry since the 1980s, it is only recently that there has been a general trend for paper mills to switch from conventional single component systems to dual systems.

Microparticle/Flocculant - In the past few years the use of microparticles in conjunction with ultra high molecular weight flocculants as drainage and retention aids has grown. The more compact, shear stable flocs produced by the cationic flocculant followed by the anionic microparticle give technical improvements in both paper formation and dewatering and retention of fibre and fines within the paper. In general terms, the former is regarded as a more powerful drainage and retention system compared to the latter.

Low Molecular Weight Flocculants - Contrary to the trend to increase molecular weight, which has resulted in the commercial availability of ultra high molecular weight (*HMW*) flocculants, there are a number of applications where low molecular weight (*LMW*) flocculants are gaining considerable credibility.

SOME USEFUL FORMULAS WHEN DEALING WITH CHEMICAL ADDITIVES

To calculate the Lbs. of chemical per gal. of solution:
Strength of Chemical (*as %*) × Specific Gravity × Lbs of Water

To calculate mL per min. of water flow in a treatment plant:
(GPD of water × 3785)/1440

To calculate the make-up requirements in gallons for a rectangular tank:
Length of tank (*in ft*) × Width × Depth × 7.47

To calculate the make-up requirements in gallons for a round tank or clarifier:
3.1417 × R^2 × Depth (*in ft*) × 7.47
Where R is the tank radius in ft.

Fermentation Broth Pre-treatment - Flocculants can be used at several stages of the fermentation product purification process, e.g. cell-broth separation, cell debris flocculation and protein precipitation. Such processes are becoming increasingly more relevant with the commercialization of new products using biotechnology routes. The trend in flocculant pre-treatment of whole cells in complex fermentation media is to use low molecular weight flocculants. Herein low molecular weight flocculants lead to more compact centrifuge cakes. This is advantageous whether the fermentation product to be recovered is extra-cellular, intracellular or the cellular material itself.

If the fermentation product is extracellular, the benefits come from the minimisation of product entrainment into the cake. Alternatively, if the product is intracellular or the cellular material itself, a compact cake minimizes the contamination that would result from entrainment of the fermentation medium in the cake.

RECOMMENDED RESOURCES FOR THE READER

The following references include Web sites that you can refer to for specific properties and discussions of applications of alum.

1. Liquid Aluminum Sulfate (Viscosity of aqueous solutions of liquid alum $[Al_2(SO_4)_2 \cdot 14.3\ H_2O]$ at different temperatures), go to the following web site: http://www.ecoservices.us.rhodia.com/alum.asp.

2. Pulp and Paper Technical Association of Canada - Standards of Sulphite Waste Liquor, C-3, Jan. 68. Viscosity of Sulphite Waste Liquor, C-4,Jan. 68.; F-2, Nov. 70. Wire Screening, F-3, July 92. Alum Solutions, F-4, Jan. 68, http://www.paptac.ca/english/engdatsh.htm.

3. The Role of Alum in Historical Papermaking ... and other Polyvalent Metals on the Viscosity of Gelatin Solutions." Journal of the Science of Food ... WF et al. "The Effect of Alum and pH on Sheet Acidity", go to the following web site: http://palimpsest.stanford.edu
4. Production of Alum From Awaso Bauxite. Go to the following web site: http://home.att.net/~africantech/GhIE/Awaso1.htm

5. GARNET - WEDC/publications/wares. Discusses soda ash and alum solutions. Go to the following web site: http://www.lboro.ac.uk

6. UCL Crystallization Research Group Publications : Composition, density and viscosity of saturated solutions", Go to the following web site for information: http://www.ucl.ac.uk/~ucec02j/Xtn_pubs.htm.
7. Montgomery College, Maryland - contains diagrams, surface tension, viscosity, intermolecular forces, types, vapor pressure of solutions and colligative properties. Also covers the synthesis of alum crystals. Go to the following web site: http://www.mc.cc.md.us/Departments/chemrv/ch101.htm

8. Discussion on Cd(II) from Aqueous Solutions by Fungal Adsorbents and Sludge Conditioning. http://poai3.ev.nctu.edu
9. Cheremisinoff, N. P. and P. N. Cheremisinoff, Filtration Equipment for Wastewater Treatment, Prentice Hall Publishers, Englewood Cliffs, New Jersey, 1993.

The following references are Web sites that you can refer to for specific properties and discussions of applications of diatomaceous earth.

10. What is Diatomaceous Earth? How does it worK? - Hydroponics ... Diatomaceous Earth has a unique combination of physical properties: High Porosity: Up to eighty-five percent of the volume of Diatomaceous Earth is made up of ... http://www.hydromall.com/happy_grower16.html

11. Results Diatomaceous Earth Insect Control ... the proven long-term insect control properties of White Mountain's proprietary brand of diatomaceous earth (DI-ATOMATE) with the "instant gratification" of an ... http://www.whitemountainnatural.com/results.html

12. Diatomaceous Earth ... then it couldn't have any significant insecticidal properties and would be labeled as something other than diatomaceous earth. http://froebuck.home.texas.net/toppage7.htm

13. Evaluation of Diatomaceous Earth Topdressing for Cyanobacterial ... ... are available, such as diatomaceous earth (DE), which is also called diatomite.Information is available regarding the physical properties of DE (Koski and ... http://www.mafes.msstate.edu/pubs/b1079.html

14. Mervat S. Hassan, Ibrahim A. Ibrahim and Ismail S. Ismail - ... ... poor due to the general high level of impurities. Technologic properties of diatomaceous earth was significantly improved due to upgrading of diatoms through ... http://www.salty2k.com/sse/abstracts/vol7/hassanetal2v7.html [More Results From: www.salty2k.com]

15. Daitomaceous Earth ... Loss on ignition, 11.50. Diatomaceous Earth - Physical Properties. Bulk Density, 0.26 gm/cc. Temperature Resistance, 1400 DC. ... http://www.infoindia.com/uff/daito.html

16. GCSAA - GCM July 2000 -- Soil amendments affect turf ... five days (9). To improve water-holding properties, another team recommended the use of diatomaceous earth as an amendment to sand (8). However, another report ... http://www.gcsaa.org/gcm/2000/july00/07soil.html

You can refer to the following references for more details on flocculants.

17. Johnson, K. M., 1996. Ind Water Treatment, 28, p. 24

18. Dahlstrom, D. A., 1990. Ind Eng Chem Res, 29, p.1020-1025

19. Moody, G. M., 1994. In Selection and Scale - Up for Solid/Liquid Separation, The Filtration Society 30th Anniversary Symposium, p.9-29

20. Moody, G. M. and McColl, P., 1997. In Proceedings of Filtech Europa Conference, The Filtration Society, Dusseldorf, Germany, p.393-406

21. Ramsden, D. K., Hughes, J. and Weir, S., 1998. Biotech Tech, 8, p.599.

QUESTIONS FOR THINKING AND DISCUSSING

1. Go to the Web and develop a list of suppliers along with unit costs for some of the various types of flocculants described in this chapter. Also obtain the MSDS (Material Safety Data Sheets) for these chemicals. Once you have collected the information, develop a relative ranking of unit cost and health risk associated with each. For question 2 you will need information of recommended doses.
2. For a 50,000 GPD filtration operation with an average loading 50 mg/L TSS (*Total Suspended Solids*) determine the optimum flocculant to use in order to achieve at least an 85 % reduction in solids. Assume that a rotary drum filter unit is used.
3. For each of the chemicals selected in question 1, calculate the recommended dosage (As Lbs/gal. solution) for 25,000 gpd of waster treatment.
4. Discuss the advantages and disadvantages of polyelectrolytes over other convention flocculants.
5. Develop a list of the issues and parameters to examine when making a selection of the proper filter aid.
6. What characteristics should the optimum filter aid possess?
7. What are the reasons for applying a precoat to a filter? List several common materials used as precoat filter aids.
8. List and describe the types of dry chemical feeders that are in common use. Go to the Web and develop a list of major vendors for these equipment.

Chapter 4

SELECTING THE RIGHT FILTER MEDIA

INTRODUCTION

The role of the filter media is to provide a substrate surface through which the suspension passes, and suspended solids are captured onto the surface and within the pore structure of the substrate. As described in Chapter 1, when pressure or a vacuum is applied over one side of the media, the liquor is drawn through the filter media, leaving behind suspended particles retained on the media, with the clean filtrate discharged from the apparatus. This simplified view is illustrated in Figure 1. In many ways the media is the workhorse of any filtration apparatus. The properties, durability and cost are key characteristics that we need to examine when making selections. In this chapter we examine criteria for selecting different filter media. The discussions focus only on conventional filter media, and other, less conventional systems are discussed in later chapters. As with other chapters you will find a series of sidebar discussions that offer additional information and highlighted factors that you should consider. And at the end of this chapter are some ***Questions for Thinking and Discussing***. Remember to refer to the ***Glossary*** at the end of the book if you run across any terms that are unfamiliar to you.

TYPES OF FILTER MEDIA TO CHOOSE FROM

There are many filter media from which to choose from; however, the optimum type depends on the properties of the suspension and specific process conditions. Filter media may be classified into several groups, however the two most common classes are the *surface-type* and *depth-media-type*.

Surface-type filter media are distinguished by the fact that the solid particles of suspension on separation are mostly retained on the medium's surface. That is, particles do not penetrate into the pores. Common examples of this type of media are filter paper, filter cloths, and wire mesh.

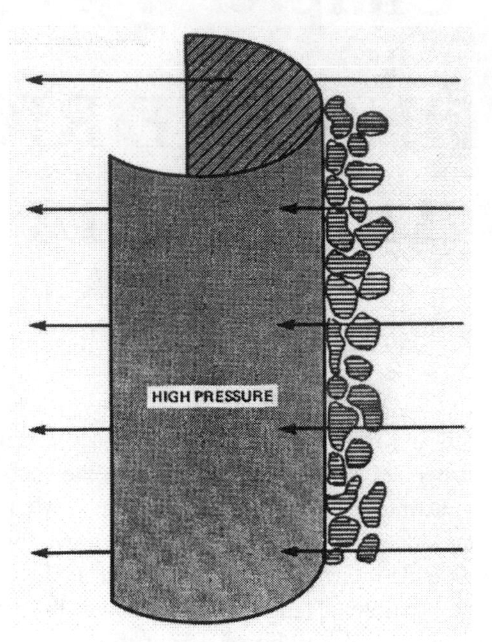

Figure 1. Pressure filtration.

Depth-type filter media are largely used for wastewater clarification purposes. They are characterized by the fact that the solid particles penetrate into the pores where they are retained. The pores of such media are considerably larger than the particles of suspension. The suspension's concentration is generally not high enough to promote particle bridging inside the pores. Particles are retained on the walls of the pores by adsorption, settling and sticking. In general, depth-type filter media cannot retain all suspended particles, and their retention capacity is typically between 90 and 99 percent. Sand and filter aids, as examples, fall into this category. Some filter media may act as either surface-type or depth-type, depending on the pore size and suspension properties (e.g., particle size, solids concentration and suspension viscosity).

It is also common practice to classify filter media by their materials of construction. Examples are cotton, wool, linen, glass fiber, porosmooth surface caused by carrying the warp (or the weft) on the fabric surface over many weft (or warp) yarns. Intersections between warp and weft are kept to a minimum, just sufficient to hold the fabric firmly together and still provide a smooth fabric surface. The percentage of open area in a textile filter indicates the proportion of total fabric area that is open, and can be determined by the following relationship:

$$\% \ open \ area = \frac{(mesh \ opening)^2}{(mesh \ opening + thread \ diameter)^2} \times 100$$

The other type of common weave is called a Dutch weave. Table 1 provides a conversion chart between the two types of weave and micron ratings. Discussions that follow focus on different types of common flexible filter media.

Table 1. Weave Conversion Chart.

Dutch Weave	Square Weave	Micron Rating
12x64	60	270-290
14x88	70	220-240
14x100	80	210-230
21x110	120	115-125
20x200	140	100-110
30x150	150	85-100
20x250	180	80-90
40x200	200	70-75
50x250	250	58-63
60x300	300	45-53
80x700	325	40-45
80x700	400	35-38
200x600	450	28-32
165x800	500	24-26
165x1400	635	16-18
200x1400	-	12-14
325x2300	-	8-9

Reverse Dutch Weave	Square Weave	Micron Rating
48x10	40	400
72x15	60	300
132x14	80	200
152x30	120	120
236x33	150	90
338x37	200	80

GLASS CLOTHS

Glass cloths are manufactured from glass yarns. They have high thermal resistance, high corrosion resistance and high tensile strength, and are easily handled; the composition and diameter of the fibers can be altered as desired. The disadvantages of glass cloth are the lack of flexibility of individual fibers, causing splits and fractures, and its low resistance to abrasion. However, backing glass cloth with a lead plate, rubber mats or other rigid materials provides for longevity. Backing with cotton or rubber provides about 50% greater life than in cases where no backing is used.

COTTON CLOTHS

Cotton filter cloths are among the most widely used filter media. They have a limited tendency to swell in liquids and are used for the separation of neutral suspensions at temperatures up to 100° C, as well as suspensions containing acids up to 3% or alkalies with concentrations up to 10% at 15-20° C. Hydrochloric acid at 90-100° C destroys cotton fabric in about 1 hour, even at concentrations as low as 1.5%. Nitric acid has the same effect at concentrations of 2.5%, and sulfuric acid at 5%. Phosphoric acid (70%) destroys the cloth in about six days. Water and water solutions of aluminum sulfate cause cotton fabrics to undergo shrinkage. Woven cotton filter cloths comprise ducks, twills, chain weaves, canton flannel and unbleached muslins. Cotton duck is a fabric weave that is a plain cloth with equal-thickness threads and texture in the "over one and under one" of the warp and woof. The twill weave is over two and under two with the next filling splitting the warp strands and giving a diagonal rib at 45° if the number of warp and filling threads are equal. Canton flannel is a twill weave in which one surface has been brushed up to give a nap finish. A muslin cloth is a very thin duck weave, which is unbleached for filtering. In chain weave one filling goes over two warp threads and under two, the next reversing this; the third is a true twill sequence, and the next repeats the cycle. A duct may be preferable to a twill of higher porosity, because the hard surface of the duck permits freer cake discharge. Under high increasing pressure a strong, durable cloth (duck) is required, since the first resistance is small as compared with that during cake building. Certain types of filters, such as drum filters, cannot stand uneven shrinkage and, in some cases, cloths must be pre-shrunk to ensure fitting during the life of the cloth.

Nitro-filter (nitrated cotton cloth) cloths are about the same thickness and texture as ordinary cotton filtration cloths, but are distinguished by a harder surface. It is claimed that the cake is easily detached and that clogging is rare. Their tensile strength is 70-80% of that of the specially manufactured cotton cloths from which they are prepared. They are resistant to the corrosive action of sulfuric, nitric, mixed nitration and hydrochloric acids. They are recommended for filtering sulfuric acid solutions to 40% and at temperatures as high as 90°C, with the advantage of removing finely divided amorphous particles, which would quickly clog most ceramic media. Nitro-filter cloths are composed of cellulose nitrate, which is an ester of cellulose. Any chemical compound that will saponify the ester will destroy the cloth. Caustic soda or potash in strengths of 2% at 70° C or over; alkali sulfides, polysulfides and sulfohydrates; or mixtures of ethyl alcohol and ether, ethyl, amyl and butyl acetates, pyridine, ferrous sulfates, and other reducing agents are detrimental to the cloth.

Cellulose nitrate is inflammable and explosive when dry, but when soaked in water it is considered entirely safe if reasonable care is taken in handling. For this reason it is colored red and packed in special containers. Users are cautioned to keep the cloths wet and to handle them carefully.

Proper selection of the filter media is often the most important consideration for assuring efficient suspension separation. A good filter medium should have the following general characteristics:

- ***The ability to retain a wide size distribution of solid particles from the suspension,***
- ***Offer minimum hydraulic resistance to the filtrate flow,***
- ***Allow easy discharge of cake,***
- ***High resistance to chemical attack,***
- ***Resist swelling when in contact with filtrate and washing liquid,***
- ***Display good heat-resistance within the temperature ranges of filtration,***
- ***Have sufficient strength to withstand filtering pressure and mechanical wear,***
- ***Capable of avoiding wedging of particles into its pores.***

WOOL CLOTHS

Wool cloths can be used to handle acid solutions with concentrations up to 5-6%. Wool cloth has a life comparable to that of cotton in neutral liquors. Wool is woven in the duck-like square cloth weave, or with a raised nap; or it may be formed as a felt. Originally the smooth cloth weave was used for filtering electrolytic slimes and similar slurries. The hairlike fibers, as in cotton cloth, ensure good filtrate clarity. Long-nap wool cloth has found wide application in sewage sludge dewatering and in cases where only ferric chloride is used for conditioning. The wool has a long life and it does not clog easily. Wool cloths are sold by weight, usually ranging 10-22 oz/yd^2 with the majority at 12 oz/yd^2. The clarity through wool cloths is considerably less than through cotton cloths.

PAPER PULP AND FIBER CLOTHS

Paper pulp and fiber cloths are excellent materials for precoats and filter aids. Paper pulp gives a high rate of flow, is easily discharged and shows little tendency to clog.

Paper pulp's disadvantage lies in its preparation. Soda or sulfate pulp, most commonly used, must be disintegrated and kept in suspension by agitation before precoating. This requires considerable auxiliary equipment. Diatomaceous earths, while they should be kept in suspension, are very easy to handle and do not undergo disintegration. Paper pulp compressed into pads is used in pressure filters for beverage clarification. After becoming dirty, as evidenced by decrease in the rate of flow, the paper may be repulped, water-washed and reformed into pads. Although this involves considerable work, excellent clarity and high flowrates are obtained. The impurities do not form a cake as such, but penetrate into the pad and can only be removed by repulping and washing the pad.

Pads of a mixture of paper pulp and asbestos fiber are used in bacteriological filtrations. In sheet form it is employed in the laboratory for all kinds of filtration. Filter papers are made in many grades of porosity for use in porcelain and glass funnels. Industrially, paper in the form of sheets is used directly or as a precoat in filter presses.

Used directly in lubricating clarification in a "blotter press", it acts much the same manner as the paper pads, but is much thinner and is not reused. As a precoat, paper protects the filter medium from slimy fines; it may be peeled off and discarded after clogging, leaving the medium underneath clean.

RUBBER MEDIA

Rubber media appear as porous, flexible rubber sheets and microporous hard rubber sheets. Commercial rubber media have 1100-6400 holes/in.^{2} with pore diameters of 0.012-0.004 in. They are manufactured out of soft rubber, hard rubber, flexible hard rubber and soft neoprene. The medium is prepared on a master form, consisting of a heavy fabric belt, surfaced on one side with a layer of rubber filled with small round pits uniformly spaced. These pits are 0.020 in. deep, and the number per unit area and their surface diameter determine the porosity of the sheet. A thin layer of latex is fed to the moving belt by a spreader bar so that the latex completely covers the pits, yet does not run into them. This process traps air in each pit. The application of heat to the under-surface of the blanket by a steam plate causes the air to expand, blowing little bubbles in the film of latex. When the bubbles burst, small holes are left, corresponding to the pits. The blown rubber film, after drying, is cooled and the process repeated until the desired thickness of sheet is obtained. The sheet is then stripped off of the master blanket and vulcanized. Approximately 95% of the pits are reproduced as holes in the rubber sheet. The holes are not exactly cylindrical in shape but are reinforced by slight constrictions which contribute to strength and tear resistance. This type is referred to as "plain," and can be made with fabric backing on one or both sides to control stretching characteristics. If the unvulcanized material is first stretched, and then vulcanized while stretched, it is called "expanded." Resulting holes are oval and have a higher porosity (sometimes up to 30%). Special compounds have been formulated for resistance to specific chemicals under high concentrations at elevated temperatures, such as 25% sulfuric acid at 180°F. The smooth surface allows the removal of thinner cakes than is possible with cotton or wool fabrics. Rubber does not show progressive binding and it can be readily cleaned and used in temperatures up to 180°F. On the other hand, because a clear filtrate is difficult to obtain when filtering finely divided solids, a precoat often becomes necessary.

SYNTHETIC FIBER CLOTHS

Cloths from synthetic fibers are superior to many of the natural cloths thus far considered. They do not swell as do natural fibers, are inert in many acid, alkaline and solvent solutions and are resistant to various fungus and bacterial growths (the degree depending on the particular fiber and use). Several synthetic fibers resist relatively high temperatures, and have a smooth surface for easy cleaning and good solids discharge. Some of the most widely used synthetic filter media are nylon, Saran, Dacron, Dynel, Vinyon, Orlon, and Acrilan. Table 2 compares the physical properties of several synthetic fiber filter media. Tightly woven, monofilament (single-strand) yarns consist of small-diameter filaments. They tend to lose their tensile strength, because their small diameters reduce their permeability; thus multifilament yarns are normally used. Monofilament yarns in loose weaves provide high flowrates, good solids discharge, easy washing and high resistance to blinding, but the turbidity of the filtrate is high and recirculation is usually necessary, initially at least.

FLEXIBLE METALLIC MEDIA

Flexible metallic media are especially suitable for handling corrosive liquors and for high-temperature filtration. They have good durability and are inert to physical changes. Metallic media are fabricated in the form of screens, wire windings, or woven fabrics of steel, copper, bronze, nickel and different alloys. Perforated sheets and screens are used for coarse separation, as supports for filter cloths or as filter aids. Metallic cloths are characterized by the method of wire weaves as well as by the size and form of holes and by the wire thickness. Metallic cloths may be manufactured with more than 50,000 holes/cm^2 and with hole sizes less than 20 μm.

Table 2. Properties of woven filter cloth fibers

Fibers	Acids	Alkalies	Solvents	Fiber Tensile Strength	Temperature Limit (°F)
Acrilan	Good	Good	Good	High	275
Asbestos	Poor	Poor	Poor	Low	750
Cotton	Poor	Fair	Good	High	300
Dacron	Fair	Fair	Fair	High	350
Dynel	Good	Good	Good	Fair	200
Glass	High	Fair	Fair	High	600
Nylon	Fair	Good	Good	High	300
Orlon	Good	Fair	Good	High	275
Saran	Good	Good	Good	High	240
Teflon	High	High	High	Fair	180
Wool	Fair	Poor	Fair	Low	300

METALLIC/NONMETALLIC CLOTH

Combination metallic and nonmetallic cloths consist of metallic wires and weak cloth or asbestos threads. There are some difficulties in weaving when attempting to maintain uniformity between wires and the cloth, and considerable dissatisfaction has been experienced with such construction. While cotton weaves well with the asbestos, the cotton fibers destroy the fabric's resistance to heat and corrosion. Its use is, therefore, quite limited, despite its resistance to high temperatures, acids and mildew.

Cotton cloths are sometimes treated with metallic salts (copper sulfate) to improve their corrosion-resistant qualities. Such cloths are in the usual cotton filter cloth grades, and while they are not equivalent to metallic cloths, the treatment does materially prolong the life of the cotton fiber. When dealing with metallic cloths, the following terms are important:

Mesh - Number of openings per lineal inch. Measured from the center of wires. The number of openings precedes the word "mesh".

Wire - The diameter of wire used in weaving cloth measured by gauge or decimal inch.

Opening - The size of clear opening between parallel wires. For a given mesh, the space is determined by the diameter of wire used.

Woven wire cloth is available in a broad range of alloys with mesh counts as coarse as 1" in most alloys and as fine as 635 in some alloys. The following is a list of alloys normally used:

Galvanized Steel	Stainless Steel Type 304
Plain Steel	Aluminum
Brass	Bronze
Copper	Exotic Alloys
Galvanized Steel	High Temperature Alloys
Monel 400	Nickel
Plain Steel	Precious Metals
Stainless Steel Type 304	Stainless Steel Type 316
Stainless Steel Type 321	Stainless Steel Type 330
Stainless Steel Type 347	Stainless Steel Type 410
Stainless Steel Type 430	

NONWOVEN MEDIA

Nonwoven media are fabricated in the form of belts or sheets from cotton, wool, synthetic and asbestos fibers or their mixtures, as well as from paper mass. They may be used in filters of different designs, for example, in filter presses, filters with horizontal discs and rotary drum vacuum filters for liquid clarification. Most of these applications handle low suspension concentrations; examples are milk, beverages, lacquers and lubricating oils. Individual fibers in nonwoven media are usually connected among them as a result of mechanical treatment. A less common approach is the addition of binding substances. Sometimes the media are protected from both sides by loosely woven cloth. Nonwoven media of various materials and weights, and in several grades of retentiveness per unit weight can be formed, in either absorbent or nonabsorbent material. These filter media retain less dispersed particles (more than 100 μm) on their surface, or close to it, and more dispersed particles within the depths of the media.

Nonwoven filter media are mostly used for filter medium filtration with pore clogging. Because of the relatively low cost of this medium, it is often replaced after pore clogging. In some cases, nonwoven media are used for cake filtration. In this case, cake removal is so difficult that it must be removed altogether from the filter medium. Nonwoven filter media can be prepared so that pore sizes decrease in the direction from the surface of the filter media contacting suspension to the surface contacting the supporting device. This decreases the hydraulic resistance of filtration and provides retention of relatively large particles of suspension over the outer layer of the nonwoven medium. Nonwoven filter media of synthetic, mechanically pressed fibers are manufactured by puncturing the fiber layer with needles (about 160 punctures/cm^2), and subsequent high temperature treatment with liquid which causes fiber contraction. Such filter media are distinguished by sufficient mechanical strength and low hydraulic resistance, as well as uniform fiber distribution. Filter media from fibers connected by a blinder are manufactured by pressing at 70N/cm^2 and 150°C. These media have sufficient mechanical strength, low porosity and are corrosion resistant. Filter media may be manufactured by lining a very thin layer of heat-resistant metal (e.g., nickel 360) over a fiber surface of inorganic or organic material. Such filter media may withstand temperatures of 200°C and higher. Of the flexible filter media described, the synthetic fabrics are perhaps the most widely relied on in industrial applications. Each filtration process must meet certain requirements in relation to flowrate, clarity of filtrate, moisture of filter cake, cake release and nonbinding characteristics. The ability of a filter fabric to help meet these criteria, and to resist chemical and physical attack depend on such characteristics as fiber type, yarn size, thread count, type of weave, fabric finish and yarn type (monofilament, multifilament or spun). Monofilament yarns consist of a single, continuous filament with a relatively smooth surface. The different sizes are specified as a measurement of the diameter in mils or in micrometers. Multifilament yarns are made from many fine filaments extruded simultaneously. The different sizes are specified by a measurement of weight known as the denier. These yarns are generally used for filter fabrics which require a smooth surface and relatively tight weave. Spun yarns are made from filaments which are chopped in short lengths and then spun or twisted together. Spun yarns are made into filter fabrics with a hairy, dense surface very suitable for filtration of very fine particles. It is necessary to select the type of fiber that will offer the most resistance to breakdown normally caused by chemical, temperature and mechanical conditions of the filter process. Tables 3 and 4 can serve as rough guides to proper media selection. Table 5 provides linear conversion units between mesh size, inches and micrometers.

RIGID FILTER MEDIA

FIXED RIGID MEDIA

Fixed rigid media are available in the forms of disks, pads and cartridges. They are composed of firm, rigid particles set in permanent contact with one another. The media formed have excellent void uniformity, resistance to wear and ease in handling as piece units. Depending on the particle size forming the filter media, temperature, pressure and time for caking, it is possible to manufacture media with different porosities. The higher the pore uniformity, the more uniform the shape of the particles. These media are distinguished by long life, high corrosion resistance and easy cake removal. However, the particles that penetrate inside the pores are very difficult to extract.

METALLIC MEDIA

Metallic filter media are widely used throughout the chemical and process industries in the form of perforated or slotted plates of steel, bronze or other materials. These designs provide for easy removal of coarse particles and for supporting loose rigid media. Powdered metal is a porous medium. The physical characteristics, chemical composition, structure, porosity, strength, ductility, shape and size can be varied to meet special requirements. The porosity ranges up to 50% void by volume, tensile strength up to 10,000 psi, varying inversely with porosity, and ductility of 3-5% in tension, and higher in compression.

Table 3. Physical properties and chemical resistance of polyester fibers. [a]

Specific Gravity	1.38
Moisture Regain	
At 65%RH and 68°F(20°C)(%)	0.4
Water Retention Power (%)	3-5
Tensile Strength	
cN/dtex	7-9.5
Wet in % of dry	95-100
Elongation at Break	
%	10-20
Wet in % of dry	100-105
Ultraviolet Light Resistance	R[b]
Resistance to Fungus, Rot, and Mildew	R
Resistance to Dry Heat	
Continuous	
°F	302
°C	150
Short-Term Exposure	
°F	392
°C	200
Chemical Resistance to	
Acids	C[c]
Acetic Acid Concentration	R
Sulfuric Acid 20%	R
Nitric Acid 10%	C

Hydrochloric Acid 25%	C
Alkalies	C
Saturated Sodium Carbonate	R
Chlorine Bleach Concentration	R
Caustic Soda 25%	U[d]
Ammonia Concentration	U
Potassium Permanganate 50%	R
Formaldehyde Concentration	R
Chlorinated Hydrocarbons	R
Benzene	R
Phenol	C
Ketones, Acetone	R

a Average properties reported as based on typical industry sources.

b R = recommended; c C = conditional; d U = unsatisfactory

Table 4. Standard fibers and micrometer ratings for bag filters.[a]

		Available Micrometer Ratings											
Construction		**1**	**10**	**25**	**50**	**100**	**150**	**200**	**250**	**300**	**400**	**600**	**800**
Felts	Polyester	X	X	X	X	X		X					
	Polypropylene	X	X	X	X	X							
Multifilament Meshes	Polyester					X	X	X	X	X	X		X
	Nylon (heavy)												X
Monofilament Meshes	Nylon				X	X	X	X	X	X	X	X	X
	Polypropylene									X		X	

Compatibility and Temperature Limits for Standard Bag Materials

Fiber	Compatibility with							**Temperature Limits (°F)**
	Organic Solvents	**Animal, Vegetable, and Petro Oils**	**Microorganisms**	**Alkalies**	**Organic Acids**	**Oxidizing Agents**	**Mineral Acids**	
Polyester	Excellent	Excellent	Excellent	Good	Good	Good	Good	300
Polypropylene	Excellent	Excellent	Excellent	Excellent	Excellent	Good	Good	225
Nylon	Excellent	Excellent	Excellent	Good	Fair	Poor	Poor	325

Bag Capacities and Dimensions

Bag Size No.	Fits Rosedale Model No.	Bag Surface Area (ft²)	Bag Volume (gal)	Bag Dimensions Length (in.)	Diameter (in.)
1	815	2.0	2.1	16.5	7
2	830	4.4	4.6	32.0	7
1 (inner)	815	1.6	1.7	14.5	5.75
2 (inner)	830	3.6	3.8	30.0	5.75

Powdered metal cannot be readily ground or machined. It is made in discs, sheets, cones or special shapes for filtering fuel oil, refrigerants, solvents, etc. The smooth surface associated with a perforated plate permits brush cleaning or scrubbing in addition to the naturally easier discharge from such surfaces. The hard metallic material has a long life, not being subject to abrasion or flexing. The size of particles filtered on such plates must be relatively large. Normally, plates are confined to free-filtering materials where there is little danger of clogging. Metallic filter media may be used either for cake filtration or depth filtration, i.e., pore clogging. Regeneration of media may be achieved by dissolving solid particles inside the pores or by back thrust of filtrate.

Table 5. Comparative particle sizes.

U.S. Mesh	in.	μm
3	0.265	6730
3½	0.223	5660
4	0.187	4760
5	0.157	4000
6	0.132	3360
7	0.111	2830
8	0.0937	2380
10	0.0787	2000
12	0.0661	1680
14	0.0555	1410
16	0.0469	1190
18	0.0394	1000
20	0.0331	841
25	0.0280	707
30	0.0232	595
35	0.0197	500
40	0.0165	420
45	0.0138	354

U.S. Mesh	in.	μm
50	0.0117	297
60	0.0098	250
70	0.0083	210
80	0.0070	177
100	0.0059	149
120	0.0049	125
140	0.0041	105
170	0.0035	88
200	0.0029	74
230	0.0024	63
270	0.0021	53
325	0.0017	44
400	0.0015	37

CERAMIC MEDIA

Ceramic filter media are manufactured from crushed and screened quartz or chamotte, which is then thoroughly mixed with a binder (for example, with silicate glass) and sintered. Quartz media are resistant to concentrated mineral acids but not resistant to low-concentration alkalies or neutral water solutions of salts. Chamotte media are resistant to dilute and concentrated mineral acids and water solutions of their salts, but have poor resistance to alkali liquids. The rough surface of ceramic filter media promotes adsorption of particles and bridging. Sintering of chamotte with a binder results in large blocks from which filter media of any desired shape can be obtained. Using synthetic polymers as binders, ceramic filter media that do not contain plugged pores are obtained.

DIATOMACEOUS MEDIA

Diatomaceous media are available in various shapes. Their relatively uniform particle size establishes high efficiency in retaining solid particles of sizes less than 1μm as well as certain types of bacteria. Media in the form of plates and cartridges are manufactured by sintering a mixture of diatomite with a binder.

COAL MEDIA

Coal media are manufactured by mixing a fraction of crushed coke with an anthracene fraction of coal tar and subsequent forming under pressure, drying and heating in the presence of a reducing flame. These media of high mechanical strength are good for use in acids and alkalies.

EBONITE MEDIA

Ebonite media are manufactured from partially vulcanized rubber, which is crushed, pressed and vulcanized. These media are resistant to acids, salt solutions and alkalies. They may be used for filtration at temperatures ranging from -10 to +110° C.

FOAM PLASTIC MEDIA

Foam plastic media are manufactured from polyvinyl chloride, polyurethane, polyethylene, polypropylene and the other polymer materials. The foam plastic media are economical.

LOOSE RIGID MEDIA

Filter media may also be composed of particles that are rigid in structure, but are applied in bulk loose form. That is, individual particles merely contact each other. This form has the advantage of being cheap and easy to keep clean by rearrangement of the particles. When the proper size and shape of particles are selected, the section of passage may be regulated over extremely wide limits. The disadvantages of a rigid medium in simple contact are that it can be used conveniently only in a horizontal position and that it does not allow removal of thick deposits or surface cleaning except by backwashing, without disturbing the filter bed.

COAL AND COKE

Coal (hard) and coke are used in water filtration, primarily for the removal of coarse suspensions, care being taken to prevent them from scouring or washing away, because of their relative lightness and fine division. Coal is principally composed of carbon, and is inert to acids and alkalies. Its irregular shapes are advantageous at times over silica sand.

Though inert to acids, sand is affected by alkalies, and its spherical particle shape allows deeper solids penetration and quicker clogging than does coal. With the lighter weight of coal (normally 50 lb/ft^3, compared with 100 lb/ft^3 for sand), a greater surface area is exposed for solids entrapment.

CHARCOAL

Charcoal, whether animal or vegetable, when used as a filter medium, is required to perform the dual services of decoloring or adsorbing and filtering. The char filters used in the sugar industry are largely decoloring agents and the activated carbons used in water clarification are for deodorizing and removal of taste. There are many types of charcoal in use as filter media, ranging from ordinary wood char to specially prepared carbons.

Activated carbon in particular is very versatile as a filter media because it not only can physically separate out suspended solids, but it can adsorb materials. The adsorption process occurs at solid-solid, gas-solid, gas-liquid, liquid-liquid, or liquid-solid interfaces. Adsorption with a solid such as carbon depends on the surface area of the solid. Thus, carbon treatment of water involves the liquid-solid interface. The liquid-solid adsorption is similar to the other adsorption mechanisms. There are two methods of adsorption: physisorption and chemisorption. Both methods take place when the molecules in the liquid phase become attached to the surface of the solid as a result of the attractive forces at the solid surface (adsorbent), overcoming the kinetic energy of the liquid contaminant (adsorbate) molecules.

Factors affecting adsorption include:

☞ ***The physical and chemical characteristics of the adsorbent, that is, sur- face area, pore size, chemical composition, and so on.***
☞ ***The physical and chemical characteristics of the adsorbate, that is, molecular size, molecular polarity, chemical composition, and so on; the concentration of the adsorbate in the liquid phase (solution).***
☞ ***The characteristics of the liquid phase, that is, pH and temperature.***
☞ ***The residence time of the system.***

Physisorption occurs when, as a result of energy differences and/or electrical attractive forces (weak van der Waals forces), the adsorbate molecules become physically fastened to the adsorbent molecules. This type of adsorption is multilayered; that is, each molecular layer forms on top of the previous layer with the number of layers being proportional to the contaminant concentration. More molecular layers form with higher concentrations of contaminant in solution. When a chemical compound is produced by the reaction between the adsorbed molecule and the adsorbent, chernisorption occurs. Unlike physisorption, this process is one molecule thick and irreversible because energy is required to form the new chemical compound at the surface of the adsorbent, and energy would be necessary to reverse the process. The reversibility of physisorption is dependent on the strength of attractive forces between adsorbate and adsorbent. If these forces are weak, desorption is readily affected. Certain organic compounds in wastewaters are resistant to biological degradation and many others are toxic or nuisances (odor, taste, color forming), even at low concentrations. Low concentrations may not be readily removed by conventional treatment methods. Activated carbon has an affinity for organics and its use for organic contaminant removal from gaseous streams and wastewaters is widespread. The effectiveness of activated carbon for the removal of organic compounds from fluids by adsorption is enhanced by its large surface area, a critical factor in the adsorption process. The surface area of activated carbon typically can range from 450 to 1,800 m^2/g, with some carbons observed to have a surface area up to 2,500 m^2/g. Some examples are given in Table 6.

The versatility of charcoal, or carbon lies in its ability to adsorb materials. Carbon has been known throughout history as an adsorbent with its usage dating back centuries before Christ. Ancient Hindus filtered their water with charcoal. In the thirteenth century, carbon materials were used in a process to purify sugar solutions. In the eighteenth century, Scheel discovered the gas adsorptive capabilities of carbon and Lowitz noted its ability to remove colors from liquids. Carbon adsorbents have been subjected to much research resulting in numerous development techniques and applications. One of these applications was begun in England in the mid-nineteenth century with the treatment of drinking waters for the removal of odors and tastes. From these beginnings, water and wastewater treatment with carbon has become widespread in municipal and industrial processes, including wineries and breweries, paper and pulp, pharmaceutical, food, petroleum and petrochemical, and other establishments of water usage. Interest in carbon use for air as well as water pollution control and traditional industrial/product applications has received increased attention since the early 1970s with the advent of more stringent environmental regulations.

Table 6. Typical Surface Areas of Activated Carbons.

Origin	Surface Area, m^2/g
Bituminous coal	1,200-1,400
Bituminous coal	800-1,000
Coconut shell	1,100-1,150

Origin	Surface Area, m^2/g
Pulp mill residue	550-650
Pulp mill residue	1,050-1,100
Wood	700-1,400

Of less significance than the surface area is the chemical nature of the carbon's surface. This chemical nature or polarity varies with the carbon type and can influence attractive forces between molecules. Alkaline surfaces are characteristic of carbons of vegetable origins and this type of surface polarity affects adsorption of dyes, colors, and unsaturated organic compounds. Silica gel, an adsorptive media that is not a carbon compound, has a polar surface which also exhibits an adsorptive preference for unsaturated organic com- pounds as opposed to saturated compounds. However, for the most part, activated carbon surfaces are nonpolar, making the adsorption of inorganic electrolytes difficult and the adsorption of organics easily effected.

Pores of the activated carbon exist throughout the particle in a manner illustrated in Figure 2. The pore structure of activated carbon affects the large surface-to-size ratio. The macropores do not add appreciably to the surface area of the carbon but provide a passageway to the particle interior and the micropores. The micropores are developed primarily during carbon activation and result in the large surface areas for adsorption to occur.

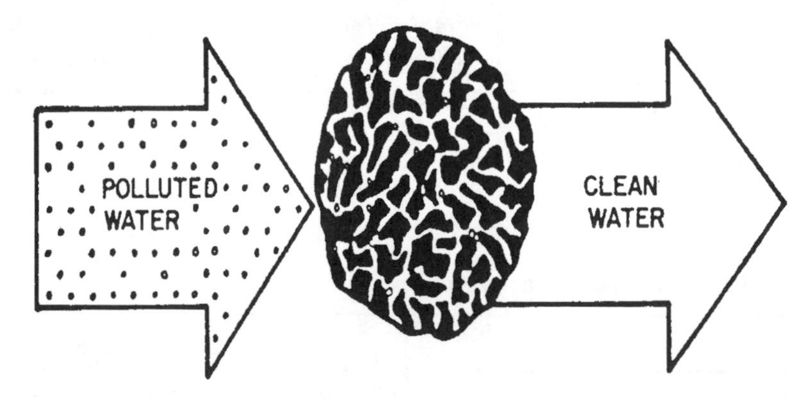

Figure 2. Carbon granule at work.

Macropores are those pores greater than 1,000 Å; micropores range between 10-1,000 Å. Pore structure, like surface area, is a major factor in the adsorption process. Pore-size distribution determines the size distribution of molecules that can enter the carbon particle to be adsorbed. Large molecules can block off micropores, rendering useless their available surface areas. However, because of irregular shapes and constant molecule movement, the smaller molecules usually can penetrate to the smaller capillaries. Since adsorption is possible only in those pores that can be entered by molecules, the carbon adsorption process is dependent on the physical characteristics of the activated carbon and the molecular size of the adsorbate. Each application for carbon treatment must be cognizant of the characteristics of the contaminant to be removed and designed with the proper carbon type in order to attain optimum results. Basically, there are two forms of activated carbon: powdered and granular. The former are particles that are less than U.S. Sieve Series No. 50, while the latter are larger. The adsorption rate is influenced by carbon particle size, but not the adsorptive capacity which is related to the total surface area. By reducing the particle size, the surface area of a given weight is not affected. Particle size contributes mainly to a system's hydraulics, filterability, and handling characteristics.

CARBON ACTIVATION

Carbon materials are activated through a series of processes which includes:

- ***Removal of all water (dehydration)***
- ***Conversion of the organic matter to elemental carbon, driving off the noncarbon portion (carbonization)***
- ***Burning off tars and pore enlargement (activation).***

Initially, the material to be converted is heated to 170° C to effect water removal. Temperatures are than raised above 170° C driving off CO_2, CO, and acetic acid vapors. At temperatures of about 275° C, the decomposition of the material results and tar, methanol, and other by-products are formed. Nearly 80 percent elemental carbon is then effected by prolonged exposure to 400- 600° C. Activation of this product follows with the use of steam or carbon dioxide as an activating agent. The superheated steam, 750-950° C, passes through the carbon burning out by-product blockages, and expanding and extending the pore network.

DIATOMACEOUS EARTH

Diatomaceous earths may resemble the forms of the charcoals. The earths are primarily filter aids, precoats or adsorbents, the function of the filter medium being secondary. Fuller's earth and clays are used for decoloring applications; diatomaceous earths are used for clarification. The adsorbtivity of diatomaceous earth works in the same fashion as activated carbon, but isotherms (affinity) for many chemical species like the hydrocarbons is weaker. For this reason, activated charcoal or carbon is much preferred in wastewater treatment applications expecially when taste and odor issues are priorities.

PRECIPITATES AND SALTS

Precipitates or salts are used when corrosive liquor must be filtered, and where there is no available medium of sufficient fineness that is corrosion-resistant and will not contaminate the cake. In these cases, precipitates or salts are used on porous supports. In the filtration of caustic liquors, ordinary salt (sodium chloride) is used as the filter medium in the form of a precoat over metallic cloth. This procedure has the advantage that the salt medium will not be detrimental to either the cake or the filtrate if inadvertently mixed with it.

SAND AND GRAVEL

Sand and gravel are the most widely used of the rigid media simple contact. Most of the sand used this way is for the clarification of water for drinking or industrial uses. Washed, screened silica sand is sold in standard grades for this work and is used in depths ranging from a few inches to several feet, depending on the type of filter and clarification requirements. Heavy, irregular grains, such as magnetite, give high rates of flow and low penetration by the solid particles, and are easily cleaned. They are, however, considerably more expensive than silica sand, so their use is limited. Sand beds are often gravel-supported, but gravel alone is seldom used as the filter medium.

CRUSHED STONE OR BRICK

Crushed stone or brick is used for coarse filtration of particularly corrosive liquors. Their use, however, is extremely limited and they are not considered important filter media.

GENERAL PROPERTIES OF LOOSE AND GRANULAR MEDIA

The physical properties of loose and granular media are important, both from the standpoint of the operation of the filtration device, but also from the standpoints of feeding and storing these materials in bins and silos. These considerations are equally important and quite pertinent to dry chemicals that are used as filtration aids (Chapter 3). In a general sense, loose solid matter is comprised of large numbers of individual particles. The physical properties and forces of attraction which exist between individual particles have important effects on their flow behavior, as fluids flow around them, and in their stagnant states in bins and silos. With many materials there exists forces of attraction between individual and clusters of particles, and hence these materials are described as being cohesive, Ideal, loose solids have no forces of attraction between them. The properties of loose materials in contrast to fluids and composite solids are characterized by several parameters that must either be measured or known prior to specifying the equipment for handling them. The first of these parameters is the material's density, of which there are several related terms, namely bulk, particle and skeletal densities. The bulk density is the overall density of the loose material and includes the interparticle distance of separation. It is simply defined as the overall mass of the material per unit volume. A material's bulk density is sensitive to the particle size, the mean particle density, moisture content, and the interparticle separation distance (better known as the *degree of solids packing* or simply the *packing density*). It can be measured by simply pouring a weighed sample of the particles into a graduated cylinder, and from the volume occupied one determines the loose bulk density. By gently vibrating the container walls, the distance between particles decreases and hence, the volume decreases. The material thus becomes denser with time and its bulk density achieves some limiting value, ρ_{max}, known as the *tapped* or *packed bulk density*.

The ratio of ρ_{max}/ρ_{min} can be as high as 1.52 depending on the material. Consequently, when bulk densities are reported it is important to note whether the value was determined under loose or tapped conditions, along with the mean particle size. Most literature values report an average bulk density that is representative of the material most often handled. Loose solids may be broadly characterized according to their bulk densities:

Light material	$\rho_b < 600\ kg/m^3$
Average	$600 < \rho_b < 2,000\ kg/m^3$
Extra heavy	$\rho_b > 2,000\ kg/m^3$

The loose bulk density (kg/m^3) can be computed as:

$$\rho_b = (G_1 - G)/V$$

where G_1, G = weights of filled and empty cylinders, respectively
V = internal volume of cylinder

Bulk density is related to particle density through the interparticle void fraction ϵ in the sample.

$$\rho_b = \rho_p (1-\epsilon)$$

The value of ϵ varies between the limits of 0 and unity; however, many particles have a loosely poured voidage of approximately 0.4 to 0.45.

Particle density, ρ_p, is the density of a particle including the pores or voids within the individual solids. It is defined as the weight of the particle divided by the volume occupied by the entire particle. Sometimes this is referred to as the material's *apparent density.*

The skeletal density, ρ_S, also called the *true density,* is defined as the density of a single particle excluding the pores. That is, it is the density of the *skeleton* of the particle if the particle is porous. For nonporous materials, skeletal and particle densities are equivalent. For porous particles, skeletal densities are higher than the particle density.

Particle and skeletal densities are related through the following expression:

$$\rho_p = (1 + \xi\rho_f)/(1/\rho_s + \xi)$$

where ρ_p = particle density

ρ_s = skeletal density

ρ_f = density of fluid within the pores of the solid

ξ = pore volume per unit mass of solids

When the particle pores are saturated with solids, $\xi\rho_f$ is negligible and the expression simplifies to the following:

$$1/\rho_p = 1/\rho_s + \xi$$

Another property of importance is the *pore volume.* It can be measured indirectly from the adsorption and/or desorption isotherms of equilibrium quantities of gas absorbed or desorbed over a range of relative pressures. Pore volume can also be measured by mercury intrusion techniques, whereby a hydrostatic pressure is used to force mercury into the pores to generate a plot of penetration volume versus pres- sure. Since the size of the pore openings is related to the pressure, mercury intrusion techniques provide information on the pore size distribution and the total pore volume.

Moisture can significantly affect loose materials, particularly their flowability. Low temperatures, particle bridging, and caking can alter interparticle void fractions and cause dramatic changes in bulk density. Moisture becomes bound to solids because of mechanical, physicochemical, and chemical mechanisms. Moisture retained between particles and on their surfaces is strictly a mechanical mechanism. Physicochemical binding results when moisture penetrates inside particle pores because of diffusion and adsorption onto pore walls. Chemically bound moisture appears as hydrated or crystalline structures. The terms *moisture or moisture content* is used to denote the degree of liquid retained on and in solids.

Moisture is defined as the ratio of the fluid's weight retained by solids to the weight of the wet material:

$$W = (G_w - G_d)/G_w$$

where G_w and G_d are the weights of the wet and absolute dry material, respectively. Moisture content, W_c, is the ratio of the moisture weight to the weight of absolute fry material:

$$W_c = (G_w - G_d)/G_d$$

Values of W and W_c can be expressed either as fractions or percentages. The presence of moisture tends to increase the relationship between moisture content and the density of loose or lump materials as follows:

$$\rho_m = \rho(1 + W_c)$$

For dusty and powdery materials:

$$\rho_m = \rho(1 + W_c)/(1 + (1/3)W_c\ \rho_f/\rho_p)$$

where ρ_m, ρ = densities of wet and dry loose materials, respectively

ρ_p = particle density

ρ_f = density of liquid filling the solid particle pores

In addition to the physical properties just described there are those properties which affect the flowability of the material. Specifically, these properties are the material's angle of repose, angle of internal friction, and the angle of slide.

The *angle of repose* is defined as the angle between a line of repose of loose material and a horizontal plane. Its value depends on the magnitude of friction and adhesion between particles and determines the mobility of loose solids, which is a critical parameter in designing conical discharge and feeding nozzles and in establishing vessel geometries. In all cases the slopes of such nozzles should exceed the angle of repose.

Materials can be roughly categorized according to their angle of repose as follow:

Free Flowing Granules	**$25° < \beta < 30°$**
Free Flowing Granules	**$30° < \beta < 38°$**
Fair to Passable Flow of Powders	**$38° < \beta < 45°$**
Cohesive Powders	**$45° < \beta < 55°$**

The angle of repose is the measured angle between a horizontal plane and the top of a pile of solids. The *poured angle* of repose is obtained when a pile of solids is formed, whereas the *drained angle* results when solids are drained from a bin. Figure 3 distinguishes between the two terms signifying the angle of repose. For monosized particles or particles with a relatively narrow particle size distribution, the drained and poured angles of repose are approximately the same. If however, the solids have a broad particle size distribution, then the drained angle can be higher than the poured angle. In general the lower the angle of repose, the more free flowing the material is.

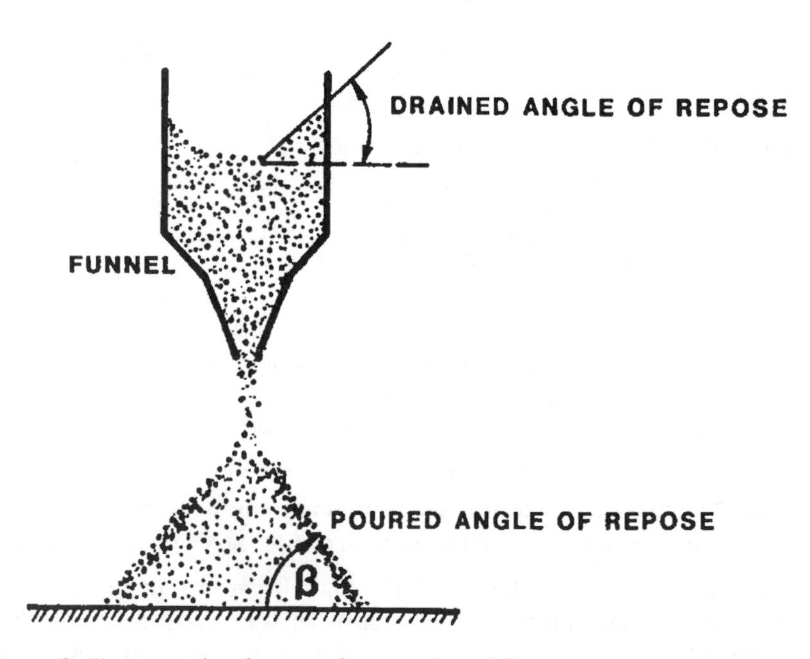

Figure 3. Shows angle of repose for granular solids.

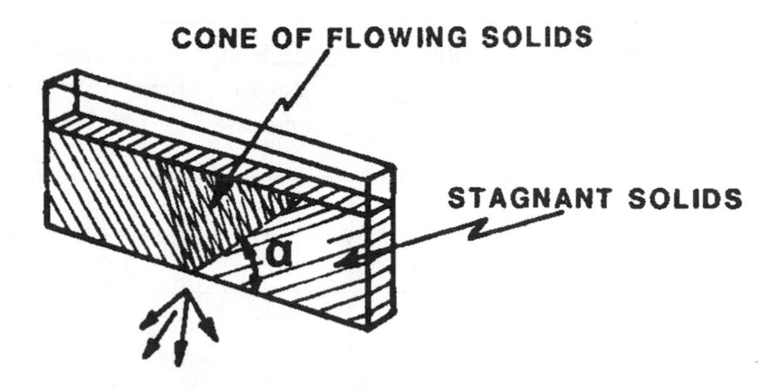

Figure 4. Illustrates angle of internal friction.

The angle of repose is sensitive to the conditions of the supporting surface; the smoother the surface, the smaller the angle. The angle may also be reduced by vibrating the supporting surface.

When handling slow moving materials having large angles of repose, well-designed, bunkers and hoppers are provided with highly polished internal surfaces and low amplitude vibrators. The angle is also sensitive to moisture. Specifically, moisture tends to increase the angle of repose. The variation of β with moisture content is likely due to the surface layer of moisture that surrounds each particle and surface tension effects which bind aggregates of solids together.

The *angle of internal friction*, α, is defined as the equilibrium angle between flowing particles and bulk or stationary solids in a bin. Figure 4 illustrates the definition. The angle of internal friction is greater than the angle of repose.

The angle of slide is defined as the angle from the horizontal of an inclined surface on which an amount of material will slide downward due to the influence of gravity. This angle is important in the sizing and operation of chutes and hoppers as well as pneumatic conveying systems. It provides a measure of the relative adhesiveness of a dry material to a dissimilar surface. This angle depends on the type or nature of solids, the physical and surface properties of equipment (e.g., roughness, cleanliness, dryness), the surface configuration (e.g., the degree of curvature), the manner in which solids are placed onto surfaces, and the rate of change of the slope of the surface during measurements.

Table 7 provides some literature reported measurements of the properties discussed for common filter media and aids as loose materials. You might try an experiment and measure the various angles. In developing your data base, obtain the measurements in triplicate, and calculate a standard deviation for each angle measurement.

Table 7. Physico-Mechanical Properties of Loose Materials.

Material	Bulk Density (g/cm^3)	Angle of Repose (degress)	Friction Coefficient	
			Inside	Outside on Steel
Phosphate powder	1.52	36	0.49	0.37
Calcium chloride	0.68	35	0.63	0.58
Carbamide (powdered)	0.54	42	0.825	0.56
Superphosphate (granulated)	1.1	31	0.64	0.46
Superphosphate (powdered)	0.8	36	0.71	0.7
Talc	0.85	40	-	-
Chalk	1.1	42	0.81	0.76
Sand (fine)	1.51	33	1.0	0.58
Coal (fine)	0.95	36	0.67	0.47

FILTER MEDIA SELECTION CRITERIA

Due to the wide variety of filter media, filter designs, suspension properties, conditions for separation and cost, selection of the optimum filter medium is complex. Filter media selection should be guided by the following rule: a filter medium must incorporate a maximum size of pores while at the same time providing a sufficiently pure filtrate. Fulfilment of this rule invokes difficulties because the increase or decrease in pore size acts in opposite ways on the filtration rate and solids retention capacity.

The difficulty becomes accentuated by several other requirements that cannot be achieved through the selection of a single filter medium. Therefore, selection is often reduced to determining the most reasonable compromise between different, mutually contradictory requirements as applied to the filter medium at a specified set of filtration conditions. Because of this, some problems should be solved before final medium selection. For example, should attempts be made to increase filtration rate or filtrate purity? Is cost or medium life more important? In some cases a relatively more expensive filter medium, such as a synthetic cloth, is only suitable under certain filtration conditions, which practically eliminates any cost consideration in the selection process.

Thus, the choice may only be made after consideration of all requirements. It is, however, not practical to analyze and compare each requirement with the hope of logically deducing the best choice. There is, unfortunately, no generalized formula for selection that is independent of the details of the intended application. Each cake requires study of the specific considerations, which are determined by the details or the separation process.

One can to outline a general approach for medium selection along with a test sequence applicable to a large group of filter media of the same type. There are three methods of filter media tests: laboratory- or bench-scale pilot-unit, and plant tests. The laboratory-scale test is especially rapid and economical, but the results obtained are often not entirely reliable and should only be considered preliminary. Pilot-unit tests provide results that approach plant data. The most reliable results are often obtained from plant trials.

Different filter media, regardless of the specific application, are distinguished by a number of properties. The principal properties of interest are the permeability of the medium relative to a pure liquid, its retention capacity relative to solid particles of known size and the pore size distribution. These properties are examined in a laboratory environment and are critical for comparing different filter media.

The permeability relative to a pure liquid, usually water, may be determined with the help of different devices that operate on the principle of measurement of filtrate volume obtained over a definite time interval at known pressure drop and filtration area. The permeability is usually expressed in terms of the hydraulic resistance of the filter medium. This value is found from:

$$V = \frac{\Delta p S}{\mu(r_o h_c + R_f)} \tau$$

When the cake thickness is 0, we may write the equation as:

$$R_f = \frac{\Delta p S \tau}{V} \mu$$

Note that:

$$\Delta p = \Delta p_t - \Delta p_f$$

where Δp_t = pressure difference accounting for the hydrostatic pressure of a liquid column at its flow through the filter medium, supporting structure and device channels, and Δp_f = same pressure drop when the flow of liquid is through the supporting structure and device channels

Analytical determination of the hydraulic resistance of the medium is difficult. However, for the simplest filter medium structures, certain empirical relationships are available to estimate hydraulic resistance. The relationship of hydraulic resistance of a cloth of monofilament fiber versus fiber diameter and cloth porosity can be based on a fixed-bed model.

In evaluation and selection of a filter medium, one should account for the fact that hydraulic resistance increases gradually with time. In particular, the relationship between cloth resistance and the number of filter cycles is defined by:

$$R = R_{in} e^{KN}$$

$$K = \frac{g' - g''}{g''}$$

The retentivity relative to solid particles (e.g., spherical particles of polystyrene of definite size) is found from experiments determining the amount of these particles in the suspension to be filtered before and after the filter media. The retentivity K is determined as follows: where g', g" =amounts of solid particles in liquid sample before and after the medium, respectively.

The pore sizes distribution, as well as the average pore size, is determined by the "bubble" method. The filter medium to be investigated is located over a supporting device under a liquid surface that completely wets the medium material. Air is introduced to the lower surface of the medium. Its pressure is gradually increased, resulting in the formation of single chains of bubbles. This corresponds to air passages through the largest-diameter pores. As pressure is increased, the number of bubble chains increases due to air passing through the smaller pores. In many cases a critical pressure is achieved where the liquid begins to "boil." This means that the filter medium under investigation is characterized by sufficiently uniform pores. If there is no "boiling," the filter medium has pores of widely different sizes. The pore size through which air passes is calculated from known relations. For those pores whose cross section may be assumed close to a triangle, the determining size should be the diameter of a circle that may be inscribed inside the triangle.

For orientation in cloth selection for a given process, the following information is essential: filtration objectives (obtaining cake, filtrate or both), and complete data (if possible) on the properties of solid particles (size, shape and density), liquid (acid, alkali or neutral, temperature, viscosity, and density), suspension (ratio of solids to liquid, particle aggregation and viscosity), and cake (specific resistance, compressibility, crystalline, friable, plastic, sticky or slimy). Also, the required capacity must be known as well as what constitutes the driving force for the process (e.g., gravity force, vacuum or pressure). Based on such information, an appropriate cloth that is resistant to chemical, thermal and mechanical aggression may be selected. In selecting a cloth based on specific mechanical properties, the process driving force and filter type must be accounted for. The filter design may determine one or more of the following characteristics of the filter cloth: tensile strength, stability in bending, stability in abrasion, and/or ability of taking the form of a filter-supporting structure. Tensile strength is important, for example, in belt filters. Bending stability is important in applications of metallic woven cloths or synthetic monofilament cloths. If the cloth is subjected to abrasion, then glass cloth cannot be used even though it has good tensile strength.

From the viewpoint of accommodation to the filter-supporting structure, some cloths cannot be used, even though the filtering characteristics are excellent. For rotary drum filters, for example, the cloth is pressed onto the drum by the "caulking" method, which uses cords that pass over the drum. In this case, the closely woven cloths manufactured from monofilament polyethylene or polypropylene fiber are less desirable than more flexible cloths of polyfilament fibers or staple cloths.

Depending on the type of filter device, additional requirements may be made of the cloth. For example, in a plate-and-frame press, the sealing properties of cloths are very important. In this case, synthetic cloths are more applicable staple cloths, followed by polyfilament and monofilament cloths. In leaf filters operating under vacuum and pressure, the cloth is pulled up onto rigid frames. Since the size of a cloth changes when in contact with the suspension, it should be pretreated to minimize shrinkage.

In selecting cloths made from synthetic materials, one must account for the fact that staple cloths provide a good retentivity of solid particles due to the short hairs on their surface. However, cake removal is often difficult from these cloths – more than from cloths of polyfilament and, especially, monofilament fibers. The type of fiber weave and pore size determine the degree of retentivity and permeability. The objective of the process, and the properties of particles, suspension and cake should be accounted for. The cloth selected in this manner should be confirmed or corrected by laboratory tests. Such tests can be performed on a single filter. These tests, however, provide no information on progressive pore plugging and cloth wear. However, they do provide indications of expected filtrate pureness, capacity and final cake wetness.

A single-plate filter consists of a hollow flat plate, one side of which is covered by cloth. The unit is connected to a vacuum source and submerged into the suspension (filtration), then suspended in air to remove filtrate, or irrigated by a dispersed liquid (washing). The filter cloth is directed downward or upward or located vertically, depending on the type of filter that is being modeled in the study.

The following is a recommended sequence of tests that can assist in cloth selection for continuous vacuum filters. If the cycle consists of only two operations (filtration and dewatering), tests should be conducted to determine the suspension weight concentration after 60 sec of filtration and 120 seconds of dewatering. The cake thickness should be measured and the cake should be removed to determine the weight of wet cake and the amount of liquid in it. The weight of filtrate and its purity are also determined. If the cake is poorly removed by the device, it is advisable to increase the dewatering time, vacuum or both. If the cake is poorly removed after an operating regime change, it should be tested with another cloth. If the cake is removed satisfactorily, filtration time should be decreased under increased or decreased vacuum. Note that compressible cakes sometimes plug pores faster at higher vacuum. After the filtration test for a certain filter cycle (which is based on the type of the filter being modeled), the suspension's properties should be examined. Based on the assumed cycle, a new filtration test should be conducted and the characteristics of the process noted. Capacity (N/m^2-hr), filtration rate (m^3/m^2-hr) and cake wetness can then be evaluated. Also, if possible, the air rate and dewatering time should be computed. The results of the first two or three tests should not be taken into consideration because they cannot exactly characterize the properties of the cloth. A minimum of four or five tests is generally needed to achieve reproducible results of the filtration rate and cake wetness to within 3 to 5%. When the cycle consists of filtration, washing and dewatering, the tests are considered principally in the same manner. The economic aspects of cloth selection should be considered after complete determination of cloth characteristics.

RECOMMENDED RESOURCES FOR THE READER

Check out these references on Pretreatment. They can all be found on the EPA Website (www.epa.gov).

1. Application and Use of the Regulatory Definition of Significant Noncompliance for Industrial Users, September 9, 1991 (Memorandum) ERIC: W986; NTIS: PB95-201786.

2. CERCLA Site Discharges to POTWs Treatability Manual, August 1990 540/2-90-007 ERIC: W570; NTIS: PB91-921269 Disk PB91-507236.

3. CERCLA Site Discharges to POTWs: CERCLA Site Sampling Program - Detailed Data Report, May 1990 540/2-90-008 ERIC: W515; NTIS: PB91-921270.

4. CERCLA Site Discharges to POTWs: Guidance Manual, August 1990 540/G-90-005 ERIC: W150; NTIS: PB90-274531.

5. Environmental Regulations and Technology: The National Pretreatment Program, July 1986 625/10-86-005 ERIC: W350; NTIS: PB90-246521.

6. Federal Guidelines: State and Local Pretreatment Programs Appendices 1-7 - Volume 2, January 1977 430/9-76-017B ERIC: W185; NTIS: PB-266782.

7. Federal Guidelines: State and Local Pretreatment Programs Volume 1, January 1977 430/9-76-017A ERIC: U041; NTIS: PB-266781.

8. Federal Guidelines: State and Local Pretreatment Programs Appendix 8 - Volume 3, January 1977 430/9-76-017C ERIC: W186; NTIS: PB-266783.

9. Guidance for Reporting and Evaluating POTW Noncompliance with Pretreatment Implementation Requirements, September 1987 ERIC: W304; NTIS: PB95-157764.

10. Guidance Manual for Conducting RCRA Facility Assessments at Publicly Owned Treatment Works, September 1987 ERIC: W830; NTIS: PB95-157715.

11. Guidance Manual for Control of Slug Loadings to POTWs, September 1988 ERIC: W111; NTIS: PB93-202745.

12. Guidance Manual for POTW Pretreatment Program Development, October 1983 833/B-83-100 ERIC: W639 ; NTIS: PB93-186112.

13. Guidance Manual for Preventing Interference at POTWs, September 1987 833/B-87-201 NSCEP: 833/B-87-201; ERIC: W106 ; NTIS. PB92-117969.

14. Guidance Manual for the Control of Wastes Hauled to Publicly Owned Treatment Works, September 1999 833/B-98-003 NSCEP: 833/B-98-003; WRC: 833/B-98-003; ERIC: C281. NTIS: PB2000-102387.

15. Guidance Manual for the Identification of Hazardous Wastes Delivered to Publicly Owned Treatment Works by Truck, Rail, or Dedicated Pipe, June 1987 833/B-87-100 ERIC: W202; NTIS: PB92-149251.

16. Guidance Manual for the Use of Production Based Pretreatment Standards and the Combined Wastestream Formula, September 1985 833/B-85-201 NSCEP: 833/B-85-201; ERIC: U095 ; NTIS: PB92-232024.

17. Guidance Manual on the Development and Implementation of Local Discharge Limitations Under the Pretreatment Program - Volume 1, November 1987 ERIC: W025; NTIS: PB95-157707.

18. Guidance Manual on the Development and Implementation of Local Discharge Limitations Under the Pretreatment Program - Volume 2 - Appendices, November 1987 ERIC: W026; NTIS: PB95-157699.

19. Guidance Manual on the Development and Implementation of Local

Discharge Limitations Under the Pretreatment Program, December 1987 833/B-87-202 ERIC: W107; NTIS: PB92-129188.

20. Guidance to POTWs for Enforcement of Categorical Standards, November 5, 1984, Memorandum ERIC: W296; NTIS: PB95-157673.

21. Guidance to Protect POTW Workers from Toxic and Reactive Gases and Vapors, June 1992 812/B-92-001 NSCEP: 812/B-92-001; ERIC: W115; NTIS: PB92-173236.

22. Guide to Discharging CERCLA Aqueous Wastes to Publicly Owned Treatment Works (POTWs), March 1991 NTIS: PB91-219364 Handbook for Monitoring Industrial Wastewater, August 1973 625/6-73-002 ERIC: W318; NTIS: PB-259146.

23. Industrial User Inspection and Sampling Manual for POTWs (Diskette Version and Printed Appendices), April 1994 ERIC: W493; NTIS: PB96-502646.

24. Industrial User Permitting Guidance Manual, September 1989 833/R-89-001 ERIC: W109; NTIS: PB92-123017.

25. Pretreatment Bulletin No. 3, November 6, 1987 ERIC: W859; NTIS: PB95-159414.

26. Pretreatment Bulletin No. 4, November 6, 1987 ERIC: W860 ; NTIS: PB95-159406.

27. Pretreatment Bulletin No. 6, June 1989 ERIC: W861 ; NTIS: PB95-159398.

28. Pretreatment Compliance Inspection and Audit Manual for Approval Authorities, July 1986 833/B-86-100 NSCEP: 833/B-86-100; WRC: 833/B-86-100; ERIC: W277; NTIS: PB90-183625.

29. Pretreatment of Industrial Wastes: Joint Municipal and Industrial Seminar, 1978 625/4-78-012 ERIC: W662 Procedures Manual for Reviewing a POTW Pretreatment Program Submission, October 1983 833/B-83-200 ERIC: W137; NTIS: PB93-209880.

30. Treatability Manual: Volume I - Treatability Data, September 1981 Revised 600/2-82-001A ERIC: W754 Treatability Manual: Volume II - Industrial Descriptions, September 1981 Revised 600/2-82-001B ERIC: W755 Treatability Manual: Volume III - Technologies for Control/Removal of Pollutants 600/2-82-001C ERIC: W756 Treatability Manual: Volume IV - Cost Estimates, April 1983 Revised 600/2-82-001D ERIC: W757 Treatability Manual: Volume V - Summary, January 1983, Change 2 600/2-82-001E ERIC: W753 U.S. EPA Pretreatment Compliance Monitoring and Enforcement System Version 3.0: User's Guide, Final, September 1992 831/F-92-001; NSCEP: 831/F-92-001; ERIC: W269; NTIS: PB94-118577.

Check out the following Web sites for more general information:

31. Provides water chemicals and equipment for potable, waste and process water in industrial, municipal and mining treatment systems.URL: http://www.tramfloc.com.

32. Tramfloc, Inc. - Home Page - Text Version - flocculant, coagulants. Provides water chemicals and equipment for potable, waste and process water in industrial, municipal and mining treatment systems. URL: http://www.tramfloc.com/indext.html.

33. Water and Wastewater Treatment Plant Operators Occupational Outlook Handbook; URL: http://stats.bls.gov/oco/ocos229.htm.

34. Control and Optimization of Wastewater Treatment Plants, Department of Systems...Research progress in wastewater treatment at Uppsala University. URL: http://www.syscon.uu.se/Research/waste.html.

35. Wastewater Treatment and Metal Finishing Equipment ... Your complete source of new and reconditioned industrial wastewater treatment, metal finishing, and biological treatment systems and equipment. URL: http://wmi-inc.com.

36. Water and Wastewater Research and Co-operation Directory - European Centres of ...The Water and Wastewater Directory is a data source of more than 750 European organisations from 31 countries. It offers an easy search for experts...URL: http://www.metra-martech.com.

37. Environmental Dynamics - Worldwide Wastewater Treatment Systems Wastewater : Environmental Dynamics, Inc. wastewater treatment. Biological wastewater treatment and advanced technology aeration-mixing systems. URL: http://www.wastewater.com.

38. Wastewater Treatment Engineering; Bureau of Land & Water Quality Last update: 03/12/01. Wastewater Treatment Engineering, Technical Assistance and Pollution Prevention. Waste Treatment; Go to the following web site: http://janus.state.me.us/dep/blwq/engin.htm.

39. International provider of water purification and wastewater treatment solution ... Waterlink is a provider of water purification and wastewater treatment solutions, carbon systems, separations technology, aeration systems, solids...URL: http://www.waterlink.com.

QUESTIONS FOR THINKING AND DISCUSSING

1. What are the characteristics of a good filter medium? Make a list of several commercially available products and determine whether or not they meet your criteria.

2. What is the % open area of the weave of a cloth filter, where the opening is on the average 50 microns and the thread diameter is 85 mills.
3. What is the micron rating of a 50 × 250 Dutch weave filter cloth?
4. Select a cloth filter for a filter press application in which the water is both alkaline and has a high content of solvents. The operating temperature could reach as high as 200° F on excursions.
5. Define the terms mesh, wire and opening.
6. Select a bag filter that must operate at temperatures above 300° F and handle concentrations of organic acids in the wastewater.
7. What are some of the factors the impact on the adsorptivity of materials like carbon and diatomaceous earth?
8. Discuss how carbon adsorption works and how it can be used in water treatment applications. Give some specific examples where this technology is used to remove specific contaminants.
9. Develop a general classification system for dry bulk chemical additives and filter aids based on ease of feeding to a filtering machine.
10. Conduct a lab test to measure the bulk densities of several filter aids. If you don't have the chemicals, then describe how you would do the tests and what specific measurements and calculations are needed.
11. The pore volume of a material is 30 % and its particle density is 1.3 g/cm^3. Calculate the skeletal density of the material.
12. What is the effect of moisture on the flowability and feeding characteristics of dry bulk filter aids?
13. Explain the terms angle of repose, angle of internal friction, and angle of slide. Why are these important to dry bulk chemical handling?
14. What is permeability and how can it be expressed as an engineering parameter.?
15. Describe the method for estimating pore size distribution.
16. Develop a checklist of items/issues you should follow when selecting a filter media.

Chapter 5

WHAT PRESSURE AND CAKE FILTRATION ARE ALL ABOUT

INTRODUCTION

This chapter provides a summary of the governing expressions describing conventional pressure-induced filtration and a description of major equipment. Standard filtration practices refers to the most common or classical method of filtration, sometimes referred to as cake filtration.

This type of filtration relies on the use of a porous bed, or more accurately - a porous media which can be cloth. With high-solids-concentration suspensions, even relatively small particles (in comparison to the pore size) will not pass through the medium, but tend to remain on the filter surface, forming "bridges" over individual openings in the filter medium. The filtrate flows through the filter medium and cake because of an applied pressure on one side of the media bed, the magnitude of which is controlled proportionally to the filtration resistance. This resistance results from the frictional drag on the liquid as it passes through the filter and cake.

The hydrostatic pressure varies from a maximum at the point where suspension enters the cake, to zero where liquid is expelled from the medium; consequently, at any point in the cake the two are complementary. That is, the sum of the hydrostatic and compression pressures on the solids always equals the total hydrostatic pressure at the face of the cake. Thus, the compression pressure acting on the solids varies from zero at the face of the cake to a maximum at the filter medium.

When solid particles undergo separation from the mother suspension, they are captured both on the surface of the filter medium and within the inner pore passages. The penetration of solid particles into the filter medium increases the flow resistance until the filtration cycle can no longer continue at economical throughput rates, at which time the medium itself must either be replaced or thoroughly cleaned.

At the end of this chapter you will find three annexes. The first of these is a list of nomenclature used in the chapter. There are quite a few design equations that are summarized in the foregoing sections and, hence, you will need to refer to this annex from time to time. The second annex is a list of recommended references that I have relied on over the years, plus some interesting Web sites for you to visit for vendor-specific information, as well as supplemental design and equipment sizing information. The final annex is the ***Questions for Thinking and Discussing***. This annex is a collection of problem-solving questions, many of which are based on actual design cases developed over the years and also developed for seminar programs that my father (*Paul N. Cheremisinoff, P.E.*) and an old colleague of mine (*Dr. David S. Azbel*) taught to process engineers. Some of these problems are quite challenging and I recommend you work on these with your colleagues. You will find that you may have to refer to some of the recommended references in addition to the design formulae presented in this chapter to answer all of these questions. But, if you can successfully get through these questions, you should have a pretty good command of the subject and be able to confidently select and develop preliminary equipment sizing for different applications. Note that I have taken a very generalized approach to the subject in that it does not restrict us to water treatment applications alone. Indeed many of the filtration machines discussed in this chapter are not necessarily the proper or best choices for water treatment applications, but they can be depending on the unique industry application. Having said this, what this chapter attempts to provide you with is a very solid foundation on sizing and equipment selection for this important class of filtration practices. Remember to refer to the ***Glossary*** at the end of the book, if you run across any terms that are unfamiliar to you.

CONSTANT PRESSURE DIFFERENTIAL FILTRATION

When the space above the suspension is subjected to compressed gas or the space under the filter plate is under a vacuum, filtration proceeds under a constant pressure differential (the pressure in the receivers is constant). The rate of filtration decreases due to an increase in the cake thickness and, consequently, flow resistance. A similar filtration process results from a pressure difference due to the hydrostatic pressure of a suspension layer of constant thickness located over the filter medium.

If the suspension is fed to the filter with a reciprocating pump at constant capacity, filtration is performed under constant flowrate. In this case, the pressure differential increases due to an increase in the cake resistance. If the suspension is fed by a centrifugal pump, its capacity decreases with an increase in cake resistance, and filtration is performed at variable pressure differentials and flowrates.

The most favorable filtration operation with cake formation is a process whereby no clogging of the filter medium occurs. Such a process is observed at sufficiently high concentrations of solid particles in suspension. From a practical standpoint this concentration may conditionally be assumed to be in excess of 1% by volume. Filtration is frequently accompanied by hindered or free gravitational settling of solid particles. The relative directions of action between gravity force and filtrate motion may be concurrent, countercurrent or crosscurrent, depending on the orientation of the filter plate, as well as the sludge location above or below the filter plate. The different orientations of gravity force and filtrate motion with their corresponding distribution of cake, suspension, filtrate and clear liquid are illustrated in Figure 1. Particle sedimentation complicates the filtration process and influences the controlling mechanisms. Furthermore, these influences vary depending on the relative directions of gravity force and filtrate motion. If the suspension is above the filter medium (Figure 1A), particle settling leads to more rapid cake formation with a clear filtrate, which can be evacuated from the filter by decanting. If the suspension is under the filter medium (Figure 1B), particle settling will prevent cake formation, and it is necessary to mix the suspension to maintain homogeneity.

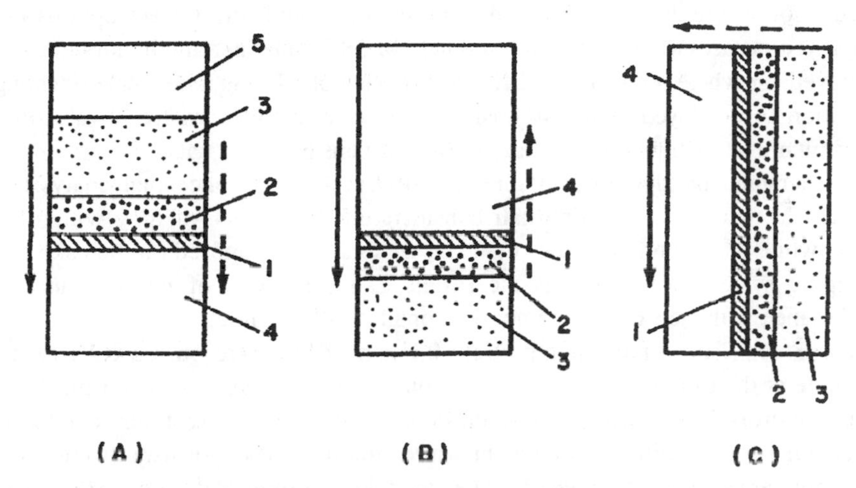

Figure 1. Direction of gravity force action and filtrate motion in filters: A-cocurrent; B-countercurrent; C-crosscurrent; solid arrow-direction of gravity force action; dashed arrow- direction of filtrate motion; 1-filter plate; 2-cake; 3-sludge; 4-filtrate; 5-clear liquid.

When the cake structure is composed of particles that are readily deformed or become rearranged under pressure, the resulting cake is characterized as being compressible. Those that are not readily deformed are referred to as sem-compressible, and those that deform only slightly are considered incompressible. Porosity (defined as the ratio of pore volume to the volume of cake) does not decrease with increasing pressure drop. The porosity of a compressible cake decreases under pressure, and its hydraulic resistance to the flow of the liquid phase increases with an increase in the pressure differential across the filter media.

Cakes containing particles of inorganic substances with sizes in excess of 100 μm may be considered incompressible. Examples of incompressible cake-forming materials are sand and crystals of carbonates of calcium and sodium. The cakes containing particles of metal hydroxides, such as ferric hydroxide, cupric hydroxide, aluminum hydroxide, and sediments consisting of easy deforming aggregates, which are formed from primary fine crystals, are usually compressible. At the completion of cake formation, treatment of the cake depends on the specific filtration objectives. For example, the cake itself may have no value, whereas the filtrate may. Depending on the disposal method and the properties of the particulates, the cake may be discarded in a dry form, or as a slurry. In both cases, the cake is usually subjected to washing, either immediately after its formation, or after a period of drying. In some cases, a second washing is required, followed by a drying period where all possible filtrate must be removed from the cake; or where wet discharge is followed by disposal: or where repulping and a second filtration occurs; or where dry cake disposal is preferable. Similar treatment options are employed in cases where the cake is valuable and all contaminating liquors must be removed, or where both cake and filtrate are valuable. In the latter, cake-forming filtration is employed, without washing, to dewater cakes where a valueless, noncontaminating liquor forms the residual suspension in the cake.

To understand the dynamics of the filtration process, a conceptual analysis is applied in two parts. The first half considers the mechanism of flow within the cake, while the second examines the external conditions imposed on the cake and pumping system, which brings the results of the analysis of internal flow in accordance with the externally imposed conditions throughout.

The characteristics of the pump relate the applied pressure on the cake to the flowrate at the exit face of the filter medium. The cake resistance determines the pressure drop. During filtration, liquid flows through the porous filter cake in the direction of decreasing hydraulic pressure gradient. The porosity (ϵ) is at a minimum at the point of contact between the cake and filter plate (i.e., where x = 0) and at a maximum at the cake surface (x = L) where sludge enters. A schematic definition of this system is illustrated in Figure 2.

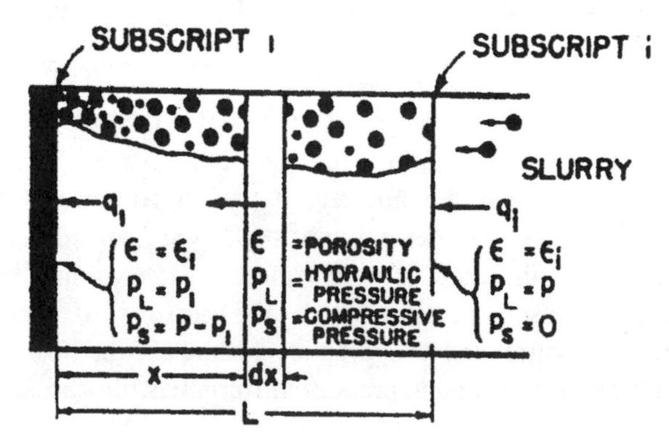

Figure 2. Important parameters in cake formation.

The drag that is imposed on each particle is transmitted to adjacent particles. Therefore, the net solid compressive pressure increases as the filter plate is approached, resulting in a decrease in porosity. Referring to Figure 3A, it may be assumed that particles are in contact at one point only on their surface, and that liquid completely surrounds each particle. Hence, the liquid pressure acts uniformly in a direction along a plane perpendicular to the direction of flow. As the liquid flows past each particle, the integral of the normal component of force leads to form drag, and the integration of the tangential components results in frictional drag. If the particles are non-spherical, we may still assume single-point contacts between adjacent particles as shown in Figure 3B.

Consider flow through a cake (Figure 3C) with the membrane located at a distance x from the filter plate. Neglecting all forces in the cake other than those created by drag and hydraulic pressure, a force balance from x to L gives:

$$F_s + Ap_L = Ap \qquad (1)$$

The applied pressure p is a function of time but not of distance x. F_S is the cumulative drag on the particles, increasing in the direction from x = L to x = 0. Since single point contact is assumed, the hydraulic pressure p_L is effectively over the entire cross section (A) of the cake; for example, against the fictitious membrane shown in Figure 3B. Dividing Equation 1 by A and denoting the compressive drag pressure by $p_S = F/A$, we obtain:

$$p_L + p_s = p \qquad (2)$$

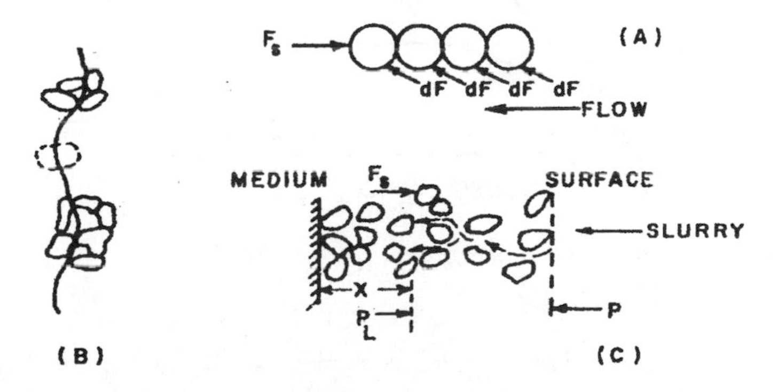

Figure 3. Frictional drag on particles in compressible cakes.

The term p_S is a fictitious pressure, because the cross-sectional area A is not equal to either the surface area of the particles nor the actual contact areas In actual cakes, there is a small area of contact A_C whereby the pressure exerted on the solids may be defined as F_S/A_C.

Taking differentials with respect to x, in the interior of the cake, we obtain:

$$dp_s + dp_L = 0 \tag{3}$$

This expression implies that drag pressure increases and hydraulic pressure decreases as fluid moves from the cake's outer surface toward the filter plate.

From Darcy's law, the hydraulic pressure gradient is linear through the cake if the porosity (ϵ) and specific resistance (α) are constant. The cake may then be considered incompressible. This is illustrated by the straight line obtained from a plot of flowrate per unit filter area versus pressure drop shown in Figure 4. The variations in porosity and specific resistance are accompanied by varying degrees of compressibility, also shown in Figure 4.

The rate of the filtration process is directly proportional to the driving force and inversely proportional to the resistance.

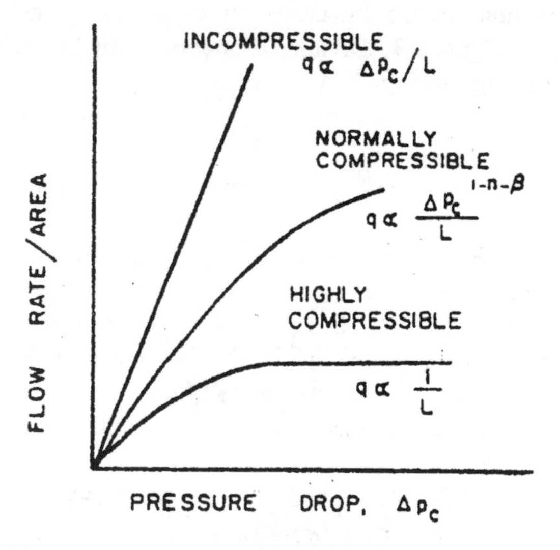

Figure 4. Flowrate/area versus pressure drop across the cake.

Because pore sizes in the cake and filter medium are small, and the liquid velocity through the pores is low, the filtrate flow may be considered laminar: hence, Poiseuille's law is applicable. Filtration rate is directly proportional to the difference in pressure and inversely proportional to the fluid viscosity and to the hydraulic resistance of the cake and filter medium. Because the pressure and hydraulic resistances of the cake and filter medium change with time, the variable rate of filtration may be expressed as:

$$u = \frac{dV}{A d\tau} \tag{4}$$

where V = volume of filtrate (m^3)
A = filtration area (m^2)
τ = time of filtration (sec)

Assuming laminar flow through the filter channels, the basic equation of filtration as obtained from a force balance is:

$$u = \frac{1}{A}\frac{dV}{d\tau} = \frac{\Delta p}{\mu(R_c + R_f)} \tag{5}$$

where Δp = pressure difference (N/m^2)
μ = viscosity of filtrate ($N\text{-}sec/m^2$)
R_C = filter cake resistance (m^{-1})
R_f = initial filter resistance (resistance of filter plate and filter channels) (m^{-1})
u = filtration rate (m/sec), i.e., filtrate flow through cake and filter plate
$dV/d\tau$ = filtration rate (m^3/sec), i.e., filtrate flow rate

Filter cake resistance (R_C) is the resistance to filtrate flow per unit area of filtration. R_C increases with increasing cake thickness during filtration. At any instant, R_C depends on the mass of solids deposited on the filter plate as a result of the passage of V (m^3) filtrate. R_f may be assumed a constant. To determine the relationship between volume and residence time τ, Equation 5 must be integrated, which means that R_C must be expressed in terms of V.

We denote the ratio of cake volume to filtrate volume as x_0. Hence, the cake volume is x_0V. An alternative expression for the cake volume is h_CA; where h_C is the cake height in meters. Consequently:

$$x_0 V = h_c A \tag{6}$$

Hence, the thickness of the cake, uniformly distributed over the filter plate, is:

$$h_c = x_0 \frac{V}{A} \tag{7}$$

The filter cake resistance may be expressed as:

$$R_c = r_0 x_0 \frac{V}{A} \tag{8}$$

where r_0= specific volumetric cake resistance (m^{-2}).

As follows from Equation 8, r_0 characterizes the resistance to liquid flow by a cake having a thickness of 1 m.

Substituting for R_C from Equation 8 into Equation 5, we obtain:

$$\frac{1}{A}\frac{dV}{d\tau} = u = \frac{\Delta p}{\mu[r_0 x_0 (V/A) + R_f]} \tag{9}$$

Filtrate volume, $x_{0,}$ can be expressed in terms of the ratio of the mass of solid particles settled on the filter plate to the filtrate volume (x_W) and instead of r_0, a specific mass cake resistance r_W is used. That is, r_W represents the resistance to flow created by a uniformly distributed cake, in the amount of 1 kg/m^2. Replacing units of volume by mass, the term r_0 x_0 in Equation 9 changes to $r_W x_W$.
Neglecting filter plate resistance (R_f = 0), and taking into account Equation 7, we obtain from Equation 3 the following expression:

$$r_0 = \frac{\Delta p}{\mu h_c u} \tag{10}$$

At μ=1 N-sec/m^2, h_c = 1 m and u = 1 m/sec, r_0 = Δp. Thus, the specific cake resistance equals the pressure difference required by the liquid phase (with a viscosity of 1 N-sec/m^2) to be filtered at a linear velocity of 1 m/sec through a cake 1 m thick. This hypothetical pressure difference, however, is beyond a practical range. For highly compressible cakes, r_0 can exceed 10^{12} m^2. Assuming V = 0 (at the start of filtration) where there is no cake over the filter plate. Equation 9 becomes:

$$R_f = \frac{\Delta p}{\mu u} \tag{11}$$

At μ = 1 N-sec/m^2 and u = 1 m/sec, R_f = Δp. This means that the filter plate resistance is equal to the pressure difference necessary for the liquid phase (with viscosity of 1 N-sec/m^2) to pass through the filter plate at a rate of 1 m/sec. For many filter plates R_f is typically 10^{10} m^{-1} .

For a constant pressure drop and temperature filtration process all the parameters in Equation 9, except V and τ, are constant. Integrating Equation 9 over the limits of 0 to V, from 0 to τ, we obtain:

$$\int_0^V \mu(r_0 x_0 \frac{V}{A} + R_f)\, dV = \int_0^\tau \Delta p A\, d\tau \tag{12a}$$

or

$$\mu_0 r_0 x_0 \frac{V^2}{2A} + \mu R_f V = \Delta p A \tau \tag{12b}$$

Dividing both sides by $\mu r_0 x_0/2A$ gives:

$$V^2 + 2\left(\frac{R_f A}{r_0 x_0}\right) V = \left(2\frac{\Delta p A^2}{\mu r_0 x_0}\right)\tau \tag{13}$$

Equation 13 is the relationship between filtration time and filtrate volume. The expression is applicable to either incompressible or compressible cakes, since at constant Δp, r_0 and x_0 are constant. If we assume a definite filtering apparatus and set up a constant temperature and filtration pressure, then the values of R_f, r_0, μ and Δp will be constant.

The terms in parentheses in Equation 13 are known as the "filtration constants", and are often lumped together as parameters K and C; where:

$$K = \frac{2\Delta p A^2}{\mu r_0 x_0} \tag{14}$$

$$C = \frac{R_f A}{r_0 x_0} \tag{15}$$

Hence, a simplified expression may be written to describe the filtration process as follows:

$$V^2 + 2VC = K\tau \tag{16}$$

Filtration constants K and C can be experimentally determined, from which the volume of filtrate obtained over a specified time interval (for a certain filter, at the same pressure and temperature) can be computed. If process parameters are changed, new constants K and C can be estimated from Equations 14 and 15. Equation 16 may be further simplified by denoting τ_0 as a constant that depends on K and C:

$$\tau_0 = \frac{C^2}{K} \tag{17}$$

Substituting τ_0 into Equation 16, the equation of filtration under constant pressure conditions is:

$$(V + C)^2 = K(\tau + \tau_0) \tag{18}$$

Equation 18 defines a parabolic relationship between filtrate volume and time. The expression is valid for any type of cake (i.e., compressible and incompressible). From a plot of V + C versus $(\tau+\tau_0)$, the filtration process may be represented by a parabola with its apex at the origin as illustrated in Figure 5. Moving the axes to distances C and τ_0 provides the characteristic filtration curve for the system in terms of volume versus time. Because the parabola's apex is not located at the origin of this new system, it is clear why the filtration rate at the beginning of the process will have a finite value, which corresponds to actual practice.

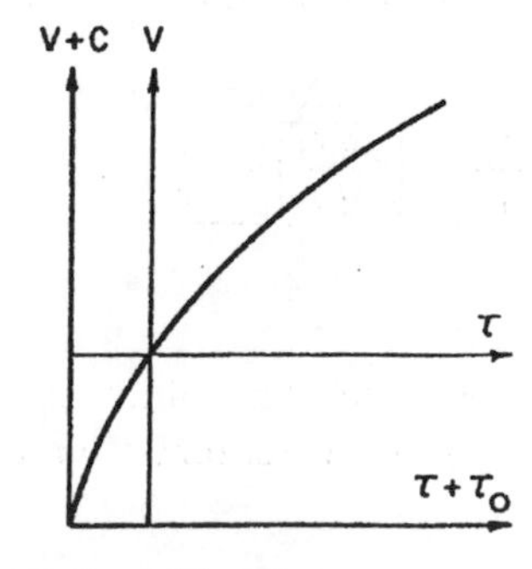

Figure 5. Typical filtration curve.

Constants C and τ_0 in Equation 18 have physical interpretations. They are basically equivalent to a fictitious layer of cake having equal resistance. The formation of this fictitious cake follows the same parabolic relationship, where τ_0 denotes the time required for the formation of this fictitious mass, and C is the volume of filtrate required. Differentiating Equation 16 gives:

$$\frac{dV}{d\tau} = \frac{K}{2(V + C)} \qquad (19)$$

And rearranging in the form of a reciprocal relationship:

$$\frac{d\tau}{dV} = \frac{2V}{K} + \frac{2C}{K} \qquad (20)$$

This form of the equation provides a linear relation like the plot in Figure 6. The expression is that of a straight line having slope 2/K, with intercept C. The experimental determination of $d\tau/dV$ is made simple by the functional form of this expression. Filtrate volumes V_1 and V_2 should be measured for time intervals τ_1 and τ_2. Then, according to Equation 16:

$$\frac{\tau_2 - \tau_1}{V_2 - V_1} = \frac{1}{V_2 - V_1}\left[\frac{V_2^2 - V_1^2}{K} + \frac{2C(V_2 - V_1)}{K}\right] =$$

$$2\frac{\left(\frac{V_1 + V_2}{2}\right)}{k} + \frac{2C}{K} \qquad (21)$$

In examining the right side of this expression, we note that the quotient is equal to the inverse value of the rate at the moment of obtaining the filtrate volume, which is equal to the mean arithmetic value of volumes V_1 and V_2:

$$\frac{\tau_2 - \tau_1}{V_2 - V_1} = \left(\frac{d\tau}{dV}\right)_{\frac{V_1 + V_2}{2}} \qquad (22)$$

Filtration constants C and K can be determined on the basis of several measurements of filtrate volumes for different time intervals.

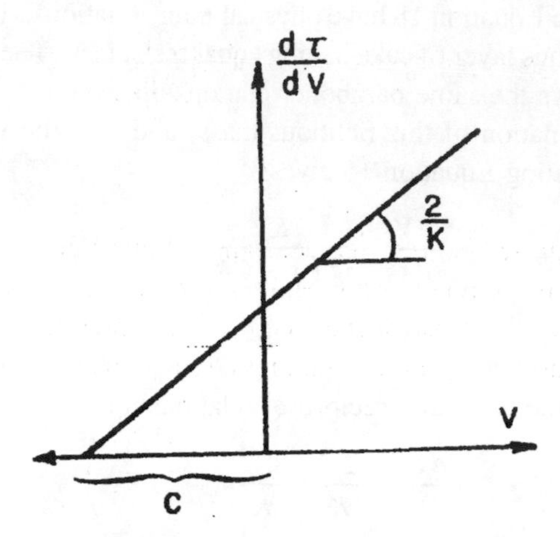

Figure 6. Plot of Equation 20.

As follows from Equations 14 and 15, values of C and K depend on r_0 (specific volumetric cake resistance), which in turn depends on the pressure drop across the cake. This Δp, especially during the initial stages of filtration, undergoes changes in the cake. When the cake is very thin, the main portion of the total pressure drop is exerted on the filter medium. As the cake becomes thicker, the pressure drop through the cake increases rapidly but then levels off to a constant value. Isobaric filtration shows insignificant deviation from Equation 16. For approximate calculations, it is possible to neglect the resistance of the filter plate, provided the cake is not too thin. Then the filter plate resistance $R_f = 0$ in Equation 15, $C = 0$ (Equation 15) and $\tau_0 = 0$ (Equation 17). Therefore, the simplified equation of filtration takes the following form:

$$V^2 = K\tau \tag{23}$$

For thick cakes, Equation 23 gives results close to that of Equation 16.

CONSTANT-RATE FILTRATION

When sludge is fed to a filter by a positive-displacement pump, the rate of filtration is nearly constant (i.e., $dV/d\tau$ = constant). During constant-rate filtration, the pressure increases with an increase in cake thickness. Therefore, the principal variables are pressure and filtrate volume, or pressure and filtration time. Equation 9 is the principal design relation, which may be integrated for a constant-rate process. The derivative, $dV/d\tau$, may be replaced simply by V/τ:

$$\Delta p = \mu r_0 x_0 \left(\frac{V^2}{A^2 \tau} \right) + \mu R_f \left(\frac{V}{A \tau} \right) \tag{24}$$

The ratios in parentheses express the constant volume rate per unit filter area. Hence, Equation 24 is the relationship between time τ and pressure drop Δp. For incompressible cakes, r_0 is constant and independent of pressure. For compressible cakes, the relationship between time and pressure at constant-rate filtration is:

$$\Delta p = \mu a x_0 \Delta p^s \left(\frac{V}{A \tau} \right)^2 \tau + \mu R_f \left(\frac{V}{A \tau} \right) \tag{25}$$

Filtration experiments are typically conducted in pilot scale equipment and generally tests are conducted either at constant pressure or constant rate to determine ax_0, as well as s and R_f, for a given sludge and filter medium. Such tests provide empirical information that will enable the time required for the pressure drop to reach the desired level for a specified set of operating conditions to be determined. In the initial stages of filtration, the filter medium has no cake. Furthermore, Δp is not zero, but has a value that is a function of the resistance of the medium for a given flowrate. This initial condition can be stated as:

$$\Delta p_0 - R_f \mu \left(\frac{V}{A \tau} \right) \tag{26}$$

For an incompressible cake (where s = 0), Equation 25 takes the form:

$$\Delta p = \mu a x_0 \left(\frac{V}{A \tau} \right)^2 \tau + \mu R_f \left(\frac{V}{A \tau} \right) \tag{27}$$

As noted earlier, for thick cakes, the resistance of the filter medium may be neglected. Hence, for $R_f = 0$, Equation 25 simplifies to:

$$\Delta p^{1-s} = \mu a x_0 \left(\frac{V}{A \tau} \right)^2 \tau \tag{28}$$

An increase in pressure influences not only coefficient r_0, but the cake's porosity as well. Since the cake on the filter plate is compressed, residual liquid is squeezed out. Thus, for constant feed, the flowrate through the medium will not be stable, but will fluctuate with time.

The weight of dry solids in a cake is:

$$W = x_0 V \tag{29}$$

where x_0 = weight of solids in the cake per unit filtrate volume.

The concentration of solids in the feed sludge is expressed by weight fraction c. It is also possible to evaluate experimentally the weight ratio of wet cake to its dry content m. Hence, a unit weight of sludge contains mc of wet cake. We denote γ as the specific weight of feed sludge. This quantity contains c amount of solids; hence, the ratio of the mass of solids in the cake to the filtrate volume is:

$$x_0 = \frac{c\gamma}{1 - mc} \tag{30}$$

Thus, from the sludge concentration c and the weight of wet cake per kg of dry cake solids m, x_0 can be computed. If the suspension is dilute, then c is small; hence, product mc is small. This means that x_0 will be approximately equal to c. According to Equations 29 and 30, the weight ratio of wet to dry cake will vary. Equation 30 shows also that because x_0 depends on the product mc, at relatively moderate suspension concentrations this effect will not be great and can, therefore, be neglected. However, when filtering concentrated sludges the above will play some role; that is, at constant feed, the filtrate changes with time.

VARIABLE-RATE AND -PRESSURE FILTRATION

The dynamics of variable-rate and -pressure filtrations can be illustrated by pressure profiles that exist across the filter medium. Figure 7 shows the graphical representation of those profiles. According to this plot, the compressed force in the cake section is:

$$p = p_1 - p_{st} \tag{31}$$

where p_1 = pressure exerted on the sludge over the entire cake thickness

p_s = static pressure over the same section of cake

p corresponds to the local specific cake resistance $(r_w)_x$. At the sludge-cake interface $p_{st} = p_1$ and $p = 0$; and for the interface between the cake and filter plate $p_{st} = p_{st}$ and $p = p_1 - p'_{st}$. p'_{st} corresponds to the resistance of filter plate pf, and is expressed by:

$$\Delta p_f = \mu R_f W \tag{32}$$

where W = rate of filtration (m^3/m^2-sec).

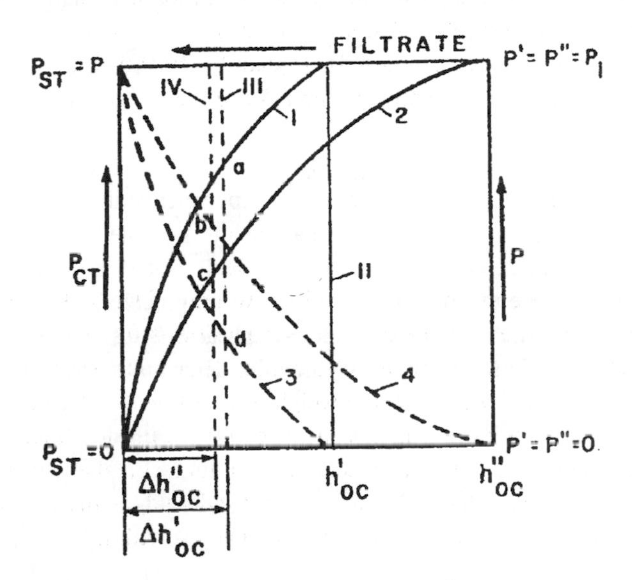

Figure 7. Distribution of static pressure p_{st} in liquid and p along the cake thickness and filter plate: I, II -boundaries between the cake and sludge at τ'' and τ'; III, IV-boundaries between cake layers or cake and filter plate at τ'' and τ'; V- boundary line between the cake and filter plate or free surface of filter plate; 1,3-curves $p_{st}=f(h_{oc})$ and $p=f(h_{oc})$ at τ'; 2, 4 -curves $p_{st}=f(h_{oc})$ and $p=f(h_{oc})$ at τ''.

Note that Δp_f is constant during the operation. Pressure p is also the driving force of the process. Therefore, starting from the governing filtration equations, the general expression for an infinitesimal increment of solid particle weight in a cake of unit of area is $x_w dq$ (q = filtrate volume obtained from 1 m^2 filtering area, m^3/m^2). The responding increment dp may be expressed as:

$$\frac{dq}{x_w dp} = \mu (r_w)_x W \tag{33}$$

x_W is not sensitive to changes in p. In practice, an average value for x_W can be assumed. Note that W is constant for any cross section of the cake. Hence, Equation 33 may he integrated over the cake thickness between the limits of p = 0 and $p = p_1 - p'_{st}$, from q = 0 to q = q:

$$q = \frac{1}{\mu x_w W} \int_0^{p_1 - p'_{st}} \frac{dp}{(r_w)_x} \qquad (34)$$

Parameters q and W are variables when filtration conditions change. Coefficient $(r_W)_X$ is a function of pressure:

$$(r_w)_x = f(p) \qquad (35)$$

The exact relationship can be derived from experiments in a device called a compression-permeability cell. Once this relationship is defined, the integral of the right side of Equation 34 may be evaluated analytically (or if the relationship is in the form of a curve, the evaluation may be made graphically). The interrelation between W and P_i is established by the pump characteristics, which define q = f(W) in Equation 34. Filtration time may then be determined from the following definition:

$$\frac{dq}{d\tau} = W \qquad (36)$$

Hence,

$$\tau = \int_0^q \frac{dq}{W} \qquad (37)$$

CONSTANT-PRESSURE AND -RATE FILTRATION

This mode of operation is achieved when a pure liquid is filtered through a cake of constant thickness at a constant pressure difference. Cake washing by displacement when the washing liquid is located over the cake may be considered to be filtration of washing liquid through a constant cake thickness at constant pressure and flowrate. The rate of washing is related to the rate of filtration during the last stages and may be expressed by Equation 9, where Δp is the pressure at the final moment and V is the filtrate volume obtained during filtration, regardless of the filtration method used (i.e., constant-pressure or constant-rate operation). In the final stages, filtration usually is performed under constant pressure. Then, the rate of this process may be calculated from Equation 19. From filtration constants C and K, at constant pressure for a given system, the filtration rate for the last period is determined. If the washing liquid passes through the filter in the same pore paths as the sludge and filtrate, then the difference between the washing rate and filtration rate for this last period will be mostly due to a difference in the viscosities of the wash liquor and filtrate. Therefore, Equation 19 is applicable using the viscosity of the washing liquid, μ_w. Denoting the rate of filtration in the last period as $(dV/d\tau)$, the washing rate is:

$$\left(\frac{dV}{d\tau}\right)_w = \left(\frac{dW}{d\tau}\right)_f \frac{\mu}{\mu_w} \tag{38}$$

FILTER-MEDIUM FILTRATION FORMULAS

Solid particles undergoing separation from the mother suspension may be captured both on the surface of the filter medium and within the inner pore passages. This phenomenon is typical in the separation of low-concentration suspensions, where the suspension consists of viscous liquids such as sugar liquors, textile solutions or transformer oils, with fine particles dispersed throughout. The penetration of solid particles into the filter medium increases the flow resistance until eventually the filtration cycle can no longer proceed at practical throughputs and the medium must be replaced. In this section standard filtration formulas are provided along with discussions aimed at providing a working knowledge of the filter-medium filtration process.

CONSTANT-PRESSURE DROP FILTRATION

Constant-pressure drop filtration can result in saturation or blockage of the filter medium. The network of pores within the filter medium can become blocked because of one or a combination of the following situations:

- Pores may become blocked by the lodging of single particles in the pore passage.
- Gradual blockage can occur due to the accumulation of many particles in pore passages.
- Blockage may occur during intermediate-type filtration.

Proper filter medium selection is based on understanding these mechanisms and analyzing the impact each has on the filtration process.

In the case of single-particle blockage, we first consider a 1 m^2 surface of filter medium containing N_p number of pores. The average pore radius and length are r_p and ℓ_p, respectively. For laminar flow, the Hagen-Poiseuille equation may be applied to calculate the volume of filtrate V′ passing through a pore in a unit of time:

$$V' = \frac{\pi r_p^4 \Delta p}{8\mu \ell_p} \tag{39}$$

Consequently, the initial filtration rate per unit area of filtration is:

$$W_{in} = V' N_p \tag{40}$$

Consider 1 m^3 of suspension containing n number of suspended particles. If the suspension concentration is low, we may assume the volume of suspension and filtrate to be the same. Hence, after recovering a volume q of filtrate, the number of blocked pores will be nq, and the number unblocked will be (N_p - nq). Then the rate of filtration is:

$$W = V' (N_p - nq) \tag{41}$$

or

$$W = W_{in} - k' q \tag{42}$$

where

$$k' = V' n \tag{43}$$

k′ is a constant having units of sec^{-1}. It characterizes the decrease in intensity of the filtration rate as a function of the filtrate volume. For constant V′, this decrease depends only on the particle number n per unit volume of suspension. The total resistance R may be characterized by the reciprocal of the filtration rate. Thus, W in Equation 42 may be replaced by 1/R (sec/m). Taking the derivative of the modified version of Equation 42 with respect to q, we obtain:

$$\frac{dR}{dq} = \frac{k'}{(W_{in} - k'q)^2} \tag{44}$$

Comparison with Equation 42 reveals:

$$\frac{dR}{dq} = \frac{k'}{W^2} \tag{45}$$

or

$$\frac{dR}{dq} = k'R^2 \tag{46}$$

Equation 46 states that when complete pore blockage occurs, the intensity of the increase in the total resistance with increasing filtrate volume is proportional to the square of the flow resistance.
In the case of multiparticle blockage, as the suspension flows through the medium, the capillary walls of the pores are gradually covered by a uniform layer of particles. This particle layer continues to build up due to mechanical impaction, particle interception and physical adsorption of particles. As the process continues, the available flow area of the pores decreases. Denoting x_0 as the ratio of accumulated cake on the inside pore walls to the volume of filtrate recovered, and applying the Hagen-Poiseuille equation, the rate of filtration (per unit area of filter medium) at the start of the process is:

$$W_{in} = BN_p r_p^4 \tag{47}$$

where

$$B = \frac{\pi \Delta p}{8 \mu \ell_p} \tag{48}$$

When the average pore radius decreases to r, the rate of filtration becomes:

$$W = BN_p r^4 \tag{49}$$

For a finite filtrate quantity, dq the amount of cake inside the pores is $x_o dq$, and the cake thickness is dr. That is:

$$x_o dq = -N_p 2\pi r \ell_p dr \qquad (50)$$

Note that the negative sign indicates that as q increases, the pore radius r decreases. Integrating this expression over the limits of 0 to q, for r_p to r we obtain:

$$q = \frac{N_p \pi \ell_p}{x_o}(r_p^2 - r^2) \qquad (51)$$

And from Equations 47 and 49, we may define the pore radii as follows:

$$r_p^2 = \left(\frac{W_{in}}{BN_p}\right)^{1/2} \qquad (52)$$

or simply:

$$r^2 = \left(\frac{W}{BN_p}\right)^{1/2} \qquad (53)$$

Substituting these quantities into Equation 51 and simplifying terms, we obtain:

$$W = [(W_{in})^{1/2} - Cq]^2 \qquad (54)$$

where

$$C = \frac{x_o}{\pi \ell_p}\left(\frac{B}{N_p}\right)^{1/2} \qquad (55)$$

It is convenient to define the following constant:

$$K = \frac{2C}{(W_{in})^{1/2}} \qquad (56)$$

From which Equation 54 may be restated as:

$$W = W_{in}(1 - 1/2Kq)^2 \tag{57}$$

Since W=1/R, we may write:

$$R = \frac{1}{W_{in}(1 - 1/2Kq)^2} \tag{58}$$

The derivative of this expression with respect to q is:

$$\frac{dR}{dq} = \frac{k}{W_{in}(1 - 1/2Kq)^3} \tag{59}$$

On some rearranging of terms, we obtain:

$$\frac{dR}{dq} = K(W_{in})^{1/2} R^{3/2} \tag{60}$$

or

$$\frac{dR}{dq} = K'' R^{3/2} \tag{61}$$

where

$$K'' = K(W_{in})^{1/2} \tag{62}$$

Equation 61 states that the intensity of increase in total resistance with increasing filtrate amount is proportional to resistance to the 3/2 power. In this case, the total resistance increases less sensitively than in the case of total pore blockage.

As follows from Equations 56 and 62:

$$K'' = 2C \tag{63}$$

Substituting Equation 55 for C and using Equation 48 for B, the above expression becomes:

$$K'' = 2(W_{in})^{1/2}\left(\frac{x_o}{N_p \pi r_p \ell_p}\right) \tag{64}$$

Note that for constant W_{in}, parameter K'' is proportional to the ratio of the settled volume of cake in the pores to the filtrate volume obtained, and is inversely proportional to total pore volume for a unit area of filter medium.

Replacing W by dq/dt in Equation 57, we obtain:

$$d\tau = \frac{1}{W_{in}}\left(1 - \frac{1}{2}Kq\right)^{-2} dq \tag{65}$$

Integration of this equation over the limits from 0 to τ for 0 to q we obtain:

$$\tau = \frac{2q}{W_{in}(2 - Kq)} \tag{66}$$

and on simplification:

$$\frac{K}{2}\tau = \frac{\tau}{q} - \frac{1}{W_{in}} \tag{67}$$

Equation 67 may be used to evaluate constants K (m^{-1}) and W_{in}.

Finally, for the case of intermediate filtration, the intensity of increase in total resistance with increasing filtrate volume is less than that occurring in the case of gradual pore blocking, but greater than that occurring with cake filtration. It may be assumed that the intensity of increase in total resistance is directly proportional to this resistance:

$$\frac{dR}{dq} = K'''R \tag{68}$$

Integration of this expression between the limits of 0 to q, from R_f to R gives:

$$\frac{R}{R_f} = e^{K'''q} \tag{69}$$

Substituting l/W for R and l/W_{in} for R_f, the last expression becomes:

$$\frac{W_{in}}{W} = e^{K'''q} \tag{70}$$

or

$$W = W_{in} e^{-K'''q} \tag{71}$$

Substituting dq/dτ for W_{in} Equation 71 and integrating over the limits of 0 to τ between 0 and q we obtain:

$$\tau = \frac{1}{W_{in}} \frac{e^{K'''q} - 1}{K'''} \tag{72}$$

Hence,

$$K'''\tau = \frac{e^{K'''q}}{W_{in}} - \frac{1}{W_{in}} \tag{73}$$

Accounting for Equation 70, the final form of this expression becomes:

$$\frac{1}{W} = \frac{1}{W_{in}} + K'''\tau \tag{74}$$

To compare the different mechanisms of filtration, the governing equation of filtration must be rearranged. The starting expression is:

$$\frac{dV}{A d\tau} = \frac{\Delta p}{\mu [r_o x_o (V/A) + R_r]} \tag{75}$$

Replacing V by q, and denoting the actual filtration rate (dq/dτ) as W, the governing filtration equation may be rewritten for a unit area of filtration as follows:

$$W = \frac{\Delta p}{\mu(r_o x_o q + R_f)} \tag{76}$$

At the initial moment when q = 0, the filtration rate is

$$W_{in} = \frac{\Delta p}{\mu R_f} \tag{77}$$

From Equations 76 and 77 we have:

$$W = \frac{W_{in}}{1 + K''' W_{in} q} \tag{78}$$

where

$$K''' = \frac{\mu r_o x_o}{\Delta p} \tag{79}$$

The numerator of Equation 79 characterizes the cake resistance. The denominator contains information on the driving force of the operation. Constant K''' (sec/m^2) characterizes tile intensity at which the filtration rate decreases as a function of increasing filtrate volume.

Substituting l/R for W in Equation 78 and taking the derivative with respect to q, we obtain:

$$\frac{dR}{dq} = K''' \tag{80}$$

The expression states that the intensity of increase in total resistance for cake filtration is constant with increasing filtrate volume. Replacing W by dq/dτ in Equation 78 and integrating over the limits of 0 to q between 0 and τ we obtain:

$$\frac{K'''}{2} q = \frac{\tau}{q} - \frac{1}{W_{in}} \tag{81}$$

Note that this expression reduces to Equation 74 on substituting expressions for W_{in} (Equation 77) and K''' (Equation 79).

Examination of Equations 46, 61, 68 and 80 reveals that the intensity of increase in total resistance with increasing filtrate volume decreases as the filtration process proceeds from total to gradual pore blocking, to intermediate type filtration and finally to cake filtration. Total resistance consists of a portion contributed by the filter medium plus any additional resistance. The source of the additional resistance is established by the type of filtration. For total pore blockage filtration, it is established by solids plugging the pores; during gradual pore blockage filtration, by solid particles retained in pores; and during cake filtration, by particles retained on the surface of the filter medium.

The governing equations (Equations 42, 67, 74 and 81) describing the filtration mechanisms are expressed as linear relationships with parameters conveniently grouped into constants that are functions of the specific operating conditions. The exact form of the linear functional relationships depends on the filtration mechanism. Table 1 lists the coordinate systems that will provide linear plots of filtration data depending on the controlling mechanism.

In evaluating the process mechanism (assuming that one dominates) filtration data may be massaged graphically to ascertain the most appropriate linear fit and, hence, the type of filtration mechanism controlling the process, according to Table 1. If, for example, a linear regression of the filtration data shows that $q = f(\tau/q)$ is the best linear correlation, then cake filtration is the controlling mechanism. The four basic equations are by no means the only relationships that describe the filtration mechanisms.

Table 1. Coordinates for representing linear filtration relationships.

Type of Filtration	Equation	Coordinates
With Total Pore Blocking	42	q vs W
With Gradual Pore Blocking	67	τ vs τ/q
Intermediate	74	τ vs $1/W$
Cake	81	q vs τ/q

All the mechanisms of filtration encountered in practice have the functional form:

$$\frac{dR}{dq} = KR^{b} \tag{82}$$

where b typically varies between 0 and 2.

CONSTANT RATE FILTRATION

Filtration with gradual pore blocking is most frequently encountered in industrial practice. This process is typically studied under the operating mode of constant rate. We shall assume a unit area of medium which has N_p pores, whose average radius and length are r_p and ℓ_p, respectively. The pore walls have a uniform layer of particles that build up with time and decrease the pore passage flow area. Filtration must be performed in this case with an increasing pressure difference to compensate for the rise in flow resistance due to pore blockage. If the pores are blocked by a compressible cake, a gradual decrease in porosity occurs, accompanied by an increase in the specific resistance of the deposited particles and a decrease in the ratio of cake-to-filtrate volumes. The influence of particle compressibility on the controlling mechanism may be neglected. The reason for this is that the liquid phase primarily flows through the available flow area in the pores, bypassing deposited solids. Thus, the ratio of cake volume to filtrate volume (x_o) is not sensitive to the pressure difference even for highly compressible cakes. From the Hagen-Poiseuille relation (Equation 39) replacing W_{in} in Equation 40 with constant filtration rate W and substituting ΔP_{in} for constant pressure drop ΔP we obtain:

$$W = B'\Delta P_{in} N_p r_p^4 \quad (83)$$

where

$$B' = \frac{\pi}{8\mu\ell_p} \quad (84)$$

The mass of particles deposited on the pore walls will be $x_o dq$, and the thickness of this particle layer in each pore is dr. Hence

$$x_o dq = -N_p 2\pi r \ell_p d\tau \quad (85)$$

Integration over the limits of 0, q from r_p to r yields

$$q = \frac{N_p \pi \ell_p}{x_o}(r_p^2 - r^2) \quad (86)$$

Radii r_p and r are defined by Equations 83 and 85, respectively, from which we obtain the following expressions:

$$q = \frac{N_p \pi \ell_p}{x_o}\left[\left(\frac{W}{B' \Delta p_{in} N_p}\right)^{1/2} - \left(\frac{W}{B' \Delta p N_p}\right)\right] \quad (87)$$

or

$$q = \frac{\pi \ell_p}{x_o}\left(\frac{W N_p}{B'}\right)^{1/2}\left[\left(\frac{1}{\Delta p_{in}}\right)^{1/2} - \left(\frac{1}{\Delta p}\right)^{1/2}\right] \quad (88)$$

Since q = Wτ, Equation 88 may be stated in a reduced form as:

$$C\tau = \left(\frac{1}{\Delta p_{in}}\right)^{1/2} - \left(\frac{1}{\Delta p}\right)^{1/2} \quad (89)$$

where

$$C = \frac{x_o}{\pi \ell_p}\left(\frac{W B'}{N_p}\right)^{1/2} \quad (90)$$

A plot of Equation 89 on the coordinate of τ vs $(1/\Delta p_{in})^{½} - (1/\Delta p)^{½}$ results in a straight line, passing through the origin, with a slope equal to C. Thus, if experimental data correlate using such coordinates, the process is gradual pore blocking. Note that at τ = 0, $\Delta p = \Delta p_{in}$, which is in agreement with typical process observations.

The filtration time corresponding to total pore blockage, when $\Delta p \rightarrow \infty$ may be estimated from:

$$\tau = \frac{1}{C}\left(\frac{1}{\Delta p_{in}}\right)^{1/2} \quad (91)$$

To express the relationship between ΔP and τ more directly. Equation 89 is restated in the form:

$$\Delta p = \frac{1}{(A - C\tau)^2} \quad (92)$$

where

$$A = \left(\frac{1}{\Delta p_{in}} \right)^{1/2} \tag{93}$$

It is important to note that pore blocking occurs when suspensions have the following characteristics:

1. relatively small particles;
2. high viscosity; and
3. low solids concentrations.

Both particle size and the liquid viscosity affect the rate of particle settling. The rate of settling due to gravitational force decreases with decreasing particle size and increasing viscosity. The process mechanisms are sensitive to the relative rates of filtration and gravity sedimentation.

Examination of the manner in which particles accumulate onto a horizontal filter medium assists in understanding the influences that the particle settling velocity and particle concentration have on the controlling mechanisms. "Dead zones" exist on the filter medium surface between adjacent pores. In these zones, particle settling onto the medium surface prevails. After sufficient particle accumulation, solids begin to move under the influence of fluid jets in the direction of pore entrances. This leads to favorable conditions for bridging. The conditions for bridge formation become more favorable as the ratio of particle settling to filtration rate increases. An increase in the suspension's particle concentration also enhances accumulation in "dead zones" with subsequent bridging. Hence, both high particle settling velocity increases and higher solids concentrations create favorable conditions for cake filtration. In contrast, low settling velocity and concentration results in favorable conditions for gradual pore blocking.

The transition from pore-blocked filtration to more favorable cake filtration can therefore be achieved with a suspension of low settling particles by initially feeding it to the filter medium at a low rate for a time period sufficient to allow surface accumulation. This is essentially the practice that is performed with filter aids.

CAKE FILTRATION EQUIPMENT

There are three general categories of filters used. These are:

Clarifying Filters - Usually cartridge-type or bag filters designed to remove small amounts of particles from a solution. Laboratory personnel typically use these types of filters. These types of filters are also used for 0.2 micron terminal sterilization.

Crossflow Filters - These are usually membrane-type filters used for ultrafiltration. In the field of biotechnology these types of filters are used in ultrafiltration devices used in concentrating solutions, and performing buffer exchanges.

Cake Filters - These are filters that are used to remove large amounts of solids from a slurry solution. They would normally be seen in biotechnology in the primary clarification of fermentation batches and in a variety of solids removal steps seen in the production of drugs via organic synthesis.

The first two categories, clarifying and crossflow filters, have been very well developed and optimized for use in biotechnology and standard wastewater treatment applications. Equipment is easily available for these applications, whether as small 0.2 micron sterilizing filter used to terminally sterilize 100 ml of product solution, or a small 500 ml crossflow filter used to concentrate a small amount of antibody solution. Many vendors of this equipment to wastewater treatment applications have their origins in the CPI (Chemical Process Industries), and have incorporated many of the scale-up and optimization properties developed in much larger units used in large scale chemical production. As a result, these two filtration unit operations are one of the most optimized and efficient used in wastewater treatment.

The third category, cake filters, although well developed in many wastewater treatment applications, are the least developed of the filtration equipment use by the Biotech Industry. In the organic synthesis laboratory sometimes very simple equipment like a funnel and filter paper is used to accomplish this operation. Some other operations used for this filtration step in the lab are more sophisticated, but many are very labor intensive and limit the capacity of the overall production process itself. As a result, there is a need for optimization of the cake filtration equipment used in biotechnology. Cake filtration equipment is available in batch and continuous modes. Following are several examples of cake filtration units:

Batch or Semi-Batch Equipment

- Plate-and-Frame Filter Press
- Pressure Leaf Filter
- Agitated/Monoplate Nutsche Filter/Dryer
- Horizontal Plate Filter/Dryer

Continuous Equipment

- Rotary Drum Filter
- Centrifugal Filter
- Horizontal Belt Filter

THE PLATE-AND-FRAME FILTER PRESS

This type of filter allows pressurized filtration of a slurry mixture to remove solids. A set of filter plates is sandwiched together in series with a configuration similar to the plates on a plate-and-frame heat exchanger. After the plates are compressed together hydraulically, a channel throughout all the plates allows the slurry to be pumped into one side of each of the plates. The filtrate goes through the filter media on the plate leaving the solid cake on the media. Filtrate on the other side of the filter plate enters a channel, joins the filtrate coming off the other plates in the filter, and leaves the equipment. When the filter plates are filled with solids, the solids can be washed and the press opened to discharge the solids off each plate. The filter press consists of a head and follower that contain in between a pack of vertical rectangular plates that are supported by side or overhead beams. The head serves as a fixed end to which the feed and filtrate pipes are connected and the follower moves along the beams and presses the plates together during the filtration cycle by a hydraulic or mechanical mechanism. Each plate is dressed with filter cloth on both sides and, once pressed together, they form a series of chambers that depend on the number of plates. The plates have generally a centered feed port that passes through the entire length of the filter press so that all the chambers of the plate pack are connected together. Likewise, four corner ports connect all the plates and collect the mother and wash filtrates in a "closed discharge" towards outlets that are located on the same side as the feed inlet. Some filter presses have plates that are fitted with cocks at their lower side so that the filtrate flows in an "open discharge" to a trough and serve as "tell tales" on the condition of the filter cloth by the clarity of the filtrate that passes through each chamber. The disadvantage of this arrangement is that it cannot be used with filtrates that are toxic, flammable or volatile. A typical flow scheme is illustrated in Figure 8.

Filterpresses were introduced nearly 100 years ago and have been applied extensively in dewatering waste sludges. They were considered labor intensive machines hence they did not find much acceptance in the sophisticated and highly automated process industries. It was not until sometime in the 1960's that this image changed by the introduction of advanced mechanisms that were oriented towards obtaining low moisture cakes that discharge automatically and enable the washing of the cloth at the end of the filtration cycle.

Often special measures are taken to ease cake discharge and enhance filtration. The measures taken include precoating and the addition of body aid. Precoating the plates prior to introducing the feed is done only in the following cases:

- When the contaminants are gelatinous and sticky it forms a barrier that avoids cloth blinding. Likewise the interface between the precoat and the cloth departs readily so the cake discharges leaving a clean cloth.
- When a clear filtrate is required immediately after the filtration cycle commences otherwise recirculation must be employed until a clear filtrate is obtained.

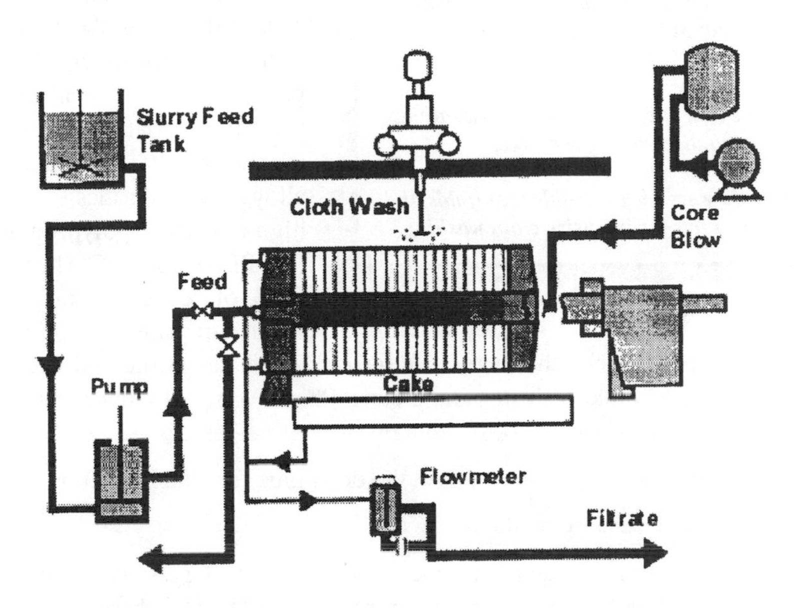

Figure 8. Process flow scheme for plate-and-frame filter press.

Once the precoating stage is completed the process slurry is pumped into the filter, the forming cake is retained on the plates and the filtrate flows to further processing. When the solids are fine and slow to filter a body-aid is added to the feed slurry in order to enhance cake permeability. However, it should be kept in mind that the addition of body-aid increases the solids concentration in the feed so it occupies additional volume between the plates and increases the amount of cake for disposal. Likewise, for all those applications when the cake is the product, precoat and filter-aid may not be used since they mix and discharge together with the cake.

For many years the plate-and-frame systems have used flush plates with separate frames to contain the cake. These filter presses had many sealing surfaces which were the main cause for leakages. As a result *recessed plates* were introduced in order to cut the number of surfaces in half and reduced the problem of drippings. The development of recessed plates has gone hand in hand with advances in cloth technology which enabled three-dimensional stretching as opposed to simple plate-and-frame where the cloth remains in one plain. Present recess depths are 16, 20 and 25 mm so the corresponding cake thicknesses are 32, 40 and 50 mm at

MAIN FEATURES OF A MODERN-DAY FILTER PRESS INCLUDE:

➡ ***Shuttle shifters that separate the plates one by one for cake discharge at a rate of 5-6 seconds per plate. A special design of the shifting mechanism ensures that two adjacent plates are not pulled together due to sticky cakes.***

➡***Shakers that subject the plate to vibrations and assist in discharging the cake.***

➡***Cloth showers with movable manifolds and high impact jets for intensive cloth washing.***

maximum filling. Filter presses are built for operating pressures of 7, 10 and 15 bar for cake squeezing and the largest available plates are 2 by 2 meters so the hydraulic pressure system that holds the closing force of the plates is designed accordingly. Filter press plates are available in various materials of construction such as cast iron, aluminum alloys, and plastics such as high-density polypropylene (HDPE) and PVDF. The major area of development, apart from automation, was in the design of the plates since thermoplastics have enabled new structural concepts which were not possible with metallic plates. Plastic composite plates have the following unique features:

- Lower plate weight has reduced the downtime for shuttle shifting during the cake discharge mode.
- Effective filtration area has gone up since with the largest available plates of 2 by 2 meters, having a 20 mm recess and 150 chambers, the area is about 1000 m^2 with a cake capacity of 20 m^3.
- The introduction of water, or air to a lesser extent, from the backside of flexible membranes reduces chamber volume and squeezes the cake yielding a further lowering of the moisture content. The filter press may be arranged as a mixed pack of flush and membrane plates, full flush or full membrane pack depending on the application.

Most plates are extruded in polypropylene which withstands temperatures of 80 to 85° C. Operating at higher temperatures will warp the plates and leakage or even squirts can be dangerous at such high temperatures.

Careful consideration to selection and sizing this equipment is required when dealing with any of the following cases: (1) When filtering saturated brines since the plates cool-off during cake discharge and require preheating prior to feeding the process slurry. For such brines autoclaved filters such as Horizontal Plates, Vertical Leaf or Candle Filters are better suited as they can be steam jacketed; (2) When there is a risk of environmental hazard from toxic, flammable or volatile cakes when the plates are opened for discharge at the end of each cycle. Again, the autoclaved filters are better suited; (3) When efficient washing is required since with a chamber filled with cake the wash water may not reach all its surface causing an uneven displacement. This, however, should present no problem when a gap is left between the formed cakes within a chamber so that the wash water is distributed evenly over the cake and reaches its entire surface. Filterpress without membrane plates have the following operational sequence (refer to Figure 9):

- Slurry is pumped and fills the chambers at a high flow rate and low pressure which gradually builds-up as the cake gets thicker. The drip trays which are positioned below the filterpress for the collection of drippings closed.
- When pressure reaches 6-7 bars wash water is pumped through the filter cake at a predetermined wash ratio to displace the adhering mother solution.
- Air blowing is applied to reduce cake moisture.
- The wet core that remains in the feed port is blown back with air for 20-30 seconds to ensure that the discharged cake is completely dry.
- The drip trays open and are ready for cake discharge.
- The hydraulic plate closing piston retracts together with the follower.
- The shuttle shifter moves the plates one by one towards the follower and the cake discharges.
- The drip trays close and are ready for the next cycle.

IT MAKES SENSE TO SELECT A FILTER PRESS

☞ **When a very low moisture content is required for thermal cake drying or incineration.**

☞ **When high filtrate clarity is required for polishing applications.**

☞ **When good cake release assisted by squeezing is required.**

☞ **When the cake is disposed as land fill for spreading with a bulldozer provided it is hard enough to carry its weight.**

☞ **When large filtration areas are required in a limited space.**

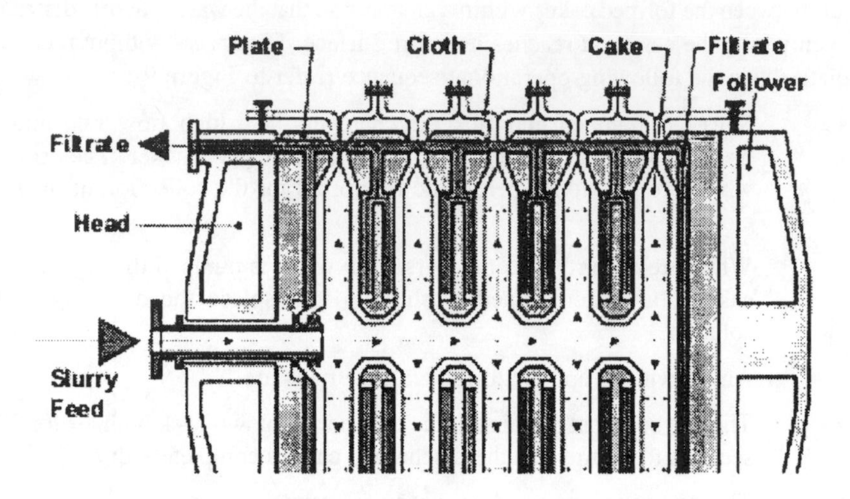

Figure 9. Feature of the operating press.

The shuttle shifter moves the plates back one by one towards the fixed header. When each plate parks the cloth is washed at 100 bar with a mechanism that lowers and lifts a pair of symmetrical manifolds with high impact nozzles.

A filter press with membrane plates has the following operational sequence:

- Slurry is pumped and fills the chambers at a high flow rate and low pressure which gradually builds-up as the cake gets thicker. The drip trays are closed.
- The membranes, of empty chamber type plates, are pressed back to allow cake formation.
- When pressure reaches 6-7 bars the cake is presqueezed for even distribution by pumping water to the backside of the membranes.
- Wash water is pumped through the filter cake at a predetermined wash ratio to displace the adhering mother solution.
- Air blowing is applied to reduce cake moisture.
- More water is pumped to the backside of the membranes for final squeezing up to 15 bar to further reduce moisture.
- The wet core that remains in the feed port is blown back with air at 6-7 bars for 20-30 seconds to ensure that the discharged cake is completely dry.
- The drip trays open and are ready for cake discharge.
- The hydraulic plate closing piston retracts together with the follower.

- The shuttle shifter moves the plates one by one towards the follower and the cake discharges.
- The drip trays close and are ready for the next cycle.
- The shuttle shifter moves the plates back one by one towards the fixed header. When each plate parks the cloth is washed at 100 bar with a mechanism that lowers and lifts a pair of symmetrical manifolds with high impact nozzles.

Cake disposal is relatively straightforward. Cakes may be discharged into bins that are trucked away or transported with a belt conveyor. With very large filter presses a well formed cake may weigh 200-300 kg per chamber and when it falls into a bin or onto a belt conveyor in one solid piece the impact is very high. Hence, special measures are required to break and de-lump the sole hard cake and, for belt conveyors, it is also recommended to increase the number of belt support rollers below the discharge chute at the point of impact.

The filterpress by itself requires little maintenance however the automation features that accompany the machine should be checked regularly and with particular attention to safety devices such as:

- The infra-red curtain that protects the operator during the closure of the plate pack should stop the hydraulic pump within 2 milliseconds.
- The switch that warns when a loss of pressure in the hydraulic plate closing system detects leakage between the plates.
- The filtrate flow meter microswitch that stops the slurry feed pump when the chambers are full.
- The microswitch that is attached to the drip trays is interlocked so that the doors are fully open during cake discharge.
- The pressure switch that permits squeezing of the membranes only when the plate pack is compressed with the hydraulic closing system.
- The zippered bellows that protect the hydraulic piston against drippings should be checked for wear and tear.
- The two manifolds that wash the cloth on both sides of the plate have high impact nozzles at a pressure of 100 bar. When some nozzles of one manifold are plugged the jet impact is uneven and the plates tend to swing.
- The cloth must be checked for holes and the optional cocks on the filtrate port of each plate help in identifying damaged cloths.
- The impregnated edges that surround the cloth and seal between adjacent plates should be checked for leakage.

Body-aid (i.e., the filter-aid) and precoating are often mentioned in connection with pressure filtration and the difference in their application is: (1) Body aid is used when the slurry is low in solids content with fine and slimy particles that are difficult to filter. To enhance filtration coarse solids with large surface area are added to the slurry and serve as a body-aid that captures and traps in its interstices

The following materials can serve as body-aid or are used to form a precoat:

Diatomaceous Earth (also called Diatomite) consisting of silicaceous skeletal remains of tiny aquatic unicellular plants.

Perlite consisting of glassy crushed and heat-expanded rock from volcanic origin.

Cellulose consisting of fibrous light weight and ashless paper like medium.

·Special groundwood is becoming popular in recent years since it is combustible and reduces the high cost of disposal. There are nowadays manufacturers that grind, wash and classify special timber to permeabilities which can suit a wide range of applications.

the slow filtering particles and produce a porous cake matrix. The amount added depends on the nature of the solids and varies from ? for non-compressible and up to 5 times for gelatinous solids; (2) Precoating the plates with a 2-3 mm thick medium of a known permeability and its application requires skills since it takes-up effective cake volume, lengthens the cycle time and an over consumption can be quite costly. Precoating prior to filtration serves two main purposes:

- When the contaminants are gelatinous and sticky it forms a barrier that avoids cloth blinding. Likewise the interface between the precoat and the cloth parts readily so the cake discharges leaving a clean cloth.
- When a clear filtrate is required immediately after the filtration cycle commences otherwise recirculation must be employed until a clear filtrate is obtained.

THE PRESSURE LEAF FILTER

This device is similar to a plate-and-frame filter press but the whole plate assembly is also housed in a tank or pressure vessel. This design allows higher pressures to be used, and also allows the filtration operation to be done more efficiently in many applications. There are two basic configurations, namely a horizontal plate and a vertical pressure leaf filter.

Horizontal plate pressure filters were commonly applied to the fine chemical process industries such as antibiotics , pesticides or pigments when the load of impure insolubles is low and polishing is required to obtain a high product clarity. In more recent years they may be seen more and more in heavier industries such as fertilizers or precious metals when the product is the cake and efficient washing and low moistures are required. These units are well suited for handling flammable, toxic and corrosive materials since they are autoclaved and designed for hazardous environments when high pressure and safe operation are required. Likewise, they may be readily jacketed for applications whenever hot or cold temperatures are to be preserved. These features are not possible on filter presses which require the opening of plates to the atmosphere and shifting them one by one to allow cake discharge at the end of each cycle.

Normally the filter structure consists of a stack of plates attached to a hollow shaft which are mounted inside a pressure vessel with each plate covered with a suitable filter medium. The slurry is fed under pressure into the vessel and the cake, which is retained by the filter medium, forms on the top of each plate whilst the filtrate passes through the hollow shaft further to the process. Filter sizes may vary but generally the maximum is 60 m^2 area and designed for a 6 bar operating pressure. Each circular plate in the stack is constructed with radial ribs that are welded to the bottom and support a horizontal coarse mesh screen which is covered with a finer woven metal screen or filter cloth to retain the cake. The bottom of the plate slopes towards the hollow central shaft which lets the filtrate flow freely through circumferential holes and further down the shaft to the filtrate outlet. The clearance between the plates is maintained by special spacers

HORIZONTAL PLATE FILTERS ARE BEST SELECTED IN THE FOLLOWING SITUATIONS:

- **When minimum floor space for large filtration areas is required.**
- **When the liquids are volatile and may not be subjected to vacuum.**
- **When there is a risk of environmental hazard from toxic, flammable or volatile cakes specially secured discharge mechanisms may be incorporated.**
- **When high filtrate clarity is required for polishing applications.**
- **When handling saturated brines that require elevated temperatures the tank may be steam jacketed.**
- **When efficient washing is required.**
- **When the cake is heavy and must be supported as opposed to a Vertical Leaf Filter where the cake forms on a vertical surface and may fall-off once the pressure drops.**
- **When the cake may be discharged either dry or as a thickened slurry.**

with "o" rings to positively seal between the slurry that surrounds the plates and the shaft that collects the filtrate. The height of the spacers determine the clearance for cake build-up and may be replaced to meet various process conditions.
One of the differences between polishing and cake filtration is the space between the plates. For polishing applications the clearances are about 20 mm as opposed to cake filtration applications where, depending on the percentage of solids and cake build-up properties, clearances may reach 100 mm. Hence, polishing filters accommodate more plates than cake filters so for the same vessel size more effective area is available with polishers. The vessels that are employed with horizontal plate filters are, as opposed to vertical leaf filters, always constructed vertically to accommodate the plates stack. All have removable dished heads but there are two options for bottom design; namely a conical bottom and a dished bottom. The selection depends largely on the cake discharge arrangement. The head of the larger vertical vessels is often pivoted so that it is swung away to allow the upwards removal of the plates stack. The layout ususally provides sufficient headroom for raising the stack over the vessel and additional floor space next to the filter for stack maintenance and replacement of damaged plates. It is good practice to design a special rig that will support the removed stack. The vessels at their bottoms are fitted with highly secured cake discharge openings to ensure safe sealing of the tank under pressure. The concept of cake filtration, as opposed to polishing, was enabled by substantial improvements in the cake discharge mechanisms since such filters are operating on a short cycle time. There are two types of cake discharge mechanisms and both use centrifugal force to throw the cake against the cylindrical wall which then falls to the bottom of the tank:

- The rotating disc stack.
- The vibrating disc stack.

The rotating type may be driven from either the top or the bottom whilst the vibrating type is always driven from the top. The removal of the tank head cover from top driven filters is generally more complex than those driven from the bottom. On the other hand bottom driven filters are more susceptible to slurry leaks. The position of the cake outlet depends on the construction of the tank bottom. There are two types available:

- With a conical bottom and a central outlet.
- With a dished bottom and a side outlet.

Tanks with conical bottoms discharge cakes by gravity and those with dished bottoms have a spade that rakes and conveys the cake towards the outlet. Hence, the conical types require more headroom as compared to the dished type having the same filtration area. Conical tanks also have often an additional scavenging plate at the lower part of the cone to filter the residual slurry heel that remains below the main plates. The slurry heel that remains at the very bottom of the tank is removed through a special dip pipe to avoid discharging a wet cake. To facilitate better cake discharge there are designs with sloping plates. With this concept the cake, owing to the centrifugal force, flies off the plate in a horizontal trajectory without being dragged and subjected to the frictional radial shear over the surface as with conventional flat plates. The cake that accumulates on the plates may be discharged as a wet thickened sludge or as a dry cake. For wet cakes the vessel will normally have a small outlet that is fitted with a valve whilst for dry cakes the opening is large and the closure locks up electrically or hydraulically with a bayonet wedge.

ADVANTAGES of a Horizontal Pressure Leaf Filter are:

The removal of the plate stack on bottom driven filters is simpler than on top driven machines since on the later the entire drive has to be removed to allow access to the stack.

Plates with the screens mounted on the topside, as opposed to two sided plates, provide good support for the forming cake and therefore are always used on applications with thick and heavy cakes.

The operation of a horizontal plates filter is labor intensive and requires a complex manipulation of valves so present day installations are in most cases fully automated. The operational steps are as follows:

Step 1. Precoating - The precoating stage is done only in the following cases: (a) When the contaminants are gelatinous and sticky it forms a barrier that avoids cloth blinding. Likewise the interface between the precoat and the cloth departs readily so the cake discharges leaving a clean cloth; (b) When a clear filtrate is required immediately after the filtration cycle commences otherwise recirculation must be employed until a clear filtrate is obtained.

Step 2. Filtration - Once the precoating stage is completed the process slurry is pumped into the filter, the forming cake is retained on the plates and the filtrate flows to further processing. When the solids are fine and slow to filter a body-aid is added to the feed slurry in order to enhance cake permeability. However, it should be kept in mind that the addition of body-aid increases the solids concentration in the feed so it occupies additional volume between the plates and increases the amount of cake for disposal. Likewise, for all those applications when the cake is the product, precoat and filter-aid may not be used since they mix and discharge together with the cake.

Step 3. Heel Removal - Once the filtration cycle is completed it is necessary to remove the slurry heel that surrounds the plates. This is done by blowing air into the tank which displaces the slurry down to the lowest plate and further to the scavenger plate if one exists. The remaining slurry at the very bottom is recirculated through a dip pipe back to the feed tank until the entire slurry has been evacuated.

Step 4. Cake Drying - The air then continues to pass through the cake until the captive moisture is reduced to a minimum and the cake is in practical terms considered to be dry.

Step 5. Cake Discharge - At this point the air pressure is released, the cake outlet is opened and the plate stack is rotated to discharge the cake. The cake outlet opening must be interlocked with the motor drive since its spinning is conditional to the outlet being open. On some filters the cloth or mesh screen may backwashed with water after cake discharge to dislodge and remove any cake residue that adhered to the medium.

DISADVANTAGES of a Horizontal Pressure Leaf Filter are:

High headroom is required for dismantling the entire plate stack.

The bearing of top and bottom driven filters, that supports the rotating plate stack and its sealing, is complex since it has to withstand the internal pressure and the side forces imposed by the mechanical drive. However, side loads on some machines are eliminated by the use of hydraulic motors.

The horizontal plate filter requires attention on a regular basis to safety devices and automation features that accompany modern filters. The space above the filter should have a hoisting device and sufficient headroom to lift the entire disc stack and move it horizontally to a location adjacent to the filter tank. It is recommended to have a special rig that will hold the plate stack for maintenance since the bigger ones may reach a length 3 meters or more. Space must also be allocated for the cover which may be either if it is hinged or removed.

Vertical Pressure Leaf Filters are essentially the same as Horizontal Plate Filters except for the orientation of the filter elements which are vertical rather than horizontal. They are applied for the polishing slurries with very low solids content of 1-5% or for cake filtration with a solids concentration of 20-25%. As with the horizontal plate filter the vertical leaf filters are also well suited for handling flammable, toxic and corrosive materials since they are autoclaved and designed for hazardous environments when high pressure and safe operation are required. Likewise, they may be readily jacketed for applications whenever hot or cold temperatures are to be preserved.The largest leaf filters in horizontal vessels have a filtration area of 300 m^2 and vertical vessels 100 m^2 both designed for an operating pressure of 6 bar.

During operation the slurry is pumped under pressure into a vessel that is fitted with a stack of vertical leaves that serve as filter elements. Each leaf has a centrally located neck at its bottom which is inserted into a manifold that collects the filtrate. The leaf is constructed with ribs on both sides to allow free flow of filtrate towards the neck and is covered with coarse mesh screens that support the finer woven metal screens or filter cloth that retain the cake. The space between the leaves may vary from 30-100 mm depending on the cake formation properties and the ability of the vacuum to hold a thick and heavy cake to the vertical leaf surface.

The two types of vessel geometries employed are vertical and horizontal. In most of the fine chemicals processes the leaves are fitted into vertical vessels whereas horizontal vessels are used in the heavier process industries such as the preparation of sulfur in phosphoric acid plants. The leaves inside horizontal tanks may be positioned either along the tank axis or perpendicular to the axis. In order to utilize the tank volume for maximum filtration area the width of the leaves is graduated so they fit to the circular contour of the tank. This also reduces the slurry heel volume that surrounds the leaves. The vessels are fitted with highly secured cake discharge openings to ensure safe sealing of the tank under pressure.

Selection Criteria

Select a Vertical Leaf Filter for an application when

Minimum floor space for large filtration areas is required.

Liquids are volatile and may not be subjected to vacuum.

There is a risk of environmental hazard from toxic, flammable or volatile cakes specially secured discharge mechanisms may be incorporated.

High filtrate clarity is required for polishing applications.

Handling saturated brines that require elevated temperatures the tank may be steam jacketed.

The cake may be discharged either dry or as a thickened slurry.

The cake that accumulates on the leaves may be discharged as a wet thickened sludge or as a dry cake. For wet cakes the vessel will normally have a small outlet that is fitted with a valve whilst for dry cakes the opening is large and the closure locks up electrically or hydraulically with a bayonet wedge. The head cover of vertical vessels is often pivoted so that it is swung away to allow the upwards removal of the leaves in the stack. It is good practice to design a special rig that will support a leaf that is removed from the vessel. Special quick opening bolts are fitted around the cover so that tightness is secured during operation but enable easy opening when access to the stack is required. Figure 10 illustrates a vertical leaf filter. An advantages of the vertical leaf filter compared to the horizontal plate filter is when cakes depart easily from the filtering medium. In such cases it is not necessary to incorporate means to assist discharge since gravity will release the cake and let it drop towards the discharge opening. For such cakes that do not discharge readily a special mechanism that vibrates the entire stack is incorporated and this will in most instances release the cake. However, with this method care must be taken so that the cake does not bridge between the two adjacent plates since this will impair cake discharge. There are instances when the cake is disposed to ponds or repulped for further treatment. In these situations special oscillating high impact jet headers sweep the medium surface and sluice the cake through the discharge outlet. These headers also serve to wash the filtering medium and dislodge particles that clog the metal screen or cloth.

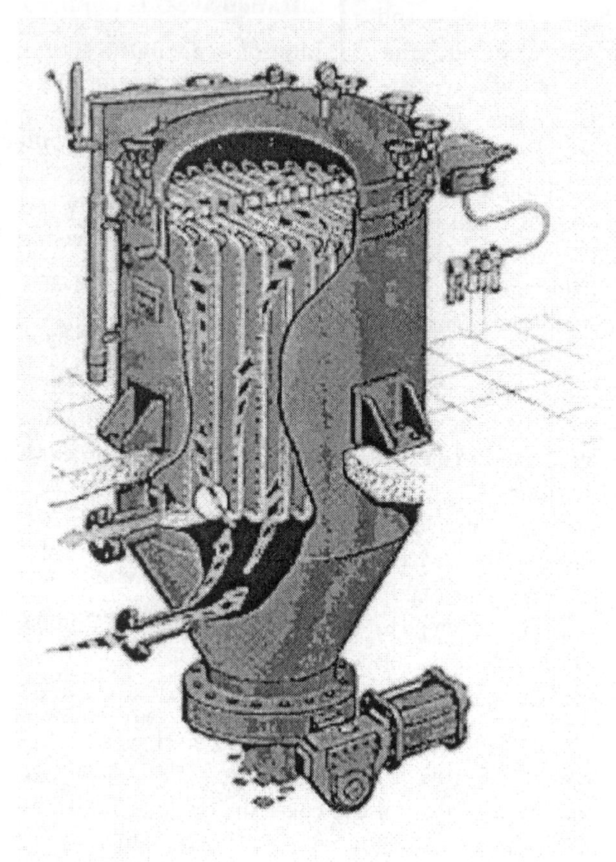

Figure 10. Vertical leaf filter machine.

The operation of a vertical pressure leaf filter is labor intensive and requires a complex manipulation of valves so present day installations are in most cases fully automated. The operational steps are as follows:

Step 1. Precoating - The precoating stage is done only in the following cases: (a) When the contaminants are gelatinous and sticky the precoat layer forms a barrier that avoids cloth blinding. Likewise the interface between the precoat and the cloth departs readily so the cake discharges leaving a clean cloth; (b) When a clear filtrate is required immediately after the filtration cycle commences otherwise recirculation must be employed until a clear filtrate is obtained.

Step 2. Filtration - Once the precoating stage is completed the process slurry is pumped into the filter, the forming cake is retained on the leaves and the filtrate flows to further processing. When the solids are fine and slow to filter a body-aid is added to the feed slurry in order to enhance cake permeability. However, it should be kept in mind that the addition of body-aid increases the solids concentration in the feed so it occupies additional volume between the leaves and increases the amount of cake for disposal. Likewise, for all those applications when the cake is the product, precoat and filter-aid may not be used since they mix and discharge together with the cake.

Step 3. Heel Removal - Once the filtration cycle is completed it is necessary to remove the slurry heel that surrounds the leaves otherwise the cake will be wet while being discharged. For this purpose a special dip pipe at the very bottom of the tank evacuates the remaining slurry heel which is recirculated back to the feed tank.

Step 4. Cake Drying - The air then continues to pass through the cake until the captive moisture is reduced to a minimum and the cake is in practical terms considered to be dry.

Step 5. Cake Discharge - At this point the air pressure is released, the cake outlet is opened and the leaf stack is vibrated to discharge the cake. The cake outlet opening must be interlocked with a pressure sensor to avoid opening under pressure. On some filters the cloth or mesh screen may be backwashed with water after cake discharge to dislodge and remove any cake residue that adhered to the medium.

The maintenance requirements on these machines is labor intensive and mirrors those of its horizontal counterpart.

THE AGITATED, MONOPLATE, NUTSCHE FILTER/DRYER

This family of filters consist of a vertical pressure vessel with a horizontal filter plate at the bottom. The filtrate from this equipment flows out a nozzle on the bottom of the filter. These devises are usually used for slurries where large amounts of solids are being collected. Variations of this equipment include equipment with removable lower heads for easy cake removal, ability to pressure or vacuum filter, ability to wash the filter cake, an agitator to break-up and rewash the filter cake, and heating or cooling jackets for the whole vessel. The Nutsche filter is the industrial version of the well known laboratory scale Buchner Funnel with the exception that it is designed to operate under either on vacuum or pressure.

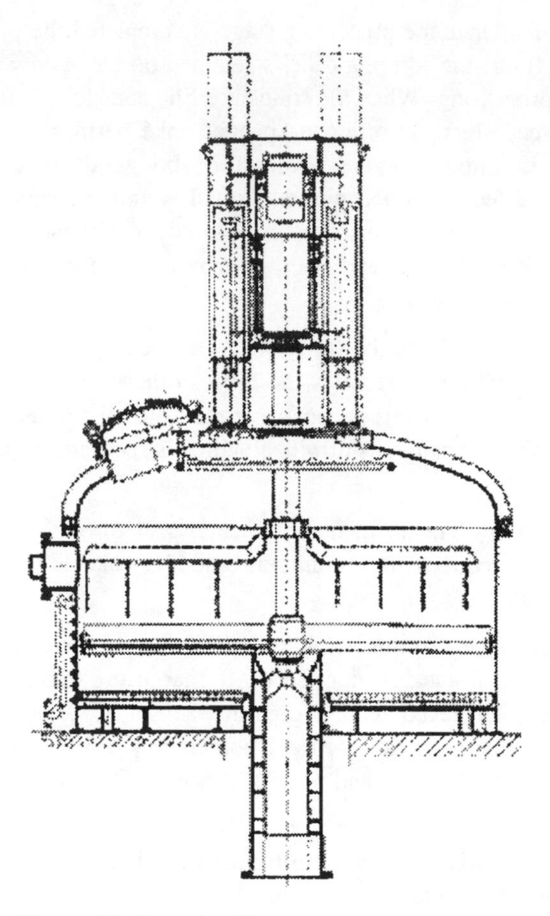

Figure 11. Nutsche filter.

Nutsche filters are constructed to perform a multitude of tasks including reaction, filtration, cake washing and thermal drying on a single unit. As such these are very sophisticated machines with tight process control on parameters such as pressure, temperature and pH.

Nutsche filters are well suited for handling flammable, toxic, corrosive and odor-noxious materials since they are autoclaved and designed for use in hazardous and ex-proof environments when extremely safe operation is required. They are available in almost any size with the larger machines for a slurry filling batch of 25 m^3 and a cake volume of 10 m^3. Such filters have a filtration area of 15 m^2 and are suitable for fast filtering slurries that produce readily 0.5 m thick cakes. The basic configuration is shown in Figure 11.

Nutsche filters are usually operated as part of a batch system , and hence the vessel's volume is designed to accept an entire charge of slurry from the upstream equipment. Therefore, so that the idle time of the filter is kept to a minimum,

The Nutsche in its full configuration consists of four major components:
- ☞ **The vessel**
- ☞ **The filter floor and cloth, woven mesh screen or sintered metal plate medium**
- ☞ **The slurry arms for cake washing and smoothing**
- ☞ **The cake discharge mechanism**

sufficient holding volume is required for fast charging and emptying of the vessel . The vessels are generally designed for an operating pressure of 2-3 bars but higher pressures may be specified if required. The vessel's cover supports the hydraulic system that controls the raising and lowering of the rotating auger and paddle arms as well as the various flanged connections and maintenance manhole. With larger units the dished head cover is bolted on its top. With this arrangement it is necessary to enter the vessel for the replacement of the filter medium. On smaller filters the head and cylinder are in one part that bolts to the filter floor featuring easy access to the filter medium. This arrangement has usually swing type fast locking bolts which are attached to the circumference of the filter floor. For applications that require preservation of temperature the vessel may be double-walled or half-pipe jacketed to provide effective heat transfer to the product. The filter floor consists of a densely perforated plate sufficiently strong to hold the cake weight and the pressure that is exerted on the cake's surface. On the larger filters additional support to the plate is given by ribs between the vessel's floor and the filter plate. There are three types of filter medium that cover the filter plate: woven metallic mesh screen, synthetic filter cloth, and sintered metallic plate. Selection depends entirely on the characteristics of the solids, the liquid and the appropriate materials of construction. Medium selection cannot be determined on a bench scale leaf test and it is only pilot testing that can establish long term effects such as medium blinding and fluctuations in the feed properties.

The unique design feature of this machine is the rake arms which are used for cake washing, smoothing and discharge of the cake. These tasks are done by paddle or paddle and auger systems: The paddle system consists of two arms with slanted blades that rotate in one direction to re-slurry the cake during washing and discharge it at end of cycle. The paddle arms are rotated in the opposite direction for cake smoothing to seal cracks prior to cake drying. In this system one hydraulic arrangement raises and lowers the paddles depending on the filtration cycle.

The auger system consists of two paddle arms that rotate in one direction to perform re-slurrying as described for the paddle system. In addition two auger arms smoothen the cake by rotating in one direction and convey the cake towards the center for discharge by rotating in the opposite direction. In this system two separate hydraulic arrangements raise and lower the paddle arms and the auger arms independent-ly depending on the filtration cycle. The cake may be conveyed for discharge towards a piped chute that is located in the center of the vessel or near the outlet located on the vessel's cylinder just above the filter plate. The operational sequence of this machine is complex, requiring a combination of manipulation of valves and hydraulics. Systems range from automatic to semi-automated. The operational sequence involves the following steps:

1. **Filtration** - The filter is charged with slurry and pressure is applied to displace the filtrate leaving the cake retained over the filter medium. For slurries with a wide distribution of coarse fast settling solids and slow settling fines there is a risk of segregation with the finer fraction settling over the coarse fraction. When this happens the fines seal the cake and slow down the cake formation so keeping the slurry in suspension with rotating arms during filtration assists in forming a homogeneous cake.

2. **Cake Washing** - In the washing stage a spray ring or connections on top of the cover introduce the wash liquid over the cake. This displaces the mother solution with the wash liquid but with such in-situ washing the efficiency may be quite low if the cake forms with an uneven thickness. One of the advantages of the Nutsche Filter is the ability to smoothen the cake's surface prior to applying spray wash so that the entire bed is washed evenly. Washing efficiency may be further improved if air or gas are not allowed to enter the cake in a multi-washing system so the wash liquids always displace the solutions in a "piston" like manner. This is achieved by a special detector that monitors the surface of the cake for moisture and once air or gas start entering the bed a signal is transmitted to close the filtrate valve and reopen it once next washing commences.

3. **Cake Repulping** - Many processes require high washing efficiency to remove the contaminating liquid from the product and washing the cake by repulping yields the most efficient product purity. This is done by resuspending the cake with the paddle arms for thorough mixing with the wash solution. During resuspension the rotating arms are moving slowly downwards and are "shaving" the bed gradually layer after layer until the entire cake enters the slurry.

4. **Pressure Drying** - In the drying stage air or gas purges the cake until the captive moisture is reduced to an asymptotic level and in practical terms the cake is considered to be as dry as possible. To obtain minimum moisture the cake is smoothened by reversing the rotation of the paddle or auger arms and exerting controlled pressure on its surface with the hydraulic system. This seals cracks in the cake so that air or gas will not bypass the bed.

5. **Vacuum Drying** - Further reduction in cake moisture may be obtained by slowly rotating and lowering the paddle arms to scrape and delump the cake. To take advantage of the drying ability of the Nutsche Filter it is worth considering the option of heating components such as the vessel, filter floor and paddles to enhance drying.

6. **Cake Discharge** - Once all the stages are completed the cake discharge valve opens and the paddle arms on the smaller machines or the auger arms on the larger ones are rotated and lowered to convey the dry cake towards the center. The same procedure also applies to side discharge machines however it should be noted that in this case the cake comes out intermittently and not continuously. This may have a layout impact on the downstream facility such as the conveyor that handles the product to storage.

THE HORIZONTAL PLATE FILTER/DRYER

These filters are similar in configuration to a Nutsche filter, but instead of one filter plate there is a series of plates inside the vessel. The filtrate is pulled through the filter media in the center of each plate to a central pipe that discharges out the bottom of the devise. The bottom plate of the filter usually discharges its filtrate thought a separate nozzle. These filters are usually used with slurries where a smaller quantity of solids is to be collected.

These devices have the same options as the Nutsche filter described above. Some configurations have variations on the method of cake removal from manual to totally automated. Automatic discharge is usually via scrapers or centrifugal action. Variations of this type of filter are mounted in horizontal and vertical vessels. These machines were commonly applied to the fine chemical process industries in such applications as antibiotics , pesticides or pigments manufacturing when the load of impure insolubles is low and polishing is required to obtain a high product clarity. However, in recent years they may be seen more and more in heavier industries such as fertilizers or precious metals when the product is the cake and efficient washing and low moisture is required. They are very well suited for

Selection Criteria

Nutsche Filters are best selected when:

➪ Minimum floor space is required

➪ There are several unit operations upstream and downstream filtration, such as reaction and thermal drying, are required by the process flow-scheme

➪ There is a risk of environmental hazard from toxic, flammable or volatile cakes

➪ Handling saturated brines or process conditions require elevated temperatures the vessel, filter floor and paddles may be heated

➪ Reslurry washing, being more efficient than in-situ displacement washing, is required.

➪ Sharp separation between the mother and wash solutions is required

➪ The cake tends to crack smoothing avoids the wash liquid, air or gas purge from by- passing.

handling flammable, toxic and corrosive materials since they are autoclaved and designed for hazardous environments when high pressure and safe operation are required. Likewise, they may be readily jacketed for applications whenever hot or cold temperatures are to be preserved.

The filter structure consists of a stack of plates attached to a hollow shaft which are mounted inside a pressure vessel with each plate covered with a suitable filter medium. The slurry is fed under pressure into the vessel and the cake, which is retained by the filter medium, forms on the top of each plate whilst the filtrate passes through the hollow shaft further to the process. Filter sizes may vary but generally the maximum is 60 m^2 area and designed for a 6 bar operating pressure. Each circular plate in the stack is constructed with radial ribs that are welded to the bottom and support a horizontal coarse mesh screen which is covered with a finer woven metal screen or filter cloth to retain the cake. The bottom of the plate slopes towards the hollow central shaft which lets the filtrate flow freely through circumferential holes and further down the shaft to the filtrate outlet. The clearance between the plates is maintained by special spacers with "o" rings to positively seal between the slurry that surrounds the plates and the shaft that collects the filtrate. The height of the spacers determine the clearance for cake build-up and may be replaced to meet various process conditions.

One of the differences between polishing and cake filtration is the space between the plates. For polishing applications the clearances are about 20 mm as opposed to cake filtration applications where, depending on the percentage of solids and cake build-up properties, clearances may reach 100 mm. Hence, polishing filters accommodate more plates than cake filters so for the same vessel size more effective area is available with polishers.

THE ROTARY DRUM FILTER

This continuous filter is used when the solids in the slurry can be easily suspended in solution and remain there. In this filter a cylindrical drum sets horizontally half submerged in a trough holding the filtration slurry. The drum is coated with a filter media and a vacuum is pulled on the inside of it. The filtrate passes through the filter media leaving the solids on the outside of the drum. The drum is rotated continuously allowing the solids to be washed and dried before they are removed by a knife blade (doctor blade) as the drum moves past it. The cleaned filter media then is rotated through the trough to pick up more solids. An illustration of a rotary drum filter is shown in Figure 12. This machine to the bottom feed group of filtering devices and is one of the oldest filters applied to the chemical process industry. The filter consists of the following subassemblies:

- A drum that is supported by a large diameter trunnion on the valve end and a bearing on the drive end and its face is divided into circumferential sectors each forming a separate vacuum cell. The internal piping that is connected to each sector passes through the trunnion and ends up with a wear plate having ports which correspond to the number of sectors. The drum deck piping is arranged so that each sector has a leading pipe to collect the filtrate on the rising side of the drum and a trailing pipe to collect the remaining filtrate from the descending side to ensure complete evacuation prior to cake discharge. However, in some instances, only leading or trailing pipes are provided and this depends on process requirements. The drum is normally driven with a variable speed drive at cycle times of between 1 rpm 10 rom.

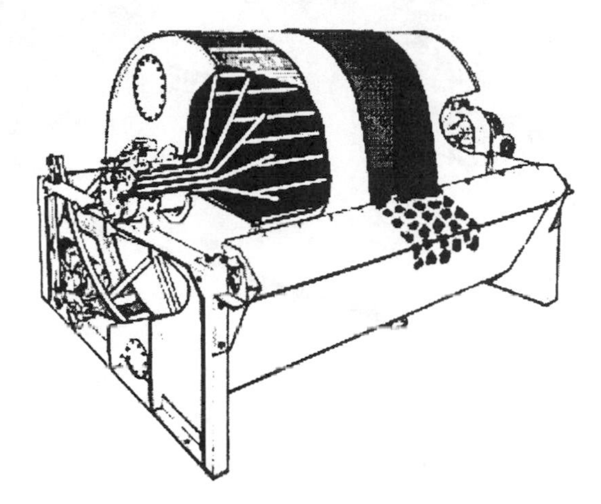

Figure 12. Rotary drum filter.

- A valve with a bridge setting which controls the sequence of the cycle so that each sector is subjected to vacuum, blow and a dead zone. When a sector enters submergence vacuum commences and continues through washing, if required, to a point that it is cut-off and blow takes place to assist in discharging the cake. The valve has on certain filters adjustable blocks and on others a fixed bridge ring. Adjustable bridge blocks enable the optimization of form to dry ratio within the filtration cycle as well as the effective submergence of the drum when the slurry level in the tank reaches a maximum.
- A cake discharge mechanism that can be either a scraper, belt, roll and in rare cases a string discharge. Blow is applied only to filters with scraper and roll discharge mechanisms but not to filters with a belt or string discharge.

The following sketches (Figures 13 through 16) illustrate key features of the operation of this machine.

Selecting the suitable mechanism largely depends on the release characteristics of the cake from the filter media. Scraper discharge mechanisms generally tend to suit cakes that release readily whereas roller discharge mechanisms are best suited for cakes that exhibit thixotropic behavior.

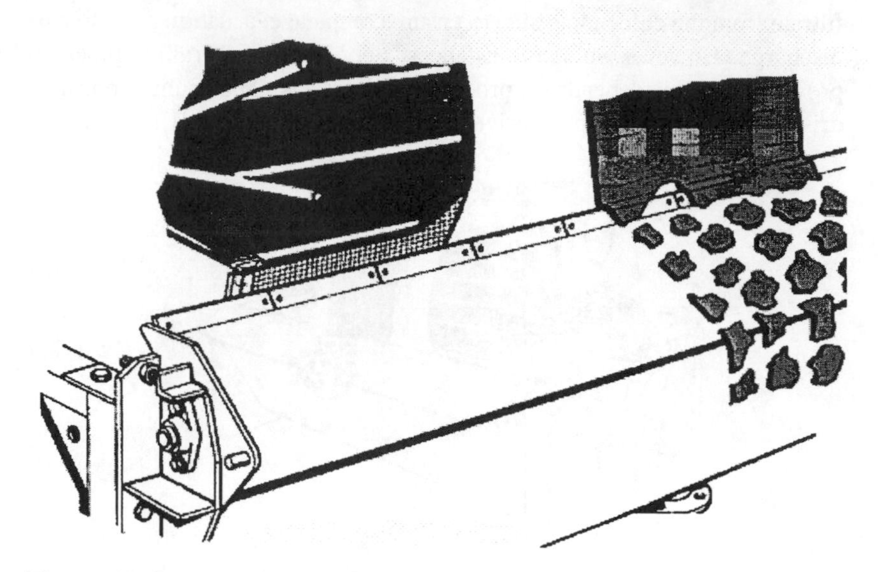

Figure 13. Details of scraper discharge.

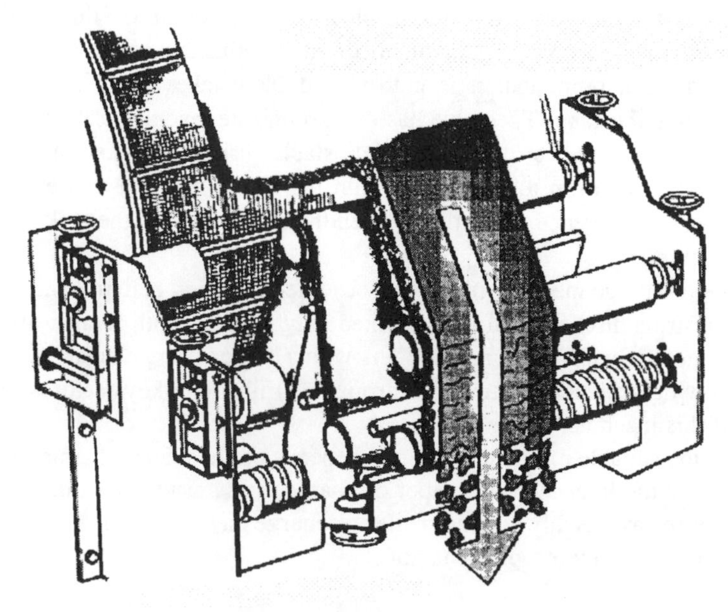

Figure 14. Details of belt discharge.

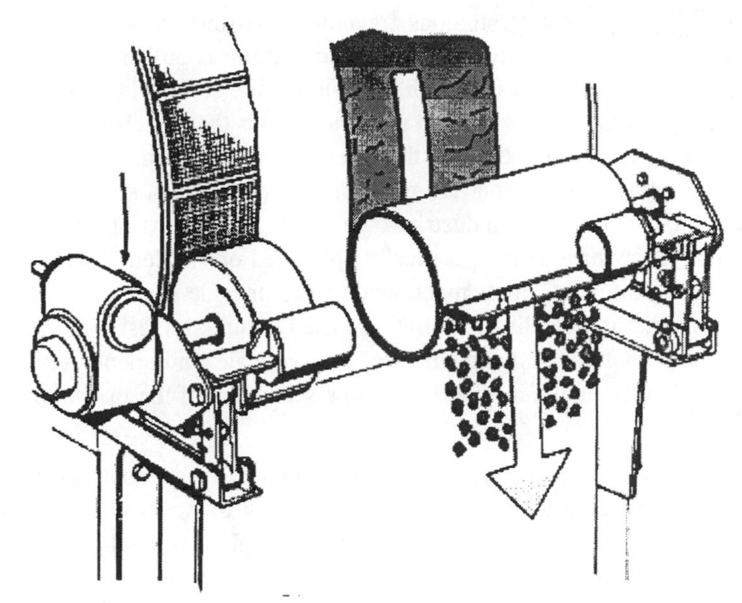

Figure 15. Details of roll discharge.

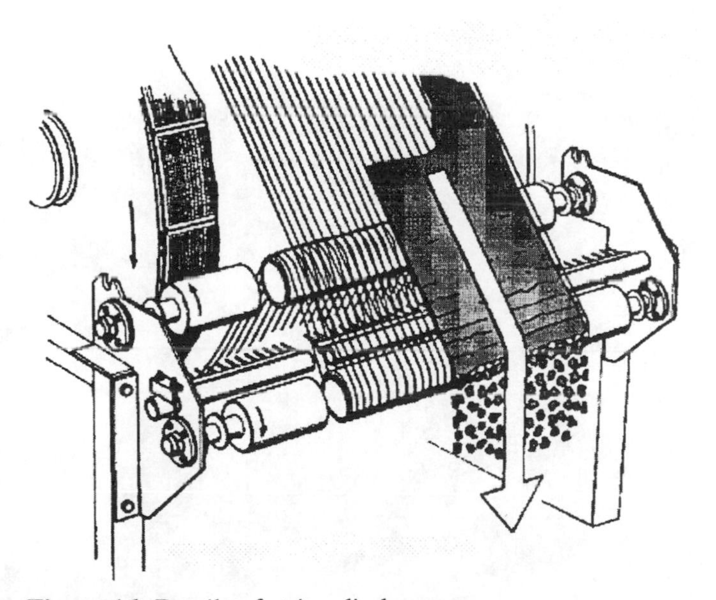

Figure 16. Details of string discharge.

THE HORIZONTAL BELT FILTER

This devise consists of a continuous traveling horizontal belt that looks like a conveyor belt. Slurry is loaded at one end and vacuum is pulled on the underside of the belt. The filtrate passes through the belt and the solid cake remains on top. As the belt moves the cake is washed and dried and is then discharged off the end of the belt as is wraps around under the machine. This equipment is usually used for slurries that have solids that are not easily suspendible in solution. Although there are many of these industrial-type cake filters in operation outside biotechnology, few have been specifically optimized or "scaled-down" use in the Biotech lab or pilot plant. Some manufacturers do provide small scale versions of their equipment, but typically these units are used to run pilot tests on a customer's slurry for the purpose of designing the proper large scale equipment. A photograph of an installation is shown in Figure 17 and a schematic of the operation is given in Figure 18.

Horizontal belt filters are perhaps the most commonly used vacuum filters in the chemical processing industry due to their flexibility of operation, adaptation to corrosive slurries and suitability to handle large throughputs. Applications to water treatment have been liomited, but still this is an important machine worth mentioning. The development of the horizontal belt filters for the chemical process industries was closely associated with the progress in rubber technology since they incorporate an endless and thick rubber belt of a complex design to support the cake retained by the filter cloth.

Figure 17. Photo of a horizontal vacuum belt filter installation.

The first known filters were the Landskrona and Lurgi built in the 1920's and the Giorgini which was a belt filter but with attached trays. The belts were very narrow and short, with a 30 cm wide by 4-5 meters length, and were primarily applied to the washing of phosphate rock.

Later, being top feed filters that facilitated multi-washing stages, they were applied in phosphoric acid plants to replace the chains of 3 or 4 internal feed rubber covered drum filters used for gypsum washing. As the demand for area has gone up filters were manufactured with three and four 30 cm wide belts running in parallel since the rubber manufacturers were unable to catch-up with the growth of the chemical plants. For this reason the main rivals over the years to belt filters were the Tilting Pan and Table Filters so when rubber belts were the constraint to filtration area growth these filters were in demand and vice versa. Belts 4 meter wide for 120 m^2 filters weigh more than 10 tons and are manufactured in one piece from sophisticated rubber compounds are common in the industry. The filter cloth retains the cake and moves together with the belt. With some exceptions, they are made from synthetic materials such polypropylene or polyester with monofilament or multifilament yarns and with sophisticated weaves. A vacuum box below the belt that is mounted along the filter and collects the filtrate through a manifold to the receivers. The box at its topside has two lips covered with low friction synthetic strip liners that seal through intermediate wear belts between the bottom side of the

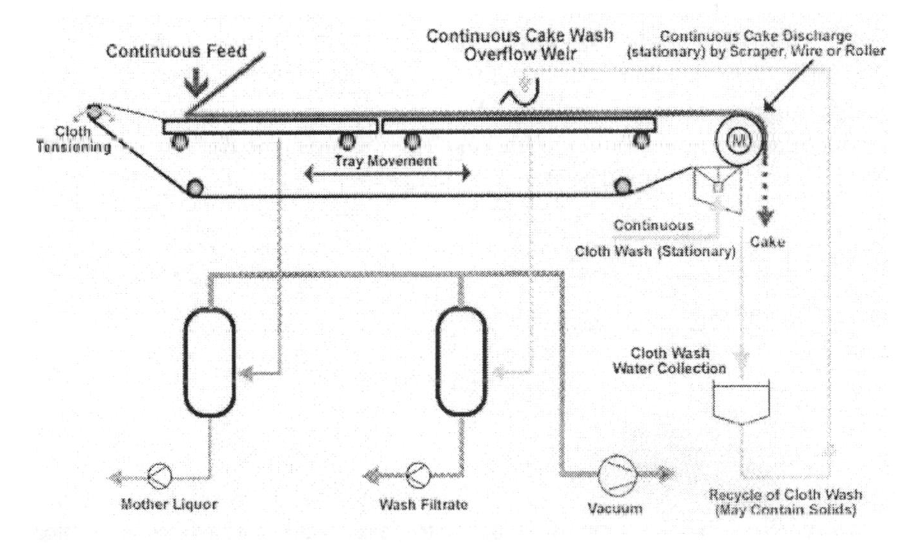

Figure 18. Schematic of operation.

belt and the surface of the strips. Since the belt is the most expensive part of the filter these endless narrow belts serve as a sacrificial component that takes the wear between the surfaces, protects the rubber belt and secures against vacuum leaks. A special mechanism allows parallel lowering or swinging of the vacuum box for

Select a Horizontal Belt Filter:

➪ ***For solids that are fast settling and cannot be kept as a homogenous slurry in bottom or side feed filters such as Drum or Disc Filters.***

➪ ***When long drying time is required to reach asymptotic moisture in the cake. On Drum Filters, for example, the ratio of dry to form cannot normally exceed 1.5 since it is determined by its geometry and the number of circumferential compartments.***

➪ ***When very short cycle times are required for fast dewatering cakes such as phosphate slurry.***

➪ ***If a clear filtrate is required right from the start it is good practice to form a thin heel that serves as filter medium over the exposed cloth. This is done by either a "cloudy port outlet" that is recirculated or, if solids are settling fast, by allocating the first 20-30 cm to act as a "sedimentation pool" prior to entering the vacuum zone.***

cleaning from fines that may have settled inside. The mechanism is designed to accurately seal between the underside of the main belt and the two narrow wear belts that move together along the slide strips attached to the top shoulders of the vacuum box. A feed box and one or more wash boxes are mounted over the filter and designed to distribute evenly the slurry and wash water across the belt. Once the belt reaches the end of the vacuum box the cake drying portion of the cycle terminates and the cloth leaves the rubber belt. The cloth continues moving, changes direction over the discharge roll and the cake drops through a chute for further handling. A deck attached to the frame and mounted underneath the belt is designed to support the heavy rubber belt and the cake load. The friction between the surfaces is reduced by injecting water for lubrication and blowing air that floats the belt or by a moving floor constructed of narrow endless belts that move together with the main rubber belt. A filtrate manifold collects the mother and wash liquids to one or more vacuum receivers. It should be kept in mind that a short path of filtrate between the vacuum box and the receivers reduces to a minimum the losses of vacuum for both the single phase flow of the mother filtrate and the two phase flow of air and wash filtrates. A pneumatic or electrical tracking mechanism controls the filter cloth from slipping sideways by guiding it to the left or to the right. There are several types of mechanisms but the following are very common: Two pairs of rolls that pinch the cloth alternatively and are positioned on both sides, and A roll is that spans across the cloth, is hinged at one end and swings forwards or backwards on the other end.

These machines are designed to meet a wide range of process requirements many of which are subjecting its components to severe and demanding conditions. Systems run at high speeds, handle thick and heavy cakes, operate at high temperatures and often in an unfriendly environment hence, they are of a sturdy design. From a maintenance standpoint, the following are areas that often require attention: (1) Cracks in the rubber belt may cause separation of the plies which are encapsulated between the rubber layers. This weakens the belt and should be repaired on site without delay; (2) The shrouds on both sides of the belt are subjected to high tension while going over the head and tail pulleys. Their duty is to contain the incoming feed and if the edges tear slurry may pour all over so inspection and their repair is essential; (3) The vacuum box is hinged and swings to one side so as to enable the periodical cleaning of its internals from settled fines. The repositioning of the box is one of the main reasons for loss of vacuum and special care must be taken to seal the box's anti-friction liners against the sacrificial wear belts and the bottom side of the main belt; (4) The endless wear belts must be inspected to ensure that they are in good condition otherwise the main belt may be damaged. Likewise, the wear belts should be checked if they seal properly between the stationary vacuum box and the moving belt; (5) The life of the belt and the main drive depend largely on the water lubrication between the surfaces of the moving and stationary parts hence, the tubes leading to those parts must be kept clean; (6) It is recommended that the alignment of the filter is inspected from time to time. This applies mainly to large filters since misalignment due to differential settling of the building foundations during the first years after start-up can cause difficulties in sealing the long and segmented vacuum box.. Also, across the filter, the thickness of the cake may taper in one direction causing uneven cake washing. The alignment across the filter is particularly important for thin cakes since a 0.5% slope on a 2 meter wide belt and a 20 mm cake reduces cake thickness on one side from 20 to 10 mm.

Select a Horizontal Belt Filter:

➪ ***When intensive cake washing is required since belt filters make it possible to apply countercurrent washing.***

➪ ***When cakes tend to crack under vacuum measures such as a flapper, compression blanket or pressure roll may assist in sealing the cracks thus avoiding loss of vacuum. When such measures are used it is necessary to make sure that the belt supporting system can take these extra vertical loads.***

➪ ***When scale formation due to flash evaporation is a problem or filtrate temperature must be maintained a vacuum box steam jacketing may be provided.***

➪ ***When the cake tends to clog the cloth its continuous removal after cake discharge enables dislodging of particles by thorough washing of the cloth on both sides with high impact nozzles.***

THE DISC FILTER

The Disc Filters belong to the side feed group of filtering machines. They are generally used in heavy duty applications such as the dewatering of iron ore, hematite, coal, aluminum hydrate, copper concentrate, pyrite flotation concentrates and other beneficiation processes. The filter consists of several discs, up to 15 in the larger machines, each made up from sectors which are clamped together to form the disc. The sectors are ribbed towards the neck and designed for a high capacity drainage rate. One of the main features is that the required floor space taken up by disc filters is minimal and the cost per m^2 of filtration area is the lowest when compared to other vacuum filters. During operation each sector enters submergence and a cake is formed on the face of the discs. It then emerges to the drying zone, the liquid drains to a central barrel and from there through a valve to the vacuum receiver. The valve with its bridge setting controls the timing so that once the sector leaves the drying zone it moves over a separating bridge and a snap or low pressure blow is applied to discharge the cake. Scraper blades on the side of each disc guide the cake to discharge chutes which are positioned between adjacent discs and are wide enough to avoid their clogging by the falling cake. A paddle type agitator located at the bottom of the tank maintains the slurry in suspension which in most of the metallurgical applications contains solids with high specific gravity which are fast settling and abrasive.

The main features of this machine include:

- Discs and sectors which may be made in injection molded polypropylene, metal or special redwood.
- A center barrel supported by the main bearings and consisting of piped or trapezoidal filtrate passages. The sectors are attached to the barrel through "o" ring sealed connections in a number equal to the number of disc sectors.
- A valve with bridges and internal compartments for form and dry under vacuum and cake discharge under pressure with 2-2.5 bar snap or 0.2-0.25 bar constant blow. Most disc filters are fitted with one valve only however two valves are often mounted on both drive and non-drive ends with long barreled filters or when the hydraulic loadings are high.
- An agitator with paddles that are positioned between the discs and far enough not to interfere with the forming cake.
- A tank which, on its discharge side, has separated slurry compartments for the discs and discharge chutes for the blown-off cake.
- Two cake discharge blades on both sides of each disc are suspended from a frame mounted on the tank. These serve to deflect and guide the cake to the discharge chutes.
- An overflow trough that spans across the entire tank length and ensures full submergence of the sectors in the cake formation zone (an exposed sector in the 6-o-clock position will cause immediate loss of pressure).

NOMENCLATURE

A = area (m^2)

B, B' = empirical parameters

C = filtration parameter

c = concentration (kg/m^3)

F_s = force (N)

h_c = cake height (m)

K, K'', K''' = filtration constants

L = cake thickness (m)

ℓ_p = pore length (m)

n = number of suspended particles

N_p = number of pores

p = pressure (N/m^2)

q = filtrate volumer per unit area of filter (m^3/m^3) or filtrate volume (m^3)

r = specific resistance (m^{-1})

r_0 = specific volumetric cake resistance (kg/m^2)

r_p = pore radius (m)

r_w = specific mass cake resistance (kg/m^2)

R = resistance (m/sec)

R_c, R_f = cake and filter resistance, respectively (m^{-1})

u = average velocity (m/sec)

V = filtrate volume (m^3)

W = mass of dry solids (kg), or rate of filtration (m^3/m^2-sec)

x_0 = ratio of cake to filtrate volume.

Greek Symbols

ϵ = porosity

μ = viscosity (P)

Π = ratio of filtration rate to gravity setting

τ = time (sec)

τ_0 = time constant (sec)

RECOMMENDED RESOURCES FOR THE READER

The following references offer some good general reading and design-specific information.

1. Cheremisinoff, P.N., Wastewater Treatment Pocket Handbook, Pudvan Publishing, Northbrook, IL, 1987.

2. Cheremisinoff, P.N., Pocket Handbook for Solid-Liquid Separations, Gulf Publishing Co., Houston, TX, 1984.

3. Noyes, R., Unit Operations in Environmental Engineering, Noyes Publishers, NJ, 1994.

4. Kirkpatrick, J., Mathematics for Water and Wastewater Treatment Plant Operators, Ann Arbor Science Pub., Ann Arbor, MI, 1976.

5. Wells, S. A. and Gordon, J. A. (1982) "Geometric Variations in Reservoir Water Quality," *Water Resources Bulletin*, Vol. 18, No. 4, 661-670

6. Adams, E. E. and Wells, S. A. (1984) "Field Measurements on Side Arms of Lake Anna, Virginia," *Journal of Hydraulic Engineering*, ASCE, Vol. 110, No. 6, 773-793

7. Adams, E. E., Wells, S. A. and Ho, E. K. (1987) "Vertical Diffusion in a Stratified Cooling Lake," *Journal of Hydraulic Engineering*, ASCE, Vol. 113, No. 3, 293-307

8. Dick, R. I., Wells, S. A. and Bierck, B. R. (1988) "A Note on the Role of Capillary Forces in Compressible Cake Filtration," *Fluid/Particle Separation Journal*, Vol. 1, No. 1, 32-34

9. Bierck, B. R., Wells, S. A., and Dick, R. I. (1988) "Compressible Cake Filtration: Monitoring Cake Formations Using X-Rays from a Synchrotron Source," *Water Pollution Control Federation Journal*, Vol. 60, No. 5, 645-650

10. Wells, S. A. and Dick, R. I. (1989) "Mathematical Modeling of Compressible Cake Filtration," *Proceedings ASCE National Environmental Engineering Conference*, Austin, TX, 788-795

11. Wells, S. A. (1990) "Effect of Winter Heat Loss on Treatment Plant Efficiency," *Research Journal of the Water Pollution Control Federation*, Vol. 62, No. 1, 34-39

12. Wells, S. A. (1990) "Determination of Sludge Properties for Modeling Compressible Cake Filtration from Specific Resistance Tests," *Proceedings A.S.C.E. National Environmental Engineering Conference*, Washington, D.C., 125-131

13. Wells, S. A. (1991) "Two-Dimensional, Steady-State Modeling of Cake Filtration in a Laterally Unconfined Domain," *Fluid/Particle Separation Journal*, Vol. 4, No. 2, June, 107-116

14. Wells, S. A. and Plaskett, J. H. (1992) "Modeling Compressible Cake Filtration with Uncertainty," in *Advances in Filtration and Separation Technology : Separation Problems and the Environment*, Volume 5, ed. by B. Scheiner, American Filtration Society, pp. 351-354

15. Wells, S. A. and Dick, R. I. (1992) "Synchrotron Radiation Measurements of Degree of Saturation in Porous Matrix," A.S.C.E., *Journal of Engineering Mechanics*, Vol. 118, No. 8, 1738-1744

16. Wells, S. A. and Dick, R. I. (1993) "Permeability, Solid and Liquid Velocity, and Effective Stress Variations in Compressible Cake Filtration," Proceedings, American Filtration Society Conference on System Approach to Separation and Filtration Process Equipment, Chicago, Illinois, May 3-6, pp. 9-12

17. Karl, J. and Wells, S. A. (1995) "Modeling of Gravity Sedimentation in One Dimension," in *Advances in Filtration and Separation*, Vol. 9, ed. by K. Choi, American Filtration and Separations Society, pp. 400-410

18. Berger, C. and Wells, S. A. (1995) "Effects of Management Strategies to Improve Water Quality in the Tualatin River, Oregon," in *Water Resources Engineering*, Vol. 2, ed. by W, Espey Jr. and P. Combs, ASCE, 1360-1364

19. A.Rushton, A.S. Ward and R.G. Holdich, Solid-Liquid Filtration and Separation Technology, 1996, VCH, Weinheim, Germany

20. R.G. Holdich, 1990, Rotary Vacuum Filter Scale-up Calculations - and the use of Computer Spreadsheets, *Filtration and Separation, 27*, pp 435-439

21. R.G. Holdich, 1994, Simulation of Compressible Cake Filtration, *Filtration and Separation, 31*, pp 825-829

22. R.G. Holdich and G. Butt, 1997, *Separation Science and Technology, 32*, pp 2129-2151

The following are Web sites to visit for additional information and calculation methods

23. Cake Filtration Equipment for Biotechnology. There are three general categories of filters used ... http://members.aol.com/jmk7/feb97.htm

24. Kongressprogramm - FILTECH EUROPA 2001... TUESDAY - 16.10.2001. 9.00 - 10.15, Plenary Session. ... TUESDAY - 16.10.2001. 9.00-10.15, Plenary Session. 10.45-12.00, L 1, Cake Filtration I, Chair: E. Vorobiev, F. ... http://www.filtecheuropa.com/programme_full_e.htm

25. Scott Wells ... processes. Research includes modeling the dynamics of cake filtration and the dynamics of liquid/particle flow in water and wastewater

clarifiers. Specific ... http://www.ce.pdx.edu/faculty/wells.htm [More Results From: www.ce.pdx.edu]

26. http://202.155.39.179/weblm/pdii-lipi/milne1/maj4.htm ... Incompressive cake filtration: Mechanism, parameters, and modeling / F Civan Page: 2379-2387 Abstract : Improved models developed verify interpretation and ... http://202.155.39.179/weblm/pdii-lipi/milne1/maj4.htm [More Results From: 202.155.39.179]
27. Filtration ... Section, Title. 1. Terminology. 2. Filtration Models.2.1. Cake Filtration: Calculation of the Pressure Drop. ... http://www.wiley-vch.de/contents/ullmann/ull_10295.html [More Results From: www.wiley-vch.de]
28. Filtration and Separation ... RVF.zip (8k): to investigate operating conditions on rotary vacuum filter throughput, assuming incompressible cake filtration; see reference 1 and 2. ... http://www-staff.lboro.ac.uk/~cgrgh/page2.htm [More Results From: www-staff.lboro.ac.uk]
29. Filtration Behaviors of Rod-shaped Bacterial Broths in Unsteady ... ... membrane surface. The permeation flux followed the cake filtration law at the initial stage of the crossflow filtration of the broths of A subtilis and E. coli ...http://www.scej.org/ronbun/JCEJe/e29p0973.html [More Results From: www.scej.org]
30. Ruech, Wolfgang: An experimental and theoretical study of ... An experimental and theoretical study of compressible cake filtration. Verfasser: Ruech ... http://www.tugraz.at/forschung/diplomarbeiten/1996/17-37.html [More Results From: www.tugraz.at]
31. Filtration Basic ... separation layer. The longer the filtration takes, the higher the separation effect, since after the initial phase the actual cake filtration will start. The ... http://www.filterpressen.de/html_e/filtration/basic.htm [More Results From: www.filterpressen.de]
32. Staff profile page - the Engineering Faculty at Loughborough ... Broad Interests and Expertise. Compressible cake filtration Selection, scale-up and process simulation of solid/liquid separation equipment Washing and ...http://www.lboro.ac.uk/departments/eng/research/staff/html/tarleton.html [More Results From: www.lboro.ac.uk]
33. Filtration and Separation ... in terms of filtrate volume with time as well as dry cake yield with filtration time. Both are outputted in tabular form suitable for importing into a ... http://www.filtration-and-separation.com/ [More Results From: www.filtration-and-separation.com]
34. Preprint 2000-041 ... which together with initial and boundary conditions determines a dynamic cake filtration
35. Process. In the case of a prescribed applied pressure function, we ...http://www.math.ntnu.no/conservation/2000/041.html [More Results From: www.math.ntnu.no]

36. DJ Mullan - Extended Abstract ... Internal Cylindrical Compressible Cake Filtration Model. To date no one has developed a model which ... http://www.und.ac.za/und/prg/posters/daveabs.html [More Results From: www.und.ac.za]
37. CHE 440: Filtration Experiment ... find the parameters (discussed below) that govern constant pressure drop cake filtration
for the filtration equipment in the Unit Operations laboratory. These ... http://chemeng.uah.edu/courses/che440sp97/filter.html [More Results From: chemeng.uah.edu]
38. AFS: Table of Contents Vol. 12 ... On Parameters Affecting Flow Behavior During Compressible Cake Filtration in the Centrifugal Field. ... http://www.afssociety.org/publications/Contents/vol12.shtml [More Results From: www.afssociety.org]
39. Diesel Trap Concepts [subscriber access] ... in the surface-type filters. That layer is commonly referred to as "filtration cake" and the process is called "cake filtration". http:// www.dieselnet.com/tech/dpf_concept.html.
40. Mining Technology - Outokumpu Technology GmbH - Filtration ... ... plates with rubber membranes • Endless filter belt • Residue-free cake discharge • Filtration capacity compares more than favourably with other systems on ... http://www.mining-technology.com/contractors/filtering/outo/ [More Results From: www.mining-technology.com]

QUESTIONS FOR THINKING AND DISCUSSING

1. A suspension of aluminum hydroxide in water is to be filtered under constant pressure in a batch Nutsch filter having a filtering area of 1 m^2. Each filter cycle is estimated to separate out 0.5 m^3 of suspension. The operating temperature is 25° C. The following expression for the cake resistance was empirically determined from pilot tests:

$$r_0 = r_0' (\Delta P)^{S'}$$

where r_0 = specific cake resistance (m^{-2})

r_0', S' = constants determined to be 0.5×10^9 and 0.93, respectively

The volume ratio of the cake to filtrate (x_o) is 0.01. Assuming that the resistance of the filter plate is negligible, determine the time required for filtering over the same pressure range.

2. A suspension of solid particles in water is to be separated by a Nutsch filter, where A = 1 m^2, μ = 10^{-3} N-sec/m^2 and the temperature of the suspension is 23° C. The allowable pressure drop is 20×10^4 N/m^2. The filter plate resistance is $R_f = 5 \times 10^{10}$ m^{-1} and is independent of the pressure drop. The volume ratio of the cake to filter is x_o = 0.025, and the filter capacity is 0.2 ×

10^{-3} m^3/sec of filtrate. The relationship between cake resistance and pressure drop is the same as described in question 1. Determine the following: (a) the filtration time; (b) the filtration volume; (c) the cake thickness.

3. A suspension is filtered using a filter cloth media. The resistance of the cloth is negligible. The following data is available: $R'_{c.in}$ = 0.050 m; $R'_{c.in}$ = 0.100 m; $r_o = 6.5 \times 10^{10}$ N-sec/m^2; $x_0 = 0.20$; $\Delta P_t = 20 \times 10^4$ N/m^2. Determine the difference in filtration times between a cylindrical and flat surface filter medium.

4. An aqueous solution of sodium bicarbonate (t = 20° C, $\mu^w = 10^{-3}$ N-sec/m^2) is being filtered under vacuum in a Nutsch filter. The filtration area is 1.5 m^2. The operation is performed in two stages: The first filtration stage is at constant rate and the second is at constant pressure drop. The allowable pressure drop is 80,000 N/m^2 and the permissible cake thickness is 0.2 m. The filter capacity at constant rate is 0.5×10^{-3} m^3/s of filtrate. The cake is incompressible, with the following characteristics: $x_0 = 0.10$; $r_o = 2 \times 10^{12}$ m^{-2} ; $R_f = 10^{10}$ m^{-1}. Calculate the total filtration time of both steps and the volumes of filtrate, cake and separated slurry. Assume for these calculations that the operation is performed without any washing and dewatering stages.

5. The system described in question 4 has an auxiliary time of τ_{aux} = 1800 sec. Determine the total filtration time at constant pressure drop, cake thickness and filtrate volume.

6. A filter press with a hydraulic drive has 35 frames with internal dimensions of 1900 × 1200 mm. The unit operates under the following set of conditions:

- Filtrate viscosity, μ (N-sec/m^2) = 1.8×10^{-3}
- Washing liquid viscosity, $\mu_{w\ell}$ (N-sec/m^2) = 1.0×10^{-3}
- Specific cake resistance, r_o (m^{-2}) = 9×10^{14}
- Volume ratio of cake to filtrate, x_o = 0.059
- Allowable pressure drop, ΔP (N/m^2) = 7×10^6
- Cake porosity, ϵ = 0.49
- Weight ratio of dissolved solids extracted to material before washing, G/G_o = 0.85
- Auxiliary operating time, τ_{aux} (min) = 20

The cake is incompressible and the average plate resistance, R_f, is 2×10^9 m^{-1}. The suspension passes through all the frames and in one direction. The cake is not dewatered. Determine the following:

(a) the rate of filtration and washing under conditions of maximum filter press capacity;

(b) the required times of filtration and washing,

(c) the capacity of the filter press,

(d) the rate of washing liquor; and

(e) the cake thickness.

7. Determine the blower requirements for a rotary vacuum filter used for filtering solids from a liquid suspension under the following set of conditions:

Properties of Slurry:

- Weight fraction of suspended solids, x' = 0.3
- Filtrate density, ρ = 90 Lbs/ft^3
- Density of solids in slurry ρ_p = 101 Lbs/ft^3
- Filtrate viscosity, μ_f = 7 cP
- Viscosity of air, μ_a = 0.018 cP
- Surface tension of filtrate, σ = 0.006 Lb$_f$/ft

Properties of Cake:

- Cake porosity, ϵ = 0.55
- Particle sphericity, ψ= 0.80
- Mean particle size, d_p = 2 10 μm

The drum has a diameter of 4.5 ft and is 2.7 ft wide. The washing ratio is 0.12. The unit operates at 2 rpm and under 5 in. Hg vacuum pressure. The cake thickness, h_c, is 3 in.

8. The operating cycle of a belt vacuum filter consists of filtration, washing and dewatering stages. The solids suspension in water is separated under the following set of conditions:

- Liquid-phase viscosity, μ (N-sec/m^2) = 0.9 $\times$ 10^{-3}
- Specific volume cake resistance, r_o (m^{-2}) = 4 x 10^2
- Volume ratio of cake to filtrate, x_o = 0.25
- Pressure difference, ΔP (N/m^2) = 5 $\times$ 10^4
- Minimum allowable cake thickness, h_c (m) = 0.008
- Washing time, τ_w (sec) = 15
- Dewatering time, τ_d (sec) = 25

- Belt effective length, ℓ (m) = 6.0
- Belt effective width, b (m) = 0.5

The resistance of the filter medium is negligible. Determine the filter's capacity in filtrate and the velocity of the belt. *To solve this problem take a look at the design criteria for a rotary drum filter – they are very similar.*

9. A suspension containing 50 % solids is to be filtered in a filter press, achieving a concentrated cake of 40 % water (on a weight basis). The filter capacity for the cake is G_c = 700 kg/hr. Determine the amount of suspension, the filtrate and the ratio x_o. The density of the solid phase is 1,600 kg/m^3 and that of the liquid phase is 900 kg/m^3.

10. For question 9, determine the washing time if 2 kg of water is used per 1 kg of cake. The filtrate viscosity is 0.7 cP and the viscosity of the washing liquor is 1 cP.

11. Determine the required surface area of a Nutsch filter operating under the following conditions:

- Amount of suspension is Q_s = 2500 kg
- Solid phase concentration = 10 %
- Cake wetness = 60 %
- Filtrate density is ρ = 1040 kg/m^3
- Cake density is ρ_c = 1,100 kg/m^3
- Specific cake resistance is $r_o = 1.324 \times 10^{14}$ m^{-2}
- Filter plate resistance is $R_f = 5.69 \times 10^{11}$ rn^{-1}
- Amount of washing water = 1.75 m^3/m^3 of wet cake
- Filtrate viscosity, μ_f = 1.1 cP
- Washing water viscosity, μ_w = 1.0 cP
- Pressure drop, $\Delta P = 1.96 \times 10^5$ N/m^2

The total time for cake removal and filter preparation is 20 min.

12. Determine the amount of wet cake obtained from filtering 10 m^3 of suspension with a relative specific weight of 1.12. The suspension is 20% by weight solids and the cake wetness is 25%.

13. Approximately 15 m^3 of filtrate was removed from a suspension containing 20% by weight solids. The cake wetness is 30%. Calculate the amount of dry cake obtained.

14. The filtration time of 20 m^3 of a suspension in a filter press is 2.5 hours. Determine the approximate washing time of the cake for 2 m^3 of water, assuming that the washing rate is 4 times less than that of filtration at the end

of the process. The resistance of the filter cloth may be neglected. The viscosity of the filtrate is approximately the same as that of the washing liquor.

15. For question 14, how much will the washing time change if the filtrate viscosity is double and the viscosity of the wash water is 1 cP?

16. Determine the required washing time under the following set of conditions: Washing Intensity = 6 Liters/m^2-min., Cake Thickness = 40 mm, Initial Solids Concentration in the Filtrate of the Wash Water = 120 g/Liter; Final Concentration = 2 g/Liter. Assume a wash rate constant, K = 375 cm^3/Liter.

17. Calculate the required filtration time for 10 Liters of liquid through 1 m^2 of filter area if tests have shown that 1 Liter of filtrate is obtained after 2.25 min. and 3.2 Liters after 14.5 min. from the start of the process.

18. For the conditions described in the above question, determine the washing time if the amount of wash water is 2.4 Liter/m^2 and the washing is cocurrent with the filtrate.

19. During the filtration test of a water suspension containing 14.5 % calcium carbonate in a filter press (A = 0.1 m^2; 50 mrn cake thickness; temperature 23° C), the following data were obtained:

Pressure (atm)	Filtrate Obtained (Liter)	Time Elapsed (sec.)
0.35	2.92	1.46
	7.80	888
1.05	2.45	50
	9.80	660

Determine the filtration constants, K (m^2/hr) and C (m^3/m^2), at the two pressures.

20. For the conditions described in the above question, compute the specific cake resistance of calcium carbonate. The wetness of the cake at 0.35 atm is 37%, and at 1.05 atm, 32% of wet cake by weight.

21. Determine the filter cloth resistance for the filtration of calcium carbonate under conditions given in the above questions.

22. A suspension is to be processed in a filter press to obtain 6 m^3 of filtrate in 3 hours. A filtration test in the lab under the same pressure and cake thickness showed that the filtration constants for 1 m^2 of filter area are as follows: K = 20.7 × 10-4 m^2/hr and C = 1.45 × 10^{-3} m^3/m^2. Determine the size of the filter.

23. A rotary drum vacuum filter is fed with 8.5 m^3/hr of a suspension containing 17.6% solids by weight. The final wetness of the cake must be 34%. The

vacuum is 600 mm Hg. During a filtration test on a model at a vacuum of 510 mm Hg, it was found that the desired wetness is achieved in 32 sec. when operating in the filtering zone. The filtration constants related to 1 m^2 of area are K = 11.2 $Liter^2/m^4$-sec. and C = 6 $Liter/m^2$. The specific weight of the suspension is 1,120 kg/m^3 and that of the filtrate is 1,000 kg/m^3. Determine the filter area and its number of rotations.

24. How long will it take to wash a cake of NaCl in a filter press to achieve 5 g/Liter NaCl in washing water? Washing is performed with clean water at an intensity of 0.33 m^3/m^2-hr. The cake thickness is 35 mm and the washing constant, K, is 520 cm^3/Liter. At the beginning of the washing, the NaCl concentration in the wash water is 143 g/Liter.

25. For the conditions in the above question, calculate the NaCl concentration in the wash water after 50 min. of washing.

26. The following data were obtained in a filtration test using a $CaCO_3$ suspension on a filter of area A = 500 cm^2 at constant pressure:

Time (sec.)	6.8	19.0	36.4	53.4	76.0	102	131.2	163
Filtrate Volume (Liter)	0.5	1.0	1.5	2	2.5	3	3.5	4

Determine the filtration constants **a** and **b**, where **a** is the constant characterizing cake resistance related to 1 m^2 of filter area in sec/m^2, and **b** characterizes the resistance of the filter medium related to 1 m^2 of filter area in sec./m^2.

27. Using the filtration constants determined from the above question, determine the filtration time of a 5 m^3 $CaCO_3$ suspension containing 5% solids on the filter with 10 m^2 area. The cake wetness is 40% and the solids density is 2,200 kg/m^3. The density of the liquid is 1,000 kg/m ^{3}. Also determine the final filtration rate after 2 hrs. of operation.

28. A water suspension is filtered in a media of 50 m^2 area at 20° C. Approximately 7 m^3 of filtrate was obtained. The filtration constants are a = 1.44 × 10^6 sec./m2 and b = 9 × 10^3 sec./m. Determine the washing time at 40° C when the water wash rate is 10 Liter/m^2.

29. A suspension of 18 % solids is filtered over an area of 50 m^2. The cake's moisture content (by weight) is 40 %. Determine the cake thickness if the filtration and washing time is 10 hrs. The wash water's volume is 1/5th that of the filtrate volume, and the filtration constants are as follows: a = 1.10 × 10^5 sec./m^2 and b = 2.15 × 10^3 sec./m. The densities of the solid and liquid phases are 3,100 kg/m^3 and 1,000 kg/m^3, respectively.

30. Size a Nutsch filter for processing a suspension containing 15 % solids. The desired cake wetness is 45 % and the filter capacity for the filtrate should be 5 m^3/hr. The allowable pressure drop is 500 mm Hg. The cake specific resistance is r_o = 9 × 1011 m^{-2} and the filter plate resistance is R_f = 2 × 10^9

m^{-1}. Solid and liquid phase densities are 2,100 kg/m^3 and 1,000 kg/m^3, respectively. The cake can be washed with water at 20° C using 1 kg of water per kg of cake.

31. For the above question, identify a suitable vendor and develop a cost estimate for this machine, including controls and installation.

Chapter 6

CARTRIDGE AND OTHER FILTERS WORTH MENTIONING

INTRODUCTION

In some ways this chapter is a sidebar discussion, although it would be remiss to treat these systems as miscellaneous. Cartridge filters are so widely used throughout the chemical process industries that, if you are a chemical engineer, it would be amazing that at some point in your career you did not specify or use this type of filter in an application. The other types of filter machines discussed in this chapter are important, but perhaps are not nearly as frequently employed in process and water treatment applications as the cartridge device. In all cases the systems are operated as batch, although there are semi-continuous to continuous operations cropping up all the time.
As with other chapters, you will find a section at the end dealing with ***Questions for Thinking and Discussing***. Some of the questions are problem-solving and rely on formulae and reference material provided to you in the last chapter. Remember to refer to the ***Glossary*** at the end of the book, if you run across any terms that are unfamiliar to you.

CARTRIDGE FILTERS

Cartridge filters are used extensively throughout the chemical process industries in applications from laboratory-scale to commercial operations ranging to more than 5,000 gpm. Figure 1 shows a photograph of a series of units that serve as a pretreatment stage to an RO unit for water treatment. Quite often cartridge filters are used as a pretreatment stage, but occasionally they may serve as polishing filters and even stand-alone systems.
Table 1 lists typical filtering applications and operating ranges. The simplest and oldest types of designs include a series of thin metal disks that are 3-10 inches in diameter and set in a vertical stack with very narrow uniform spaces between them. The disks are supported on a vertical hollow shaft and fit into a closed cylindrical

casing. Liquid is fed to the casing under pressure, from where it flows inward between the disks to openings in the central shaft and out through the top of the casing. Te suspended solids are captured between the disks and remain on the filter. As most of the solids are removed at the periphery of the disks, the unit is referred to as *an edge filter*. The accumulated solids are removed periodically from the cartridge.

Figure 1. Battery of cartridge filters used for water pretreatment in an RO process.

Table 1. Typical Operating Ranges of Cartridge Filters.

Liquid	Filtration Range	Liquid	Filtration Range
Alum	60 mesh - 60 μm	Gasoline	1 - 3 μm
Brine	100-400mesh	Hydrocarbon wax	25 - 30 μm
Ethyl alcohol	5-10 μm	Isobutane	250 mesh
Ferric chloride	30-250 mesh	MEA	200 mesh - 10 μm
Herbicides and pesticides	100-700 mesh	Naptha	25 - 30 μm

Liquid	Filtration Range	Liquid	Filtration Range
Hydrochloric acid	100 mesh to 5-10 μm	Residual oil	25 - 50 μm
Mineral oil	1-10 μm	Sea water	5 - 10 μm
Nitric acid	40 mesh - 10 μm	Vacuum gas oil	25 - 75 μm
Phosphoric acid	100 mesh - 10 μm	Steam injection	5 - 10 μm
Sodium hydroxide	1-10 μm	Calcium carbonate	30 - 100 mesh
Sodium hypochlorite	1-10 μm	Dyes	60 - 400 mesh
Sodium sulfate	5-10 μm	Fresh water	30 - 200 mesh
Sulfuric acid	250 mesh - 3 μm	Latex	40 - 100 mesh
Synthetic oils	25 30 μm	River water	20 - 400 mesh

Other designs are simpler, experience lower pressure drop and have fewer maintenance problems. In an upflow cartridge filter, the unfiltered liquid enters the inlet (bottom) port. It flows upward, around and through the filter media, which is a stainless steel or fabric screen reinforced by a perforated, stainless steel backing. Filtered liquid discharges through the outlet (top port). Because of the outside-to-inside flow path, solids deposit onto the outside of the element so that screens are easy to clean. This external gasketing design prevents solids from bypassing the filter and contaminating the process downstream. There are no "o"-ring seals that can crack, channeling media that can fail or cartridges that can collapse or allow bypassing. As with any filter, careful media selection is critical. Media that are too coarse, for example, will not provide the needed protection. However, specifying finer media than necessary can add substantially to both equipment maintenance and operating costs. Factors to be considered in media selection include solids content, type of contaminant, particle size and shape, amount of contaminant to be removed, viscosity, corrosiveness, abrasiveness, adhesive qualities of the suspended solids, liquid temperature and the required flowrate. Typical filter media are wire mesh (typically 10 mesh to 700 mesh), fabric (30 mesh to 1 μm), slotted screens (10 mesh to 25 μm) and perforated stainless steel screens (10 mesh to 30 mesh). Table 2 provides some typical particle retention sizes for different filter media. Single filtration units can be piped directly into systems that require batch to semi-continuous services. Using quick-connector couplings, the media can be removed from the filter housing, inspected and cleaned in a matter of minutes. Filtering elements are also interchangeable. This means that while one unit is being cleaned, the other can be in service. Multiple filter units in a single housing are also quite common. These consist of two or more single filter units valved in parallel to a common header. The distinguishing feature of this type of configuration is the ability to sequentially backwash each unit in place while the others remain onstream. Hence, these systems serve as continuous operations.

These units can be automated fully to eliminate manual backwashing. Backwashing can be controlled by changes in differential pressure between the inlet and outlet headers. One possible arrangement consists of a controller and solenoid valves that supply air signals to pneumatic valve actuators on each individual filter unit. As solids collect on the filter elements, flow resistance increases. This increases the pressure differential across the elements and, thus, between the inlet and outlet headers on the system. When the pressure drop reaches a preset level, an adjustable differential pressure switch relays information through a programmer to a set of solenoid valves, which, in turn, send an air signal to the pneumatic valve actuator. This rotates the necessary valve(s) to backwash the first filter element. When the first element is cleaned and back onstream, each successive filter element is backwashed in sequence until they are all cleaned. The programmer is then reset automatically until the rising differential pressure again initiates the backwashing cycle.

Table 2. Typical Filter Retentions.

	Mesh Size	Nominal Particle Retention		% Open Area
		(in.)	(μm)	
Wire Mesh	10	0.065	1650	56
	20	0.035	890	46
	30	0.023	585	41
	40	0.015	380	36
	60	0.009	230	27
	80	0.007	180	32
	100	0.0055	140	30
	150	0.0046	115	37
	200	0.0033	84	33
	250	0.0024	60	36

	Mesh Size	Nominal Particle Retention		% Open Area
		(in.)	(μm)	
	400	0.0018	45	36
	700	0.0012	30	25
Perforated	10	0.063	1600	50
	20	0.045	890	36
	30	0.024	610	30
	40	0.015	380	20
	60	0.009	230	18
Slotted	80	0.007	180	25
	100	0.006	150	13
	120	0.005	125	11
	150	0.004	100	9
	200	0.003	75	7
	325	0.002	50	5
Fabric	500	0.0016	40	Percentage of open area not applicable to fabric media.
	-	0.0010-0.0012	25-30	
	-	0.0006-0.0008	15-20	
	-	0.0002-0.0004	5-10	
	-	0.00004-0.00012	1-3	

THE TILTING PAN FILTER

The Tilting Pan Filter is predominantly employed in the phosphoric acid industry and, to some extent, in the washing of phosphate rock. There was a traditional rivalry over the years between the Pan and Belt Filter which now, owing to the substantial progress in rubber technology, swings in the favor of the later. The operation of Tilting Pan Filters is based on a series of horizontal independent trapezoidal pans mounted on a circular supporting structure that rotates under vacuum during the filtration cycle and invert under blow to discharge the cake. During filtration the cake may be washed countercurrently, generally in three stages, while the mother and wash filtrates are flowing to separate vacuum receivers. After the final drying portion of the cycle the cake is discharged dry for piling or sluiced as a slurry for disposal to ponds. Following cake discharge the pans pass over radial manifolds with high impact jets that clean the cloth and dislodge any cake leftovers that remained on the pan's surface. At the last stage before relevelling, suction is applied to evacuate the wash water that accumulates inside the pan, and at the same time it dries the cloth to avoid dilution of the mother filtrate. Since during this stage air passes freely through the exposed cloth, applying suction with the main vacuum pump will cause a loss of vacuum to the entire system, hence, an auxiliary small vacuum pump is incorporated to separate between the vacuum zones.

The main valve, which is called the distributor, controls the cycle segments with bridges that open and close the port of each pan as it passes from high vacuum through cake blow-off to low vacuum. Compartments within the valve separate the various zones to ensure that the separation is sharp and no mixing of mother, strong, middle and weak filtrates occurs during the cycle. To speed up the evacuation of the various filtrates there are valves with two bridge circles, one for the main stream and the another that opens to atmosphere and purges the pan below the deck just before entering a new zone. The speed of rotation establishes the cycle time and this, being an important design parameter, is limited mechanically by the mass inertia of the swinging pan and its wet cake load at the point of discharge. Typical cycle times for pan filters are 2.5 to 3 minutes depending on the size of the filter and the design of the peripheral cam that controls the tilting velocity. Pan Filters are available for dry or wet cake discharge depending on the method of disposal. The discharge zone for dry discharge consists of two chutes, one for the dry cake and the second for cloth washing. The zone for wet discharge has one chute only in which the cake is sluiced and then the cloth is washed. Dry

Operational Sequence

The cycle of a Tilting Pan Filter that includes three counter-current washing stages consists of the following zones:

With pans under vacuum during filtration (with short atmospheric purge on certain valves):

- **Cloudy Port Recycle or Sedimentation Pool (before applying vacuum)**
- **Cake Formation**
- **Cake Predrying**
- **First Washing**
- **First Predrying**
- **Second Washing**
- **Second Predrying**
- **Third Washing**

Operational Sequence

With pans purged to atmosphere during inversion:

- **Cake discharge combined with cloth wash and sluicing for wet cake discharge**
- **Cake discharge separated from cloth wash for dry discharge.**

With pans under low vacuum during relevelling

- **Evacuation of leftover cloth wash water**
- **Cloth drying**

cake discharge consumes about 3-4% more effective filtration area than the wet discharge so this should be taken in account while calculating the required area. The machine is constructed from the following subassemblies:

➢ Trapezoidal pans with sloping bottoms for fast evacuation of filtrates and each fitted with a roller arm assembly that tilts and relevels the pan as it passes trough the inverting track. Special circumferential fasteners enable the quick replacement of filter cloths.

➢ A central valve with bridge setting to control the various zones throughout the filtration cycle, and in some instances, two bridge circles with the second serving to purge the pan. The valve has internal compartments for the mother and wash filtrates.

➢ A feed and wash boxes that may be set in a slight slope so that the distribution over the trapezoidal surface will be proportional to the pan's area. This will ensure that the formed cake is even in thickness and that the applied wash liquids are distributed evenly.

➢ A structure that rotates over rollers, supports the pans and is fitted with a toothed rim that drives the filter.

The maintenance of these machines can be labor intensive. Major components that require frequent inspection for preventive maintenance include:

- Evacuation of leftover cloth during relevelling
- Cloth drying
- The pans for corrosion
- The wear plate of the main valve for erosion since its large diameter makes it vulnerable to a loss of vacuum
- The twisting rubber hoses or “o” ringed elbows which connect the rotating pan necks with pipe and the wear plate to the face of the stationary valve
- The condition of the high impact nozzles mounted on the cloth wash manifold and, for wet cake discharge, the sluicing manifold
- The support rollers which take the vertical load of the entire machine and the horizontal thrust rolls that maintain the rotating frame concentric
- The toothed rim and sprocket which drives the pan filter

THE TABLE FILTER

These machines were first introduced to industry nearly 60 years ago and were of rather small and of a simple design. Early designs were greatly flawed. Their main limitation was at the discharge zone since the cake was contained in a fixed rim and special sealing arrangements must be provided in order to avoid the spillage of brine at the table's circumference. Another problem was that the thin heel left between the scroll and the surface of the table was dislodged by applying a back blow but not removed from the surface of the passing cell. So, as it reached the feed zone it was mixed with the incoming slurry without the cloth being washed. This can caused progressive media blinding which effects filtration rate and requires frequent stoppage of the operation for cloth washing. As the demand for more area and higher throughputs has gone up, table filters were redesigned and upgraded.

Special attention was given to the redesign of the discharge zone and a radically new engineering concept was developed to solve the circumferential sealing problem and allowing continuous cloth washing. The new table filter features a rubber belt rim that rotates together with the cells but is removed from the table just after leaving the final wash and drying zone and before reaching the cake discharge hopper. The rim then stays away from the table and returns once it passes the wash water hopper and enters the cloth drying zone. Special rollers are diverting the rim and a tension take- up roll ensures that the belt stays tight against the table and seals the circumference against leakages.

The filtration area of large table filters is more than 200 m^2 and having few moving parts can rotate at a cycle time of 1.5 minutes. These machines can handle thick cakes and may be operated at high vacuum levels. The major subassemblies of the machine include;

- A series of fixed trapezoidal cells that form a rotating table and each connected to a stationary valve in the center of the filter. The cell is designed with steep sloped bottom for fast evacuation of the filtrate.
- A valve that may be raised from the top and has a bridge setting and compartments to control the various zones.
- An internal rim fixed to the table at the inner circumference and a continuous rubber belt that surrounds the table at the periphery and confine the slurry, wash liquids and the cake during the filtration cycle.
- Rollers that support the vertical loads, centering thrust and others that move the rim away from the table in the discharge zone and maintain it under tension.
- Radial rubber dams that separate between the feed, wash stages, cake discharge and cloth wash, and cloth drying zones to prevent the mixing of filtrates.
- A variable pitch screw that transports the cake radially towards the discharge point.

The cycle of a table filter that includes three countercurrent washing stages consists of the following zones:

With cells under vacuum during filtration:

- Cloudy Port Recycle or Sedimentation Pool (before applying vacuum).
- Cake Formation.
- Cake Predrying.
- First Washing.
- First Predrying.
- Second Washing.
- Second Predrying.
- Third Washing.
- Final Drying.

With cells purged to atmosphere:

- Cake discharged dry and conveyed by the screw to the cake hopper.
- Cloth wash and sluicing for the removal of the heel.

With cells under low vacuum:

- Evacuation of cloth wash water.
- Cloth drying.

Compared to tilting pan filters, these machines have fewer moving parts and hence they are easier and less costly to maintain.

Select the Table Filter

- **When the process downstream requires a de-lumped cake since the screw disintegrates the solid lumps while conveying them to the periphery.**
- **When the solids are fast settling and cannot be kept as a homogenous slurry in bottom or side feed filters such as Drum or Disc Filters.**
- **When very short cycle times are required for fast dewatering cakes such as phosphate slurry.**
- **When a clear filtrate is required right from the start it is good practice to form a thin heel that serves as a filter medium over the exposed cloth. This is done by either a "cloudy port outlet" that is recirculated or, if solids are settling fast, by allocating a portion of the table after the cloth drying dam and prior to entering the vacuum zone to act as a "sedimentation pool".**
- **When intensive cake washing is required.**
- **When a large filtration area is required but a horizontal belt filter does not fit into the layout.**
- **When cakes tend to crack under vacuum measures, such as a flapper or pressure roll may assist in sealing the cracks thus avoiding loss of vacuum.**

QUESTIONS FOR THINKING AND DISCUSSING

The following are some challenging calculations you can try for sizing cartridge filters. The design formulae provided in Chapter 5 should get you through these. Remember, the principles for the preliminary sizing of filtering machines are essentially the same. Once the primary sizing criteria have been established, then the detailed configuration and specific operational requirements can be established by working with the equipment supplier.

1. A saturated liquor is thickened in a cartridge filter at maximum capacity and constant pressure drop. The properties of the suspension and the cake were determined from pilot scale tests as follows:

Suspension Properties

Solids concentration before thickening	725 N/m^3
Solids concentration after thickening	3,510 N/m^3
Liquid viscosity	0.6×10^{-3} N-sec./m^2
Liquid specific weight	1.04×10^4 N/m^3
Solids specific weight	2.9×10^4 N/m^3

Cake Properties

Solids concentration	0.6 N/N cake
Specific volumetric resistance	90×10^{10} m^{-2}

The cartridge filter has the following dimensions:

Length	3.4 m
Upper base diameter	0.3 m
Lower base diameter	0.21 m

And the unit will be operated under the following set of conditions:

Capacity in suspension	60×10^5 N/hr
Pressure drop at cake removal stage	15×10^5 N/m^2
Time of cake removal	8 sec.
Ratio of filtration time to total thickness cycle	0.99
Filter medium resistance	0.4×10^{12} m^{-1}

Using the above information, determine the following sizing information:

(a) the capacity of the thickener in the filtrate and the thickened suspension;

(b) the thickening cycle time that corresponds to the minimum filtering area;
(c) the minium filtration area required; and
(d) the cake thickness.

2. In the above design case, explain why each of the parameters calculated is important not only from an operational standpoint, but from selecting a specific cartridge filter. Then go to the Web and identify one ore more vendors that have machines you feel might be applicable to this case.

3. There are a lot variations of cartridge filters. This chapter only provides the basics. Take some time to develop a list of vendors and the different types of cartridge filters on the market today. Organize the list into batch and continuous type devices.

Chapter 7

WHAT SAND FILTRATION IS ALL ABOUT

INTRODUCTION

When people think of sand filtration, they automatically relate to municipal water treatment facilities. In general that's the arena that this classical filter has ruled, but it most certainly has found applications in pure industrial settings, oftentimes for niche applications where suspended solids and organic matter persist in process waters. Very common applications that I have seen in Eastern Europe have been part of cooling and process water treatment plants, particularly for large cooling tower applications in refineries and coke chemical plants, where biological growth problems can adversely impact on heat exchange equipment. Also, in many former Soviet block republics, municipal water treatment plants were almost always part of large industrial complexes, so that both the communities and plant water treatment requirements were met by a single operation. This leads to a very distinct set of problems that we in the U.S. and parts of Western Europe don't face, because of the separation of operations, and one which is way beyond the scope of this volume.

Sand filtration is almost never applied as the primary treatment method. Most often it is a pretreatment or final stage, but sometimes intermediate stage of water treatment, and is most often used along with other filtration technologies, carbon adsorption, sedimentation and clarification, disinfection, and biological methods. The term sand filtration is somewhat misleading and stems from older municipal wastewater treatment methods. While there is a class of filtration equipment that relies principally on sand as the filter media, it is more common to employ multiple media in filtration methods and equipment, with sand being the predominant media. In this regard, the terms sand filtration and granular media filtration are considered interchangeable in our discussions. The design, operation and maintenance of these systems are very straightforward, and indeed may be viewed as the least complex or simplest filtration practices that exist. It is a very old technology and much of the operational criteria is "art" in my opinion - although I have gotten into arguments with my colleagues on the use of this term. Be as it may - let's take a look at this technology and then try some ***Questions for Thinking and Discussing***. Remember to refer to the ***Glossary*** at the end of the book if you run across any terms that are unfamiliar to you.

WASTEWATER TREATMENT PLANT OPERATIONS

Before getting into the subject of sand filtration, we should first attempt to put the technologies of municipal wastewater treatment into some perspective. Wastewater treatment plants can be divided into two major types: biological and physical/chemical. Biological plants are more commonly used to treat domestic or combined domestic and industrial wastewater from a municipality. They use basically the same processes that would occur naturally in the receiving water, but give them a place to happen under controlled conditions, so that the cleansing reactions are completed before the water is discharged into the environment.

Physical/chemical plants are more often used to treat industrial wastewaters directly, because they often contain pollutants which cannot be removed efficiently by microorganisms-- although industries that deal with biodegradable materials, such as food processing, dairies, breweries, and even paper, plastics and petrochemicals, may use biological treatment. Biological plants generally use some physical and chemical processes also.

A physical process usually treats suspended, rather than dissolved pollutants. It may be a passive process, such as simply allowing suspended pollutants to settle out or float to the top naturally-- depending on whether they are more or less dense than water. Or the process may be aided mechanically, such as by gently stirring the water to cause more small particles to bump into each other and stick together, forming larger particles which will settle or rise faster-- a process known as *flocculation*. Chemical flocculants may also be added to produce larger particles. To aid flotation processes, dissolved air under pressure may be added to cause the formation of tiny bubbles which will attach to particles.

Filtration through a medium such as sand as a final treatment stage can result in a very clear water. In contrast -- ultrafiltration, nanofiltration, and reverse osmosis (RO) are processes which force water through membranes and can remove colloidal material (very fine, electrically charged particles, which will not settle) and even some dissolved matter. Absorption (adsorption, technically) on activated charcoal is a physical process which can remove dissolved chemicals. Air or steam stripping can be used to remove pollutants that are gasses or low-boiling liquids from water, and the vapors which are removed in this way are also often passed through beds of activated charcoal to prevent air pollution. These last processes are used mostly in industrial treatment plants, though activated charcoal is common in municipal plants, as well, for odor control.

Examples of chemical treatment processes, in an industrial environment, would be:

1. converting a dissolved metal into a solid, settleable form by precipitation with an alkaline material like sodium or calcium hydroxide. Dissolved iron

or aluminum salts or organic coagulant aids like polyelectrolytes can be added to help flocculate and settle (or float) the precipitated metal.

2. converting highly toxic cyanides used in mining and metal finishing industries into harmless carbon dioxide and nitrogen by oxidizing them with chlorine.

3. destroying organic chemicals by oxidizing them using ozone or hydrogen peroxide, either alone or in combination with catalysts (chemicals which speed up reactions) and/or ultraviolet light.

In municipal treatment plants, chemical treatment-- in the form of aluminum or iron salts-- is often used for removal of phosphorus by precipitation. Chlorine or ozone (or ultraviolet light) may be used for disinfection, that is, killing harmful microorganisms before the final discharge of the wastewater. Sulfur dioxide or sulfite solutions can be used to neutralize (reduce) excess chlorine, which is toxic to aquatic life. Chemical coagulants are also used extensively in sludge treatment to thicken the solids and promote the removal of water. A conventional treatment plant is comprised of a series of individual unit processes, with the output (or effluent) of one process becoming the input (*influent*) of the next process. The first stages will usually be made up of physical processes that take out easily removable pollutants. After this, the remaining pollutants are generally treated further by biological or chemical processes. These may convert dissolved or colloidal impurities into a solid or gaseous form, so that they can be removed physically, or convert them into dissolved materials which remain in the water, but are not considered as undesirable as the original pollutants. The solids (residuals or *sludges*) which result from these processes form a side stream which also has to be treated for disposal.
Common processes found at a municipal treatment plant include:

Preliminary treatment to remove large or hard solids that might clog or damage other equipment. These might include grinders (comminuters), bar screens, and *grit* channels. The first chops up rags and trash; the second simply catches large objects, which can be raked off; the third allows heavier materials, like sand and stones, to settle out, so that they will not cause abrasive wear on downstream equipment. Grit channels also remove larger food particles (*i.e.*, garbage).

Primary settling basins, where the water flows slowly for up to a few hours, to allow organic suspended matter to settle out or float to the surface. Most of this material has a density not much different from that of water, so it needs to be given enough time to separate. Settling tanks can be rectangular or circular. In either type, the tank needs to be designed with some type of scrapers at the bottom to collect the settled sludge and direct it to a pit from which it can be pumped for further treatment-- and skimmers at the surface, to collect the material that floats to the top (which is given the rather inglorious name of "scum".) The diagram below shows the operation of a typical primary settling tank.

Secondary treatment, usually biological, tries to remove the remaining dissolved or colloidal organic matter. Generally, the biodegradation of the pollutants is allowed to take place in a location where plenty of air can be supplied to the microorganisms. This promotes formation of the less offensive, oxidized products. Engineers try to design the capacity of the treatment units so that enough of the impurities will be removed to prevent significant oxygen demand in the receiving water after discharge.
There are two major types of biological treatment processes: *attached* growth and *suspended* growth.

In an attached growth process, the microorganisms grow on a surface, such as rock or plastic. Examples include open *trickling filters*, where the water is distributed over rocks and trickles down to underdrains, with air being supplied through vent pipes; enclosed *biotowers*, which are similar, but more likely to use shaped, plastic media instead of rocks; and so-called *rotating biological contacters*, or RBC's, which consist of large, partially submerged discs which rotate continuously, so that the microorganisms growing on the disc's surface are repeatedly being exposed alternately to the wastewater and to the air. The most common type of suspended growth process is the so-called *activated sludge* system. This type of system consists of two parts, an *aeration tank* and a settling tank, or *clarifier*. The aeration tank contains a "sludge" which is what could be best described as a "mixed microbial culture", containing mostly bacteria, as well as protozoa, fungi, algae, etc. This sludge is constantly mixed and aerated either by compressed air bubblers located along the bottom, or by mechanical aerators on the surface. The wastewater to be treated enters the tank and mixes with the culture, which uses the organic compounds for *growth*-- producing more microorganisms-- and for *respiration,* which results mostly in the formation of carbon dioxide and water. The process can also be set up to provide biological removal of the nutrients nitrogen and phosphorus. Refer to Figure 1 for a simplified process flow sheet.

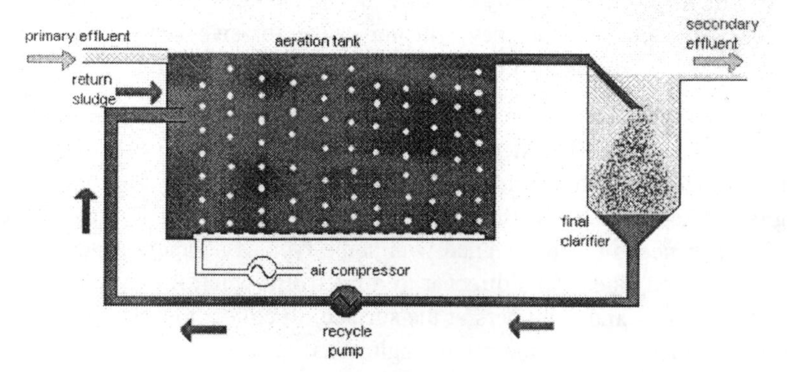

Figure 1. Simplified process flow sheet of activated sludge process.

After sufficient aeration time to reach the required level of treatment, the sludge is carried by the flow into the settling tank, or clarifier, which is often of the circular design. (An important condition for the success of this process is the formation of a type of culture which will flocculate naturally, producing a settling sludge and a reasonably clear upper, or *supernatant* layer. If the sludge does not behave this way, a lot of solids will be remain in the water leaving the clarifier, and the quality of the effluent wastewater will be poor.) The sludge collected at the bottom of the clarifier is then recycled to the aeration tank to consume more organic material. The term "activated" sludge is used, because by the time the sludge is returned to the aeration tank, the microorganisms have been in an environment depleted of "food" for some time, and are in a "hungry", or activated condition, eager to get busy biodegrading some more wastes. Since the amount of microorganisms, or *biomass*, increases as a result of this process, some must be removed on a regular basis for further treatment and disposal, adding to the solids produced in primary treatment. The type of activated sludge system that has just been described is a continuous flow process. There is a variation in which the entire activated sludge process take place in a single tank, but at different times. Steps include filling, aerating, settling, drawing off supernatant, etc. A system like this, called a sequencing batch reactor, can provide more flexibility and control over the treatment, including nutrient removal, and is amenable to computer control.)

Nutrient removal refers to the treatment of the wastewater to take out nitrogen or phosphorus, which can cause nuisance growth of algae or weeds in the receiving water. Nitrogen is found in domestic wastewater mostly in the form of ammonia and organic nitrogen. These can be converted to nitrate nitrogen by bacteria, if the plant is designed to provide enough oxygen and a long enough "sludge age" to develop these slow-growing types of organisms. The nitrate which is produced may be discharged; it is still usable as a plant nutrient, but it is much less toxic than ammonia. If more complete removal of nitrogen is required, a biological process can be set up which reduces the nitrate to nitrogen gas (and some nitrous oxide). There are also physical/chemical processes which can remove nitrogen, especially ammonia; they are not as economical for domestic wastewater, but might be suited for an industrial location where no other biological processes are in use. (These methods include alkaline air stripping, ion exchange, and "breakpoint" chlorination.)

Phosphorous removal is most commonly done by chemical precipitation with iron or aluminum compounds, such as ferric chloride or alum (aluminum sulfate). The solids which are produced can be settled along with other sludges, depending on where in the treatment train the process takes place. "Lime", or

The chemical formula for limestone is $CaCO_3$ and upon burning forms calcium oxide (CaO), which is known as burnt lime. Calcium oxide, when mixed with water, forms calcium hydroxide ($Ca(OH)_2$). Calcium hydroxide is used to treat water as a coagulation aid along with aluminum sulfate.

calcium hydroxide, also works, but makes the water very alkaline, which has to be corrected, and produces more sludge. There is also a biological process for phosphorus removal, which depends on designing an activated sludge system in such a way as to promote the development of certain types of bacteria which have the ability to accumulate excess phosphorus within their cells. These methods mainly convert dissolved phosphorus into particulate form. For treatment plants which are required to discharge only very low concentrations of total phosphorus, it is common to have a sand filter as a final stage, to remove most of the suspended solids which may contain phosphorus.

Disinfection, usually the final process before discharge, is the destruction of harmful *(pathogenic)* microorganisms, *i.e.* disease-causing germs. The object is not to kill every living microorganism in the water-- which would be *sterilization*-- but to reduce the number of harmful ones to levels appropriate for the intended use of the receiving water.

The most commonly used disinfectant is **chlorine**, which can be supplied in the form of a liquefied gas which has to be dissolved in water, or in the form of an alkaline solution called sodium hypochlorite, which is the same compound as common household chlorine bleach. Chlorine is quite effective against most bacteria, but a rather high dose is needed to kill viruses, protozoa, and other forms of pathogen. Chlorine has several problems associated with its use, among them 1) that it reacts with organic matter to form toxic and carcinogenic chlorinated organics, such as chloroform, 2) chlorine is very toxic to aquatic organisms in the receiving water-- the USEPA recommends no more than 0.011 parts per million (mg/L) and 3) it is hazardous to store and handle. Hypochlorite is safer, but still produces problems 1 and 2. Problem 2 can be dealt with by adding sulfur dioxide (liquefied gas) or sodium sulfite or bisulfite (solutions) to neutralize the chlorine. The products are nearly harmless chloride and sulfate ions. This may also help somewhat with problem 1.

FILTERING OUT WHAT'S LEFT

- ***Either slow or rapid filtration (depends on size of plant/volume of water considerations).***
- ***Rapid-sand filters force water through a 0.45-1m layer of sand (d=0.4-1.2mm) and work faster, needing a smaller area. They need frequent back-washing.***
- ***Slow-sand filters (d=0.15-0.35mm) require a much larger area but reduce bacteriological and viral levels to a greater degree. The top 1 inch must be periodically scraped off and the filter occasionally back-washed.***

A more powerful disinfectant is **ozone**, an unstable form of oxygen containing three atoms per molecule, rather than the two found in the ordinary oxygen gas which makes up about 21% of the atmosphere. Ozone is too unstable to store, and has to be made as it is used. It is produced by passing an electrical discharge through air, which is then bubbled through the water. While chlorine can be dosed at a high enough concentration so that some of it remains in the water for a considerable time, ozone is consumed very rapidly and leaves no *residual*. It may also produce some chemical byproducts, but probably not as harmful as those produced by chlorine. The other commonly used method of disinfection is **ultraviolet light.** The water is passed through banks of cylindrical, quartz-jacketed fluorescent bulbs. Anything which can absorb the light, such as fouling or scale formation on the bulbs' surfaces, or suspended matter in the water, can interfere with the effectiveness of the disinfection. Some dissolved materials, such as iron and some organic compounds, can also absorb some of the light. Ultraviolet disinfection is becoming more popular because of the increasing complications associated with the use of chlorine.

Sludge from primary settling basins, called primary or "raw" sludge, is a noxious, smelly, gray-black, viscous liquid or semi-solid. It contains very high concentrations of bacteria and other microorganisms, many of them pathogenic, as well as large amounts of biodegradable organic material. Because of the high concentrations, any dissolved oxygen will be consumed rapidly, and the odorous and toxic products of anaerobic biodegradation (*putrefaction*) will be produced. The greasy floatable skimmings from primary treatment are another portion of this *putrescible* solid waste stream. In addition to the primary sludge, wastewater plants with secondary treatment will produce a "secondary sludge", consisting largely of microorganisms which have grown as a result of consuming the organic wastes. While not quite so objectionable, due to the biodegradation which has already taken place, it is still very high in pathogens and contains much material which will decay and produce odors if not treated further. Ultimately, the sludge must all be disposed of. The way in which this is done depends on the quality of the sludge, and determines how it needs to be treated. The most desirable final fate for these solids would be for beneficial use in agriculture, since the material has organic matter to act as a soil

THE FINAL TOUCHES TO WATER

➪ ***Disinfection - water completely free of suspended sediment, is treated with a powerful oxidizing agent usually chlorine, chlorine and ammonia (chloramine), or ozone. A residual disinfectant is left in the water to prevent reinfection. Chlorine can form harmful byproducts and has suspected links to stomach cancer and miscarriages.***

➪ ***pH adjustment - so that treated water leaves the plant in the desired range of 6.5 to 8.5 pH units.***

conditioner, as well as a some fertilizer value. This requires the highest quality "biosolids", free of contamination with toxic metals or industrial organic compounds, and low in pathogens. At a somewhat lower quality, it can be used for similar purposes on non-agricultural land and for land reclamation (e.g., strip mines). Poorer quality sludge can be disposed of by landfilling or incineration.

One commonly used method of sludge treatment, called *digestion*, is biological. Since the material is loaded with bacteria and organic matter; why not let the bacteria eat the biodegradable material? Digestion can be either aerobic or anaerobic. *Aerobic* digestion requires supplying oxygen to the sludge; it is similar to the activated sludge process, except no external "food" is provided. In *anaerobic* digestion, the sludge is fed into an air-free vessel; the digestion produces a gas which is mostly a mixture of methane and carbon dioxide. The gas has a fuel value, and can be burned to provide heat to the digester tank and even to run electric generators. Some localities have compressed the gas and used it to power vehicles. Digestion can reduce the amount of organic matter by about 30 to 70 percent, greatly decrease the number of pathogens, and produce a liquid with an inoffensive, "earthy" odor. This makes the sludge safer to dispose of on land, since the odor does not attract as many scavenging pests, such as flies, rodents, gulls, etc., which spread pathogens from the disposal site to other areas-- and there are fewer pathogens to be spread.

A liquid sludge, which might contain 3 to 6% dry weight of solids, can be *dewatered* to form a drier sludge *cake* of maybe 15 to 25 percent solids, which can be hauled as a solid rather than having to be handled as a liquid. Equipment used to dewater sludge includes centrifuges, vacuum filters, and belt presses or plate-and-frame presses. Chemical coagulants are commonly added to help form larger aggregates of solids and release the water. Further processes such as composting and heat drying can produce a drier product with lower pathogen levels. Another approach involves treatment with lime (calcium oxide), which kills pathogens due to its highly alkaline nature as well as the heat that is generated as it reacts with the water in the sludge; this also results in a drier product. A final disposal method which eliminates all of the pathogens and greatly reduces the volume of the sludge is *incineration*. This is not considered a beneficial use, however, and is becoming less popular due to public concerns over air emissions.

Sludges from physical-chemical treatment of industrial waste streams containing heavy metals

ADDITIONAL STEPS

➪ ***Heavy metal removal: most treatment plants do not have special stages for metals but rely on oxygenation, coagulation and ion exchange in filters to remove them. Where metals persist, additional treatment would be needed.***

➪ ***Troublesome organics: Activated carbon filters are required where soluble organic constituents are present because many will pass straight through standard plants, e.g. pesticides, phenols, MTBE and so forth.***

and non-biodegradable toxic organic compounds often must be handled as hazardous wastes. Some of these will end up in hazardous waste landfills, or may be chemically treated for detoxification-- or even for recovery of some components for recycling. Recalcitrant organic compounds can be destroyed by carefully controlled high-temperature incineration, or by other innovative processes, such as high-temperature hydrogen reduction.

GRANULAR MEDIA FILTRATION

Granular media filtration is used for treating aqueous waste streams. The filter media consists of a bed of granular particles (typically sand or sand with anthracite or coal). The anthracite has adsorptive characteristics and hence can be beneficial in removing some biological and chemical contaminants in the wastewater. This material may also be substituted for activated charcoal.

The bed is contained within a basin and is supported by an underdrain system which allows the filtered liquid to be drawn off while retaining the filter media in place. As water containing suspended solids passes through the bed of filter medium, the particles become trapped on top of, and within, the bed. The filtration rate is reduced at a constant pressure unless an increase in the amount of pressure is applied to force the wastewater through the filter bed. In order to prevent plugging of the upper surface and uppermost depth of the bed, the filter is backflushed at high velocity to dislodge the filtered particles. The backwash water contains high concentrations of solids and is sent to further treatment steps within the watsewater treatment plant.

One of the reasons why it is important to remove suspended solids in water is that the particles can act as a source of food and housing for bacteria. Not only does this make microbiological control much harder but, high bacteria levels increase the fouling of distribution lines and especially heat transfer equipment that receive processed waters (for example, in one's household hot water heater). The removal of suspended contaminants enables chemical treatments to be at their primary jobs of scale and corrosion prevention and microbial control.

The filter application is typically applied to handling streams containing less than 100 to 200 mg/Liter suspended solids, depending on the required effluent level. Increased-suspended solids loading reduces the need for frequent backwashing. The suspended solids concentration of the filtered liquid depends on the particle size distribution, but typically, granular media filters are capable of producing a filtered liquid with a suspended solids concentration as low as 1 to as high as 10 mg/Liter. Large flow variations will affect the effluent's quality.

Granular media filters are usually preceded by sedimentation in order to reduce the suspended solids load on the filter. Granular media filtration can also be installed ahead of biological or activated carbon treatment units to reduce the suspended solids load and in the case of activated carbon to minimize plugging of the carbon columns. Granular media filtration is only marginally effective in treating colloidal size particles in suspensions. Usually these particles can be made larger by flocculation although this will reduce run lengths. In cases where it is not possible to flocculate such particles (as in the case of many oil/water emulsions), other techniques such as ultrafiltration must be considered. Figure 2 illustrates a common sand filter that most people are familiar with in swimming pool applications. Such systems rely on very fine sand media that can typically remove suspended particles about 0.5 μm in size. Filtration is an effective means of removing low levels of solids from wastes provided the solids concentration does not vary greatly and the filter is backwashed at appropriate intervals during the filtration cycle. The operation can be easily integrated with other treatment steps, and further, is well suited to mobile treatment systems as well as on-site or fixed installations. In short, sand filtration technologies, although simple, are quite versatile in meeting treatment challenges.

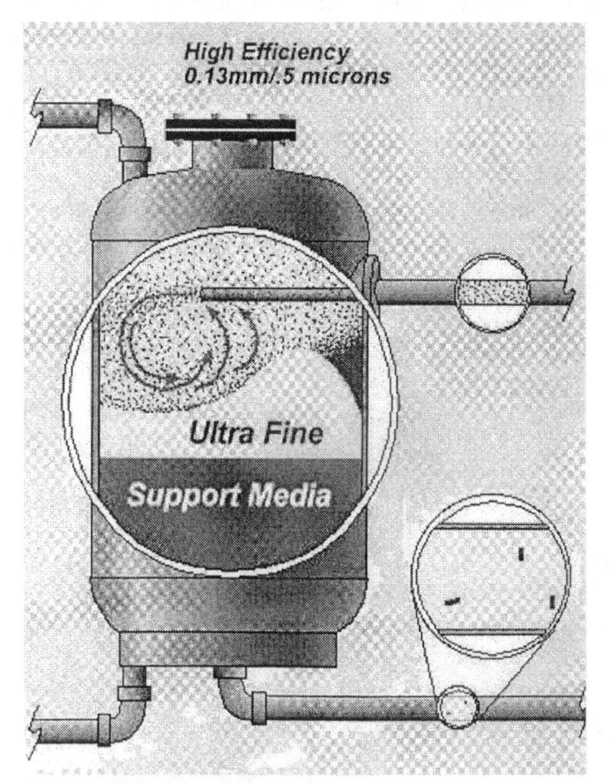

Figure 2. A simple sand filtration unit.

A typical multi-media sand filtration unit is shown in Figure 3. In this configuration a coarse layer of media is used to reduce the contaminant loading to the final layer. This allows multimedia filters to use finer media. Such units generally remove suspended solids down to about 15 μm, and they require large volumes of water to properly remove contaminant that is trapped deep within the bed. Often manufactures of these types of systems claim 90 % removal of 0.5 μm particles and larger. This can be a misleading statement as quite often only about 5 % of the 0.5 μm particles will be removed. Grouping the 0.5 μm particles with much larger particles allows the claim to be met by removing a few large volume particles from the tower sump, even though the vast majority of fine particles remain to foul heat exchange equipment.

A typical physical-chemical treatment system incorporates three "dual" medial (sand anthracite) filters connected in parallel in its treatment train. The major maintenance consideration with granular medial filtration is the handling of the backwash. The backwash will generally contain a high concentration of contaminants and require subsequent treatment.

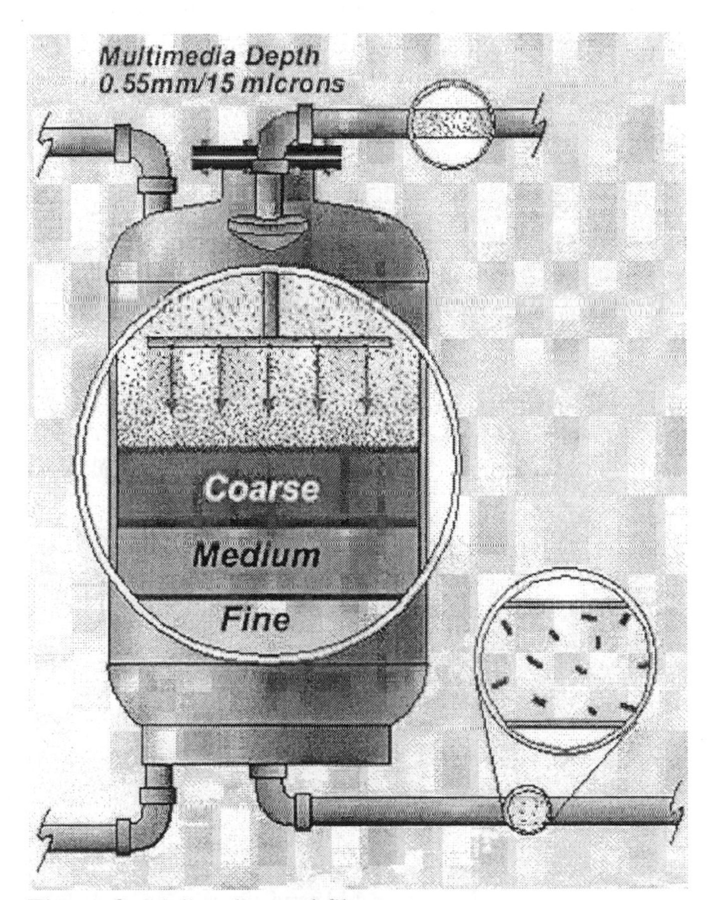

Figure 3. Mulimedia sand filter.

In this application, the operations of precipitation and flocculation play important roles. Precipitation is a physiochemical process whereby some, or all, of a substance in solution is transformed into a solid phase. It is based on alteration of the chemical equilibrium relationships affecting the solubility of inorganic species. Removal of metals as hydroxides and sulfides is the most common precipitation application in wastewater treatment. Lime or sodium sulfide is added to the wastewater in a rapid mixing tank along with flocculating agents. The wastewater flows to a flocculation chamber in which adequate mixing and retention time is provided for agglomeration of precipitate particles. Agglomerated particles are then separated from the liquid phase by settling in a sedimentation chamber, and/or by other physical processes such as filtration.

Precipitation is often applied to the removal of most metals from wastewater including zinc, cadmium, chromium, copper, fluoride, lead, manganese, and mercury. Also, certain anionic species can be removed by precipitation, such as phosphate, sulfate, and fluoride. Note that in some cases, organic compounds may form organometallic complexes with metals, which could inhibit precipitation. Cyanide and other ions in the wastewater may also complex with metals, making treatment by precipitation less efficient. A cutaway view of a rapid sand filter that is most often used in a municipal treatment plant is illustrated in Figure 4. The design features of this filter have been relied upon for more than 60 years in municipal applications.

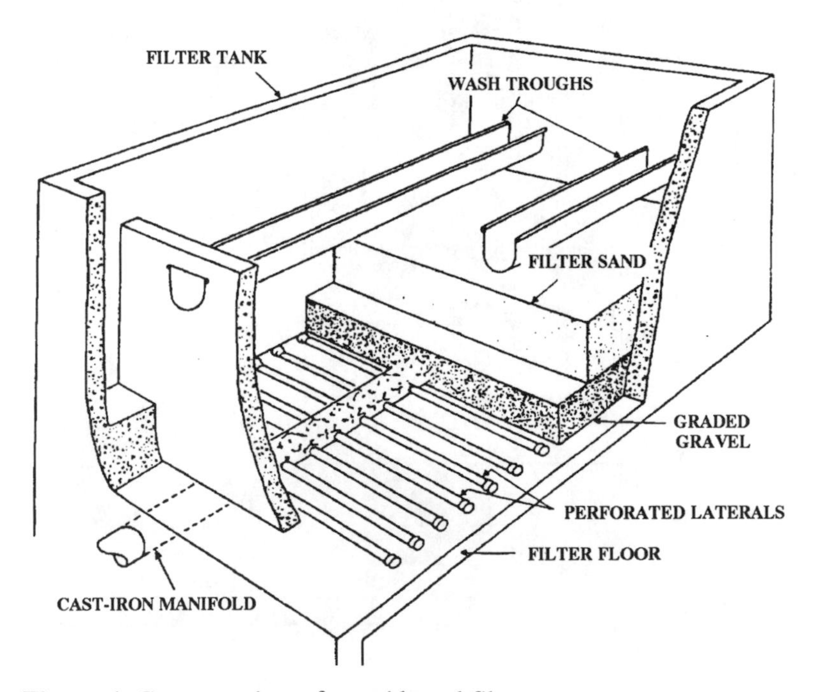

Figure 4. Cutaway view of a rapid sand filter.

LET'S TAKE A CLOSER LOOK AT SAND FILTERS

A typical sand filter system consists of two or three chambers or basins. The first is the sedimentation chamber, which removes floatables and heavy sediments. The second is the filtration chamber, which removes additional pollutants by filtering the runoff through a sand bed. The third is the discharge chamber. The treated filtrate normally is then discharged through an underdrain system either to a storm drainage system or directly to surface waters. Sand filters are able to achieve high removal efficiencies for sediment, biochemical oxygen demand (BOD), and fecal coliform bacteria. Total metal removal, however, is moderate, and nutrient removal is often low. Figure 5 illustrates one type of configuration. Typically, sand filters begin to experience clogging problems within 3 to 5 years. Accumulated trash, paper, debris should be removed every six months or as needed. Corrective maintenance of the filtration chamber includes removal and replacement of the top layers of sand and gravel as they become clogged. Table 1 provides some typical removal efficiencies for specific pollutants.

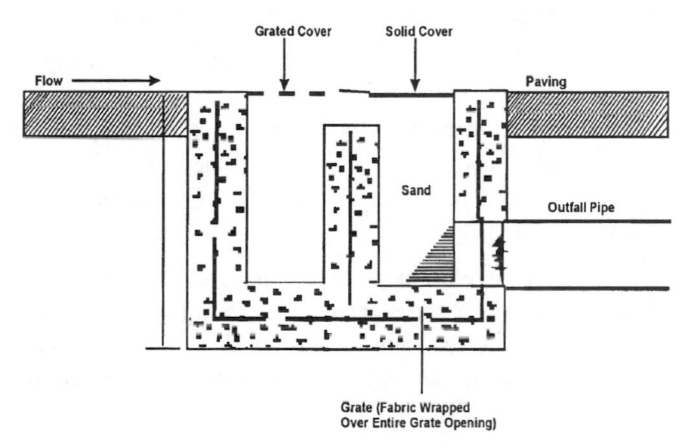

Figure 5. Example of a sand filter configuration.

Table 1. Typical removal efficiencies.

Pollutant	Percent Removal	Pollutant	Percent Removal
Fecal coliform	76	Total Organic Carbon (TOC)	48
Biochemical Oxygen Demand (BOD)	70	Total Nitrogen (TN)	21
Total Suspended Solids (TSS)	70	Iron, Lead, Zinc	45

PRECIPITATION, FLOCCULATION AND AGGLOMERATION

We will be examining these subjects in a little more detail in the next chapter. But for now, we should cover some of the basics because of their importance to sand filtration. The process of ***flocculation*** is applicable to aqueous waste streams where particles must be agglomerated into larger more settleable particles prior to sedimentation or other types of treatment. Highly viscous waste streams will inhibit the settling of solids. In addition to being used to treat waste streams, precipitation can also be used as an in situ process to treat aqueous wastes in surface impoundments. In an in-situ application, lime and flocculants are added directly to the lagoon, and mixing, flocculation, and sedimentation are allowed to occur within the lagoon.

Note that flocculation is a purely physical process in which the treated water is gently stirred to increase interparticle collisions and, thus, promote the formation of large particles. After adequate flocculation, most of the aggregates will settle out during the 1 to 2 hours of sedimentation.

Precipitation and flocculation can be integrated into more complex treatment systems. The performance and reliability of these processes depends greatly on the variability of the composition of the waste being treated. Chemical addition must be determined using laboratory tests and must be adjusted with compositional changes of the waste being treated or poor performance will result.

Precipitation is nonselective in that compounds other than those targeted may be removed. Both precipitation and flocculation are nondestructive and generate a large volume of sludge which must be disposed of. Coagulation, flocculation, sedimentation, and filtration, are typically followed by chlorination in municipal wastewater treatment processes.

Coagulation involves the addition of chemicals to alter the physical state of dissolved and suspended solids. This facilitates their removal by sedimentation and filtration. The most common primary coagulants are alum ferric sulfate and ferric chloride. Additional chemicals that may be added to enhance coagulation include activate silica, a complex silicate made from sodium silicate, and charged organic molecules called polyelectrolytes, which include large-molecular-weight polyacrylamides, dimethyl-diallylammonium chloride, polyamines, and starch.

These chemicals ensure the aggregation of the suspended solids during the next treatment step-flocculation. Sometimes polyclectrolytes (usually polyacrylamides) are also added after flocculation and sedimentation as an aid to the filtration step.

Coagulation may also remove dissolved organic and inorganic compounds. The hydrolyzing metal salts may react with the organic matter to form a precipitate, or they may form aluminum hydroxide or ferric hydroxide floc particles on which the organic molecules adsorb. The organic substances are then removed by sedimentation and filtration, or filtration alone if direct filtration or inline filtration is used. Adsorption and precipitation also removes inorganic substances.

The process of ***sedimentation*** involves the separation from water, by gravitational settling of suspended particles that are heavier than water. The resulting effluent is then subject to rapid filtration to separate out solids that are still suspended in the water. Rapid filters typically consist of 24 to 36 inches of 0.5 to 1-mm diameter sand and/or anthracite. Particles are removed as water is filtered through the media at rates of 1 to 6 gallons/minute/square foot. Rapid filtration is effective in removing most particles that remain after sedimentation. The substances that are removed by coagulation, sedimentation, and filtration accumulate in sludge which must be properly disposed of.

Coagulation, flocculation, sedimentation, and filtration will remove many contaminants. Perhaps most important is the reduction of turbidity. This treatment yields water of good clarity and enhances disinfection efficiency. If particles are not removed, they harbor bacteria and make final disinfection more difficult.

HYDRAULIC PERFORMANCE

The hydraulic performances required of the sand with slow filters are inferior to those for rapid filters. In the case of slow filters, one can use fine sand, since the average filtration velocity that is usually necessary lies in the range 2 to 5 m/day.

In slow filtration, much of the effect is obtained by the formation of a filtration layer, including the substances that are extracted from the water. At the early stages of the operation, these substances contain microorganisms able to effect, beyond the filtration, biochemical degradation of the organic matter. This effect also depends on the total surface of the grains forming the filter material. The probability of contact between the undesirable constituents of the water and the surface of the filter medium increases in proportion to the size of the total surface of the grains.

Sand often contains undesirable impurities, and additionally it can have broad particle size distributions. Sand that is used in filtration must be free of clay, dust, and other impurities. The ratio of lime, lime-stone, and magnesium oxide will have to be lower than 5 weight percent. The standard guide value of the quality of fresh sand is to be below 2% soluble matter at 20 °C within 24 hours in hydrochloric acid of a 20 weight percent concentration.

The actual diameter of the sands used during slow filtration typically lies between 0.15 and 0.35 mm. It is not necessary to use a ganged sand. The minimum thickness of the layer necessary for slow filtration is 0.3 to 0.4m, and the most efficient filtration thickness typically is at 2 to 3 cm.

The actual requirements for the sand in slow filtration are chemical in nature. Purity and the absence of undesirable matters are more important than grain-size distribution in the filtration process. On the other hand, the performance of rapid filters requires sands with quite a higher precise grain size. In the case of rapid filtration, the need for hydraulic performances is greater than in slow nitration. This means that the grain-size distribution of the medium is of prime concern in the latter case.

In wastewater treatment plants, the purity of the sand media used must be examined regularly. In addition, both the head loss of the filter beds and an analysis of the wash water during the operation of washing the filters must be checked regularly. Special attention must also be granted to the formation of agglomerates. The presence of agglomerates is indicative of insufficient washing and the possible formation of undesirable microbiological development zones within the filter bed.

The primary mechanisms that control the operation of sand filtration are:

- Straining
- Settling
- Centrifugal action
- Diffusion
- Mass attraction, or the effect of van der Waals forces
- Electrostatic attraction

Straining action consists of intercepting particles that are larger than the free interstices left between the filtering sand grains. Assuming spherical grains, an evaluation of the interstitial size is made on the basis of the grains' diameter (specific diameter), taking into account the degree of nonhomogeneity of the grains.

Porosity constitutes a important criterion in a description based on straining. Porosity is determined by the formula V_L/V_C, in which V_C is the total or apparent volume limitated by the filter wall and V_L is the free volume between the particles. The porosity of a filter layer changes as a function of the operation time of the filters. The grains become thicker because of the adherence of material removed from the water, whether by straining or by some other fixative mechanism of particles on the filtering sand. Simultaneously the interstices between the grains diminish in size. This effect assists the filtration process, in particular for slow sand filters, where a deposit is formed as a skin or layer of slime that has settled on the bed making up the active filter. Biochemical transformations occur in this layer as well, which are necessary to make slow filters efficient as filters with biological activity.

Filtration occurs correctly only after buildup of the sand mass. This formation includes a "swelling" of the grains and, thus, of the total mass volume, with a corresponding reduction in porosity. The increases and swellings are a result of the formation of deposits clinging to the empty zones between grains.

The porosity of a filter mass is an important factor. This property is best defined by experiment. A general rule of thumb is that for masses with the effective size greater than 0.4 – 0.5 mm and a specific maximum diameter below 1.2 mm the porosity is generally between 40 and 55 % of the total volume of the filter mass. Layers with spherical grains are less porous than those with angular material.

The second important mechanism in filtration is that of settling. From Stoke's law of laminar particle settling, the settling velocity of a particle is given by :

$$U = \frac{1}{18}\frac{g}{\nu}\frac{\Delta\rho}{\rho}D^2 \qquad (1)$$

where :

ρ	=	volumetric mass density of the water
$\rho, \Delta\rho$	=	volumetric mass density of the particles in suspension
D	=	diameter of the particles
g	=	$9.81 m/s^2$
ν	=	kinematic viscosity (e.g., 10^{-4} m/s at 20°C)

In sedimentation zones the flow conditions are laminar. A place is available for the settling of sludges contained in the water to be filtered.

Although the total inner surface that is available for the formation of deposits in a filter sand bed is important, only a part of this is available in the laminar flow zones that promote the formation of deposits. Usually material with a volumetric mass slightly higher than that of water is eliminated by sedimentation during filtration. Such matter could be, for example, organic granules or particles of low density. In contrast, colloidal material of inorganic origin-sludge or clay, for instance—with a diameter of 1 – 10 μm is only partially eliminated by this process, in which case the settling velocities in regard to the free surface become insufficient for sedimentation.

The trajectory followed by water in a filter mass it is not linear. Water is forced to follow the outlines of the grains that delineate the interstices. These changes in direction are also imposed on particles in suspension being transported by the water. This effect leads to the evacuation of particles in the dead flow zones. Centrifugal action is obtained by inertial force during flow, so the particles with the highest volumetric mass are rejected preferentially.

Diffusion filtration is another contributor to the process of sand filtration. Diffusion in this case is that of Brownian motion obtained by thermal agitation forces. This compliments the mechanism in sand filtration. Diffusion increases the contact probability between the particles themselves as well as between the latter and the filter mass. This effect occurs both in water in motion and in stagnant water, and is quite important in the mechanisms of agglomeration of particles (e.g., flocculation).

The next mechanism to consider is the mass attraction between particles which is due to van der Waals forces. These are universal forces contributing to the transport and fixation mechanism of matter. The greater the inner surface of the filters, the higher is the probability of attractive action. *Van der Waals forces* imply short molecular distances, and generally play a minor role in the filtration process. Moreover, they decrease very quickly when the distance between supports and particles increases. Nevertheless, the indirect effects, which are able to provoke an agglomeration of particles and, thus, a kind of flocculation, are not to be neglected and may become predominant in the case of flocculation-filtration, or more generally in the case of filtration by flocculation. Electrostatic and electrocinetic effects are also factors contributing to the filtration process. Filter sand has a negative electrostatic charge. Microsand in suspension presents an electrophoretic mobility. The value of the electrophoretic mobility, or of the corresponding zeta potential, depends on the pH of the surrounding medium. Usually a coagulation aid is used to condition the surface of microsand. In filtration without using coagulant aids, other mechanisms may condition the mass more or less successfully. For instance, the formation of deposits of organic matter can modify the electrical properties of the filtering sand surfaces. These modifications promote the fixation of particles by electrokinetic and electrostatic processes, especially coagulation. Also, the addition of a neutral or indifferent electrolyte tends to reduce the surface potential of the filtering sand by compression of the double electric layer. This is based on the principles of electrostatic coagulation. The sand, as the carrier of a negative charge spread over the surface of the filter according to the model of the double layer, will be able to fix the electropositive particles more exhaustively. This has a favorable effect on the efficiency of filtration of precipitated carbonates or of flocs of iron or aluminum hydroxide-oxide. Optimal adherence is obtained at the isoelectric point of the filtrated material. In contrast, organic colloidal particle carriers of a negative charge such as bacteria are repulsed by the electrostatic mechanism in a filter with a fresh filter mass. In this case, the negative charges of the sand itself appear unchanged. With a filter that is conditioned in advance, there are sufficient positively charged sites to make it possible to obtain an electrochemical fixation of the negative colloids.

BED REGENERATION

In addition to washing the bed, a degradated mass containing agglomerates or fermentation zones (referred to as mud balls) can be regenerated by specific treatment techniques. Among the regeneration techniques that are usually used are sodium chloride, regeneration through application of chlorine, and treatment with potassium permanganate, hydrogen peroxide, or caustic soda. Cleaning methods based on the use of caustic soda are aimed at eliminating thin clay, hydrocarbons, and gelatinous aggregates that form in filtration basins. After the filter has been carefully washed with air and water or only with water, according to its specific operating scheme, a quantity of caustic soda is spread over a water layer approximately 30 cm thick above the filter bed. The solution is then diffused in the mass by slow infiltration. After about 6 to 12 hours, the filter is washed very carefully. Sodium chloride is used specifically for rapid filters. The cleaning solution is spread in solution form in a thin layer of water above the freshly washed sand bed. After 2 or 3 hours of stagnation, slow infiltration in the mass is achieved by opening an outlet valve for the filtered water. The brine is then allowed to work for about a 24 hour period. The filter is placed back into service after a thorough washing. Sodium chloride works on proteinic agglomerates, which are bacterial in origin.

The use of potassium permanganate ($KMnO_4$) is applied to filters clogged with algae. A concentrated solution containing potassium permanganate is spread at an effective concentration over the surface of the filters to obtain, a characteristic pink-purple color on the top of the mass and allowed to infiltrate the bed for a 24 hour period. After this operation, the filter is carefully washed once again.

PPM is the abbreviation for parts per million. A part per million is equal to a milligrams per liter. Parts per million refers to a ratio of weights. For example, one part per million is one pound in a million pounds, one ounce in a million ounces, one gram in a million grams, etc.

Hydrogen peroxide is typically used in the range of 10 to 100 ppm. The cleaning method is similar to that used for permanganate. The addition of phosphates or polyphosphates makes it easier to remove ferruginous deposits. This method can be used in situ for surging the isolation sands of the wells. Adjunction of a reductor as bisulfite can be useful to create anaerobic conditions for the elimination of nematodes and their eggs when a filter has been infected. Hydrochloric acid solution is applied to the recurrent cleaning of rapid filters for sand, iron, and manganese removal. This operation has the advantage of causing the formation of chlorine in situ which acts as a disinfectant. Instantaneous cleaning of a filtering sand bed can be accomplished by the use of chlorine. A water layer is typically used as a dispersion medium. Further infiltration of the solution is obtained by percolation into the bed. The action goes on for several hours, after which the filter is washed. Chlorine is used from concentrated solutions of sodium hypochlorite. An alternative method involves the application of dioxide. This method has the advantage of arresting the formation of agglomerates of biological origin by permanent treatment of the filter wash water with chlorine.

FLOCCULATION AND FILTRATION TOGETHER

The sand filtration process is normally comprised of a clarification chain including other unit operations which precede filtration in the treatment sequence and can not be conceived of completely independent of the filtration stage. The conventional treatment scheme consists of coagulation-flocculation-settling followed by filtration. When the preceding process, in this case flocculation and/or settling, becomes insufficient, subsequent rapid filtration can be used to ensure a high quality of the effluent treated. However, this action is achieved at the expense of the evolution of filter head loss. Problems in washing and cleanliness of the mass may arise. Filtration is often viewed as serving as a coagulant flocculator. This is referred to as flocculation-filtration. The presence of thin, highly electronegative colloids (e.g., activated carbons) introduced in the form of powder in the settling phase may be a problem for the quality of the settled effluent. The carbon particles, which are smaller than 50 μm, penetrate deeply into the sand filter beds. They may rapidly provoke leakage of rapid filters. The same holds for small colloids other than activated carbon. Activated silica, which may have a favorable or an unfavorable effect on filtration, is composed of ionized micella formed by polysilicic acid-sodium polysilicate. This become negatively charged colloidal micella. The behavior of activated silicas depends on the conditions of neutralization and the grade of the silicate used in the preparation of the material. Activated silica is a coagulant aid that contributes to coalescence of the particles. Hence, it brings about an improvement in the quality of settled or filtrated water, depending on the point at which it is introduced. Preconditioning of the sand surface of filters by adding polyelectrolytes is an alternative use of sand filters as a coagulator-flocculator. In the treatment of drinking water the method depends on the limitations of these products in foodstuffs. The addition of polyphosphates to a water being subjected to coagulation usually has a negative effect; specifically the breaking of the agglomeration velocity of the particles during flocculation will occur in sand filtration. The addition of polyphosphates simultaneously with phosphates can be of value in controlling corrosion. This sometimes makes it possible to avoid serious calcium carbonate precipitation at the surface of filter grains when handling alkaline water. The application concerns very rapidly incrusting water while maintaining high hardness in solution. The addition of polyphosphates involves deeper penetration of matter into the filter mass. Hence, the breaking of flocculation obtained by the action of polyphosphates enables the thinner matters to penetrate the filters more deeply. These products favor the "in-depth effects" of the filter beds. Their use necessitates carefully checking that they are harmless from a hygienic point of view. The depth penetration of material in coagulation-filtration is almost opposite to the concept of using the filter as a screen. Precipitation initiated by germs plays a significant role. Empirical relations are normally relied on in the design of filters as a function of the penetration in depth of coagulated material. The concentration of those residual matters in filtered water (C_f) depends on several factors: the linear infiltration rate (ν_f), the effective size of the filter medium (ES), the porosity of the filter medium (ϵ), the final loss of head of the filter bed (Δh), the depth of penetration of the coagulated matter (l), the concentration of the particles in suspension in the water to be filtered (C_0), and the water height (H). The following generalized relation is often found among the filtration engineer's notes.

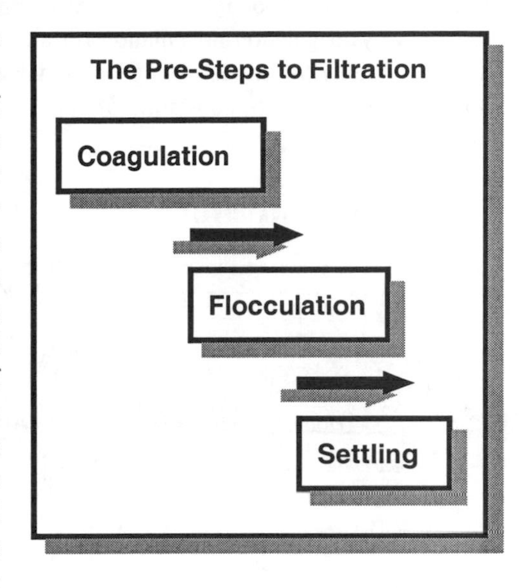

$$C_f = f\left(\nu_f \times (ES) \times \epsilon^4 \times \frac{\Delta h}{l} \times C_0 \times H\right) \tag{2}$$

It should be noted that the total loss of head of a filter bed is in inverse ratio to the depth of penetration of the matter in suspension. In a normal wastewater treatment plant, the water is brought onto a series of rapid sand filters and the impurities are removed by coagulation-flocculation-filtration. Backwashing is typically performed in the counterflow mode, using air and water. One type of common filter is illustrated in Figure 6, consisting of closed horizontal pressurized filters.

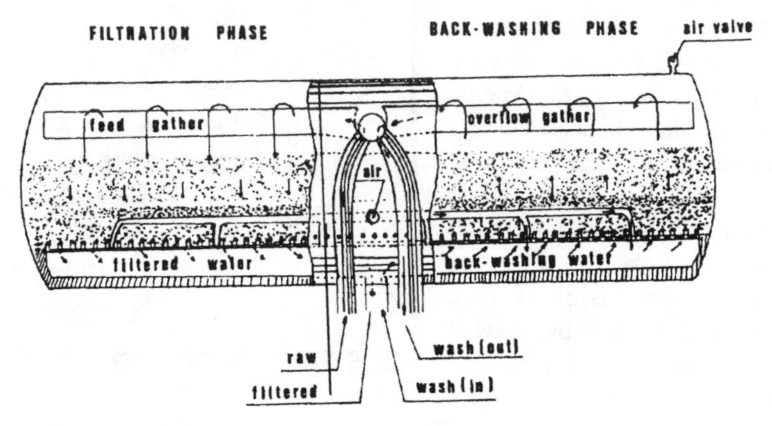

Figure 6. Cross section of a typical filtration unit.

SLOW SAND FILTRATION

Slow sand filtration involves removing material in suspension and/or dissolved in water by percolation at slow speed. In principle, a slow filter comprises a certain volume of areal surface, with or without construction of artificial containment, in which filtration sand is placed at a sufficient depth to allow free flow of water through the bed. When the available head loss reaches a limit of approximately 1 m, the filter must be pulled out of service, drained, and cleaned. The thickness of the usual sand layer is approximately of 1 to 1.50 m, but the formation of biochemically active deposits and clogging of the filter beds takes place in the few topmost centimeters of the bed. The filter mass is pored onto gravels of increasing permeability with each layer having a thickness of approximately 10 to 25 cm. The lower-permeability layer can reach a total thickness of 50 to 60 cm. So-called gravels 18 to 36 cm in size are used and their dimensions are gradually diminished to sizes of 10 to 12 cm or less for the upper support layer. The sand filter must be cleaned by removal of a few centimeters of the clogged layer. This layer is washed in a separate installation. The removal of the sand can be done manually or by mechanical means. The removed sand may not be replaced entirely by fresh sand. Placing preconditioned and washed sand is recommended as this takes into account the biochemical aspects involved in slow filtration. An alternative to manual or mechanical removal involves cleaning using a hydraulic system. Sometimes slow filtration is used without previous coagulation. This is generally practiced with water that does not contain much suspended matter. If the water is loaded (periodically or permanently) with clay particles in suspension, pretreatment by coagulation-flocculation is necessary. Previous adequate oxidation of the water, in this case preozonization producing biodegradable and metabolizable organic derivatives issuing from dissolved substances, can be favorable because of the biochemical activity in slow filters. There are several disadvantages to the use of slow filters. They may require a significant surface area and volume, and may therefore involve high investment costs. They are also not flexibile — mainly during the winter, when the open surface of the water can freeze. During the summer, if the filters are placed in the open air, algae may develop, leading to rapid clogging during a generally critical period of use. Algae often cause taste and odor problems in the filter effluent. Additional construction costs to cover slow filters are often necessary.

RAPID SAND FILTRATION

Rapid filtration is performed either in open gravitational flow filters or in closed pressure filters. Rapid pressure filters have the advantage of being able to be inserted in the pumping system, thus allowing use of a higher effective loading. Note that pressure filters are not subject to development of negative pressure in a lower layer of the filter. These filters generally support higher speeds, as the available pressure allows a more rapid flow through the porous medium made up by the filter sand. Pressure filtration is generally less efficient than the rapid open type with free-flow filtration. Pressure filters have the following disadvantages. The injection of reagents is complicated, and it is more complicated to check the efficiency of backwashing. Work on the filter mass is difficult considering the assembly and disassembly required. Also, the risk of breakthrough by suction increases. Another disadvantages is that pressure filterts need a longer filtration cycle, due to a high loss of head available to overcome clogging of the filter bed.

Another option is to use open filters, which are generally constructed in concrete. They are normally rectangular in configuration. The filter mass is posed on a filter bottom, provided with its own drainage system, including bores that are needed for the flow of filtered water as well as for countercurrent washing with water or air.

There are several types of washing bottoms. One type consists of porous plates which directly support the filter sand, generally without a layer of support gravel. Even if the system has the advantage of being of simple construction, it nevertheless suffers from incrustation. This is the case for softened water or water containing manganese. Porous filters bottoms are also subject to erosion or disintegration upon the filtration of aggressive water.

The filter bottom is often comprised of pipes provided with perforations that are turned toward the underpart of the filter bottom and embedded in gravel. The lower layers are made up of gravel of approximate diameter 35 - 40 mm, decreasing up to 3 mm. The filter sand layer, located above this gravel layer, serves as a support and equalization zone. Several systems of filter bottoms comprise perforated self-supporting bottoms or false bottoms laid on a supporting basement layer. The former constitutes a series of glazed tiles, which includes bores above which are a series of gravels in successive layers.

All these systems are surpassed to some extent by filter bottoms in concrete provided with strainers. The choice of strainers should in part be based on the dimensions of the slits that make it possible to stop the filter sand, which is selected as a function of the filtration goal. Obstruction or clogging occurs only rarely and strainers are sometimes used.

Strainers may be of the type with an end that continues under the filter bottom. These do promote the formation of an air space for backwashing with air. If this air space is not formed, it can be replaced by a system of pipes that provide for an equal distribution of the washing fluids.

Pressure filters are worth noting. These are usually set up in the form of steel cylinders positioned vertically. Another variation consists of using horizontal filtration groups. This has the drawback that the surface loading is variable in the different layers of the filter bed; moreover, it increases with greater penetration in the filter bed (the infiltration velocity is lowest at the level of the horizontal diameter of the cylinder). The filter bottom usually consists of a number of screens or mesh sieves that decrease in size from top to bottom or, as an alternative, perforated plates supporting gravel similar to that used in the filter bottoms of an open filter system.

Filter mass washing can influence the quality of water being filtered. Changes may be consequent to fermentation, agglomeration, or formation of preferential channels liable to occur if backwashing is inadequate.

Backwashing requires locating a source that will supply the necessary flow and pressure of wash water. This water can be provided either by a reservoir at a higher location or by a pumping station that pumps treated water. Sometimes an automated system is employed with washing by priming of a partial siphon pumping out the treated water stored in the filter itself. An example is shown in Figure 7. The wash water must have sufficient pressure to assure the necessary flow. Washing of the filter sands is accomplished followed by washing with water and in most cases including a short intermediate phase of simultaneous washing with air and water. Due to greater homogenization of the filter layer and more efficient washing, the formation of fermentation areas and agglomerates in the filter mass of treatment plants for surface water (mud balls) is diminished. The formation of a superficial crust on the filter sand is avoided by washing with air. After washing with air, water flow is gradually superimposed on the air flow. This operational phase ends at the same time that the wash air is terminated, to avoid the filter mass being blown away. The wash water contains materials that eventually require treatment in a sludge treatment plant. Their concentration varies as a function of the washing cycle. Accounting for the superficial load in filtration, velocity of the wash water, and length of the filtration cycle, it may be assumed that the water used for washing will not attain 5 % of the total production.

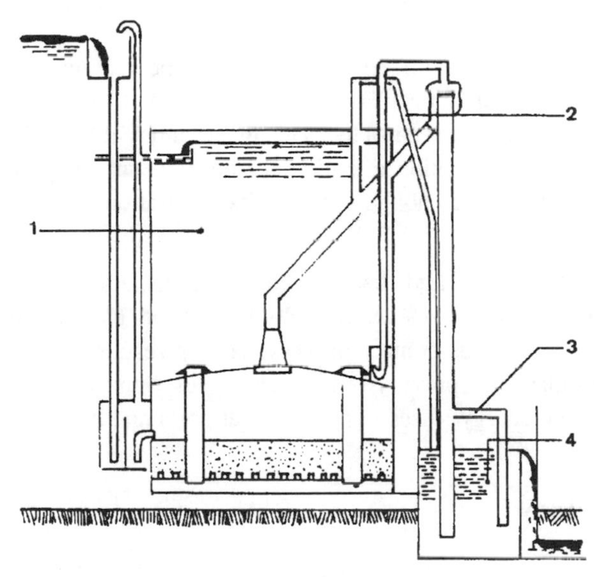

Figure 7. Automatic backwashing filter with a partial siphon system: 1-filtered water (reserve); 2-partial siphoning; 3-initiation; 4-restitution.

For new installations the first washing cycles result in the removal of fine sand as well as all the other materials usually undesirable in the filter mass, such as particles of bitumen on the inner surface of the water inlet or other residuals from the crushing or straining devices of the filter media. Consequently, it is normal that at the beginning of operation of a filter sand installation, dark colored deposits appear at the surface of the filter mass. In the long term they have no consequence and disappear after a few filtration and wash cycles. If, after several weeks of filtration, these phenomena have not disappeared, it will be necessary to examine the filter sand. The elimination of fine sand must stop after 1 or 2 months of activity. If this sand continues to be carried away after the first several dozen washings it is necessary to reexamine the hydraulic criteria of the washing conditions, the granulometry of the filter mass, and the filter's resistance to shear and abrasion.

CHEMICAL MIXING, AND SOLIDS CONTACT PROCESSES

Chemical mixing and flocculation or solids contact are important mechanical steps in the overall coagulation process. Application of the processes to wastewater generally follows standard practices and employs basic equipment. Chemical mixing thoroughly disperses coagulants or their hydrolysis products so the maximum possible portion of influent colloidal and fine supracolloidal solids are absorbed and destabilized. Flocculation or solids contact processes increase the natural rate of contacts between particles. This makes it possible, within reasonable detention periods, for destabilized colloidal and fine supracolloidal solids to aggregate into particles large enough for effective separation by gravity processes or media filtration.

These processes depend on fluid shear for coagulant dispersal and for promoting particle contacts. Shear is most commonly introduced by mechanical mixing equipment. In certain solids contact processes shear results from fluid passage upward through a blanket of previously settled particles. Some designs have utilized shear resulting from energy losses in pumps or at ports and baffles.

CHEMICAL MIXING

Chemical mixing facilities should be designed to provide a thorough and complete dispersal of chemical throughout the wastewater being treated to insure uniform exposure to pollutants which are to be removed. The intensity and duration of mixing of coagulants with wastewater must be controlled to avoid overmixing or undermixing. Overmixing excessively disperses newly-formed floc and may rupture existing wastewater solids. Excessive floc dispersal retards effective flocculation and may significantly increase the flocculation period needed to obtain good settling properties. The rupture of incoming wastewater solids may result in less efficient removals of pollutants associated with those solids. Undermixing inadequately disperses coagulants resulting in uneven dosing. This in turn may reduce the efficiency of solids removal while requiring unnecessarily high coagulant dosages. In water treatment practice several types of chemical mixing units are typically used. These include high-speed mixers, in-line blenders and pumps, and baffled mixing compartments or static in-line mixers (baffled piping sections). An example of a high-speed mixer is shown in Figure 8. Designs usually call for a 10 to 30 second detention times and approximately 300 fps/ft velocity gradient. Variable-speed mixers are recommended to allow varying requirements for optimum mixing. In mineral addition to biological wastewater treatment systems, coagulants may be added directly to mixed biological reactors such as aeration tanks or rotating biological contactors. Based on typical power inputs per unit tank volume, mechanical and diffused aeration equipment and rotating fixed-film biological contactors produce average shear intensities generally in the range suitable for chemical mixing. Localized maximum shear intensities vary widely depending on the speed of rotating equipment or on bubble size for diffused aeration.

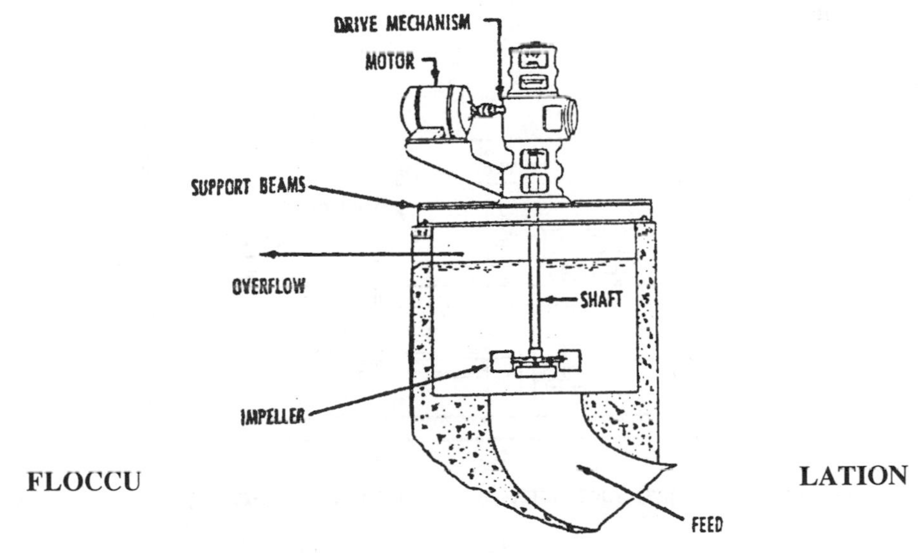

FLOCCU **LATION**

Figure 8. Example of an impeller mixer.

The proper measure of flocculation effectiveness is the performance of subsequent solids separation units in terms of both effluent quality and operating requirements, such as filter backwash frequency. Effluent quality depends greatly on the reduction of residual primary size particles during flocculation, while operating requirements relate more to the floc volume applied to separation units.

Flocculation units should have multiple compartments and should be equipped with adjustable speed mechanical stirring devices to permit meeting changed conditions. In spite of simplicity and low maintenance, non-mechanical, baffled basins are undesirable because of inflexibility, high head losses, and large space requirements. Mechanical flocculators may consist of rotary, horizontal-shaft reel units as shown in Figure 9.

Rotary vertical shaft turbine units as shown in Figure 10 and other rotary or reciprocating equipment are other examples. Tapered flocculation may be obtained by varying reel or paddle size on horizontal common shaft units or by varying speed on units with separate shafts and drives. In applications other than coagulation with alum or iron salts, flocculation parameters may be quite different. Lime precipitates are granular and benefit little from prolonged flocculation.

Polymers which already have a long chain structure may provide a good floc at low mixing rates. Often the turbulence and detention in the clarifier inlet distribution is adequate.

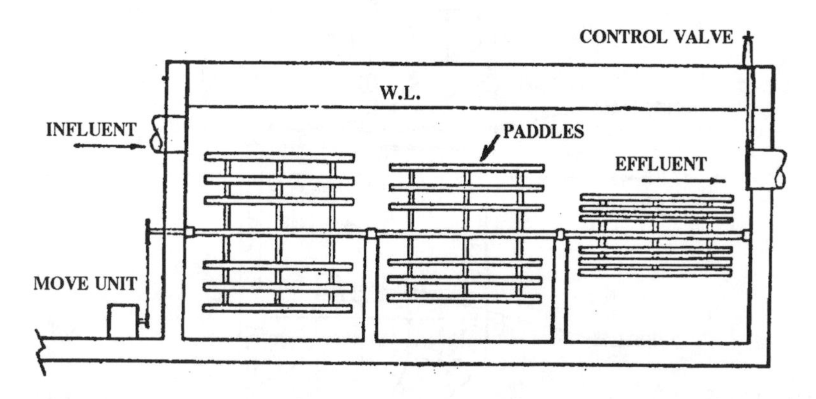

Figure 9. Mechanical flocculation basin horizontal shaft-reel type.

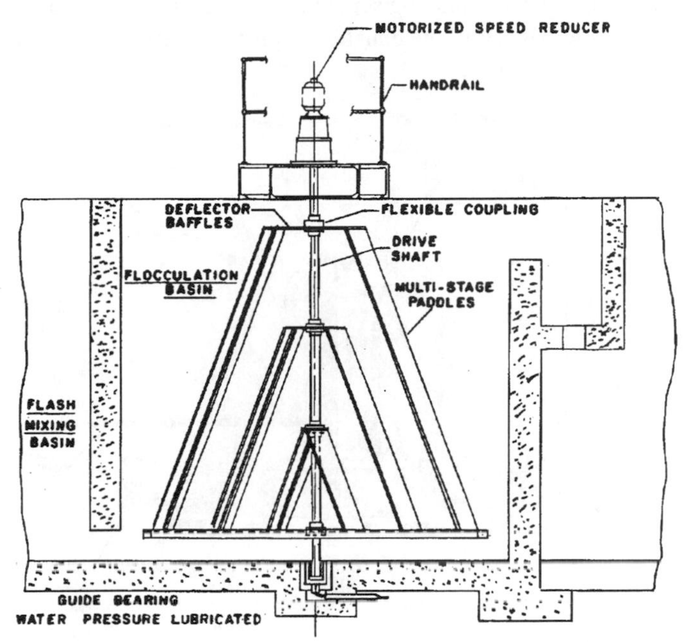

Figure 10. Mechanical flocculator vertical shaft-paddle type.

SOLIDS CONTACTING

Solids contact processes combine chemical mixing, flocculation and clarification in a single unit designed so that a large volume of previously formed floc is retained in the system. The floc volume may be as much as 100 times that in a "flow-through" system. This greatly increases the rate of agglomeration from particle contacts and may also speed up chemical destabilization reactions. Solids contact units are of two general types: slurry-recirculation and sludge-blanket. In the former, the high floc volume concentration is maintained by recirculation from the clarification to the flocculation zone, as illustrated in Figure 11. In the latter, the floc solids are maintained in a fluidized blanket through which the wastewater under treatment flows upward after leaving the mechanically stirred-flocculating compartment, as illustrated in Figure 12. Some slurry-recirculation units can also be operated with a sludge blanket. Solids contact units have the following advantages: Reduced size and lower cost result because flocculation proceeds rapidly at high floc volume concentration.; Single-compartment flocculatlon is practical because high reaction rates and the slurry effects overcome short circuiting; Units are available as compact single packages, eliminating separate units; Even distribution of inlet flow and the vertical flow pattern in the clarifier improve clarifier performances.

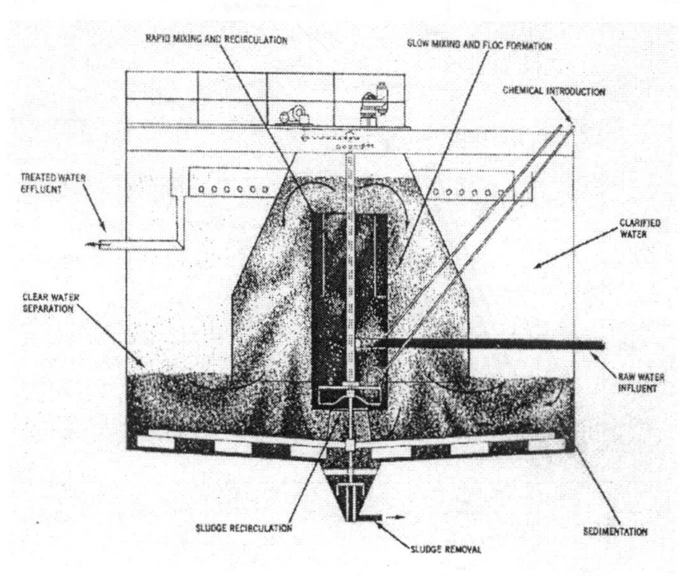

Figure 11. Solids contact clarifier without sludge blanket filtration .

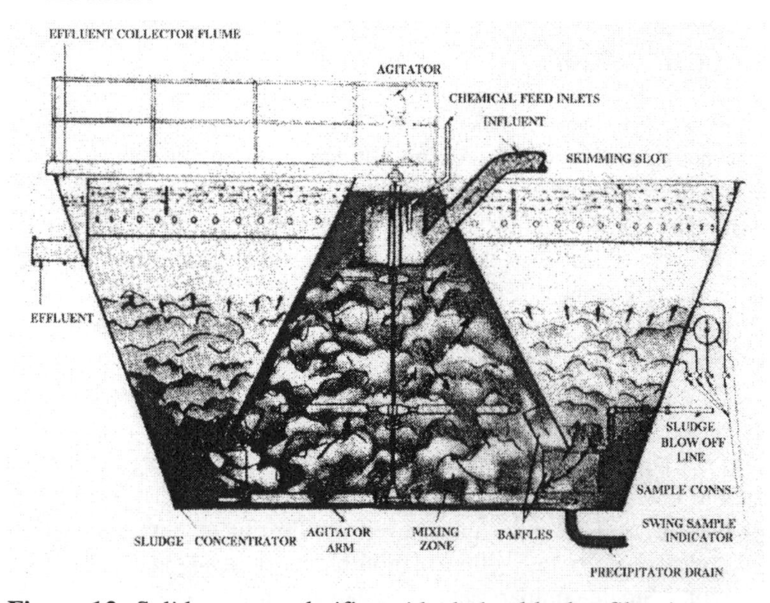

Figure 12. Solids contact clarifier with sludge blanket filtration.

Equipment typically consists of concentric circular compartments for mixing, flocculation and settling. Velocity gradients in the mixing and flocculation compartments are developed by turbine pumping within the unit and by velocity dissipation at baffles. For ideal flexibility it is desirable to independently control the intensity of mixing and sludge scraper drive speed in the different compartments.

Operation of slurry-recirculation solids contact units is typically controlled by maintaining steady levels of solids in the reaction zone. Design features of solids contact clarifiers should include:

1. Rapid and complete mixing of chemicals, feedwater and slurry solids must be provided. This should be comparable to conventional flash mixing capability and should provide for variable control, usually by adjustment of recirculator speed.
2. Mechanical means for controlled circulation of the solids slurry must be provided with at least a 3:1 range of speeds. The maximum peripheral speed of mixer blades should not exceed 6 ft/sec.
3. Means should be provided for measuring and varying the slurry concentration in the contacting zone up to 50 % by volume.
4. Sludge discharge systems should allow for easy automation and variation of volumes discharged. Mechanical scraper tip speed should be less than 1 fpm with speed variation of 3:1.
5. Sludge-blanket levels must be kept a minimum of 5 feet below the water surface.
6. Effluent launders should be spaced so as to minimize the horizontal movement of clarified water.

Further considerations include skimmers and weir overflow rates. Skimmers should be provided on all units since even secondary effluents contain some floatable solids and grease. Overflow rates and sludge scraper design should conform to the requirements of other clarification units.

RECOMMENDED RESOURCES FOR THE READER

There may be some golden oldies among these references, but the key word is "golden". Check out references 13 through 15 in particular.

1. Anon., Water Sewage Works, 6, 266 (1968).
2. Maeckelburg, D., G.W.F.,119,23 (1978).

3. O'Mella, Ch. R., and D.K. Crapps, J. AWWA, 56,1326 (1964).

4. Drapeau, A.J., and R.A. Laurence, Eau Quebec, 10, 314 (1977).

5. Burman, N.P., H_2O, 11, 348, (1978).

6. Cleasby, J.L., J. Arboleda, D.E. Burns, P.W. Prendiville, and E.S. Savage, J. AWWA, 69,115 (1977).

7. Cheremisinoff, P.N., Pollution Engineering Flow Sheets: Wastewater Treatment, Pudvan Publishing Co., Northbrook, IL, 1988.

8. Cheremisinoff, N.P., Biotechnology for Waste and Wastewater Treatment, Noyes Publication, Park Ridge, NJ, 1996.

9. Cheremisinoff, N.P. and P.N. Cheremisinoff, Carbon Adsorption for Pollution Control, Prentice Hall Publishers, Inc., Englewood, NJ ,1993.

10. Cheremisinoff, N.P. and P.N. Cheremisinoff, Liquid Filtration for Process and Pollution Control, SciTech Publishers, Inc., Morganville, NJ, 1981.

11. Cheremisinoff, N.P. and P.N. Cheremisinoff, Chemical and Non-Chemical Disinfection, Ann Arbor Science Publishers, Ann Arbor, MI, 1981.

12. Cheremisinoff, P.N. and R.B. Trattner, Fundamentals of Disinfection for Pollution Control, SciTech Publishers, Inc., Morganville, NJ, 1990.

13. Belfort, Georges "Evaluation of a Rapid Sand Filter", Filtration Experiment, Rensselaer Polytechnic Institute, 1990.

14. Ives, K.J. "Capture Mechanisms in Filtration", The Scientific Basis Of Filtration, Noordhoff Int. Publish. Co., Leyden, pp.55,93, 1975.

15. Yao, K-M., Habibian, M.T., and O'Melia, C.R., "Water and waste water Filtration: Concepts and Applications", Env. Sci. Tech., 5(11), 1105 (1971).

QUESTIONS FOR THINKING AND DISCUSSING

1. Calculate the settling velocity of coal particles in water at 20° C. The average size of the particles is 225 μm.

2. Develop a detailed process flow scheme for a wastewater treatment plant. The contaminants that are to be removed are TSS, BOD, E. coli, various nutrients, heavy grit, oil and grease.

3. Explain how the activated sludge process works.

4. Explain the difference between, and compare the advantages and disadvantages between rapid- sand and slow-sand filtrations.
5. Explain the process of digestion and when it should be used.
6. Make a list of chemical flocculating agents and relate each of these chemicals to the specific types of contaminants they are best suited to handle in wastewater. Obtain MSDS (Material Safety Data Sheets) for each of these chemicals.
7. Define hydraulic performance.
8. List the mechanisms and forces that are important to sand filtration.
9. What chemicals can be used for bed regeneration and how is this operation performed?
10. Define a colloidal suspension and give some examples.
11. Explain what is meant by a sludge displaying thixotropic behavior.
12. How does one control slurry recirculation of solids in the reaction zone?
13. Under what conditions would skimmers be used in wastewater treatment?

Chapter 8

SEDIMENTATION, CLARIFICATION, FLOTATION, AND COALESCENCE

INTRODUCTION

Suspended solid matter in a fluid media represents a heterogeneous system that from a very general viewpoint is a fundamental fluid dynamics problem that virtually all chemical engineers are familiar with. It is so frequently encountered in numerous industry applications that considering only the problem of water treatment simply does not do the subject justice. Having noted this, the principles covered in this chapter are general enough to be applied to the class of unit operations aimed at separation, handling or processing heterogeneous or two-phase systems, including those dealing with gas-solid suspensions.

Examples of operations where heterogeneous systems are encountered include sedimentation of dust in chambers and cyclone separators, separation of suspensions in settlers, separation of liquid mixtures by settling and centrifuging, hydraulic and pneumatic transport, hydraulic and air classification, flotation, mixing by air and others. Each of these operations involves the simultaneous flow of gas and solid, or liquid and solid phases. The widespread and successful application of these hydrodynamic processes to a large number of industrial problems is based on our ability to take advantage of one or a combination of five primary forces: gravity, centrifugal , buoyancy, pressure, and electric. Gravity is the controlling force for separations achieved in settlers; centrifugal force is applied to cyclone separators, dryers and mixers; and pressure forces are employed in sprayers, pneumatic transport and filters. Electrical forces are employed in special techniques, such as precipitators. Buoyancy is related to gravity and takes advantage of density differences.

You will find a very fundamental and in part, empirical approach to this subject. My experience has been for the majority of water treatment assignments over the

years that a "first-principles" approach to sedimentation and clarification techniques generally tends to get the job done. The techniques and methodologies covered have been around for quite a few years and still seem to be applied with a high degree of success in preliminary and even detailed equipment sizing and overall plant designs. There are certainly more sophisticate methods, including numerical modeling that enables one to develop reliable designs, but our intent is to look at rather simple calculation methods that enable us to define the type of equipment needed and just enough information so that we can work with an equipment vendor to define the hardware in greater detail. More sophisticated approaches will be needed for the detailed design and operation.
At the end of this chapter you will find three annexes. The first of these is a list of nomenclature used in the chapter. There are quite a few equations that are summarized in the foregoing sections and hence, you will need to refer to this annex from time to time. The second annex is a list of recommended references that I have relied on over the years, plus some interesting Web sites for you to visit for vendor-specific information as well as supplemental design and equipment sizing information. The final annex is the ***Questions for Thinking and Discussing***. Remember to refer to the ***Glossary*** at the end of the book if you run across any terms that are unfamiliar to you.

LET'S LOOK AT HOW A SINGLE PARTICLE BEHAVES IN A SUSPENSION

During the motion of viscous flow over a stationary body or particle, certain resistances arise. To overcome these resistances or drag and to provide more uniform fluid motion, a certain amount of energy must be expended. The developed drag force and, consequently, the energy required to overcome it, depend largely on the flow regime and the geometry of the solid body. Laminar flow conditions prevail when the fluid medium flows at low velocities over small bodies or when the fluid has a relatively high viscosity. Flow around a single body is illustrated in Figure 1. As shown in Figure 1(A), when the flow is laminar a well-defined boundary layer forms around the body and the fluid conforms to a streamline motion. The loss of energy in this situation is due primarily to fiction drag. If the fluid's average velocity is increased sufficiently, the influence of inertia forces becomes more pronounced and the flow becomes turbulent.
Under the influence of inertia forces, the fluid adheres to the particle surface, forming only a very thin boundary layer and generating a turbulent wake, as shown in Figure 1(B). The pressure in the wake is significantly lower than that at the stagnation point on the leeward side of the particle. Hence, a net force, referred to as the *pressure drag*, acts in a direction opposite to that of the fluid's motion. Above a certain value of the Reynolds number, the role of pressure drag becomes significant and the friction drag can be ignored.

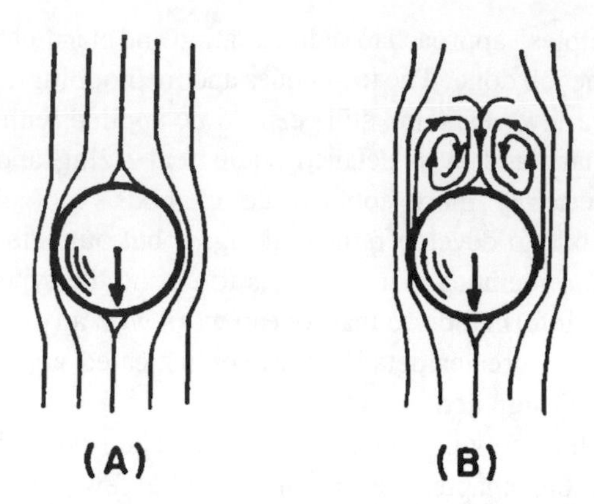

Figure 1. Flow around a single particle.

We shall begin discussions by analyzing a dilute system that can be described as a low concentration of noninteracting solid particles carried along by a water stream. In this system, the solid particles are far enough removed from one another to be treated as individual entities. That is, each particle individually contributes to the overall character of the flow. Let's consider the dynamics of motion of a solid spherical particle immersed in water independent of the nature of the forces responsible for its displacement. A moving particle immersed in water experiences forces caused by the action of the fluid. These forces are the same regardless whether the particle is moving through the fluid or whether the water is moving over the particle's surface. For our purposes, assume the water to be in motion with respect to a stationary sphere. The fluid shock acting against the sphere's surface produces an additional pressure, P. This pressure is responsible for a force, R (called the drag force) acting in the direction of fluid motion. Now consider an infinitesimal element of the sphere's surface, dF, having a slope, α, with respect to the normal of the direction of motion (Figure 2). The pressure resulting from the shock of the fluid against the element produces a force, dF, in the normal direction. This force is equal to the product of the surface area and the additional pressure, PdF_0. The component acting in the direction of flow, dR, is equal to $d\tau\cos\alpha$. Hence, the force, R, acting over the entire surface of the sphere will be:

$$R = \int PdF_0\cos\alpha = \int PdF = PF \tag{1}$$

where dF is the projection of dF_0 on the plane normal to the flow. The term F refers to a characteristic area of the particle, either the surface area or the maximum cross-sectional area perpendicular to the direction of flow. The pressure P represents the ratio of resistance force to unit surface area (R/F), and it depends on several factors, namely the diameter of the sphere (d), its velocity (u), the fluid density (ρ) and the fluid viscosity (μ).: i.e., $P = f(d, u, \rho, \mu)$. Applying dimensional analysis, the following dimensionless groups are identified:

$$Eu = \Phi(Re) \tag{2}$$

where Eu is the dimensionless Euler number, defined as $P/u^2\ \rho$, and Re is the Reynolds number ($Re = du\ \rho/\mu$). By substituting for density using the ratio of specific gravity to the gravitational acceleration, an expression similar to the well-known Darcy-Weisbach expression is obtained:

$$R/F = C_D(u^2/2g)\gamma \tag{3}$$

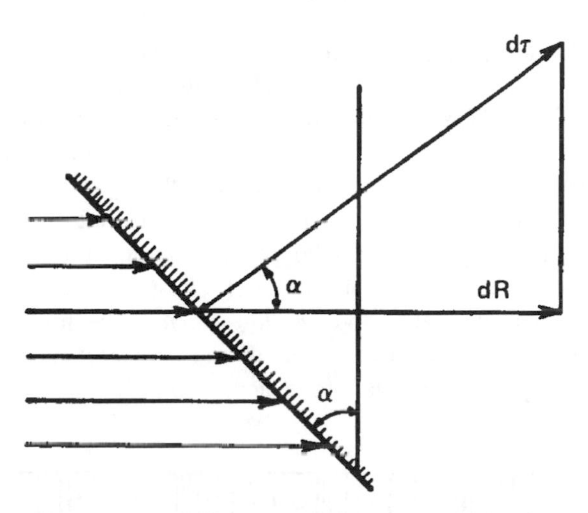

Figure 2. Infinitesimal element of a sphere's surface inclined at angle α from the direction of flow.

where C_D is the drag coefficient, which is a dimensionless parameter that is related to the Reynolds number. The relationship between C_D and Re for flow around a smooth sphere is given by the plot shown in Figure 3. As shown in this plot, there are three regions that can be approximated by expressions for straight lines. These three regions are the Stokes law region, Newton's law region, and the intermediate region. Refer to the sidebar discussion for these expressions. By substituting the expressions for the drag coefficient into equation (3), we obtain a convenient set of expressions that will enable us to calculate settling velocities. The details are left to you. There are plenty of empirical correlations in the literature for the drag

DRAG COEFFICIENTS

Laminar Regime (Stokes Law Region)
$C_D = 24/Re$; for Re < 2

Transitional Flow Regime (Intermediate Law Region)
$C_D = 18.5/Re^{0.6}$; for 2 <Re < 500

Turbulent Flow Regime (Newton's Law Region)
$C_D = 0.44$; for 500 < Re < 200,000

coefficient for different geometry objects, but practice is to use a simple sphere and then account for geometry effects by means of a correction or shape factor. We will cover this further on. For now, let's examine the phenomenon of particle settling more closely.

HOW PARTICLES SETTLE

If a particle at rest (with mass 'm' and weight 'mg') begins to fall under the influence of gravity, its velocity is increased initially over a period of time. The particle is subjected to the resistance of the surrounding water through which it descends. This resistance increases with particle velocity until the accelerating and resisting forces are equal. From this point, the solid particle continues to fall at a constant maximum velocity, referred to as the *terminal velocity,* u_t. You should recall from the last chapter that we called the *settling velocity*. The force responsible for moving a spherical particle of diameter 'd' can be expressed by the difference between its weight and the buoyant force acting on the particle.

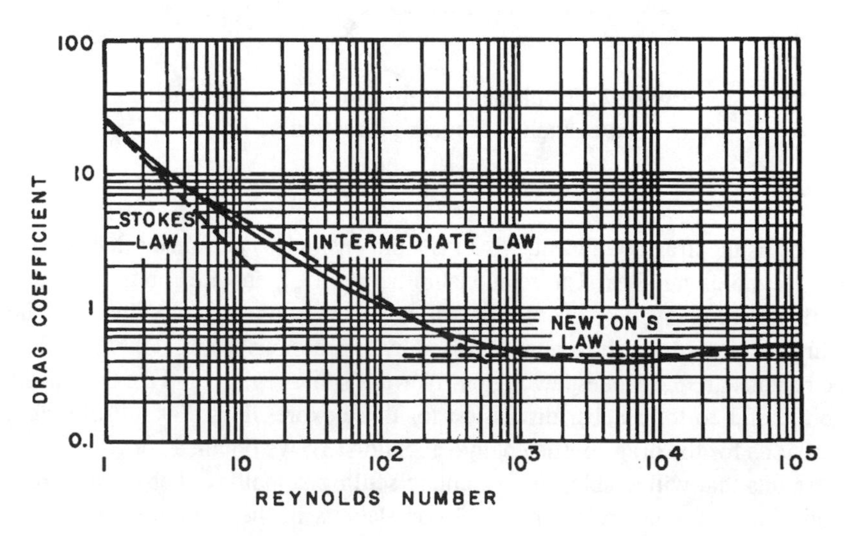

Figure 3. Drag coefficients for spheres.

The buoyant force is proportional to the mass of fluid displaced by the particle, that is, as the particle falls through the surrounding water, it displaces a volume of fluid equivalent to its own weight:

$$u = (\pi d^3/6)\, g\, (\rho_p - \rho) \quad (4)$$

where ρ_p = density of the solid particle
ρ = density of the fluid

We now have all the information necessary to develop some working expressions for particle settling. Look back at equation 3 (the resistance force exerted by the water), and the expressions for the drag coefficient (sidebar discussion on page 261). The important factor for us to realize is that the settling velocity of a particle is that velocity when accelerating and resisting forces are equal:

$$(\pi d^3/6)\, g\, (\rho_p - \rho) = C_D\, (\pi d^2/4)(\rho u_s^2/2) \quad (5)$$

From this point on, it's some very simple algebra. You can solve equation 5 for u_s, and then substitute an expression in for C_D for each of the three flow regimes (laminar, turbulent and intermediate). You can work through the details, but the working expressions are summarized for you in the sidebar discussion on this page. Remember that to apply these equations you need to know the flow regime, and so you need to make some assumptions when applying any one of these expressions. We will get to this in a moment.

THE THREE REGIMES OF SETTLING

<u>The Laminar Regime</u> (Best known as *Stokes Law*; for Re < 2)

$$u_s = d^2 g\, (\rho_p - \rho)/18\mu$$

<u>The Intermediate Regime</u> (Best known as the *Transition Regime Law*, for 2 <Re < 500)

$$u_s \approx 0.78\{d^{0.43}\, (\rho_p - \rho)^{0.715}\}/(\rho^{0.285}\mu^{0.43})$$

<u>The Turbulent Regime</u> (Best known as *Newton's Law*; for 500 < Re < 200,000)

$$u_s \approx 5.46[g d_p\, (\rho_p - \rho)/\rho]^{0.5}$$

One point we can make is that the expressions can be further developed to give us an idea of the maximum size particles that will settle out in the first two regimes of settling. If we take Stokes law for example, the maximum size particle whose velocity follows Stokes' law can be found by substituting $\mu Re/d_p\rho$ for the settling velocity into the first sidebar equation, and then setting Re = 2 (the limiting Reynolds number value for the flow regime). This then gives us the following useful expression:

$$d_{max} \approx 1.56[\mu^2/(\rho\, (\rho_p - \rho))]^{1/3} \quad (6)$$

The minimum size particles that do not follow Stokes' law occurs at Re ≈ 10^{-4}. The settling velocity in this lower bound regime is less than that computed by the Stokes' Law expression, and generally an empirical correction factor is applied to account for particle slippage. This correction factor, which is applied by dividing the value into the Stokes' law calculated u_s value is: K= 1 + A λ/d, where λ is defined as the mean free path of a fluid molecule, and constant A varies between 1.4 and 20 (as a point of reference, A =1.5 for air). But this is a correction factor we will likely never have to consider in a conventional water treatment assignment. A more convenient set of expressions for settling velocity can be derived by expressing the three settling regime equations in terms of dimensionless groups. We won't get tangled in the derivations, although they are reasonably straightforward, but rather just list these expressions. The relationships are based on the dimensionless Archimedes number, defined as:

$$Ar = [d^3\rho^2 g/\mu^2][(\rho_p - \rho)/\rho] \quad (7)$$

Note that the Archimedes number is a dimesnionless group that describes the physical properties of the heterogeneous system. It can be related to the Reynolds number (and hence the settling velocity, u_s) for each settling regime as follows: For the Stoke's settling regime:

$$Re = Ar/18 \quad (8)$$

Note that the upper limiting or critical value of the Archimedes number for this range occurs at Re = 2, and hence $Ar_{cr,1} = 18 \times 2 = 36$. This means that the laminar settling regime corresponds to Ar < 36.

For the intermediate settling regime, where 2 < Re < 500, we have the following expression:

$$Re = 0.152Ar^{0.715} \quad (9)$$

For the critical value Re = 500, the limiting value of Ar for the intermediate settling regime is $Ar_{cr,2}$ = 83,000. In other words, the intermediate settling regime corresponds to 36 < Ar < 83,000.

For the Newton's law (*turbulent settling regime*) region, where Ar > 83,000, the expression of interest is:

$$Re = 1.74Ar^{1/2} \quad (10)$$

The usefulness of these relationships lies in the recognition that by evaluating the Archimedes number, we can establish the theoretical settling range for the particles we are trying to separate out of a wastewater stream. This very often gives us a starting point for evaluating the settling characteristics of suspended solids for dilute systems. Note that from the definition of the Reynolds number, we can readily determine the settling velocity of the particles from the application of the above expressions ($u_s = \mu Re/d_p\rho$). The following is an interpolation formula that can be applied over all three settling regimes:

$$Re = Ar/[18 + 0.575Ar^{1/2}] \tag{11}$$

For low values of Ar, the second term in the denominator may be neglected, and equation 11 simplifies to equation 8; at high Ar values, we may neglect the first term in the denominator and the expression simplifies to equation 10, which corresponds to the Newton's law range.

The settling velocity of a nonspherical particle is less than that of a spherical one. A good approximation can be made by multiplying the settling velocity, u_s, of spherical particles by a correction factor, ψ, called the *sphericity factor*. The sphericity, or shape factor is defined as the area of a sphere divided by the area of the nonspherical particle having the same volume:

$$u'_s = \psi u_s \tag{12}$$

The factor $\psi < 1$ must be determined experimentally for particles of interest. Typical values are $\psi = 0.77$ for particles of rounded shape; $\psi = 0.66$ for particles of angular shape; $\psi = 0.43$ for particles of a flaky geometry.

The above analysis applies only to the free settling velocities of single particles and does not account for particle-particle interactions. Hence, the application of these formulas only applies to very dilute systems. At high particle concentrations, mutual interference in the motion of particles exists, and the rate of settling is considerably less than that computed by the given expressions. In the latter case, the particle is settling through a suspension of particles in a fluid, rather than through a simple fluid medium.

Note - In designing a system based on the settling velocity of nonspherical particles, the linear size in the Reynolds number definition is taken to be the equivalent diameter of a sphere, d, which is equal to a sphere diameter having the same volume as the particle.

The above provides us with a theoretical staring basis for particle settling. Let's now take a closer look at some of the standard hardware.

GRAVITY SEDIMENTATION

THICKENERS AND CLARIFIERS

Sedimentation involves the removal of suspended solid particles from a liquid stream by gravitational settling. This unit operation is divided into *thickening*, i.e., increasing the concentration of the feed stream, and *clarification,* removal of solids from a relatively dilute stream. A thickener is a sedimentation machine that operates according to the principle of gravity settling. Compared to other types of liquid/solid separation devices, a thickener's principal advantages are:

- simplicity of design and economy of operation;
- its capacity to handle extremely large flow volumes; and
- versatility, as it can operate equally well as a concentrator or as a clarifier.

Standard Bridge-Support Thickener

Typical diameters are up to 150 ft. and machines have operating torques to 750,000 ftr-Lbs. The "Bridge" - or "Beam" or "Truss" - spans the diameter of the tank and supports the drive and rake mechanisms. The underflow is removed from the discharge cone at the bottom center.

In a batch-operating mode, a thickener normally consists of a standard vessel filled with a suspension. After settling, the clear liquid is decanted and the sediment removed periodically. The operation of a continuous thickener is also relatively simple. Figure 4 illustrates a cross-sectional view of a standard thickener. A drive mechanism powers a rotating rake mechanism. Feed enters the apparatus through a feed well designed to dissipate the velocity and stabilize the density currents of the incoming stream. Separation occurs when the heavy particles settle to the bottom of the tank. Some processes add flocculants to the feed stream to enhance particle agglomeration to promote faster or more effective settling. The clarified liquid overflows the tank and is sent to the next stage of a process. The underflow solids are withdrawn from an underflow cone by gravity discharge or pumping. Various common configurations can be found in the sidebar discussions over the next several pages.

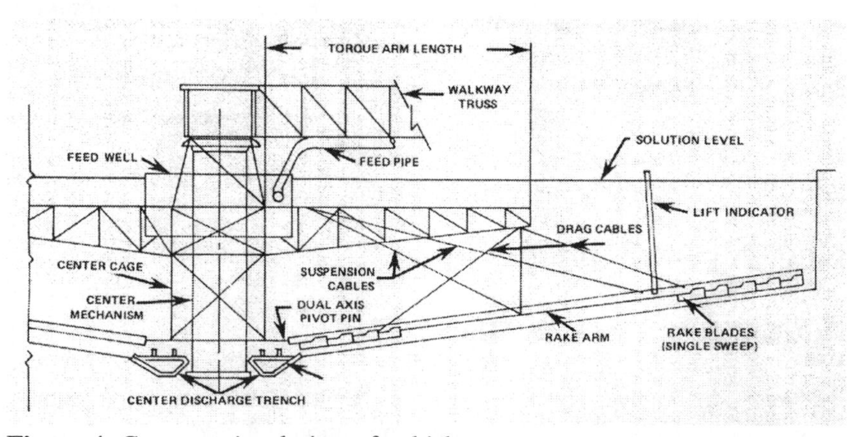

Figure 4. Cross-sectional view of a thickener.

Thickeners can be operated in a countercurrent fashion. Applications are aimed at the recovery of soluble material from settleable solids by means of continuous countercurrent decantation (CCD). The basic scheme involves streams of liquid and thickened sludge moving countercurrently through a series of thickeners. The thickened stream of solids is depleted of soluble constituents as the solution becomes enriched. In each successive stage, a concentrated slurry is mixed with a solution containing fewer solubles than the liquor in the slurry and then is fed to the thickener. As the solids settle, they are removed and sent to the next stage. The overflow solution, which is richer in the soluble constituent, is sent to the preceding unit. Solids are charged to the system in the first-stage thickener, from which the final concentrated solution is withdrawn. Wash water or virgin solution is added to the last stage, and washed solids are removed in the underflow of this thickener. The flow scheme for a three-stage CCD system is illustrated in Figure 5. The feed stream, F, is mixed with overflow O_2 (from thickener 2) before entering stage 1. The overflow of concentrated solution, O_1, is withdrawn from the first stage. The underflow from the first stage, U_1, is mixed with third-stage overflow, O_3, and fed to the second stage. Similarly, the second-stage underflow, U_2, is mixed with wash water and fed to thickener 3.

Standard Center-Pier Thickener

The diameters on these machines are over 400 ft., with operating torques to 2,400,000 ft.-Lbs. The stationary center pier supports the drive and rake mechanisms. The truss extends from the center pier to the tank periphery supports walkway, power lines and feed lasunder.

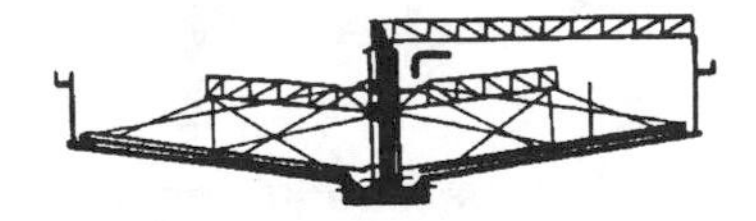

Caisson Center-Pier Thickener

(Diameters are up to 600 ft with operating torques of 4,000,000 ft.-Lbs. The center pier forms the control and pumping station, as well as support for the rake assembly. The underflow is pumped up through the caisson.

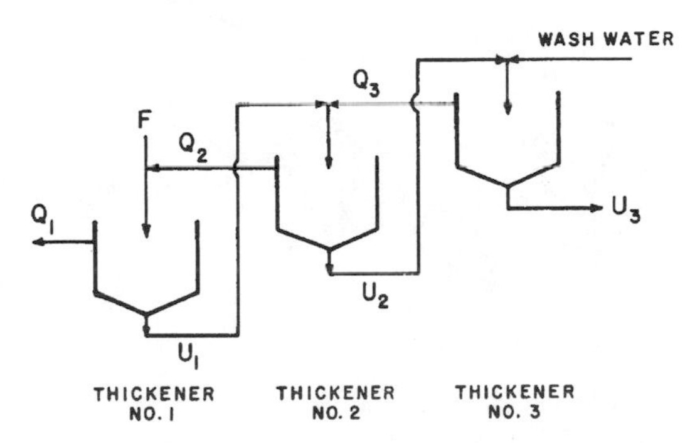

Figure 5. Flow scheme for three-stage CDD.

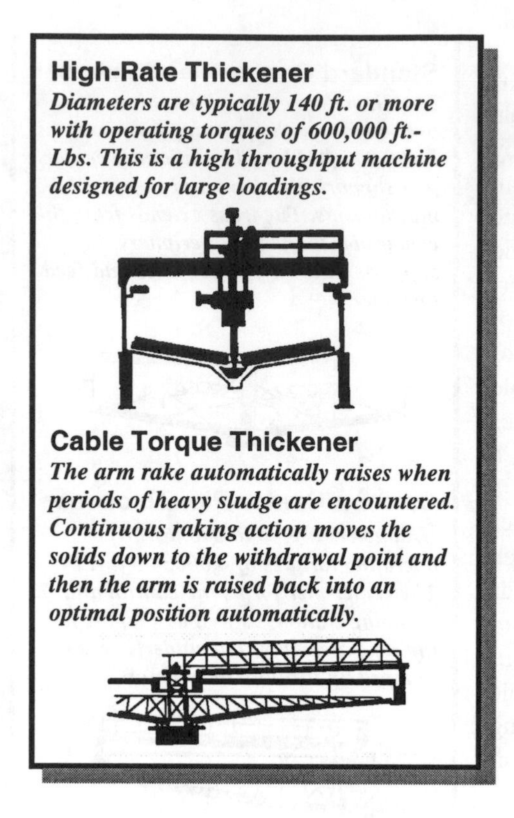

The washed solids are removed from the third stage as the final underflow, U_3. *Continuous clarifiers* handle a variety of process wastes, domestic sewage and other dilute suspensions. They resemble thickeners in that they are sedimentation tanks or basins whose sludge removal is controlled by a mechanical sludge-raking mechanism. They differ from thickeners in that the amount of solids and weight of thickened sludge are considerably lower. Figures 6 and 7 show examples of cylindrical clarifiers. In this type of sedimentation machine, the feed enters up through the hollow central column or shaft, referred to as a *siphon feed system.* The feed enters the central feed well through slots or ports located near the top of the hollow shaft. Siphon feed arrangements greatly reduces the feed stream velocity as it enters the basin proper. This tends to minimize undesirable cross currents in the settling region of the vessel.

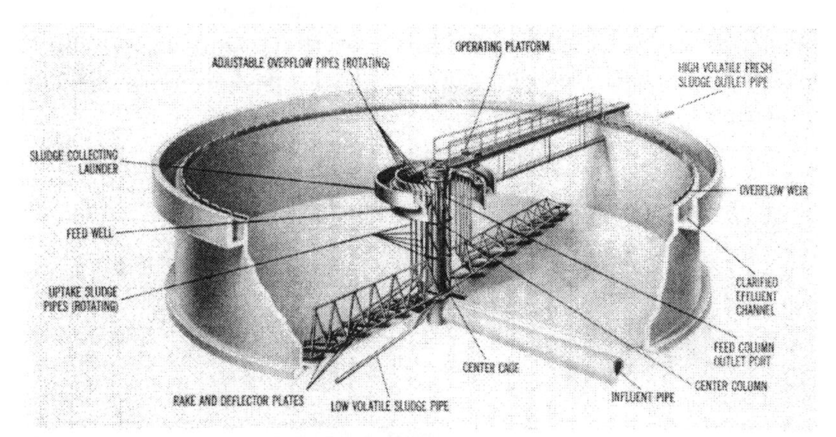

Figure 6. A cylindrical clarifier and its features.

Figure 7. Clarifier used in a wastewater treatment plant.

Figure 8. Rectangular settling basins in operation.

Most cylindrical units are equipped with peripheral weirs; however, some designs include radial weirs to reduce the exit velocity and minimize weir loadings. The unit shown also is equipped with adjustable rotating overflow pipes. Although there are a fixed number of U.S. suppliers for these unique machines, the designs are pretty much universal, and nearly identical configurations and comparable operating parameters can be found throughout the world. The photograph shown in Figure

7 is an installation at a Russian municipal wastewater treatment plant, where in fact, the operation was a part of an auto-making facility that also provided sanitary and drinking water supplies to the city. Figure 8 shows a battery of rectangular settling basins in the same treatment plant.

SOME MORE IMPORTANT COMMENTS ON THICKENING

Gravity Thickening - This process involves the concentration of thin sludges to more dense sludge in special circular tanks designed for this purpose. Its use is largely restricted to the watery excess sludge from the activated sludge process, and in large plants of this type where the sludge is sent direct to digesters instead of to the primary tanks. It may also be used to concentrate sludge to primary tanks or a mixture of primary and excess activated sludge prior to high rate digestion. The thickening tank is equipped with slowly moving vertical paddles built like a picket fence. Sludge is usually pumped continuously from the settling tank to the thickener which has a low overflow rate so that the excess water overflows and the sludge solids concentrate in the bottom. A blanket of sludge is maintained by controlled removal which may be continuous at a low rate. A sludge with a solids content of ten percent or more can be produced by this method. This means that with an original sludge of two percent, about four-fifths of the water has been removed, and one of the objectives in sludge treatment has been attained.

Flotation Thickening - Flotation thickening units are becoming increasingly popular at sewage treatment plants, especially for handling waste activated sludges. With activated sludge they have the advantage over gravity thickening tanks of offering higher solids concentrations and lower initial cost for the equipment.

Dissolved Air-Pressure Flotation - Although we haven't covered this per se, we should mention that the objective of flotation-thickening is to attach a minute air bubble to suspended solids and cause the solids to separate from the water in an upward direction. This is due to the fact that the solid particles have a specific gravity lower than water when the bubble is attached. Dissolved air flotation depends on the formation of small diameter bubbles resulting from air released from solution after being pressurized to 40 to 60 psi. Since the solubility of air increases with pressure, substantial quantities of air can be dissolved. In current flotation practice, two general approaches to pressurization are used: (1) air charging and pressurization of recycled clarified effluent or some other flow used for dilution, with subsequent addition to the feed sludge; and (2) air charging and pressurization of the combined dilution liquid and feed sludge. Air in excess of the decreased solubility, resulting from the release of the pressurized flow into a chamber at near atmospheric pressures, comes out of of solution to form the minute air bubbles. Sludge solids are floated by the air bubbles that attach themselves to and are enmeshed in the floc particles. The degree of adhesion depends on surface properties of the solids. When released into the separation area of the thickening tank, the buoyed solids rise under hindered conditions analogous to those in gravity settling and can be called hindered separation or flotation. The upward moving particles form a sludge blanket on the surface of the flotation thickener. Parameters: The primary variables for flotation thickening are: (1) pressure, (2) recycle ratio, (3) feed solids concentration, (4) detention period, (5) air-to-solids ratio, (6) type and quality of sludge, (7) solids and hydraulic loading rates, and (8) use of chemical aids. Similar to gravity sedimentation, the type and quality of sludge to be floated affects the unit performance. Flotation thickening is most applicable to activated sludges but higher float concentrations can be achieved by combining primary with activated sludge. Equal or greater concentrations may be achieved by combining sludges in gravity thickening units.

Centrifugation - Centrifugation has been demonstrated to be capable of thickening a variety of wastewater sludges. Centrifuges are a compact, simple, flexible, self-contained unit, and the capital cost is relatively low. They have the disadvantages of high maintenance and power costs and often a poor, solids-capture efficiency if chemicals are not used.

THE LAMELLA CLARIFIER

Cross-flow lamellar clarification is a technology used in industrial environments to remove oils and solids from residual water. It takes advantage of the natural tendency of oils to float, and the decantation principle for suspended solids that are denser than water. The originality of this process is the combination of natural flotation and clarification techniques in one system. The strip decanter performs as well as conventional clarifiers, but is more compact and occupies a smaller area. The process is mainly used for dealing with oils and grease present in residual liquids emanating from industrial activities in the petrochemical, chemical, mechanical, metallurgical and food-processing sectors. Depending on the specific design criteria (available space, quantity and quality of water to be treated, structural and hydraulic constraints, budget limits and need for mobility), the equipment can be housed in a circular or rectangular reservoir, or can consist of mechanical components attached to a below-grade concrete tank. If the oils are partially or completely emulsified, the cross-flow lamellar clarifier can be equipped with a coalescer, which uses a physical process to trigger separation of the oil and water phases. The coalescer is filled with various elements (rings, plastics, honeycombs, other appropriate materials ...) which maximize the potential contact surface. The accretion of microglobules to these elements leads to phase separation. System performance depends on the specific nature of the effluent to be treated and varies according to the type of industry. Depending on the particular situation, removal of free oils and greases, and of suspended solids, varies from 90 to 99%. With no chemical amendment (i.e. demulsifying agent), 20-40% removal of emulsified oils and greases can be achieved. The addition of an agent enables the process to achieve 50-99% removal, depending on the application.

This technology is designed to handle effluents containing a maximum of 10,000 mg/l of grease, and 3,000 mg/l of solids. Hydraulic load is not a limiting factor. The cross-flow lamellar clarifier consists of the following basic units: primary screening chamber, separator plate cell, sludge silo, oil and grease storage chamber. First, water is pretreated in the primary screening chamber to remove part of the floating oil and grease and to allow sedimentation of large solid particles (>500mm). Then the effluent feeds through the plate cell where separation of phases is accomplished as follows: oils are deflected upwards by the plates to form a film on the surface of the water, sludges settle to the bottom and the purified water flows horizontally to the reservoir outlet. Before leaving the system, water passes through a calibrated opening which controls the unit's hydraulic load. Sludges are recovered in a conical extraction silo which aids their compaction and provides easier handling. Dryness rates of 1 to 5% are achieved depending on sludge type. The sludge is kept apart from water to be treated, so as not to draw it back into the process flow. Floating oils are recovered from the water surface and channeled into a storage reservoir located beneath the sludge silo.

Equipment installation and start-up take less than a week. Running the system requires no energy input except for the effluent pump. Maintenance is limited to monthly testing, and cleaning of the separator plates once every six months. If needed, various products (polymers, demulsifiers and coagulants) can be added to the process to improve its performance. Running the system requires no special safety measures. If required, the equipment can be designed to provide a safe environment for treating effluents containing volatile compounds with risk of explosion.

Various parameters are taken into account in designing a cross-flow lamellar clarifier. These include:

- hydraulic load;
- suspended matter load;
- hydrocarbon load;
- desired performance;
- space available for clarifier's installation.

The investment required to set up a treatment unit complete with coalescer varies from $750 to $2,500 per cubic meter of water to be treated, depending on unit size and available space. Operating costs are low.

THE SEDIMENTATION PROCESS IN GREATER DETAIL

To examine sedimentation in greater detail, let us examine the events occurring in a small-scale experiment conducted batchwise, as illustrated in Figure 9. Particles in a narrow size range will settle with about the same velocity. When this occurs, a demarcation line is observed between the supernatant clear liquid (zone A) and the slurry (zone B) as the process continues. The velocity at which this demarcation line descends through the column indicates the progress of the sedimentation process.

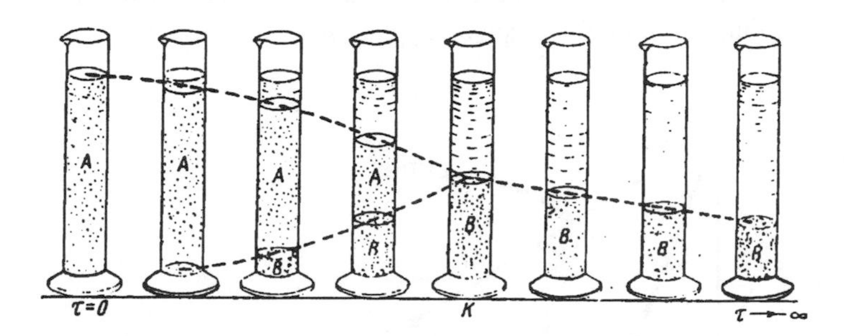

Figure 9. batch sedimentation in glass cylinder test.

The particles near the bottom of the cylinder pile up, forming a concentrated sludge (zone D), whose height increases as the particles settle from zone B. As the upper interface approaches the sludge buildup on the bottom of the container, the slurry appears more uniform as a heavy sludge (zone D); the settling zone B disappears; and the process from then on consists only of the continuation of the slow compaction of the solids in zone D. By measuring the interface height and solids concentrations in the dilute and concentrated suspensions, a graphic representation of the sedimentation rates can be prepared as shown in Figure 10.

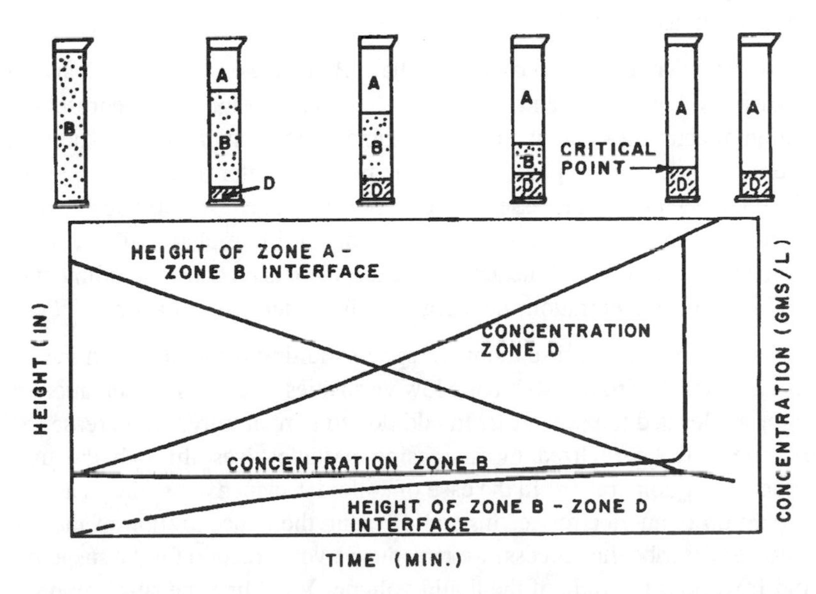

Figure 10. Plot of interface height and solids concentration versus time.

The plot shows the difference in interface height plotted against time, which is proportional to the rate of settling as well as to concentration. Examining these data in more detail as a plot of sediment height, Z, versus time, t, in Figure 11, we note that $Z \propto t$, meaning that the sedimentation rate is, and continues to be, constant. Then the sedimentation rate of the heavy sludge decreases with time, which corresponds to the curve on the graph after point K.

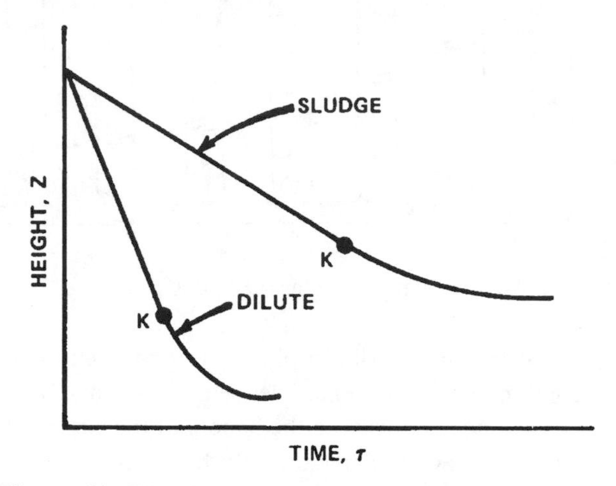

Figure 11. Plot showing the kinetics of sedimentation.

Naturally, the higher the concentration of the initial suspension, the slower the sedimentation process. Observations show that the solids concentration in the dilute phase is constant up to the point of complete disappearance of phase A. This is illustrated by the plot in Figure 11, and corresponds to a constant rate of sedimentation in the phase.

Note, however, that the concentration in phase B changes with height Z and time t (as shown by Figure 11) and, hence, each curve in Figure 12 represents the distribution of concentrations at any Oven moment. The initial concentration is C_1, which remains in the dilute phase during the process. After a sufficient period of time, the concentration increases to C_2, but in zone D. Obviously, if the concentration of the feed suspension is too high, no dilute phase will exist, even during the initial period of sedimentation. Hence, there is no constant sedimentation rate. In this case, concentration, not height, will change with time only.

As follows from an earlier discussion on spheres falling through a fluid medium, sedimentation is faster in liquids having low viscosities. Hence, sedimentation rates are higher at elevated temperatures. In addition to temperature, an increase in the process rate may be realized by increasing particle sizes through the use of coagulation or agglomeration. In the case of colloidal suspensions, this is achieved by the addition of an electrolyte. Instead of using the concentration of the initial suspension to describe the process, we introduce a void fraction for the suspension. The void fraction is the ratio of the liquid volume, V_f, filling the space among the particles, to the total volume which is the sum of the liquid volume and the actual volume of the solid particles, V_P:

$$\epsilon = V_f/(V_p + V_f) \tag{13}$$

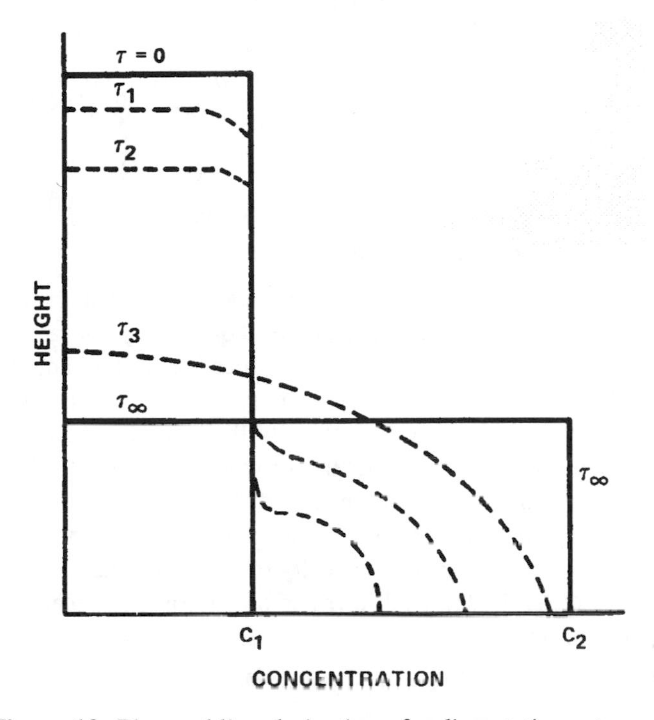

Figure 12. Plot enabling derivation of sedimentation rates.

As particles settle, forming a thickened zone, the void fraction, ϵ, decreases. At the total settling of the slurry, the void fraction is at a minimum, its value depending on the shape of the particles. For example, the minimum void fraction of spherical particles is $\epsilon_{min} = 0.2\ 15$; for small crystals, $\epsilon_{min} = 0.4$. For most systems, the void fraction of a thickened sludge is approximately $\epsilon_{min} = 0.6$; however, values should be determined experimentally for the specific system.

As the sludge compacts, its void fraction and height, X, decrease. If the initial void fraction, ϵ_0, is known when the sludge height is X_0, an average void fraction, ϵ, can be estimated assuming that the height of the sludge decreases to X. For a vertical cylinder of cross-sectioned area F, the initial volume of the sediment is FX_0 and, hence, the volume of solid particles in the sediment is $(1 - \epsilon_0)X_0 F$. Similarly, the volume of solid particles for voidage ϵ is $(1 - \epsilon)XF$. Consequently,

$$(1 - \epsilon_0)X_0 F = (1 - \epsilon)XF \tag{14}$$

Hence:

$$\epsilon = (X_0/X)(1 - \epsilon_0) \quad (15)$$

For a unit volume of slurry, its weight, γ, is the sum of the weights of the solid particles, $\gamma_p(1 - \epsilon)$, and of the liquid, $\gamma_f \epsilon$, where γ_f = specific weight of liquid and γ_p = specific weight of particles:

$$\gamma = \gamma_p(1 - \epsilon) + \gamma_f \epsilon \quad (16)$$

or

$$\epsilon = (\gamma_p - \gamma)/(\gamma_p - \gamma_f) \quad (17)$$

This expression can be used to compute the void fraction from experimentally determined values of the specific weights.

Let's now direct our attention to the sedimentation process in the zone of constant settling velocity, i.e., in the dilute phase. To simplify the analysis w assume spherical particles of the same size. The process may be simplified further by viewing sedimentation as fixed particles in an upward-moving stream of viscous liquid, whose average velocity is u_f. Due to the viscosity of the liquid, a certain velocity gradient exists relative to the distance from the surface of spherical particle, du/dx. This velocity also depends on the average distance among particles, which is determined at any moment by a void fraction, ϵ, of the slurry and the particle diameter, d. The average velocity of the liquid may be presented in this case as $u_f = f(d, \epsilon, du/dx)$. And rewritten on the basis of dimensional analysis in the following form:

$$u_f = K_1 d (du/dx)\Phi_1(\epsilon) \quad (18)$$

where K_1 = constant

$\Phi_1(\epsilon)$ = dimensionless function of the void fraction

The resistance to liquid flow around particles may be presented by an equation similar to the viscosity equation but with considering the void fraction. Recall that the shear stress is expressed by the ratio of the drag force, R, to the active surface, $K_2\pi d^2$. The total sphere surface is πd^2 and K_2 is the coefficient accounting for that part of the surface responsible for resistance. Considering the influence of void fraction as a function $\Phi_2(\epsilon)$, we obtain:

$$(R/K_2)\pi d^2 = \mu(du/dx)\Phi_2(\epsilon) \quad (19)$$

Dividing equation 19 by Equation 18, we obtain:

$$R = (K_2/K_1)\pi d\mu u_f[\Phi_2(\epsilon)/\Phi_1(\epsilon)] \quad (20)$$

For very dilute suspensions, in which the void fraction does not influence the sedimentation process, the function $\Phi_2(\epsilon)/\Phi_1(\epsilon) = \Phi(\epsilon)$ reduces to unity. It is known also that in dilute suspensions the sedimentation of small particles follows Stokes' law:

$$R_\infty = 3\pi d\mu u_f \tag{21}$$

Equating equations 20 and 21, we find that (K_2/K_1) is equal to 3. Hence, the resistance of the liquid relative to a spherical particle in the sedimentation process is

$$R = 3\pi d\mu u_f / \Phi(\epsilon) \tag{22}$$

This resistance is balanced by the gravity force acting on a particle:

$$W = (\pi d^3/6)(\gamma_p - \gamma) \tag{23}$$

where γ_p is the actual specific weight of a particle, and γ is the average specific weight of the sludge, which depends on void fraction ϵ. Using equation 17, we replace $(\gamma_p - \gamma)$ with $(\gamma_p - \gamma_f)\epsilon$, where γ_f is the specific weight of the liquid:

$$W = (\pi d^3/6)(\gamma_p - \gamma_f)\epsilon \tag{24}$$

By comparing the gravity force acting on the particle (equation 24) with the resistance to liquid flow (equation 22), we obtain the average liquid velocity relative to the particles:

$$u_f = [d^2(\gamma_p - \gamma_f)/18\mu]\epsilon\,\Phi(\epsilon) \tag{25}$$

In practice, however, the liquid velocity relative to fixed particles, u_f, is not very useful. Instead, the velocity of settling relative to the walls of an apparatus, $u_f - u$, is of practical importance. The volume of the solid phase moving downward should be equal to that of liquid moving upward. This means that volume rates of these phases must be equal. Consider a column of slurry having a unit cross section and imagine the liquid and solid phases to have a well defined interface. The column of solid phase will have a base $1 - \epsilon$, and the liquid column phase will have a base ϵ. Hence, the volumetric rate of the solid column will be $(1 - \epsilon)u$, and that of the liquid column will be $(u_f - u)\epsilon$. Because these flowrates are equal to each other, we obtain

$$(1 - \epsilon)u = (u_f - u)\epsilon \tag{26}$$

Therefore, the settling velocity of the solid phase relative to the wall of an apparatus, depending on the average liquid velocity relative to the sludge with void fraction ϵ, will be

$$u = u_f\,\epsilon \tag{27}$$

Substituting this expression into equation 25, we obtain the actual settling velocity:

$$u = [d^2 (\gamma_p - \gamma_f)/18\mu]\epsilon^2 \Phi(\epsilon) \quad (28)$$

Note that the term in parentheses expresses the velocity of free failing, according to Stokes' law:

$$u_p = d^2 (\gamma_p - \gamma_f)/18\mu \quad (29)$$

or

$$u = u_p \epsilon^2 \Phi(\epsilon) \quad (30)$$

For a very dilute suspension, i.e., $\epsilon = 1$ and $\Phi(\epsilon) = 1$, the settling velocity will be equal to the free-fall velocity. As no valid theoretical expression for the function $\Phi(\epsilon)$ is available, common practice is to rely on experimental data. Note that a unit volume of thickened sludge contains ϵ volume of liquid and $(1 - \epsilon)$ volume of solid phase, i.e., a unit volume of particles of sludge contains $\epsilon/(1 - \epsilon)$ volume of liquid. Denoting σ as the ratio of particle surface area to volume, we obtain the hydraulic radius as the ratio of this volume, $\epsilon/(1 - \epsilon)$, to the surface, σ, when both values are related to the same volume of particles:

$$r_h = \epsilon/[(1 - \epsilon)\sigma] \quad (31)$$

For spherical particles, σ is equal to the ratio of the surface area, πd^2 , to the volume $\pi d^3/6$, i.e., $\sigma = 6/d$. Hence,

$$r_h = \epsilon/[(1 - \epsilon)6] \quad (32)$$

For a specified void fraction, the diameter of the sphere is a measure of the distance between sludge particles ($u_f = f(d, \epsilon, du/dx)$). However, it is more practical to introduce the hydraulic radius, and instead of $\Phi_1(\epsilon)$ and $\Phi_2(\epsilon)$, according to equation 30, we assume the following value:

$$\Phi(\epsilon) = [\epsilon/(1 - \epsilon)]\theta(\epsilon) \quad (33)$$

where $\theta(\epsilon)$ is the new experimental function of the void fraction. Hence, the settling velocity equation may be rewritten in the following form:

$$u = u_p [\epsilon/(1 - \epsilon)]\theta(\epsilon) \quad (34)$$

By representing the velocity in this manner, we can anticipate a small change in the function $\theta(\epsilon)$ because the influence of the flow pattern is, to a large extent, accounted for in the hydraulic radius. Laboratory and pilot scale testing have shown that the function $\Phi(\epsilon)$ may be presented by the following empirical equation:

$$\Phi(\epsilon) = 10^{-1.82(1-\epsilon)} \tag{35}$$

Multiplying equation 35 by $(1 - \epsilon)/\epsilon$, we obtain the function $\theta(\epsilon)$. For $\epsilon \leq 0.7$, i.e., for thickened sludges, this function is practically constant and equal to $\theta(\epsilon) = 0.123$. The settling velocity of spherical particles is therefore:

$$u = u_p[\epsilon^2 \times 10^{-1.82(1-\epsilon)}] \tag{36}$$

For more thickened sludges:

$$u = [0.123\ \epsilon^3/(1 - \epsilon)]u_p \tag{37}$$

Thus far, the analysis has been based on independently settling spherical particles. To relate to the design of the unit operations, we now must consider the kinetics of nonspherical particle settling and the sedimentation of flocculent particles. In contrast to single-particle settling, such systems form a certain structural unity similar to tissue. The sludge is compacted under the action of gravity force, i.e., the void fraction decreases and the liquid is squeezed out from the pore structure. The formation of a regular sediment from a flocculent may be achieved by the addition of electrolytes, as described earlier. The general characteristic of normal settling of nonspherical particles (as well as flocculent ones) is that the sediment carries along with it a portion of the liquid by trapping it between particle cavities. This trapped volume of liquid flows downward with the sludge and is proportional to the volume of the sludge. That is, it can be expressed as $a(1 - \epsilon)$, where a is a coefficient and $(1 - \epsilon)$ is the volume of particles. Consequently, a portion of the liquid remains in a layer above the sludge, and a portion is carried along with the sludge corresponding to the modified void fraction:

$$\epsilon' = \epsilon - a(1 - \epsilon) \tag{38}$$

This is the difference of the total relative liquid volume and liquid moving together with particles. Substituting ϵ' for ϵ in equation 37, we obtain the settling velocity at $\epsilon \leq 0.7$:

$$u = 0.123(1 + a)^2 u_p\ [(\epsilon - a/(1 + a)]^3/(1 - \epsilon) \tag{39}$$

Denoting $a/(1 + a) = \beta$, the above expression can be written as follows:

$$u = [0.123 u_p(\epsilon - \beta)^3]/[(1 - \beta)^2(1 - \epsilon)] \tag{40}$$

Similarly equation 36 for slurries with nonspherical particles is

$$u = u_p[(\epsilon - \beta)^2/(1 - \beta)]10^{-1.82 (1 - \epsilon)/(1 - \beta)} \quad (41)$$

Parameter β is equal to the ratio of the liquid volume entrained and the sum of the volumes of this liquid and particles. Values of β are determined experimentally from measured settling velocities. In general, the smaller the effective particle size, the more liquid is entrained by the same mass of solids phase. For example, particles of carborundum with d = 12.2 μm have β = 0.268; d = 9.6 μm, β = 0.288; and d = 4.6 μm, β = 0.35.

AN ALTERNATIVE ANALYSIS - THE FORGOTTEN METHOD OF DIMENSIONAL ANALYSIS

I often refer to dimensional analysis as "lost art" - because it is usually not heavily emphasized in engineering education today. However, for well over 100 years its has provided simply a wealth of practical design correlations that are still relied upon in virtually all aspects of chemical engineering, ranging from classes of problems dealing with heat and mass transfer, reaction kinetics, momentum exchanges in flow dynamics. Much of sedimentation theory, and indeed the basis for more sophisticated analyses beyond this book, are based on relating dimensionless groups that have been correlated with experimental observation. The following discussion will walk you through the approach of dimensional analysis as applied to the general theory of sedimentation, providing us with some expressions that will give us more than a head start in analyzing specific separation problems in water treatment.

To accomplish this, let's take a few steps back and start by examining the forces acting on a single particle settling through a continuous fluid medium. These forces are gravity, $\boldsymbol{G}$, buoyant or Archimedes forces, $\boldsymbol{A}$, centrifugal field, $\boldsymbol{C}$, and an electrical field, $\boldsymbol{Q}$. The system diagram for defining this is shown in Figure 13. Geometrically summing all the forces, we obtain:

$$\boldsymbol{P} = \boldsymbol{G} + \boldsymbol{A} + \boldsymbol{C} + \boldsymbol{Q} \quad (42)$$

If force $\boldsymbol{P}$ is greater than zero, the particle will be in motion relative to the continuous phase at a certain velocity, w. At the beginning of the particle's motion, a resistance force develops in the continuous phase, $\boldsymbol{R}$, directed at the opposite side of the particle motion. At low particle velocity (relative to the continuous phase), fluid layers running against the particle are moved apart smoothly in front of it and then come together smoothly behind the particle (Figure 14). The fluid layer does not intermix (a system analogous to laminar fluid flow in smoothly bent pipes). The particles of fluid nearest the solid surface will take the same time to pass the body as those at some distance away.

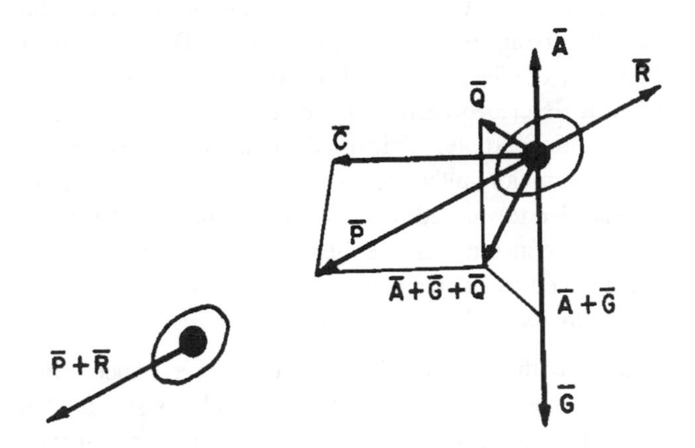

Figure 13. System of forces acting on a single particle.

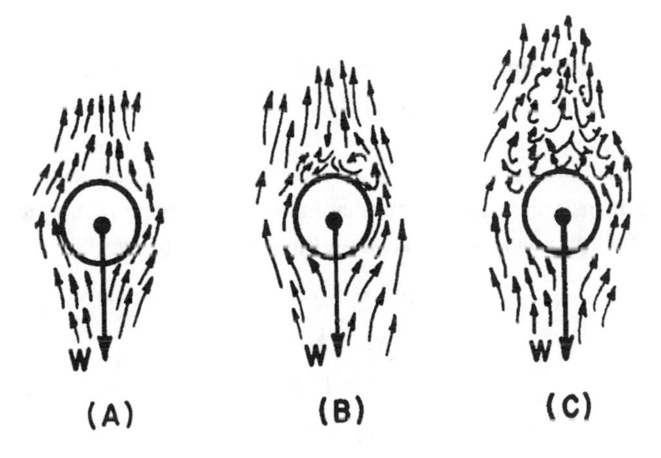

Figure 14. Flow around a particle: (A) laminar; (B) transional; (C) turbulent.

Because the liquid layers move at different velocities relative to each other, planes of slip exist between them and, from Newton's law, the forces of viscous friction arise. Consequently, the resistance force depends on the viscosity of the medium as is determined by the viscosity coefficient, μ. At higher particle velocities (or higher medium velocities relative to the particle) the flow around the object is broken, forming swirling fluid patches (Figure 14(B)). The formation of vortices is influenced by the relative flow velocity, the shape of the particle and the smoothness of the object's surface. The higher the velocity, the more complicated the particle shape; and/or the greater the roughness, the more intense the vortex formation. Eventually, this leads to the generation of eddies along the downstream surface of the particle (Figure 14(C)).

The eddies are entrained by the flow and, at a certain distance from the particle, they disappear while being replaced by new eddies. Due to eddy formation and their breaking away from the particle, a low-pressure zone forms at the front of the particle. Hence, as described earlier, a pressure gradient is formed between the front and rear of the particle. This gradient is responsible primarily for the resistance to particle motion in the medium. The amount of this resistance depends on the energy expended toward eddy formation: the more intensive this formation, the greater the energy consumption and, hence, the greater the resistance force. The inertia forces generated by eddies play an important role. They are characterized by the mass and velocity of the fluid relative to the particle.

The total resistance is the sum of friction and eddy resistances. Both factors act simultaneously, but their contribution in the total resistance depends on the conditions of the flow in the vicinity of the particle. Hence, for the most general case the resistance force is a function of velocity, w, density, ρ, viscosity, μ, the linear size of a particle, ℓ, and its shape, ψ. Thus,

$$\boldsymbol{R} = (w, \rho, \mu, \ell, \psi) \tag{43}$$

Assuming this relationship as an exponential complex, we obtain

$$\boldsymbol{R} = A' \, w^x \rho^y \mu^z \ell^\alpha \tag{44}$$

where A' is dimensionless coefficient that includes the shape factor, ψ Noting the dimensions of all parameters appearing in this expression:

$$[R] = LMT^{-2}$$

$$[w] = LT^{-1}$$

$$[\rho] = L^{-3}M$$

$$[\ell] = L$$

$$[\mu] = L^{-1}MT^{-1}$$

where L, M and T are the principal unit measures-length, mass and time. Expressing this in terms of its dimensions:

$$LMT^{-2} = (LT^{-1})^x(L^{-3}M)^y L^z (L^{-1}MT^{-1})^\alpha$$

or

$$LMT^{-2} = L^{x-3y+z-\alpha} M^{y+\alpha} T^{-x-\alpha} \tag{45}$$

For the dimensions on the LHS of this expression to satisfy the RHS, the exponents on the principal units of measure must be equal. Thus, we have the following system of three equations (corresponding to the number of values with independent dimensions):

$$x - 3y + z - \alpha = 1$$

$$y + \alpha = 1$$

$$-x - \alpha = -2$$

Because this system of equations cannot be solved (there are fewer equations than variables), we express all exponents in terms of α:

$$x = 2 - \alpha;\ y = 1 - \alpha;\ z = 2 - \alpha$$

We may now write:

$$\boldsymbol{R} = A' w^{2-\alpha} \rho^{1-\alpha} \ell^{1-\alpha} \mu^{\alpha} \tag{46}$$

Equation 46 is a general expression that may be applied to the treatment of experimental data to evaluate exponent α. This, however, is a cumbersome approach that can be avoided by rewriting the equation in dimensionless form. Equation 42 shows that there are n = 5 dimensional values, and the number of values with independent measures is m = 3 (m, kg, sec.). Hence, the number of dimensionless groups according to the π-theorem is $\pi = 5 - 3 = 2$. As the particle moves through the fluid, one of the dimensionless complexes is obviously the Reynolds number: $Re = w\,\ell\rho/\mu$. Thus, we may write:

$$\boldsymbol{R} = A' (w\ell\rho/\mu)^{-\alpha} \rho w^2 \ell 2 \tag{47}$$

As one of two possible dimensionless numbers is now known, the second one can be obtained by dividing both sides of the equation through by the remaining values:

$$Eu = \boldsymbol{R} / (\rho w^2 \ell^2) \tag{48}$$

The result is a modified Euler number. You can prove to yourself that the pressure drop over the particle can be obtained by accounting for the projected area of the particle through particle size, ℓ, in the denominator. Thus, by application of dimensional analysis to the force balance expression, a relationship between the dimensionless complexes of the Euler and Reynolds numbers, we obtain:

$$Eu = A' Re^{-\alpha} \tag{49}$$

Coefficient A' and exponent α must be evaluated experimentally. Experiments have shown that A' and α are themselves functions of the Reynolds number. Equation 47 shows that the resistance force increases with increasing velocity. If the force field (e.g., gravity) has the same potential at all points, a dynamic equilibrium between forces $\boldsymbol{P}$ and $\boldsymbol{R}$ develops shortly after the particle motion begins. As described earlier, at some distance from its start the particle falls at a constant velocity. If the acting force depends on the particle location in space, in a centrifugal field, for example, it will move with uniformly variable speed until it travels outside of the boundary of the field's action or runs into an obstacle such as a vessel wall. We now shall define the relationship between particle motion and the acting factors. Moving under the action of force ***P***, with acceleration a_g at infinitesimal distance $\delta\ell$, the particle with mass m_p performs work $m_p a_g\, \delta\ell$. This work is spent on overcoming the resistance force and the displacement of fluid mass, m_c, in a volume equal to the volume of the particle, V, at the same distance, but in opposite direction, and with the same acceleration, a_g:

$$m_p a_g \delta\ell = m_c a_g\, \delta\ell + \boldsymbol{R}\, \delta\ell \qquad (50)$$

Dividing through by $\delta\ell$ and expressing the masses of the particle and medium in terms of their volumes and densities, we obtain

$$V(\rho_p - \rho_c)a_g = \boldsymbol{R} \qquad (51)$$

Consider again the simple motion of a sphere. In this case, the equivalent diameter of a sphere, d_{eq}, is equal to its geometric diameter, d. Equating the above expressions and replacing ℓ by d (and denoting the Euler umber, Eu, by Υ), we obtain an expression for the resistance force:

$$\boldsymbol{R} = \Upsilon\, \rho_c w^2 d^2 \qquad (52)$$

where

$$\Upsilon = A' Re^{-\alpha} \qquad (53)$$

In sedimentation, the Eule number is often referred to as the *resistance number*. Multiplying and dividing the RHS of equation 52 by $\pi/8$, we obtain

$$\boldsymbol{R} = C_D F\, \rho_c w^2/2 \qquad (54)$$

where $C_D = (8/\pi)\Upsilon$ = drag coefficient

$F = \pi d^2/4$ = the cross-sectional area of the spherical particle

Equation 54 is Newton's resistance law. Substituting Equation 53 into the definition for C_D, we obtain

$$C_D = B Re^{-\alpha} \qquad (55)$$

where $B = 8\, A'/\pi$ since $\Upsilon = f(Re)$ and $C_D = f_1(Re)$.

The Reynolds number for a sphere is

$$Re = w d \rho_c/\mu = wd/\nu \qquad (56)$$

Substituting the resistance force into equation 51 and expressing *F* and V in terms of d, the basic equation of sedimentation theory is obtained:

$$d^3(\rho_p - \rho_c)a_g = \tfrac{3}{4}\, C_D w^2 d^2 \qquad (57)$$

or

$$d^3\,(\rho_f/\rho_c)a_g = \tfrac{3}{4} C_D w^2 d^2 \qquad (58)$$

where $\rho_f = \rho_p - \rho_c$ is the effective density of the system. In separation calculations for heterogeneous systems, the important parameter is settling velocity:

$$w = \{(4/3) \times (d/C_D)(\rho_f/\rho_c)a_g\}^{1/2} \qquad (59)$$

Application of the above formulas is difficult because the drag coefficient, C_D, is a function of velocity and particle geometry. A generalized calculation procedure for settling under the influence of gravity was developed many years ago. The method, however, also may be applied to settling due to the influence of any force field, provided the relationship between particle acceleration and the coordinates of field is defined. The procedure is based on expressing the basic equation of sedimentation (equation 58) in terms of a relationship of criteria. For this, both sides of equation 58 are 2 divided by v^2 , and the RHS multiplied and divided by the acceleration due to gravity, g:

$$(gd^3/v^2)(\rho_f/\rho_c)(a_g/g) = \tfrac{3}{4} C_D w^2 d^2/v^2 \qquad (60)$$

The LHS of this expression contains a dimensionless group known as the *Galileo number*, defined as $Ga = gd^3/v^2$.

Multiplying by a simplex composed of densities results in the Archimedes number:

$$Ar = Ga\,(\rho_f/\rho_c) \qquad (61)$$

And introducing the ratio of accelerations, $K_s = a_g/g$, where K_s indicates the relative strength of acceleration, a_g, with respect to the gravitational acceleration g. This is known as the *separation number*. The LHS of equation 60 contains a Reynolds number group raised to the second power and the drag coefficient. Hence, the equation may be written entirely in terms of dimensionless numbers:

$$Ar\,K_s = \tfrac{3}{4} C_D Re^2 \qquad (62)$$

The Archimedes number contains parameters that characterize the properties of the heterogeneous system and the criterion establishing the type of settling. The criterion of separation essentially establishes the separating capacity of a sedimentation machine. The product of these criteria is:

$$S_1 = Ar\ K_s \tag{63}$$

This product contains information on the properties of the suspension and characterizes the settling process as a whole. Substituting equation 62 into 63 gives

$$S_1 = \tfrac{3}{4} C_D Re^2 \tag{64}$$

This expression represents the first form of the general dimensionless equation of sedimentation theory. As the desired value is the velocity of the particle, equation 64 is solved for the Reynolds number:

$$Re = [(4/3)S_1/C_D]^{1/2} \tag{65}$$

To determine the size of a particle having a velocity, w, in the gravitational field, both sides of equation 63 are multiplied by the complex Re/ArK_s:

$$(Re^3/Ar(1/K_s) = (4/3)\ Re/C_D \tag{66}$$

The dimensionless complex Re^3/Ar is expressed simply as:

$$\zeta = Re^3/Ar = (w^3/g\nu)(\rho_f/\rho_c) \tag{67}$$

Denoting $S_2 = \zeta/K_s$, then equation 62 may be rewritten as

$$S_2 = (4/3)Re/C_D \tag{68}$$

This is the second form of the dimensionless equation for sedimentation. The Reynolds number also may be calculated from this equation:

$$Re = \tfrac{3}{4} C_D\ S_2 \tag{69}$$

As with S_1, the Reynolds number is the dependent variable and S_2 is the determining one. For settling under the influence of gravity, we note that $a_g = g$, $K_s = 1$ and, hence, $S_1 = Ar$ and $S_2 = \zeta$. Therefore, the general dimensionless equations for sedimentation are applicable in any force field. They need be transformed only into the appropriate dimensionless groups describing the type of force field influencing the process. Again, for gravity settling, $Ar = \tfrac{3}{4} C_D Re^2$, and $\zeta = 4Re/3C_D$. The dimensionless numbers of sedimentation, S_1 and S_2, as well as C_D and Υ are all functions of Re. The parameter Υ must be determined experimentally. Equation 53 can be written as a straight line when expressed in terms of its logarithms:

$$\log(\Upsilon) = \log(A') - \alpha \log(Re) \tag{70}$$

Coefficient A' and exponent α can be evaluated readily from data on Re and Υ. The dimensionless groups are presented on a single plot in Figure 15. The plot of the function $C_D = f_l(Re)$ is constructed from three separate sections. These sections of the curve correspond to the three regimes of flow. The laminar regime is expressed by a section of straight line having a slope $\beta = 135°$ with respect to the x-axis. This section corresponds to the critical Reynolds number, $Re'_{cr} < 0.2$. This means that the exponent a in equation 53 is equal to 1. At this α value, the continuous-phase density term, ρ_c, in equation 46 vanishes.

Therefore, the inertia forces have an insignificant influence on the sedimentation process in this regime. Theoretically, their influence is equal to zero. In contrast, the forces of viscous friction are at a maximum. Evaluating the coefficient B in equation 55 for $\alpha = 1$ results in a value of 24. Hence, we have derived the expression for the drag coefficient of a sphere, $C_D = 24/Re$.

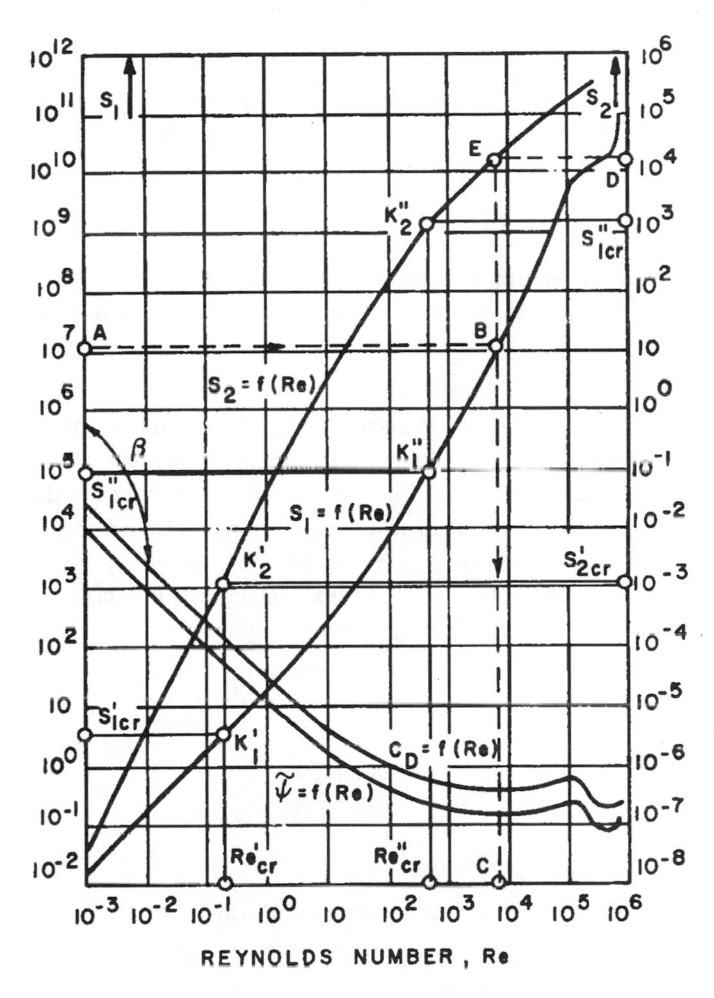

Figure 15. Dimensionless sedimentation theory plot.

The first critical values of the dimensionless sedimentation numbers, S_1 and S_2, are obtained by substituting for the critical Reynolds number value, $Re'_{cr} = 0.2$, into the above expressions:

$$S'_{1cr} < 3.6 \tag{71}$$

$$S'_{2cr} < 0.0022 \tag{72}$$

Substituting the expression for C_D, we again obtain the settling velocity of an isolated particle in larninar flow:

$$w = (d^2/18\nu)(\rho_f/\rho_c)a_g \tag{73}$$

Changing kinematic viscosity, ν, to dynamic viscosity, the velocity of particle sedimentation in the laminar regime is:

$$w = (d^2/18\mu)\rho_f a_g \tag{74}$$

From equations 64 and 60 at $S_1 \leq 3.6$, we obtair the first critical value of the particle diameter:

$$d'_{cr} \leq 1.53\{\nu^2 \rho_c/a_g \rho_f\}^{1/3} \tag{75}$$

In applying this equation it is possible to determine the maximum size particle in larninar flow, taking into account the given conditions of sedimentation (ρ_c, ρ_p, μ and a_g). However, this equation does not determine what the flow regime is when $d > d'_{cr}$.

The turbulent regime for C_D is characterized by the section of line almost parallel to the x-axis (at the $Re''_{cr} > 500$). In this case, the exponent α is equal to zero. Consequently, viscosity vanishes from equation 46. This indicates that the friction forces are negligible in comparison to inertia forces. Recall that the resistance coefficient is nearly constant at a value of 0.44. Substituting for the critical Reynolds number, $Re'_{cr} \geq 500$, into equations 65 and 68, the second critical values of the sedimentation numbers are obtained:

$$S''_{1cr} \geq 82{,}500$$

$$S''_{2cr} > 1{,}515$$

And substituting $C_D = 0.44$ into equation 59,

$$w = 1.75[d\,(\rho_f/\rho_c)a_g]^{1/2} \tag{76}$$

At $S''_{1cr} > 82{,}500$, we obtain the second critical value of particle size:

$$d''cr = 43.5\{\nu^2 \rho_c/a_g \rho_f\}^{1/2} \tag{77}$$

Those particles with sizes $d > d''_{cr}$ at a given set of conditions (ν, ρ_c, ρ_p, and a_g) will settle only in the turbulent flow regime. For particles with sizes $d'_{cr} < d$, d''_{cr} will settle only when the flow around the object is in the transitional regime. Recall that the transitional zone occurs in the Reynolds number range of 0.2 to 500. The sedimentation numbers corresponding to this zone are: $3.6 < S_1 < 82{,}500$; and $0.0022 < S_2 < 1{,}515$.

The slope of the curve in the transitional zone changes from 135 to 180°. It shows that the exponent in changes as follows: $0 \leq \alpha \leq 1$. This means that the friction and inertia forces are commensurable in the process of sedimentation. Several empirical formulas have been proposed for estimating the resistance coefficient in the transition zone. One such correlation is

$$C_D = 13/(Re)^{1/2} \tag{78}$$

Introducing this to equation 59 produces:

$$w_s = 0.22d[(a_g \rho_f)^2/\mu \rho_c]^{1/3} \tag{79}$$

When we consider many particles settling, the density of the fluid phase effectively becomes the bulk density of the slurry, i.e., the ratio of the total mass of fluid plus solids divided by the total volume. The viscosity of the slurry is considerably higher than that of the fluid alone because of the interference of boundary layers around interacting solid particles and the increase of form drag caused by particles. The viscosity of a slurry is often a function of the rate of shear of its previous history as it affects clustering of particles, and of the shape and roughness of the particles. Each of these factors contributes to a thicker boundary layer.

Experimental measurements of viscosity almost always are recommended when dealing with slurries and extrapolations should be made with caution. Most theoretically based expressions for liquid viscosity are not appropriate for practical calculations or require actual measurements to evaluate constants. For nonclustering particles, a reasonable correlation may be based on the ratio of the effective bulk viscosity, μ_B, to the viscosity of the liquid. This ratio is expressed as a function of the volume fraction of liquid x' in the slurry for a reasonable range of compositions:

$$\mu_B/\mu = (1/x')10^{1.82(1-x)} \tag{80}$$

A correction factor, R_c, incorporating both viscosity and density effects can be developed for a given slurry, which provides a more convenient expression based on the following equation:

$$u_t = (\rho_p - \rho)gd^2/18\mu \tag{81}$$

as

$$v_H = R_c(\rho_p - \rho)gd^2/18\mu \tag{82}$$

where v_H is the terminal velocity in hindered settling.

Measurements of the effective viscosity as a function of composition may be fitted to equation 80 or presented in graphic form as in Figure 16. The correction factor, R_c also may be determined by accounting for the volume fraction, η_v, of particles through the Andress formula:

$$R_c = (1 - \eta_v)^2/(1 + 2.5\,\eta_v + 7.35\,\eta_v^2) \qquad (83)$$

In summarizing sedimentation principles, we note that the particle settling velocity is the principal design parameter that establishes equipment sizes and allowable loadings for separating heterogeneous systems. However, design calculations are not straightforward because prediction of the settling velocity requires knowledge of the flow regime in the vicinity of the particles. Therefore, the following generalized design method is recommended.

From known values of d, ρ_p, ρ_c, μ and a_g, compute the first sedimentation dimensionless number (equation 63). From the plot given in Figure 17, obtain the corresponding Reynolds number, Re, and evaluate the theoretical settling velocity. If the flow regime is laminar, the settling velocity may be calculated directly from equation 74 and the regime checked by computing the Reynolds number for the flow around an individual particle. After determining w, determine the appropriate shape factor, ψ (either from literature values or measurements) and correction factor, R_c. The design settling velocity then will be:

$$w_0 = R_c\,\psi\,w \qquad (84)$$

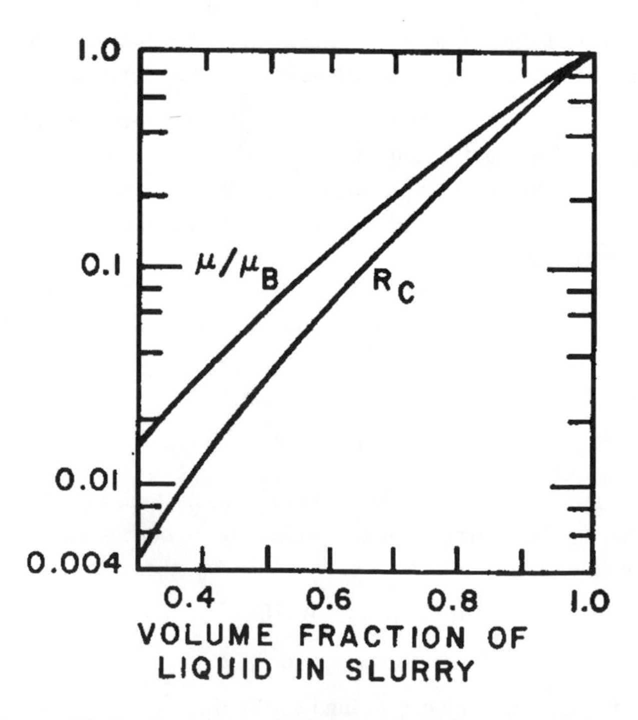

Figure 16. Settling factor for hindered settling.

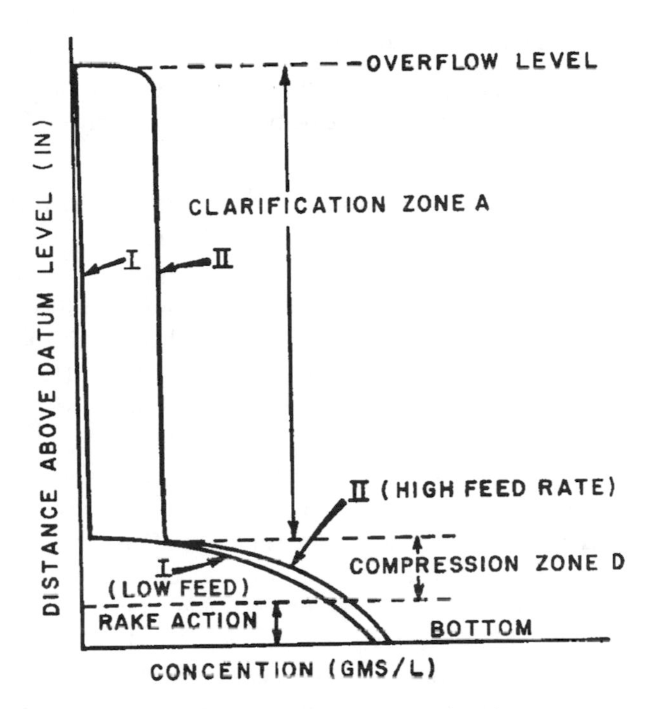

Figure 17. Plot of concentration versus height in a continuous sedimentation device. Curve (1) - low feedrate; Curve (2) - high feed rate.

Sedimentation equipment is designed to perform two operations: to clarify the liquid overflow by removal of suspended solids and to thicken sludge or underflow by removal of liquid. It is the cross section of the apparatus that controls the time needed for settling a preselected size range of particles out of the liquid for a given liquid feed rate and solids loading. The area also establishes the clarification capacity. The depth of the thickener establishes the time allowed for sedimentation (i.e., the solid's residence time) for a given feed rate and is important in determining the thickening capacity. The clarification capacity is established by the settling velocity of the suspended solids. Sedimentation tests are almost always recommended when scaling up for large settler capacities. By means of material balances, the total amount of fluid is equal to the sum of the fluid in the clear overflow plus the fluid in the compacted sludge removed from the bottom of the thickener. The average vertical velocity of fluid at any height through the thickener is the volumetric rate passing upward at that level divided by the unit's cross section. Note that if the particle settling velocity is less than the upward fluid velocity, particles will be entrained out in the overflow, resulting in poor clarification. For those size particles whose settling velocity approximately equals that of the upward fluid velocity, particles remain in a balanced suspension, i.e., they neither rise nor fall, and the concentration of solids in the clarification zone increases. This eventually results in a reduction of the settling velocity until the point where particles are entrained out in the overflow.

The thickener must be designed so that the settling velocity of particles is significantly greater than the upward fluid velocity, to minimize any increase in the solids concentration in the clarification zone.

Solids concentration varies over the thickener's height, and at the lower levels where the solution is dense, settling becomes retarded. In this region the upward fluid velocity can exceed the particle settling velocity irrespective of whether this condition exists in the upper zone or not. Figure 17 illustrates this situation, where curve II denotes a higher feed rate. A proper design must therefore be based on an evaluation of the settling rates at different concentrations as compared to the vertical velocity of the fluid. If the feed rate exceeds the maximum of the design, particulates are unable to settle out of the normal clarification zone. Hence, there is an increase in the solids concentration, resulting in hindered settling. The result is a corresponding decrease in the sedimentation rate below that observed for the feed slurry. The feed rate corresponding to the condition of just failing to initiate hindered settling represents the limiting clarification capacity of the system. That is, it is the maximum feed rate at which the suspended solids can attain the compression zone. The proper cross-sectional area can be estimated from calculations for different concentrations and checked by batch sedimentation tests on slurries of increasing concentrations. You will find some problems in the section on ***Questions for Thinking and Discussing*** that illustrate the need to check the thickener's calculated area against concentrations at various points in the vessel (including both the clarification and thickening zones). Figure 18 shows the effect of varying the underflow rate on the thickening capacity. In this example, the depth of the thickening zone (compression zone) increases as the underflow rate decreases; hence, the underflow solids concentration increases, based on a constant rate of feed.

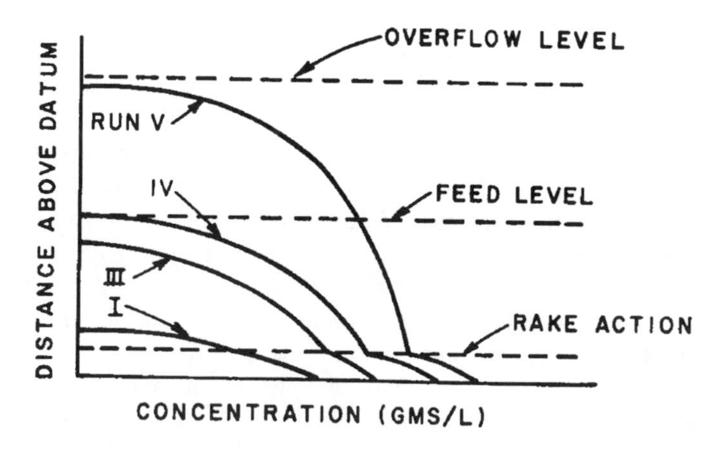

Figure 18. Shows effect of underflow rate on thickening capacity.

The curves of concentration as a function of depth in the compression zone are essentially vertical displacements of each other and are similar to those observed in batch sedimentation. When the sludge rakes operate, they essentially break up a semirigid structure of concentrated sludge. Generally, this action extends to several inches above the rakes and contributes to a more concentrated underflow. The required height of the compression zone may be estimated from experiments on batch sedimentation. The first batch test should be conducted with a slurry having an initial concentration equivalent to that of the top layer of the compression zone during the period of constant rate settling. This is referred to as the critical concentration. The time required for the sample slurry to pass from the critical concentration to the desired underflow concentration can be taken as the retention time for the solids in the continuous operation. The underlying assumption here is that the solids concentration at the bottom of the compression zone in the continuous thickener at any time is the same as the average concentration of the compression zone in the batch unit and at a time equal to the retention time of the solids in the continuous thickener. Hence, it is assumed that the concentration at the bottom of the thickener is an implicit function of the thickening time. The retention time is obtained from a batch test by observing the height of the compression zone as a function of time. The slope of the compression curve is described by the following expression:

$$-dZ/dt = k(Z - Z_\infty) \tag{85}$$

where Z, Z_∞ are the heights of compression at times t and infinity, respectively, and k is a constant that depends on the specific sedimentation system. Integrating this expression gives:

$$\ell n\,(Z - Z_\infty) = -kt + \ell n\,(Z_c - Z_\infty) \tag{86}$$

where Z_c is the height of the compression zone at its critical concentration. This expression is the equation of a straight line and normally is plotted as $\log[(Z - Z_\infty)/(Z_0 - Z_\infty)]$ versus time, where Z_0 is the initial slurry concentration. If batch tests are performed with an initial slurry concentration below that of the critical, the average concentration of the compression zone will exceed the critical value because it will consist of sludge layers compressed over varying time lengths. A method for estimating the required time to pass from the critical solids content to any specified underflow concentration can be done as follows:

1. Extrapolate the compression curve to the critical point or zero time.

2. Locate the time when the upper interface (between the supernatant liquid and slurry) is at height Z'_0, halfway between the initial height, Z_0, and the extrapolated zero-time compression zone height, Z'_0. This time represents the period in which all the solids were at the critical dilution and went into compression. The retention time is computed as $t - t_c$, where t is the time when the solids reach the specified

underflow concentration. The procedure is illustrated in Figure 19. It is recommended that you determine the required volume for the compression zone to be based on estimates of the time each layer has been in compression. The volume for the compression zone is the sum of the volume occupied by the solids plus the volume of the entrapped fluid. This may be expressed as:

$$V = Q(\Delta t)/\rho_s + \int\{m_\ell Q/m_s\ \rho_\ell\}dt \qquad (87)$$

where Q = solids mass feed per unit time; $\Delta t = t - t_c$ = retention time; m_ℓ = mass of liquid in the compression zone; m_s = mass of solids in the compression zone.

This expression is based on our earlier assumption that the time required to thicken the sludge is independent of the interface height of the compression zone. An approximate solution to this expression can be obtained if we assume m_ℓ/m_s to be constant, i.e., an average mass ratio in the thickening zone from top to bottom. Then,

$$V = Q\ \Delta t[1/\rho_s + (1/\rho_\ell)(\{m_\ell/m_s)_{avg}] \qquad (88)$$

More reliable results can be obtained by assuming average conditions over divided parts of the compression zone. That is, the above expression can be applied to divisions of the compression zone and the total volume obtained by the sum of these calculations. Try some of the problems in the section on ***Questions for Thinking and Discussing*** to strengthen your understanding of the principles covered.

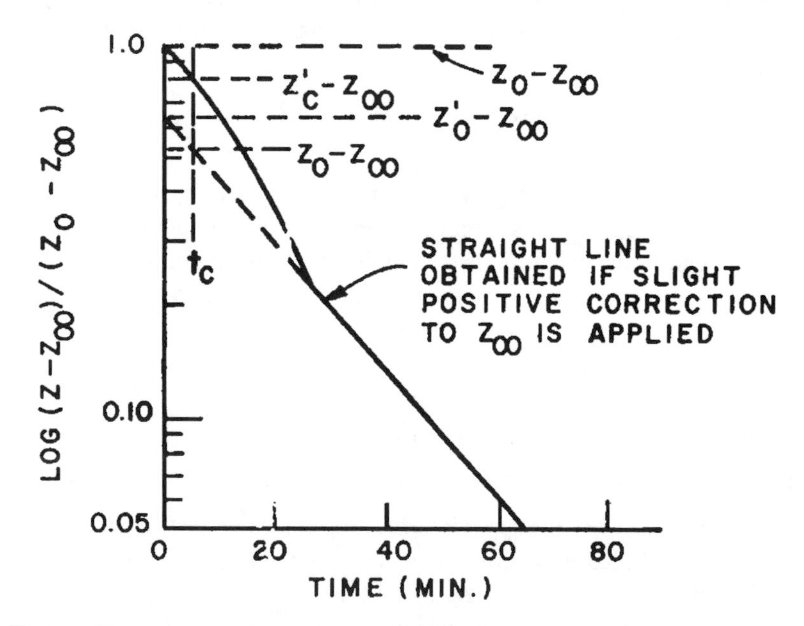

Figure 19. Extrapolation of sedimentation data to estimate time for critical concentration.

There are commercially available clarifier simulation software for comprehensive (2D) analyzes of wastewater treatment processes in circular and rectangular clarifiers. With these software, you can predict processes like:

- distribution of sludge in the settler
- flow streamlines in the settler
- vertical and horizontal flow velocities
- vertical and horizontal flow velocities
- sludge concentration in the effluent
- return sludge concentration
- total mass of sludge in the settler

Different processes like eddy turbulence, bottom current, stagnation of flows, and storm-water events can be simulated, using either laminar or turbulent flow model for simulation. All processes are displayed in real-time graphical mode (history, contour graph, surface, etc.); you can also record them to data files. Thanks to innovative sparse matrix technology, calculation process is fast and stable: a large number of layers in vertical and horizontal directions can be used, as well as a small time step. You can hunt for these on the Web.

A CLOSER LOOK AT THE MECHANICAL CLARIFICATION PROCESS AND THE CHEMISTRY OF CLARIFICATION

So by now it should be clear that what the process of clarification is all about is removing suspended solids from water. Important concepts that we have eluded to, but maybe not spelled out so clearly up to now are:

1. *Stable solids suspensions in water*- The mechanisms involved in keeping solids suspended in water,

2. *Chemical treatments -How* organic polymers and inorganic coagulants work to counteract solids stabilization mechanisms and enhance removal of solids from water, and

3. *The function of clarification unit operations*- How these units work and how chemical treatment enhances their performance.

A term that we should get into our vocabulary is "subsidence". This term essentially means settling. While a degree of clarification can be accomplished by subsidence, most industrial processes require better quality water than can be obtained from settling only. Most of the suspended matter in water would settle, given enough time, but in most cases the amount of time required would not be practical. As we have shown from our derivations of expressions describing the classical theory of sedimentation, settling characteristics depend upon the :

1. Weight of the particle,
2. Shape of the particle,
3. Size of the particle, and
4. Viscosity and/or frictional resistance of the water, which is a function of temperature.

The settling rates of various size particles at 50° F (10° C) is illustrated in Table 1.

Table 1. Some Settling Rates for Different Particles (*assumed spherical*) and Sizes

Particle Diameter (mm)	Particle Type	Time to Settle One Foot
10.0	Gravel	0.3 sec.
1.0	Coarse sand	3.0 sec.
0.1	Fine sand	38.0 sec.
0.01	Silt	33.0 minutes
0.001	Bacteria	35.0 hours
0.0001	Clay particles	230 days
0.00001	Colloidal particles	65 years

Look closely at the settling times in Table 1 – the times span from a fraction of a second to almost a lifetime! A great deal of the suspended matter found in waste waters fall into the colloidal suspension range, so obviously we cannot rely on gravitational force alone to separate out the pollutants.

WHAT COAGULATION IS ALL ABOUT

The term *coagulation* refers to the first step in complete clarification. It is the neutralization of the electrostatic charges on colloidal particles. Because most of the smaller suspended solids in surface waters carry a negative electrostatic charge, the natural repulsion of these similar charges causes the particles to remain dispersed almost indefinitely. To allow these small suspended solids to agglomerate, the negative electrostatic charges must be neutralized. This is accomplished by using inorganic coagulants, which are water soluble inorganic compounds), organic cationic polymers or polyelectrolytes. The most common and widely used inorganic coagulants are:

- Alum-aluminum sulfate-$A1_2(SO_4)_3$
- Ferric sulfate-$Fe_2(SO_4)_3$
- Ferric chloride-$FeC1_3$
- Sodium aluminate-$Na_2Al_2O_4$

Inorganic salts of metals work by two mechanisms in water clarification. The positive charge of the metals serves to neutralize the negative charges on the turbidity particles. The metal salts also form insoluble metal hydroxides which are gelatinous and tend to agglomerate the neutralized particles. The most common coagulation reactions are as follows:

$$A1_2(SO_4)_3 + 3Ca(HCO_3)_2 = 2Al(OH)_3 + 3CaSO_4 + 6CO_2$$

$$A1_2(SO_4)_3 + 3Na_2CO_3 + 3H_2O = 2Al(OH)_3 + 3Na2SO_4 + 3CO_2$$

$$A1_2(SO_4)_3 + 6NaOH = 2Al(OH)_3 + 3Na2SO_4$$

$$A1_2(SO_4)_3\,(NH_4)_2SO_4 + 3Ca(HCO_3) = 2Al(OH)_3 + (NH_4)_2SO_4 + 3CaSO_4 + 6CO_2$$

$$A1_2(SO_4)_3\,K_2SO_4 + 3Ca(HCO_3)_2 = 2Al(OH)_3 + K_2SO_4 + 3CaSO_4 + 6CO_2$$

$$Na_2Al_2O_4 + Ca(HCO_3)_2 + H_2O = 2Al(OH)_3 + CaCO_3 + Na_2CO_2$$

$$Fe(SO_4)_3 + 3Ca(OH)_2 = 2Fe(OH)_3 + 3CaSO_4$$

$$4Fe(OH)_2 + O_2 + 2H_2O = 4Fe(OH)_3$$

$$Fe_2(SO_4)_3 + 3Ca(HCO_3) = 2Fe(OH)_3 + 3CaSO_4 + 6CO_2$$

The effectiveness of inorganic coagulants is dependent upon water chemistry, and in particular -- pH and alkalinity. Their addition usually alters that chemistry. Table 2 illustrates the effect of the addition of 1 ppm of the various inorganic coagulants on alkalinity and solids concentration.

Table 2. Coagulant, Acid and Sulfate - 1 ppm Equivalents.

1 ppm Formula or Chemical	ppm Alkalinity Reduction	ppm SO_4 Increase	ppm Na_2SO_4 Increase	ppm CO_2 Increase	ppm Total Solids Increase
$Al_2(SO_4)_3 \cdot 18H_2O$	0.45	0.45	0.64	0.40	0.16
$Al_2(SO_4)_3 \cdot (NH_4)2SO_4 \cdot 24H_2O$	0.33	0.44	0.63	0.29	0.27
$Al_2(SO_4)_3 \cdot K_2SO_4 \cdot 24H_2O$	0.32	0.43	0.60	0.28	0.30
$FeSO_4 \cdot 7H_2O$	0.36	0.36	0.61	0.31	0.13
$FeSO_4 \cdot 7H_2O + (SCl_2)$	0.54	0.36	0.51	0.48	0.18
$Fe_2(SO_4)_3$	0.76	0.76	1.07	0.64	0.27
H_2SO_4 - 96%	1.00	1.00	1.42	0.88	0.36
H_2SO_4 - 93.2% (66° Be)	0.96	0.95	1.36	0.84	0.34
H_2SO_4 - 77.7% (66° Be)	0.79	0.79	1.13	0.70	0.28
$NaSO_4$	-	0.64	0.95	-	1.00
$Na_2Al_2O_4$	Increase 0.54	-	-	Reduces 0.47	0.90

An Important Note

Note that the use of metal salts for coagulation may increase the quantity of dissolved solids. One must consider the downstream impact of these dissolved solids. In addition, the impact of carryover of suspended Al ... and Fe ... compounds and their related effect on downstream processes must be considered.

Aluminum salts are most effective as coagulants when the pH range is between 5.5 and 8.0 pH. Because they react with the alkalinity in the water, it may be necessary to add additional alkalinity (called buffering) in the form of lime or soda ash. Use the values in Table 3 to guide you. Iron salts, on the other hand, are most effective as coagulants at higher pH ranges (between 8 and 10 pH). Iron salts also depress alkalinity and pH levels; therefore, additional alkalinity must be added. Sodium aluminate increases the alkalinity of water, so care must be taken not to exceed pH and alkalinity guidelines. As is evident from the reactions discussed above, a working knowledge of the alkalinity relationships of water is mandatory. By using inorganic coagulants we can wind up producing a voluminous, low-solids content sludge the is difficult to dewater and dries very slowly. The properties of the sludge to be generated and estimated quantities needs to be carefully determined, in part from pilot-scale and bench testing prior to the design and construction of a plant.

Polymers are often described as long chains with molecular weights of 1,000 or less to 5,000,000 or more. Along the chain or backbone of the molecule are numerous charged sites. In primary coagulants, these sites are positively charged. The sites are available for adsorption onto the negatively charged particles in the water. To accomplish optimum polymer dispersion and polymer/particle contact, initial mixing intensity is critical. The mixing must be rapid and thorough, Polymers used for charge neutralization cannot be over-diluted or over mixed. The farther upstream in the system these polymers can be added, the better their performance. Because most polymers are viscous, they must be properly diluted before they are added to the influent water. Special mixers such as static mixers, mixing tees and specially designed chemical dilution and feed systems are all aids in polymer dilution. Static or motionless mixers in particular are popular for this application. Refer to Figure 20 for an example of an in-line static mixer.

Names of Chemicals

Filter Alum
$Al_2(SO_4)_3 \cdot 18H_2O$

Ammonium Alum
$Al_2(SO_4)_3 \cdot (NH_4)2SO_4 \cdot 24H_2O$

Potash Alum
$Al_2(SO_4)_3 \cdot K_2SO_4 \cdot 24H_2O$

Copperas (ferrous sulfate)
$FeSO_4 \cdot 7H_2O$

Chlorinated Copperas
$FeSO_4 \cdot 7H_2O+(SCl_2)$

Ferric Sulfate
$Fe_2(SO_4)_3$

Sulfuric Acid
H_2SO_4

Salt Cake
$NaSO_4$

Sodium Aluminate
$Na_2Al_2O_4$

Table 3. Recommended Alkali and Lime 1 ppm Equivalents.

Chemical - 1ppm	Formula (1 ppm)	Alkalinity Increase (1 ppm)	Free CO_2 Reduction (1 ppm)	Hardness as $CaCO_3$ Increases
Sodium bicarbonate	$NaHCO_3$	0.60	-	-
Soda ash (56% Na_2, 99.16% Na_2CO_3)	Na_2CO_3	0.94	0.41	-

Chemical - 1ppm	Formula (1 ppm)	Alkalinity Increase (1 ppm)	Free CO_2 Reduction (1 ppm)	Hardness as $CaCO_3$ Increases
Caustic soda (76% Na_2O, 98.06% Na_2CO_3)	NaOH	1.22	1.09	
Quicklime (90% CaO)	CaO	1.61	1.41	1.61
Hydrated lime (93% $Ca(OH)_2$)	$Ca(OH)_2$	1.26	1.10	1.36

Figure 20. Example of an in-line static mixer for polyelectrolyte additions.

Once the negative charges of the suspended solids are neutralized, flocculation begins. Flocculation can be thought of as the second step of the coagulation process. Charge reduction increases the occurrence of particle-particle collisions, promoting particle agglomeration. Portions of the polymer molecules not absorbed protrude for some distance into the solution and are available to react with adjacent particles, promoting flocculation. Bridging of neutralized particles can also occur when two or more turbidity particles with a polymer chain attached come together. It is important to remember that during this step, when particles are colliding and forming larger aggregates, mixing energy should be great enough to cause particle collisions but not so great as to break up these aggregates as they are formed. In some cases flocculation aids are employed to promote faster and better flocculation. These flocculation aids are normally high molecular weight anionic polymers. Flocculation aids are normally necessary for primary coagulants and water sources that form very small particles upon coagulation. A good example of this is water that is low in turbidity but high in color (colloidal suspension).

A final are we should discuss is color removal. This is perhaps the most difficult impurity to remove from waters. In surface waters color is associated with dissolved or colloidal suspensions of decayed vegetation and other colloidal suspensions. The composition of this material is largely tannins and lignins, the components that hold together the cellulose cells in vegetation. In addition to their undesirable appearance in drinking water, these organics can cause serious problems in downstream water purification processes. For examples:

1. Expensive dernineralizer resins can be irreversibly fouled by these materials.
2. Some of these organics have chelated trace metals, such as iron and manganese within their structure, which can cause serious deposition problems in a cooling system.

There are many ways of optimizing color removal in a clarifier. The three most common methods are:

- Prechlorination (before the clarifier) significantly improves the removal of organics as well as reducing the coagulant demand.
- The proper selection of polymers for coagulation has a significant impact on organic removal.
- Color removal is affected by pH. Generally, organics are less soluble at low pH.

A REVIEW OF EQUIPMENT OPTIONS

Although we have discussed the major hardware, it is still worthwhile reviewing these in relation to the major classes of clarifier processes. The major categories of this process are:

- conventional
- upflow
- solids-contact
- sludge-blanket

Conventional clarification is the simplest form of the process. It relies on the use of a large tank or horizontal basin for sedimentation of flocculated solids. Figure 21 provides a sketch of the basic configuration. The basin normally contains separate chambers for rapid mix and settling. The first two steps critical in achieving good clarification. An initial period of turbulent mixing is needed for contact between the coagulant ans suspended solids. This is followed by a period of gentle stirring which helps to increase particle collisions and floc size. Retention times are typically between 3 and 5 minutes , 15 to 30 minutes for flocculation, and 4 to 6 hours for settling. Coagulants are added to the wastewater in the rapid mix chamber, or sometimes immediately upstream. The water passes through the mix chambers and enters the settling basin. Refer again to Figure 21, which is a classical large-tank clarifier. The water passes out to the circumference, while the flocculated particles settle to the bottom. Accumulated sludge are scraped into a sludge collection basin for removal and disposal (sometimes post processing, as discussed in Chapter 10). The clean water flows over a weir and is held in a tank , which is referred to as a clearwell. A rectangular version of a conventional clarifier is illustrated in Figure 22. This unit is referred to as a horizontal basin clarifier.

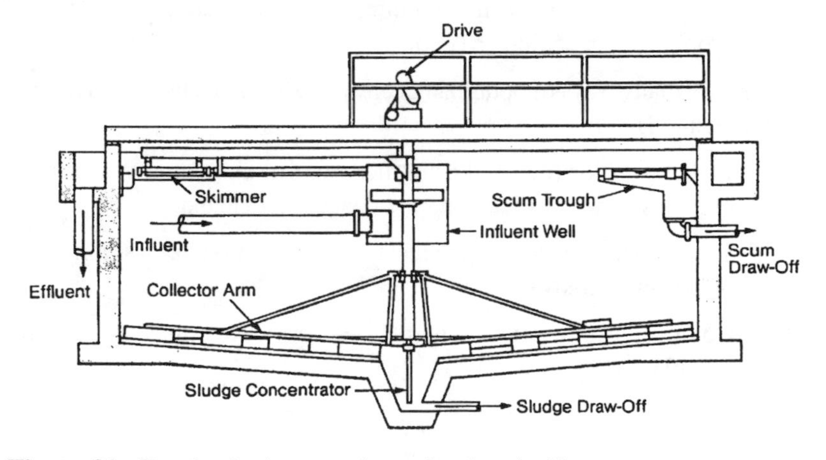

Figure 21. Sketch of a large tank or circular clarifier.

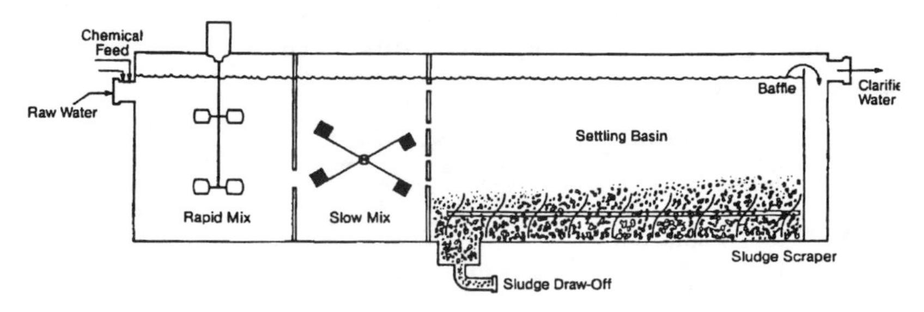

Figure 22. Sketch of a horizontal basin clarifier.

It is often advantageous to employ a zone of high solids contact to achieve a better quality effluent. This is accomplished in an upflow clarifier, so called because the water flows upward through the clarifier as the solids settle to the bottom. Most upflow clarifiers are either solids-contact or sludge-blanket type clarifiers, which differ somewhat in theory of operation. Cross-sections of these two types of units are illustrated in Figures 23 and 24. Both units have an inverted cone within the clarifier. Inside the cone is a zone of rapid mixing and a zone of high solids concentration. The coagulant is added either in the rapid mix zone or somewhere upstream of the clarifier.

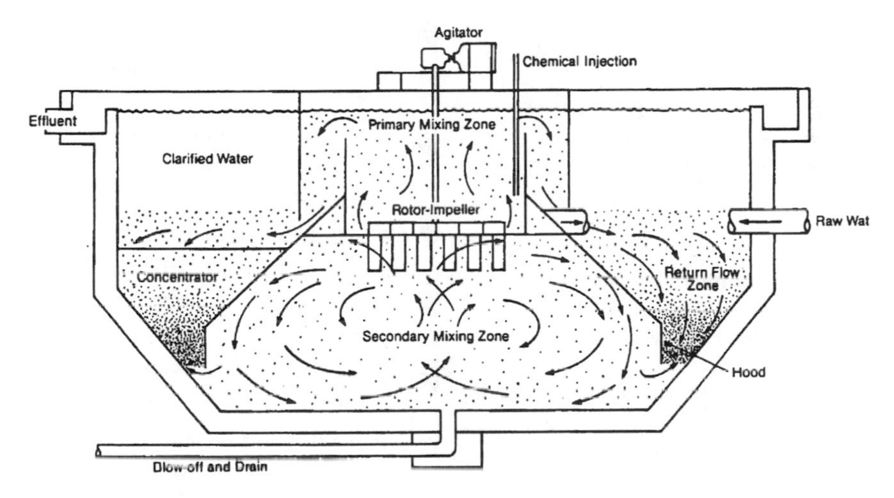

Figure 23. Sketch of a solids-contact clarifier.

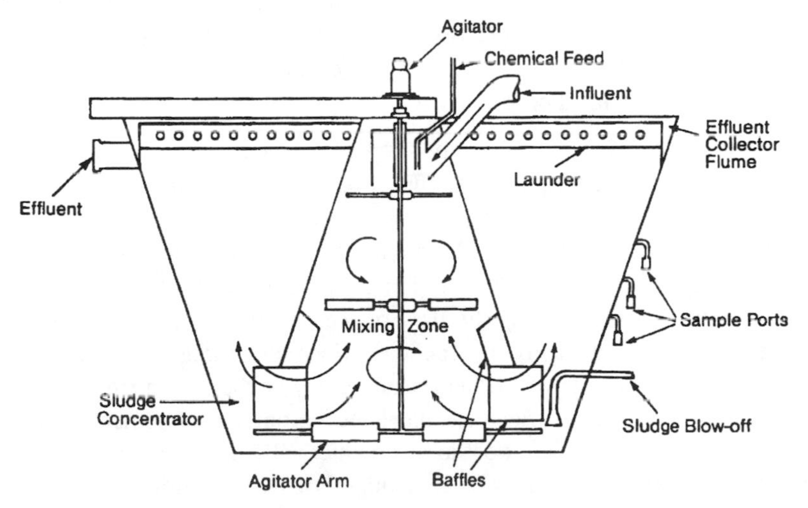

Figure 24. Sketch of a sludge-blanket clarifier.

In the solids-contact clarifier, raw water is drawn into the primary mixing zone, where initial coagulation and flocculation take place. The secondary mixing zone is used to produce a large number of particle collisions so that smaller particles are entrained in the larger floc. Water passes out of the inverted cone into the settling zone, where solids settle to the bottom and clarified water flows over the weir. Solids are drawn back into the primary mixing zone, causing recirculation of the large floc. The concentration of solids in the mixing zones is controlled by occasional or continuous blowdown of sludge.

The sludge-blanket clarifier (Figure 24) goes one step further, by passing the water up from the bottom of the clarifier through a blanket of suspended solids that acts as a filter. The inverted cone within the clarifier produces an increasing cross-sectional area from the bottom of the clarifier to the top. Thus, the upward velocity of the water decreases as it approaches the top. At some point, the upward velocity of the water exactly balances the downward velocity of a solid particle and the particle is suspended, with heavier particles suspended closer to the bottom. As the water containing flocculated solids passes up through this blanket, the particles are absorbed onto the larger floc, which increases the floc size and drops it down to a lower level. It eventually falls to the bottom of the clarifier to be recirculated or drawn off.

Some Application Pointers for Clarifiers

1. **Primary coagulants need good mixing**
2. **Carefully read over all equipment manuals and know the operating ranges and limitations**
3. **Experiment with different feed points**
4. **Don't feed polymers too close to chlorine or other oxidants**
5. **Use dilution water with polymers**
6. **Split up feeding of polymers**
7. **Watch blowdown carefully**
8. **Watch the centerwell for troubleshooting and to observe the need for any chemical dosage changes**
9. **Optimize the operating variables (e.g., sludge bed depth, turbine speed, chemical feed rate, etc.)**

Although these processes seem relatively simple, especially in relation to many chemical manufacturing operations or unit processes, there are a number of operational problems that can make the life of an operator miserable. Excessive floc carryover is a very common problem. This is most often associated with hydraulic overload or unexpected flow surge conditions. You can tackle this problem by relying on equalize flow (metering the flow of the clarifier), which will help to dampen out surges. Unfortunately, hydraulic overload conditions are not the only causes of excessive floc carryover. Other reasons many be thermal currents, short-circuiting effects, low density floc, chemical feed problems. Another common operator problem is simply no floc in the centerwell. This can result from underfeeding of chemicals or a loss of the sludge bed recirculation. Refer to the sidebar discussion on this page for some general corrective actions you can try.

You will have to investigate and apply trial and error field tests to resolve some of these problems. When new equipment are installed, it is wise to spend time during a shake-down and start-up period to explore the operational limitations of the process and train operators on how to handle these types of problems.

Corrective Actions for Floc Problems

Excessive Floc Carryover
***Cause*: Hydraulic overloading; flow surges.**
***Corrective Action*: Equalize flow to eliminate surges.**
***Cause*: Thermal currents.**
***Corrective Action*: Equalize flow.**
***Cause*: Short-circuiting.**
***Corrective Action*: Use tracer dye to confirm and identify exact nature of problem. Check clarifier internals such as mixing baffles and modify as needed.**
***Cause*: Low density floc.**
***Corrective Action*: Increase coagulant/flocculant doses.**
***Cause*: Chemcial feed problems.**
***Corrective Action*: Check pump and meter settings. Confirm thast setting correspond with anticipated chemical feed rates.**
No Floc in Centerwell
***Cause*: Chemical underfed.**
***Corrective Action*: Check operation of feed pump. Check pump settings.**
***Cause*: Sludge bed recirculation lost.**
***Corrective Action*: Increase recirculation slowly.**

RECTANGULAR SEDIMENTATION TANKS

The process concept for sedimentation tanks has hardly changed over the past 80 years. Dimensioning these vessels according to existing guidelines guarantees safe operation. With ever tightening legislation, however, the question of expansion or upgrading of existing sewage treatment plants arises. Expansion is an expensive solution and impossible if the available space is scarce so that a new construction has to be built. The basis of upgrading consists in changed process concepts which are able to exploit the unused potential of existing tanks. An essential prerequisite for upgrading plants is the question whether the settling volume for activated sludge is sufficient. Beside the clarified water discharge, the feeding method for the sludge/water mix and the skimmer system have an essential influence on the separation efficiency in tanks. The inlet height is approx. 2/3 to 1/3 of the tank depth and the skimming direction in the counter flow. This concept is based on the empirical knowledge of normally minor turbulence in the tank. However, changed process concepts with a bottom-near inlet and concurrent skimming are able to minimize such turbulences to such an extent that the sludge load and thus the separation efficiency can be increased.

It is important to recognize that the process-engineering installations in rectangular sedimentation tanks have a great influence on the performance of this final treatment stage. Of particular importance is the design of the inlet section as turbulences are generated there by mixing with the wastewater inflow which may have an intense influence on the sedimentation process. The density of flows has a strong influence on the separation efficiency. The density flow sinks to the tank bottom during inflow and passes to the tank end. The rising density flow initiates back flow of the clarified water on the surface. To increase the separation efficiency of tanks, the density flow should be minimized or the engineering process modified in a way that the density flow will be integrated with the sedimentation process.

Inlet dimensions are important sizes The density flow can be substantially influenced by the inlet structure by minimizing the potential and kinetic energy of the wastewater stream with a suitable feed design. The inlet should have bottom-near feed openings to have an as small intermixing zone as possible between the activated sludge/water mix and the tank content. The velocity gradient in the inlet section should be small to avoid floc disturbance by shearing forces. The inlet section sludge scraping also influences the separation efficiency, most notably it impacts on the degree of thickening in the sedimentation tank. For minimisation of the turbulences in the secondary clarification tank and a consequently improved separation efficiency and for sludge thickening, practice has shown that scraping the sludge in the direction of the density flow works best. The scraping velocity should be low in order to prevent re-suspension of the activated sludge flocs. To increase the degree of thickening and to minimize the volume flow of the return sludge, a minimum sludge residence time in the tank must be provided. Although a sludge hopper for thickening the activated sludge is not necessary, a sludge hopper at the tank end tends to increase the surface load of the tank.

Sludge Removal in Rectangular Tanks

The first mechanized rectangular sedimentation units in the United States were designed and installed by William M. Platt in 1920 in Gastonia, North Carolina. Since then they have found widespread application in standard designs. Sludge removal equipment of this type usually consists of a pair of endless conveyer chains running over sprockets attached to shafts, one of the shafts is connected by chain and sprocket to a drive unit at one end of the tank. Attached to the chains at about 10 foot intervals are cross pieces of wood or flights, extending the width of the tank. Linear conveyer speeds of 2-3 feet per minute are common with 1 foot per minute for activated sludge.

A major problem is underwater maintenance and repair cost. The average life of the underwater equipment is about 8 years.

Further information on the operation and design basic for rectangular settling basins can be obtained from references 1 through 3.

AIR FLOTATION SYSTEMS

Air Flotation systems for separating oil and solids from wastewater include:

⇨ Dissolved Air Flotation

⇨ Dispersed Air Flotation

⇨ Induced Air Flotation

⇨ Diffused Air Flotation

Designs include rectangular, circular, cross flow and inclined parallel plate systems.

Air flotation is one of the oldest methods for the removal of solids, oil & grease and fibrous materials from wastewater. Suspended solids and oil & grease removals as high as 99%+ can be attained with these processes. Air flotation is simply the production of microscopic air bubbles, which enhance the natural tendency of some materials to float by carrying wastewater contaminants to the surface of the tank for removal by mechanical skimming. Many commercially available units are packaged rectangular steel tank flotation systems; shipped completely assembled and ready for simple piping and wiring on site. Models typically range from 10 to over 1000 square feet of effective flotation area for raw wastewater flows to over 1000 gallons per minute. Complete systems often include chemical treatment processes. A dissolved air flotation (DAF) system can produce clean water in wash operations where reduction of oil and grease down to 2 mg/l is achievable in certain applications. In addition to municipal and heavy industry applications, DAF has found a home with commercial vehicle washing, industrial laundries, food processing. Vehicle wash applications need not be confined to automobiles and trucks, but can extend to buses, tank cars and many other types of vehicles. There is a broad and varied market for DAF. Excessive oil and grease (especially emulsified oils), plus high levels of suspended solids and metals are good candidates for our DAF systems. Another example is an industrial laundry, where there is the need for good waste treatment systems. A DAF system can be used in a variety of food processing applications, including Vegetable Oils, Animal and Seafood Fats, Red Meat Butchering, Poultry processing, and Kitchen and Equipment Washing. Oil Drilling on and offshore is under pressure from the USEPA. The removal of the oils from the dirty water which surfaces when drilling wells cannot be dumped. DAF technology has proven very effective in this industry. Some waste stream problems found in shipyards and aboard ships are oily bilge water and solids containing copper and other heavy metals used in marine paints. DAF is a proven method in these applications as well. In metal finishing operations DAF will remove cadmium, chromium, lead, zinc and other toxic heavy metals from a waste stream. Finally, when combined with other treatment processes DAF can be applied upstream of large water treatment systems to handle contaminated water before it mingles with the rest of the waste stream.

When the primary target is oil removal, we should distinguish between the forms of oil. There are two forms of oil that we find in wastewater. ***Free oil*** is oil that will separate naturally and float to the surface. ***Emulsified oil*** is oil that is held in suspension by a chemical substance (Detergents - Surfactants) or electrical energy. When making an evaluation, free oil will normally separate by gravity and float to the surface in approximately 30 minutes. Emulsified oil is held in a molecular structure called a *micelle* and will not separate on its own. Hence, there is the need for a more sophisticated method of treating suspensions containing emulsified oils.

One of the driving forces for so-called "non-conventional applications" is water usage is becoming a prime concern for many areas of the country. Impact fees for new operations and high sewer costs are common in the southern United States.

A good way to see how DAF technology works is to fill a glass full of water from the tap and observing the tiny, almost microscopic bubbles in the water. Dissolved-air flotation uses the same principle in order to introduce tiny bubbles into water. These bubbles form because the water inside the pipes, which is at high pressure, had somehow dissolved enough air that the water becomes supersaturated with air when the pressure drops before the water falls into your glass. As a result, the excess air precipitates out in the form of tiny bubbles. These bubbles are much smaller than we produce by other means of dispersing air in water.

The flotation process was developed in the mining and coal processing industries as a way of separating suspended solids from a medium such as water. As noted above, the flotation process has found uses in other fields, such as wastewater treatment. The process introduces fine air bubbles into the mixture, so that the air bubbles attach to the particles, and lift them to the surface.

Dissolved-air flotation uses a particular way of introducing the air bubbles into the flotation tank. A dissolved-air flotation machine dissolves air into the water to be treated by passing the water through a pressurizing pump, introducing air, and holding the air-water mixture at high pressure long enough for the water to become saturated with air at the high pressure. Typical pressures are 20-75 psig. After saturating the water with air at high pressure, the water passes through a pressure-relief nozzle, after which air precipitates as tiny bubbles. This process for creating air bubbles has two advantages over other processes. Dissolved-air flotation typically produces bubbles in the 40-70 micron range, whereas in normal foam fractionation, a bubble of 500 microns is considered small. The smaller bubbles have much more surface area for their volume than do the larger bubbles. A particular volume of air has 10 times the surface area when distributed as 50 micron bubbles as it does when distributed as 500 micron bubbles. Looking at this fact another way, you need 10 times the air flow with 500 micron bubbles as you need with 50 micron bubbles in order to achieve the same air-water interfacial area.

The second, and probably more important, advantage of producing bubbles by precipitation is that the process provides a more positive attachment between air bubbles and the particles or globules that you want to remove. Particles and globules in the water act as nucleation sites for the precipitation process; the precipitating air seeks out these sites to begin bubble formation. This is better than relying on chance encounters between waste particles and large bubbles introduced by some other means.

A typical DAF process is not simply a physical separation technique. One must consider the entire treatment process, which is based on chemical coagulation, clarification and rapid sand filtration. This process train is widely accepted and is very applicable to the treatment of colored and turbid surface waters for municipal and industrial applications. Normally the clarification stage employs DAF. The suspended solid matter in the chemically treated water is separated by introducing a recycle stream containing small bubbles which floats this material to the surface of the tank. This is achieved by recycling a portion of the clarified flow back to the DAF unit. The recycle flow is pumped to higher pressure and is then mixed with compressed air. The flow passes through a tank where the air dissolves to saturation at the higher pressure. When the pressure is released at the clarifier, the dissolved air precipitates as a cloud of micro bubbles which attach to the particulate matter causing it to rise to the surface.
The DAF process is particularly well suited for the removal of floc formed in the treatment of low alkalinity, low turbidity, colored water. This type of floc tends to be very fragile and voluminous, making traditional gravity sedimentation inefficient. Flotation processes do not require large, heavy floc in order to achieve efficient solids removal. This results in lower chemical dosages and reduced time required for flocculation. The compressive forces applied to the sludge by the buoyant bubble/floc agglomerates result in greatly reduced volumes of waste-water from the clarification process. This enhances the efficiency of chemical use and reduces the volume of residuals to be treated. An equally important benefit of the technology is the efficiency of the clarification process. Since the performance of the filtration process is directly affected by the amount of solids in the clarified water, the high degree of solids removal achieved in the DAF results in an overall increase in system performance. In order to meet stringent standards for turbidity removal in potable water applications, this high performance is essential.

In recapping, DAF is the process of removing suspended solids, oils and other contaminants via the use of bubble flotation. Air is dissolved into the water, then mixed with the wastestream and released from solution while in intimate contact with the contaminants. Air bubbles form, saturated with air, mix with the wastewater influent and are injected into the DAF separation chamber. The dissolved air then comes out of solution, producing literally millions of microscopic bubbles. These bubbles attach themselves to the particulate matter and float then to the surface where they are mechanically skimmed and removed from the tank. Most systems are versatile enough to remove not only finely divided suspended solids, but fats, oils and grease (FOG). Typical wastes handled include various suspended solids, food/animal production/processing wastes, industrial wastes, hydrocarbon oils/emulsions, and many others. Clarification rates as high as 97 % or more are achievable. The basic flow scheme for a DAF system is illustrated in Figure 25.

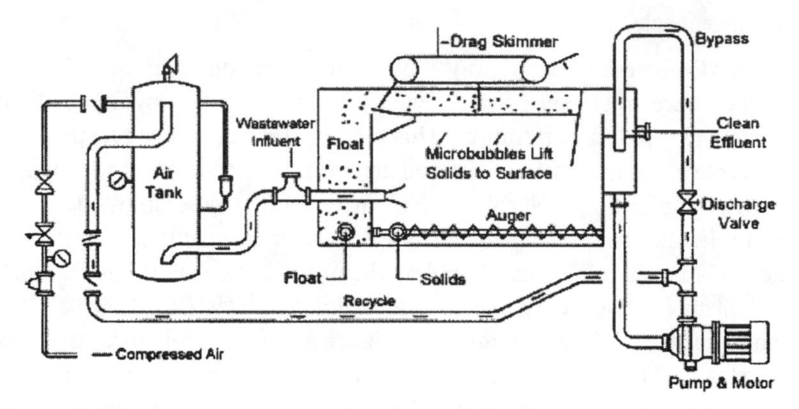

Figure 25. Conventional DAF process scheme.

The conventional DAF saturation design relies on a recycle pump combined with a saturation vessel and air compressor to dissolve air into the water. This type of system is effective, however it has the drawbacks of being labor intensive, is expensive, and can destabilize its point of equilibrium, creating "burps" due to incorrect, loss or creeping of EQ set-point in the saturation vessel. Such designs are slow to recover and can upset the flotation process. Air transfer efficiency is roughly 9 % with 80 % entrainment. This operational methodology can result in an increase in chemical use, labor costs, downtime, effluent loadings, production schedules due to the EQ loss. To overcome these shortcomings, some equipment suppliers have devised operational and control schemes that are best categorized as pollution prevention techniques.

A variation for one vendor is shown in Figure 26. The design and control of the system takes into consideration the following parameters: flow rate, water temperature, waste characteristics, chemical pretreatment options, solids loading, hydraulic loading, the air to solids ratio. Units are designed on the basis of peak flow rate expected.

Chemical pretreatment is often used to improve the performance of contaminant removal. The use of chemical flocculants is based on system efficiency, the specific DAF application and cost. Commonly used chemicals include trivalent metallic salts of iron, such as $FeCl_2$ or $FeSO_4$ or aluminum, such as $AlSO_4$. Organic and inorganic polymers (cationic or anionic) are generally used to enhance the DAF process.

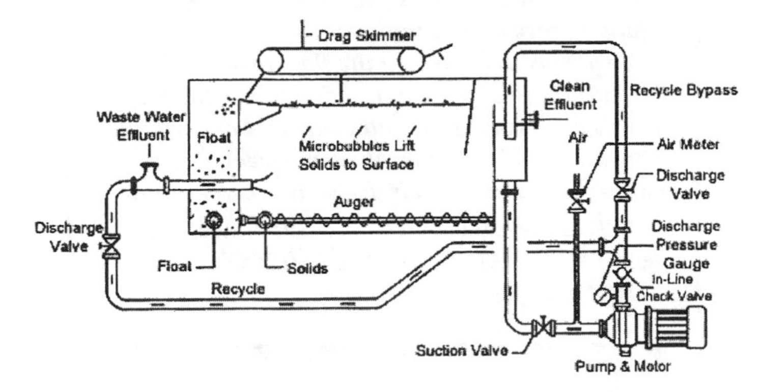

Figure 26. Variation of DAF. Vendor scheme of Pan America. *Source: Downloaded from site www.panamev.com.*

The most commonly used inorganic polymers are the polyacrylamides. Chemical flocculant concentrations employed normally range from 100 to 500 mg/Liter. The wastewater pH may require adjustment between 4.5 and 5.5 for the ferric compounds or between 5.5 and 6.5 for the aluminum compounds using an acid such as H_2SO_4 or a base such as NaOH. In many applications, the DAF effluent requires additional pH adjustment, normally with NaOH to assure that the effluent pH is within the limits specified by the POTW.. The pH range of the effluent from a DAF is typically between 6 and 9.

One mg/Liter of flocculant in 1 million gallons of water treated per day is 8.34 Lbs of material.

The mechanism by which flocculants work and enhance DAF is as follows. Attachment of most bubbles to solid particles can be effected throughsurface energies while others are trapped by solids or by hydrous oxide flocs as the floc spreads out of the water column. Colloidal solids are normally too small to allow the formation of sufficient air-particle bonding. They must first be coagulated by a chemical such as the aluminum or iron compounds mentioned above. . The solids are essentially absorbed by the hydrous metal oxide floc generated by these compounds. Often, a coagulant aid is needed in combination with the flocculant to agglomerate the hydrous oxide floc. to increase particle size and improve the rate of flotation. Mechanical/chemical emulsions can also be broken through the application of pH and polymer reactions.

The material that we recover from the surface of the DAF is referred to as the "float". The float often contains 2 to 10 % solids. These solids will need to be dewatered before ultimately finding a home for them. The subject of dewatering is covered in Chapter 12.

Note that some treatment operations choose a pollution prevention technique to dispose of the float. This involves feeding the float to animals. When this is done for the situation where the feed animals are used for human consumption, organic compounds such as chitosan, carrageenan, lignosulfonic acid,or their derivatives can be used. Use only compounds that are approved by the Food and Drug Administration (FDA) Office of Veterinary Medicine.

In general, for many applications, air flotation is the system of choice. Micro-bubbles are produced by inducing air into a vortex as the floc is formed. This controlled induction of air allows the micro-bubbles to permanently attach to the floc, resulting in the clarifier. In fully-integrated systems, the clarifier has abuilt-in sludge-holding section. Figures 27 and 28 illustrate a continuous flotation system for treating waste waters containing free-floating or emulsified oils and a typical process flow sheet, respectively.

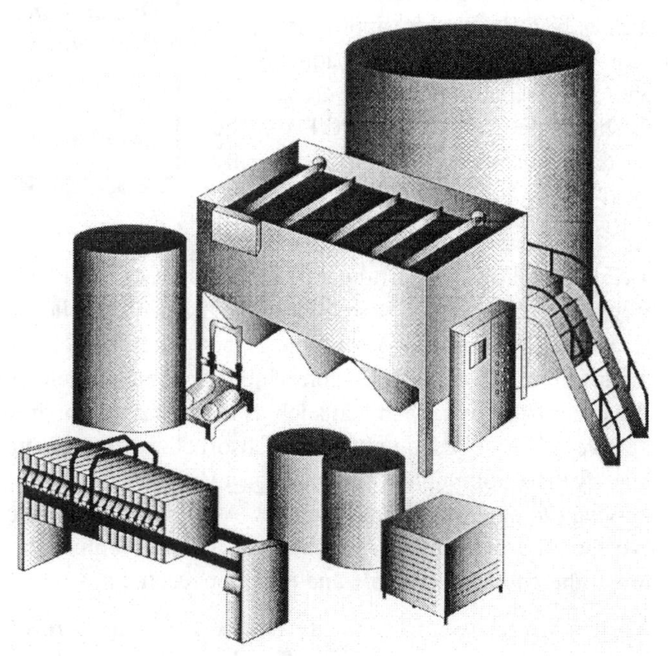

Figure 27. Continuous flotation system (Courtesy of Beckart Environmental, Inc., Kenosha, WI, email: info@beckart.com)

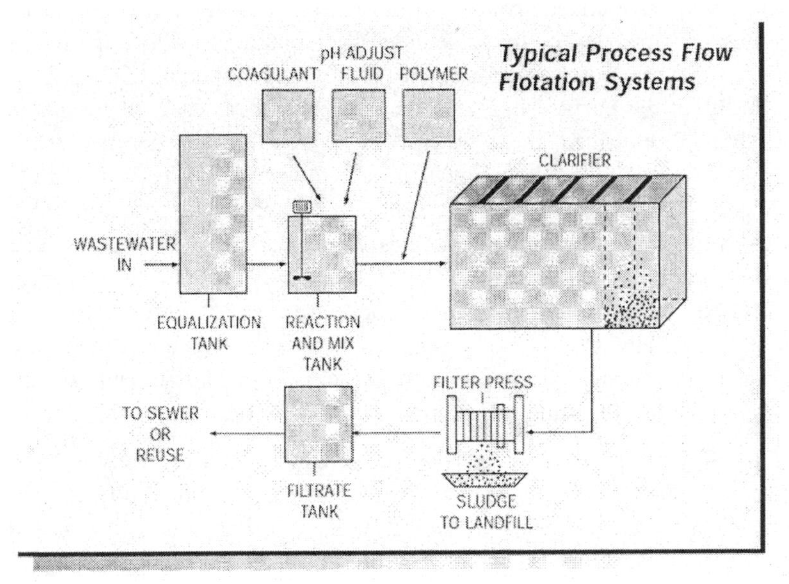

Figure 28. Continuous flotation system flow sheet. (Courtesy of Beckart Environmental, Inc., Kenosha, WI, email: info@beckart.com)

SEPARATION USING COALESCERS

A coalescer achieves separation of an oily phase from water on the basis of density differences between the two fluids. These systems obviously work best with non-emulsified oils. Applications historically have been in the oil and gas industry, and hence the most famous oil/water separator is the API separator (API being the abbreviation for the American Petroleum Institute).

Modern-day designs are more sophisticated than the early, simple separators of a few decades ago that were introduced by the petroleum industry. Commercial systems are comprised of cylindrical vessels, rectangular vessels, above and underground installations. Figure 29 shows an underground system advertised by Tank Direct.

In this diagram the key features are: **A** - Diffusion baffle: this serves four roles. First to dissipate the velocity head, thereby improving the overall hydraulic characteristics of the separator. Next, to direct incoming flow downward and outward maximizing the use of the separator volume. Third, to reduce flow turbulence and to distribute the flow evenly over the separator's cross-sectional area. Finally, to isolate inlet turbulence from the rest of the separator. **B-** Internal chambers: In the sediment chamber, heavy solids settle out, and concentrated slugs of oil rise to the surface. As the oily water passes through the parallel corrugated plate coalescer (an inclined arrangement of parallel corrugated plates) the oil rises and coalesces into sheets on the underside of the plate. The oil creeps up the plate surface, and breaks loose at the top in the form of large globules. The globules then rise rapidly to the surface of the separation chamber where the separated oil accumulates. The effluent flows downward to the outlet downcomer, where it is discharged by gravity displacement from the lower region of the separator. **C -** PetroScreen™: This part of the design focuses on improved separation efficiency. It is essentially a polypropylene coalescer (a bundle of oleophilic (i.e., oil extracting) fibers). The oleophilic fibers are layered from coarse to fine, and are encapsulated within a solid framework. These are used to intercept droplets of oil that are too small to be removed by the parallel corrugated plate coalescer section. **D-** Monitoring systems: Various monitoring and control instruments are included in a system for level sensing and pumpout control purposes.

There are variations of coalescer designs, each achieving different degrees of separation depending upon the application and properties of the influent. One fairly popular design used primarily in the oil and gas industry is a so-called deoiler cyclone vessel and cyclone system. These systems take advantage of the hydroclone (or cyclone) principle of separationMore correctly stated, the cyclonic action results in an increase in the oil droplet size, enabling an more efficient separation of the phases. Deoiling cylcones have steep droplet-cut size curves, like the one illustrated in Figure 30. The typical performance curve for a high efficiency cyclone separator (or hydroclone) shows that a small increase in the droplet size from 5 to 10 microns typically increases the separation from 15 % to over 90 %. The inlet chamber of a conventional deoiling hydroclone is usually the largest chamber in the coalescer vessel, and has a residence time up to about 20 seconds. Some commercial units employ specially designed low-intensity pre-coalescing internals and inlet vanes that take advantage of this residence time by optimizing the flow distribution. This enhances the coalescence of droplets and enables a pre-coalescing stage.

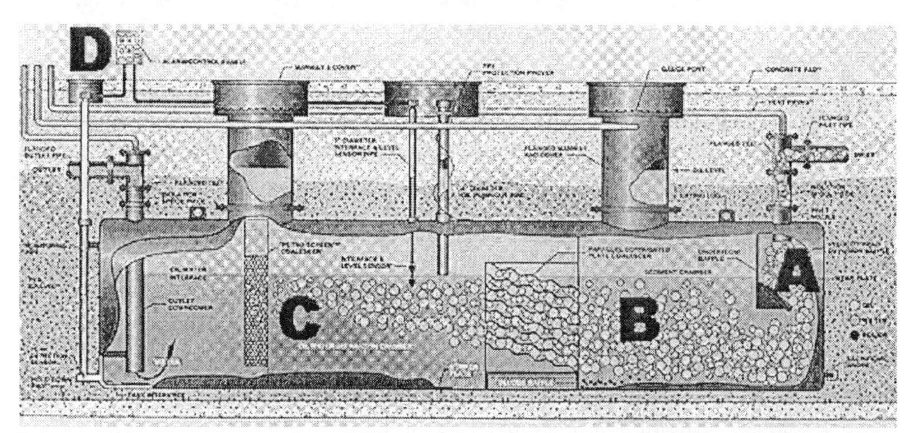

Figure 29. Underground installation of an oil/water separator. *Source: Tanks Direct from their website: http://oilwaterseparators.com/process.htm*

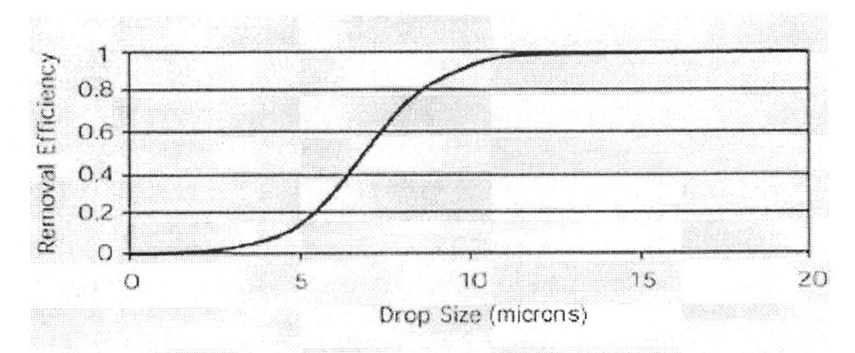

Figure 30. Cut size curve for a deoiling hydroclone.

An example of a pre-coalescer unit is illustrated in Figure 31. In general, the technology largely relies on many years of experience and empiricism. Before investing in a system, it is wise to run batch tests and perhaps pilot tests using vendor facilities. Units may range in size, complexity of internals, configuration (e.g., rectangular, slant rib designs that borrow concepts from the classical Lamella separator, to cylindrical (the classical design case). There are varying degrees of claims for removal efficiency. Cost can vary greatly with coalescers, depending not only on throughput requirements and add-on controls and monitors, but with construction as well. Because of strict environmental regulations tied to underground and above ground tanks, double wall vessel construction is often needed. Double wall vessels are normally constructed in several ways.

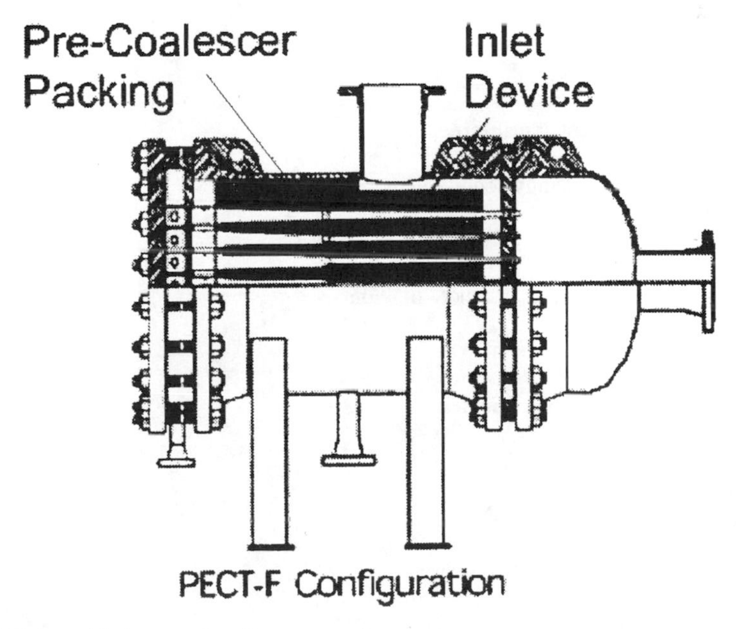

Figure 31. Example of a pre-coalescer.

One common construction is achieved by wrapping a secondary steel wall completely around the primary vessel. Each double wall vessel is constructed using the same basic fabrication techniques as used on single wall vessels. the area between the vessel walls, known as the interstice, can be monitored by a leak detection system installed in the monitor tube, located on the vessel head. Other variations of vessel construction exist and careful consideration to the advantages, disadvantages and impacts on cost must be considered. Remember to select a separator based in part on vessel construction quality meeting API, UL (Underwriter Laboratories), and Steel Tank Institute (STI) ACT-100 or STI-P3 specifications. It goes without saying that the supplier must conform to ISO 9000 quality assurance standards.

NOMENCLATURE

A	constant
A	dimensionless coefficient or Archimedes force, N
Ar	Archimedes number, dimensionless
a_s	sphericity function
a	coefficient accounting for liquid entrapped between sludge layers
a_g	acceleration, m/sec^2
B	parameter, m
C	concentration, kg/m^3
C_r	constant or centrifugal force, N
C_0	upward gas stream velocity, m/sec.
C_D	drag coefficient
d	particle diameter or distance, in., μm
E	field intensity, V/sec-m
Eu	Euler number, dimensionless
F	area, m^2
G	average gravity force, N
Ga	Galileo number, dimensionless
H	height, in. or m

K	settling velocity correction factor
K_f	time constant
K_1, K_2	coefficients
K_s	separation number
k	settling flux constant
L	length, in. or m
Ly	Lyachshenko number, dimensionless
LHS	left hand side, dimensionless
m_p	particle mass, kg
P	pressure, N/m^2 or force, N
Q	mass feed rate, kg/s or volumetric flowrate in m^3/hr
R	drag or resistance force, N
R_c	physical properties correction factor for slurries
Re	Reynolds number, dimensionless
RHS	right hand side
r_h	hydraulic radius, m
S_1	product of Archimedes number and separation number, dimensionless
S2	sedimentation number, dimensionless
T, t	time, sec. or temperature in °C or °F
u	velocity, m/sec
u_t	terminal settling velocity, m/sec
V	volume, m^3
W	total mass, kg
w	velocity, m/sec
X	sludge height, m
x	distance, m
x′	liquid volume fraction
Z	height of compression zone, m

Greek Symbols

α angle, deg.

β ratio of liquid volume entrained to total sludge volume, or angle

γ specific weight, kg_f/m^3

ϵ void fraction

η_v volume fraction of particles in a slurry

$\theta(\epsilon)$ experimental function of void fraction

λ length of mean free path, m

μ viscosity, poise

ν kinematic viscosity, m^2/sec

ν_H terminal velocity for hindered settling, m/sec potential difference, V

ρ density, kg/m^3

σ ratio of particle surface area to volume, m^{-1}

τ force, N, or stress, N/m^2

Υ dimensionless settling number or Lyashenko number

Φ parameter

ψ sphericity factor

RECOMMENDED RESOURCES FOR THE READER

It's difficult to narrow down the multitude of publications to a few for you to look at because of the exhaustive studies and design cases that have evolved in the wastewater field. However, there are three books in particular that I recommend you acquire for general information on the technologies of sedimentation, clarification and the overall relationship of these operations within a wastewater treatment plant facility. I recommend that you begin building your library with these volumes. There are plenty of design-specific case studies and sample calculations, as well as equipment scale-up methodologies provided, with each volume bringing something valuable to the subject.

1. ***Wastewater Treatment Technology***, Paul N. Cheremisinoff - editor, Published 08/01/1989, ISBN 0872012476.

2. ***Water Treatment Principles and Design***, James M. Montgomery, Published 08/01/1985, ISBN 0471043842.

3. ***Municipal Water Treatment Technology: Recent Developments***, USEPA, Published 03/01/1993, ISBN 0815513097.

For Web-sites, the selection process is even more challenging. A simple search on AltaVista for clarifiers and thickeners resulted in over 157,000 sites identified. The vast majority of these are equipment suppliers, design, engineering and consulting firms. In reviewing a few hundred of these sites while writing this volume, a few struck me as being quite useful, especially for those of you that really need to nail down the basics. These sites, along with the reasons why I think you should visit them are as follows:

4. **http://members.aol.com/erikschiff/primary.htm** - This site is actually a review of the Lynn Regional Wastewater Treatment Facility (Lynn, Massachusetts), which is a primary and secondary treatment plant with a design purpose of the primary to remove settleable and floatable material (sludge, grease, etc.). The secondary treatment is directed principally toward the removal of biodegradable organic and suspended solids. What is very nice about this site is that it provides a reasonably good technical description of the basis for the overall design, and then focuses in on some of the key unit operations, including the primary and secondary clarifiers. There are plenty of installation and equipment photographs, and some simple but very effective animated drawings that illustrate how the equipment work. You will obtain a very good understanding of the practical aspects of clarification and sedimentation practices for a municipal water treatment facility by spending some time here.

5. **http://www.baaqmd.gov/permit/handbook/sewage.htm** - This is a municipal chapter titled **SEWAGE TREATMENT FACILITIES (POTWs),** Last adopted: July 17, 1991. This chapter covers the permitting of typical unit operations at publicly owned treament works (POTW) facilities. These plants treat wastewater from sanitary and storm sewer systems prior to discharge into surface waters or reuse as reclaimed water. Typical POTW sources may be defined as a combination of the liquid or water carried wastes removed from residences, institutions, and commercial and industrial establishments, together with groundwater, surface water, and storm water runoffs. Publicly owned treatment works are typically large land intensive facilities with numerous ponds, buildings, pump stations, etc required to handle large daily flows. I recommend you spend some time reading through this chapter, not only for the sense of overall design issues, but also to gain appreciation and understanding of air emissions issues associated with these plants.

6. **http://www.state.sd.us/denr/DES/Surfacewater/clarifie.htm** - A very good

site that walks you through clarifiers, sedimentation equipment, trickling filters, aeration basins, solids thickeners, and much more. There are plenty of photographs of large-scale equipment that will give you a sense for the magnitude and complexity of these equipment, as well as a feel for the operating parameters.

7. **http://www.Treat-Wastewater.com** - This is a vendor of wastewater treatment simulation software. This is not an endorsement of the products, but I recommend you peruse the site and examine the features. You can try some sample calculations and acquire some of the products for trial periods. I cannot vouge for the usefulness of these simulations in actual design cases, however they seem to have the features needed. The range of programs includes the following: *(1) Process Advisor Pro [Carbon removal/nitrification/denitrification : Price $495]* - A software package for real time simulation, analysis and control of the biological wastewater treatment process. Visualizes unobservable processes such as activated sludge concentration, the evolution of organic matter concentration, population dynamics, and many more! The reactor can be plug flow like most of zone-aerated treatment facilities or completely mixed. Any system of sludge and water feeding can be used. Can analyze your own data; *(2) Process Advisor [Carbon removal : Price $395]* - A lesser version of Process Advisor Pro that does without nitrification and denitrification capabilities and cannot be as extensively set up to support different configurations. Otherwise supports all the features of Process Advisor Pro, including population dynamics and hydrodynamic settler model; *(3) Process Advisor MBR [Carbon removal in membrane bioreactor: Price $449]* - A software package for simulation, analysis and control of the high-concentration biological wastewater treatment processes that are using membrane filter for complete separation of the sludge from treated water and partial filtration of the dissolved solids. Supports both submerged and grossflow membrane filters; *(4) Settler Specialist [Clarifier analysis : Price $199]* - A clarifier simulation software for 2-D analyzes of wastewater treatment processes in circular and rectangular clarifiers. Can predict processes like distribution of sludge in the settler, flow streamlines in the settler, vertical and horizontal flow velocities, and much more. Supports both turbulent and laminar flow models.

The following are useful references on dissolved air flotation. This bibliography contains a number of references for equipment and process scale-up, and design methods.

8. Baeyens, J., Mochtar, Y., Liers, S., and De Wit, H. Plugflow dissolved air flotation. *Water Environment Research*. Vol. 67, Num. 7. 1995. pp. 1027-1035.

9. Fuerstenau, M. C., ed. *Flotation--A. M. Gaudin Memorial Volume*. Vols. 1 and 2, AIME, New York. 1976.

10. Gaudin, A. M. *Flotation*. Second Edition, McGraw-Hill, New York, 1957.

11. Grainger-Allen, T. J. N. Bubble generation in froth flotation machines. *Trans. IMM* Vol 79, C15-22. 1970.
12. Hedherg, T., Dahlqvist, J., Karlsson, D., Soerman, L.-O. Development of an air removal system for dissolved air flotation. *Water science and technology*. Vol. 37, No. 9, p. 81, 1998.
13. Haarhoff, J. and Steinbach, S. A model for the prediction of the air composition in pressure saturators. *Water Research*. Vol. 30, No. 12, pp. 3074-3082, 1996.
14. Liers, S, Baeyens, J, Mochtar, J. Modeling dissolved air flotation. *Water environment research*. Vol. 68, No. 6, p. 1061, 1996.
15. Klassen, V. I., and Mokrousov, V. A. *An Introduction to the Theory of Flotation*. English translation by J. Leja and G. W. Poling. Butterworths, London. 1963.
16. Leja J. *Surface Chemistry of Froth Flotation*. Plenum Press, New York. 1982.
17. Rykaart E. M. and Haarhoff J. Behaviour of air injection nozzles in dissolved air flotation. *Water Science and Technology*. Vol. 31, No. 3-4, pp. 25-35, 1995.
18. Walter, J., Wiesmann, U. Comparison of Dispersed and Dissolved Air Flotation for the Separation of Particles from Emulsions and Suspensions *Das Gas- und Wasserfach*. Wasser, Abwasser : GWF. Vol. 136, No. 2, p. 53. 1995.
19. Vrablik, E. R. Fundamental principles of dissolved air flotation of industrial wastes. *Industrial Waste Conference Proceedings*. 14th, 1959, Purdue University, Ann Arbor, USA, 743-779.
20. Zabel, T. The Advantages of Dissolved-air Flotation for Water Treatment. *J. Am. Water Works Assoc.* Vol. 77, No. 5, pp. 42-46. 1985.
21. Zlokarnic, M. Separation of activated sludge from purified waste water by induced air flotation. *Water Research*, Vol. 32, No. 4, pp. 1095-1102, 1998.

QUESTIONS FOR THINKING AND DISCUSSING

The following questions will challenge you and help to reinforce some of the principles presented in this chapter.

1. A waste stream from a pulp mill has an average concentration of 7.2 Lbs of water/Lbs of solids. A treatment plant to be designed will have a thickening stage that concentrates the stream to 1.8 Lbs of water/Lbs of solids with the production of a relatively clear overflow. Batch settling tests were conducted on different concentration slurries to ensure that the velocity of settling exceeds the upward flow of fluid at all concentrations normally encountered in the thickening of the specified feed. Tabulated results from these tests are given below Prepare an additional column for this table showing the

estimated minimum area required for clear overflow in units of 1 ton/day/ft^2 of solids feed. What is the minimum area to be used for design purposes?

Mass Ratio Fluid to Solids	Fluid Rising per Lbs Solids (Lbs/Lbs)	Calculated Fluid Rising (ft^3/hr)	Measured Rate of Settling (ft/hr)
7.2	5.9	7.3	2.9
6.3	4.1	4.8	1.5
4.3	3.2	3.9	1.1
3.4	2.6	3.1	0.85
2.5	1.7	2.6	0.59

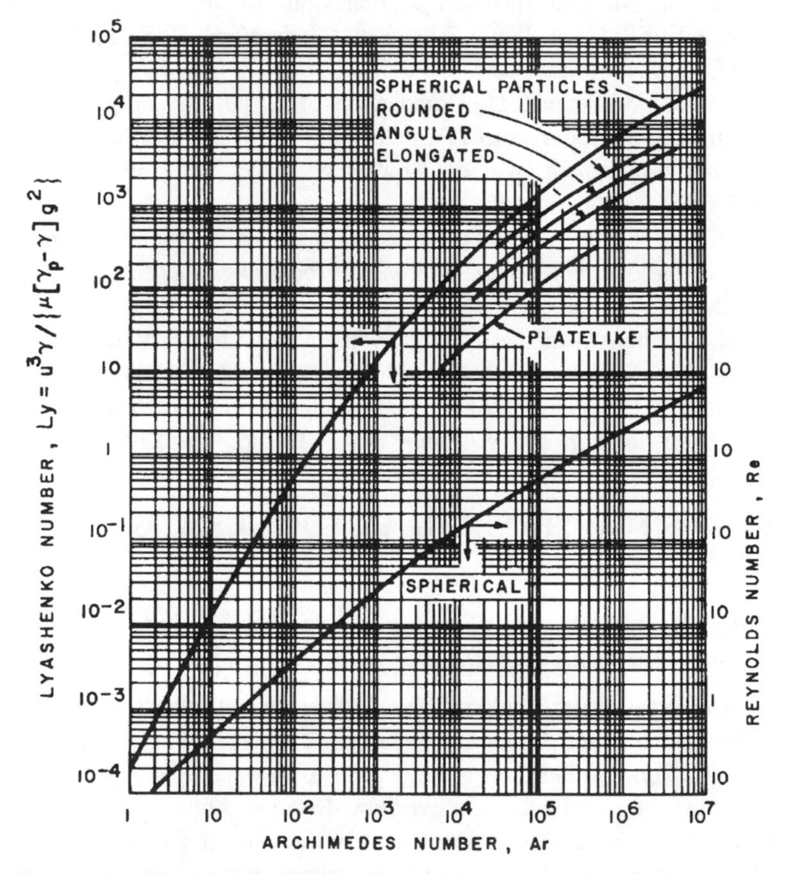

Figure for Question 2. *Plot of Reynolds number, and settling number (Lyashenko number) versus Archimedes number. Use this plot for question 3. It also useful for your own design problems.*

2. Determine the settling velocity of spherical quartz particles in water (d = 0.9 mm) using the dimensionless plot of the Lyachshenko and Reynolds numbers versus the Archimedes number in the figure above. The Lyashenko number is the same as the dimensionless settling number. The specific weight of the quartz is 2650 kg/m^3, and thc tcmpcraturc of thc watcr is 20° C.

3. Determine the maximum size of quartz particles settling in water (t = 20° C) that can be described by Stokes' law. What is this particle's settling velocity? The specific weight of quartz is 2650 kg/m^3.

4. Determine the maximum diameter of spherical chalk particles entrained by an upward-moving water stream with a velocity of 0.5 m/sec. The liquid temperature is t = 10° C, and the specific weight of the chalk is 2,710 kg/10° C.

5. Determine the settling velocity in the water (t = 20° C) for lead particles having an angular shape with d_{eq} = 1 mm. The specific weight of lead particles is 7,560 kg/10° C.

6. Calculate the sizes of elongated coal particles (ρ_{p1} = 1,400 kg/m^3) and plate-like particles of shale (ρ_{P2} = 2,200 kg/m^3) that have the same settling velocities of 0.1 m/sec through water at 20° C.

7. Determine the settling velocity of solid spherical particles if the particle diameter is d = 25 μm and particle density is 2,750 kg/m^3. The density of the liquid phase is 1,200 kg/m^3 and its viscosity is 2.4 cp.

8. Determine the velocity of hindered sedimentation of the suspension considered in question 2 if the concentration of solids in the feed is x = 30%, the density of suspension is 1,440 kg/m^3 , and the density of the solid phase is 2,750 kg/m^3.

9. Determine the capacity, cross-sectional area and diameter of a continuous sedimentation tank for liquid suspension clarification in the amount of Q_s = 20,000 kg/hr. The concentration of solids is x_1 = 50%, the settling velocity is u_0 = 0.5 m/hr, and the density of liquid phase is 1,050 kg/m^3.

10. During the spring, the mean temperature over the bottom 3 ft of the lagoon described above is 10° C, whereas the temperature at the surface is about 16° C. Would you expect good separation of solids? Substantiate your conclusions.

11. Two primary settling basins are each 100 ft in diameter with an 8-ft side water depth. The tanks are equipped with single effluent weirs located on the peripheries. For a water flow of 10 mgd, calculate the overflow rate, gpd/ft^2, detention time, hr, and weir loading, gpd/ft. The overflow rate for a clarifier

is defined as the surface settling rate, i.e., $q_0 = q/F$, where q = volumetric flowrate and F = total surface area of basin.

12. A thickener handles 80,000 gpd of sludge, increasing the solids content from 2.0 to 8.0 wt.% with 85% solids recovery. Determine the quantity of thickened sludge generated per day.

13. Develop a list of air pollution issues and discuss possible permitting requirments you may have to face with a treatment plant that relies on both primary and secondary clarifiers for municipal sewerage treatment.

Chapter 9

MEMBRANE SEPARATION TECHNOLOGIES

INTRODUCTION

There are five types of membrane processes, which are commonly used in water and wastewater treatment:

- **Electrodialysis**
- **Microfiltration**
- **Ultrafiltration**
- **Nanofiltration**
- **Reverse Osmosis**

Through these processes dissolved substances and/or finely dispersed particles can be separated from liquids. All five technologies rely on membrane transport, the passage of solutes or solvents through thin, porous polymeric membranes.

A membrane is defined as an intervening phase separating two phases forming an active or passive barrier to the transport of matter. Membrane processes can be operated as: (1) Dead-end filtration; and (2) Cross-flow filtration. Dead-end filtration refers to filtration at one end. A problem with these systems is frequent membrane clogging. Cross-flow filtration overcomes the problem of membrane clogging and is widely used in water and wastewater treatment.

The membrane itself is a polymeric coating or extrusion with inverted conical-shaped pores. Membrane filters do not plug because the pore diameter is smaller at the top, which is the point of contact with the wastewater. Material passing through the membrane passes unimpeded through the membrane structure, therefore eliminating accumulation of material within the filter. Wastewater is pumped across the membrane surface at high flow rates. This parallel fluid flow eliminates the cake-like build-up typical of conventional filters such as bags and cartridges which must be frequently replaced. Some wastewater contaminants slowly accumulate on the membrane surface, forming a thin film, during normal operating conditions. This fouling process is normal and causes the filtration rate to slowly decrease with time. When membranes no longer produce clean water at the desired rate they are cleaned in place with soap and water and returned to service. Membranes can be repeatedly cleaned for years of productive, dependable service prior to replacement.

Most of these processes are oftentimes used with ***chemical mechanical polishing*** (CMP), which is fast becoming the established technology for planarizing multilevel devices. This process requires large quantities of ultrapure water for rinsing slurry particles off the polished wafers. Treatment by this method is generally needed in order to maintain an acceptable level of total suspended solids (TSS) in industrial wastewater effluent. With large quantities of particles in the CMP wastewater stream, crossflow filtration is the most economical method for TSS removal.

The technologies discussed in this chapter are changing relatively rapidly, and hence you need to stay in touch with vendor developments. Remember to refer to the ***Glossary*** at the end of the book if you run across any terms that are unfamiliar to you.

AN OVERVIEW OF MEMBRANE PROCESSES

The following tables will give you a basic appreciation for the technologies. Table 1 provides a comparison of the factors that affect the performance of the five technologies.

Table 1. Factors Impacting on the Performance of Membrane Processes.

Technology	**Driving Force**	**Influencing Factors**			
		Size	**Diffusivity**	**Ionic Charge**	**Solubility**
Microfiltration	Pressure	+++	-	-	-
Ultrafiltration	Pressure	+++	-	+	-
Nanofiltration	Pressure	+++	+	+	-
Reverse Osmosis	Pressure	+	+++	+	+++
Electrodialysis	Electrical	+	+	+++	-

Table 2 provides a comparison of membrane structures. Between these two tables, you should get an idea of the operating conditions viz., membrane structural types, the driving forces involved in separation, and the separation mechanisms.

Common types of membrane materials used are listed in Table 3. This gets us into the concept of geometry. There are three types of modules generally used, namely: Tubular, Spiral wound, and Hollow fiber. A comparison of the various geometries is given in Table 4.

Table 2. Compares Membrane Structures.

Technology	Structure	Driving Force	Mechnism
Microfiltration	Symmetric microporous (0.02 - 10 μ	Pressure, 1 - 5 atm	Sieving
Ultrafiltration	Asymmetric microporous (1 - 20 nm)	Pressure, 2 - 10 atm	Sieving
Nanofiltration	Asymmetric microporous (0.01 - 5 nm)	Pressure, 5 - 50 atm	Sieving
Reverse Osmosis	Asymmetric with homogeneous skin and microporous support	Pressure, 10 - 100 atm	Solution diffusion
Electrodialysis	Electrostatically charged membranes (cation and anion)	Electrical potential	Electrostatic diffusion

Table 3. Summary of Common Membrane materials and Their Characteristics.

Technology	Membrane Materials	Polar Character
Microfiltration	Polypropylene (PP)	
	Polyethylene (PE)	Non polar
	Polycarbonate (PC)	Non polar
	Ceramic (CC)	Non polar

Technology	Membrane Materials	Polar Character
Ultrafiltration	Polysulfone (PSUF) Dynel Cellulose acetate (CA)	Non polar Non polar Non polar
Nanofiltration	Polyvinylidene fluoride (PVDF)	Polar
Reverse Osmosis	Cellulose acetate Polyamide Nylon	Polar Polar Polar
Electrodialysis	Styrene/vinylpyridene Divinyl benzene	-

Table 4. A Comparison Between Module Types.

Parameter	Module Type		
	Tubular	Spiral wound	Hollow fiber
Surface area, m^2/m^3	300	1000	15,000
Inner dia./spacing, mm	20 - 50	4 - 20	0.5 - 2
Feed flow rate, Liter/m^2-day	300 - 1000	300 - 1000	30 - 100
Production, m^3/m^3 of membrane per day	100 - 1000	300 - 1000	450 - 1500
Pretreatment requirements	Simple	Average	High
Extent of clogging	Little	Average	High

Parameter	Module Type		
	Tubular	**Spiral wound**	**Hollow fiber**
Mechanical Cleaning Methods	Possible	Not possible	Not possible
Chemical Cleaning Methods	Possible	Possible	Possible

WHAT ELETCRODIALYSIS IS

The principle behind electrodialysis is that electrical potential gradients will make charged molecules diffuse in a given medium at rates far greater than attainable by chemical potentials between two liquids as in conventional dialysis. When a DC electric current is transmitted through a saline solution, the cations migrate toward the negative terminal, or cathode, and the anions toward the positive terminal, the anode. By adjusting the potential between the terminals or plates, the electric current and, therefore, the flow of ions transported between the plates can be varied.

Electrodialysis can be applied to the continuous-flow type of operation needed in industry. Multi-membrane stacks can be built by alternately spacing anionic- and cationic-selective membranes. Among the technical problems associated with the electrodialysis process, concentration polarization is perhaps the most serious (discussed later). Other problems in practical applications include membrane scaling by inorganics in feed solutions as well as membrane fouling by organics. Efficient separation or pretreatment in the influent streams can include activated carbon absorption to reduce or prevent such problems. Principal applications of electrodialysis include:

As noted, the principle of ED is that electrical potential gradients will make charged molecules diffuse in a given medium at rates far greater than obtained by chemical potentials between two liquids, as in conventional dialysis. When a DC electric current is transmitted through a saline solution, most salts and minerals are dissolved in water as positively charged particles (anions, for example, Na^+) and negatively charged particles (anions, for example, Cl^-). The cations migrate toward the negative terminal, or cathode, and the anions toward the positive terminal, the anode. By adjusting the potential between the terminals or plates and the electric current, the flow of ions transported between the plates can be varied.

(1) Recovery of materials from liquid effluents, such as processes related to conservation, cleanup, concentration, and separation of desirable fractions from undesirable ones; (2) Purification of water sources; (3) Effluent water renovation for reuse or to meet point source disposal standards required to maintain suitable water quality in the receptor streams.

Cconcentration-polarization is a problem which also exists in reverse osmosis systems, and is due to of a build-up in the concentration of ions on one side of the membrane and a decrease in concentration on the opposite side. This adversely affects the operation of membranes and can even damage or destroy them. Polarization occurs when the movement of ions through the membrane is greater than the convective and diffusional movements of ions in the bulk solutions toward and away from the membrane. Along with a deleterious pH shift occurring at the membrane surface, polarization may cause solution contamination and sharply decrease energy efficiency. Commercial electrodialyzer designs incorporate various baffles or turbulence promoters and limit current densities to avoid these effects. Increased feed flow also assists mixing but requires additional power for pumping.

Treatment of brackish waters in the production of potable supplies has been the largest application of electrodialysis. Costs associated with electrodialysis processes depend on such factors as the total dissolved solids (TDS) in the feed, the level of removal of TDS (percent rejection), and the size of the plant. In brackish water treatment, operating costs for very large ED installations (on the order of millions of gallons a day) have been between 40 cents to 50 cents per 1,000 gallons for brackish feed waters, which compares favorably with RO costs.

A rough rule of thumb for the energy requirements for dernineralizing 1,000 gallons of salt water by ED in large capacity plants (4 mgd) is 5 to 7 kWh per 1,000 ppm of dissolved solids removed. Since the efficiency of electrodialytic demineralization decreases rapidly with increasing feed concentrations, this process is best utilized for treatment of weakly saline (brackish) waters containing less than 5,000 ppm of total dissolved solids. In fact, for waters at the low-concentration end of the brackish scale, ED may be the most cost-effective process of all. Electrodialysis is widely used in the United States in the dairy industry, namely in the desalting of cheese whey. Electrical requirements may vary from 5 to 14 kWh per pound of product solids. Another application of ED is the sweetening of prepared citrus juices. Other less extensive uses of electrodialysis in commercial operations in the United States include tertiary or advanced treatment of municipal sewage water and treatment of industrial wastewaters such as metal-plating baths, metal-finishing rinse waters, wood pulp wash water, and glass-etching solutions. Potential applications of ED are many. A particular advantage of the electrodialysis process is its ability to produce solutions of high concentrations of soluble salts. A combination of electrodialysis with conventional evaporation, for example, may be substantially cheaper than evaporation alone for the production of dry salt from saline solutions. Competing technologies include reverse osmosis and crystallization.

Added control of the movement of the ions can be obtained by placing sheet-type membranes of cation- or anion-exchange material between the outer plates, as shown diagrammatically in Figure 1. These sheets of cationselective resins and anion-selective resins permit the passage of the respective ions in the solution. Under an applied DC field, the cations and anions will collect on one side of each membrane through which they are transported and vacate the other side. Thus, if a NaCl solution is supplied to the central zone of the cell shown in Figure 1, the Na^+ ions will migrate through Membrane A, depleting the central zone (termed the *diluting* or *product feed stream)* of the salt ions. The two outer zones where the ions collect are commonly known as the *concentrating* or *brine streams*.

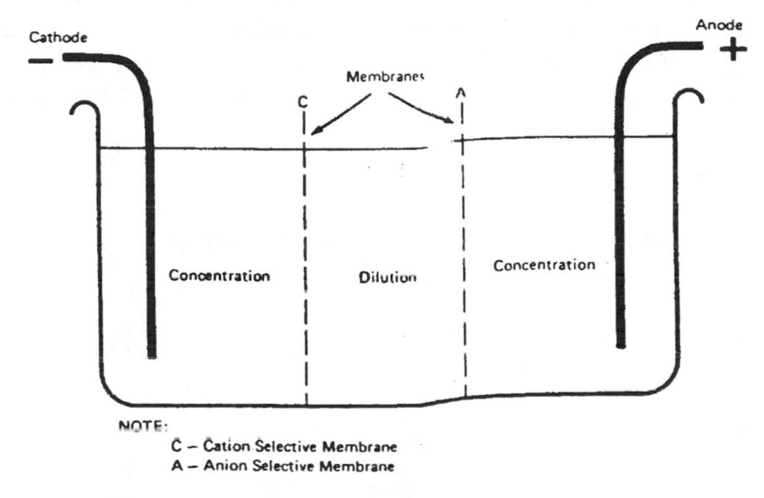

Figure 1. Electrodialysis cell diagram.

Multi-membrane stacks can be built from alternately spacing anionic- and cationic-selective membranes. Flow of solutions through specific compartments and appropriate recombination of transported ions permit desired enrichment of one stream and depletion of another. A schematic view of a typical stack based on the concept of alternating these concentrating and diluting compartments is shown in Figure 2. The feed stream enters each compartment along the top of the figure and flows downward toward the lower exit ports and manifolds. However, as the ionized streams move tangentially along the membranes, cations are transported, or attempt movement, toward the left and anions to the right, causing an alternate build-up and a depletion of ions in adjoining compartments. Thus, one resultant output stream is a diluted product water and the other is a concentrated stream of dissolved salt.

The process flow stream through a commercial demineralizer, incorporating two stacks in series dernineralized water, is shown in Figure 3. Several of the refinements required for continuous-flow operational systems are shown on this diagram, representing a two-stage demineralizer.

Although ED is more complex than other membrane separation processes, the characteristic performance of a cell is, in principle, possible to calculate from a knowledge of ED cell geometry and the electrochemical properties of the membranes and the electrolyte solution.

Another kind of electrodialysis cell configuration, shown in Figure 4, is a multiple electrodialysis system consisting of ten-unit cells, in series rather than manifolded in parallel. The feed solution is introduced at four points: It enters at both upper end points to sweep directly through both electrode chambers and is introduced into the working chambers near either end. The feed solution into the left side traverses depleted chambers and exits as depleted effluent at the right. The feed solution into the rightmost enriching cell flows in the other direction and exits as enriched effluent at the left side.

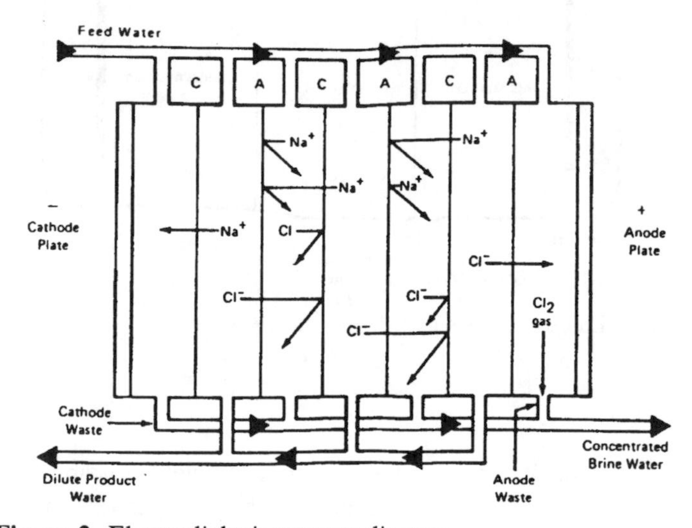

Figure 2. Electrodialysis process diagram.

In addition to the membrane stacks in electrodialysis units, various supporting equipment is essential. This includes pumps for circulation of concentrating and diluting flows; flushing streams for cathode and anode plates; injection systems for pH control; pressure concentration, pH alarms, and control systems and backflushing controls; feed strainers and filters; and grounding systems. Because of the high pH of the cathode stream, substances, such as carbonates and hydroxides could precipitate on the cathode surface and adjoining membrane; often sulfuric acid is injected to maintain the stream at pH of 2 or less. Also, recirculating concentrate requires an acid addition to yield a low pH for stability along with additional substances such as sodium hexametaphosphate. A key point to remember is that the separation is achieved by removing ions by passing them through a semipermeable membrane. The electric field applied across the membrane transports only ions. As noted the application of this technology is to desalting brackish water, to removing TDS from water and to the removal of certain heavy metals. The issue of concentration polarization results in an increase of the resistance of flow of ions across the membrane. The current must therefore be increased to overcome this resistance.

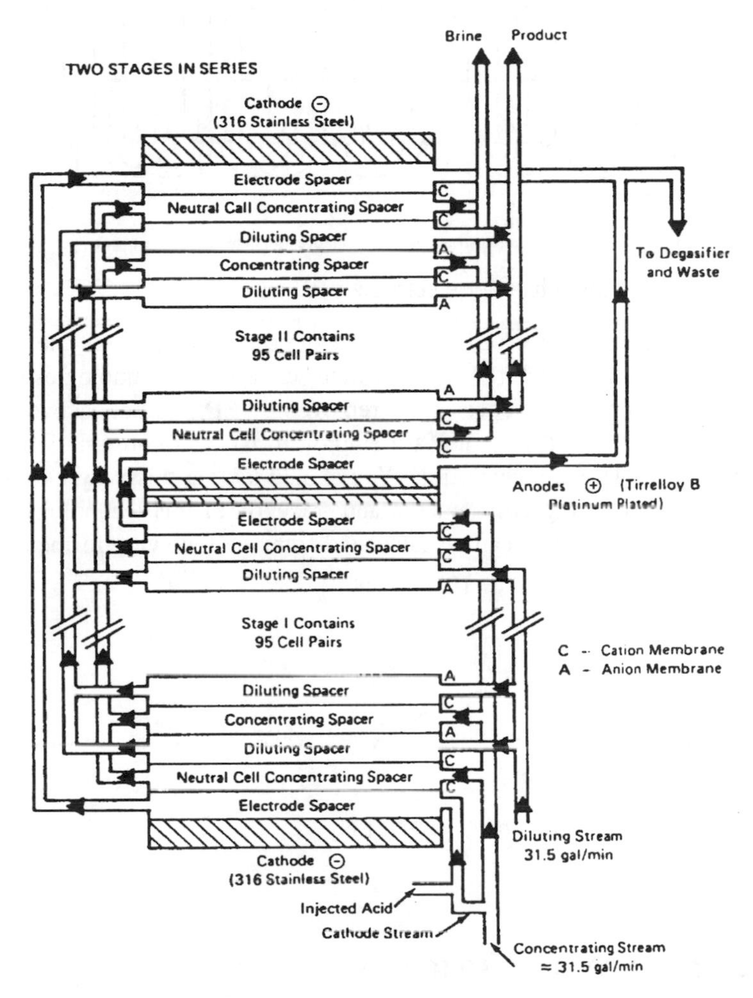

Figure 3. Flow through electrodialysis stack.

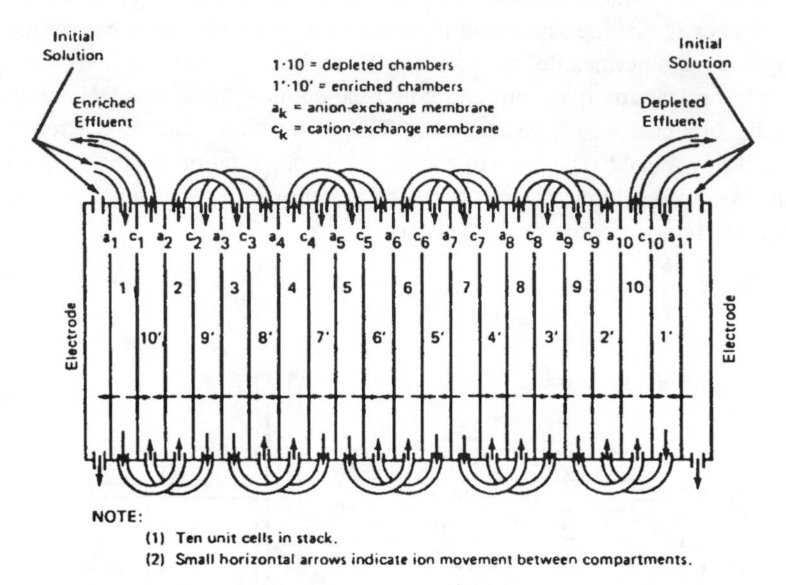

Figure 4. Multiple-chamber electrodialysis unit.

Major differences between ED and other processes are, first, the solute is transferred across the membrane against water in the other technologies discussed below, whereas only ionic species are removed by ED. As noted, two different membranes (anionic and cationic) are employed. Current consumption depends primarily on the TDS concentration. You should look at this very closely when comparing the operating cost benefits and tradeoffs of this technology to other options. Current efficiency can be calculated from the following formula:

$$E = FqN\xi/\eta I \tag{1}$$

where q = the stream flow rate

F = Faraday's constant, 96,540 amp-sec/g.eq

η = number of cells

ξ = removal efficiency, [(N-Ne)/N]

I = current, amp

WHAT ULTRAFILTRATION IS

Suspended materials and macromolecules can be separated from a waste stream using a membrane and pressure differential, called Ultrafiltration. This method uses a lower pressure differential than reverse osmosis and doesn't rely on overcoming osmotic effects. It is useful for dilute solutions of large polymerized macromolecules where the separation is roughly proportional to the pore size in the membrane selected.

Ultrafiltration membranes are commercially fabricated in sheet, capillary and tubular forms. The liquid to be filtered is forced into the assemblage and dilute permeate passes perpendicularly through the membrane while concentrate passes out the end of the media. This technology is useful for the recovery and recycle of suspended solids and macromolecules. Excellent results have been achieved in textile finishing applications and other situations where neither entrained solids that could clog the filter nor dissolved ions that would pass through are present. Membrane life can be affected by temperature, pH, and fouling.

Ultrafiltration equipment are combined with other unit operations. The unique combination of unit operations depends on the wastewater characteristics and desired effluent quality, and cost considerations.

Like normal filtration, with ultrafiltration (UF), a feed emulsion is introduced into and pumped through a membrane unit; water and some dissolved low molecular weight materials pass through the membrane under an applied hydrostatic pressure. In contrast to ordinary filtration however, there is no build-up of retained materials on the membrane filter.

A variety of synthetic polymers, including polycarbonate resins, substituted olefins, and polyelectrolyte complexes, are employed as ultrafiltration membranes. Many of these membranes can be handled dry, have superior organic solvent resistance, and are less sensitive to temperature and pH than cellulose acetate, which is widely used in RO systems.

In UF, molecular weight (MW) cutoff is used as a measure of rejection. However, shape, size, and flexibility are also important parameters. For a given molecular weight, more rigid molecules are better rejected than flexible ones. Ionic strength and pH often help determine the shape and rigidness of large molecules. Operating temperatures for membranes can be correlated generally with molecular weight cutoff. For example, maximum operating temperatures for membranes with 5,000 to 10,000 MW cutoffs are about 65° C, and for a 50,000 to 80,000 MW cutoff, maximum operating temperatures are in the range of 50° C.

Ultrafiltration is a preferred alternative to the conventional systems of chemical flocculation and coagulation followed by dissolved air flotation. Ultrafiltration provides lower capital equipment, installation, and operating costs.

The largest industrial use of ultrafiltration is the recovery of paint from water-soluble coat bases (primers) applied by the wet electrodeposition process (electrocoating) in auto and appliance factories. Many installations of this type are operating around the world. The recovery of proteins in cheese whey (a waste from cheese processing) for dairy applications is the second largest application, where a market for protein can be found (for example, feeding cattle and farm animals). Energy consumption at an installation processing 500,000 pounds per day of whey would be 0.1 kWh per pound of product. Another large-scale application is the concentration of waste-oil emulsions from machine shops, which are produced in association with cooling, lubrication, machining, rolling heavy metal operations, and so on. Ultrafiltration of corrosive fluids such as concentrated acids and ester solution is also an important application. The chemical inertness and stability of ultrafilters make them particularly useful in the cleaning of these corrosive solutions. Uses include separation of colloids and emulsions, and recovery of textile sizing chemicals. Biologically active particles and fractions may also be filtered from fluids using ultrafilters. This process is used extensively by beer and wine manufacturers to provide cold stabilization and sterilization of their products. It is also used in water pollution analysis to concentrate organisms from water samples. Food concentration applications can be applied to processing milk, egg white, animal blood, animal tissue, gelatin and glue, fish protein, vegetable extracts, juices and beverages, pectin solutions, sugar, starch, single-cell proteins, and enzymes.

Figure 5 conceptually illustrates how ultrafiltration works. Water and some dissolved low molecular weight materials pass through the membrane under an applied hydrostatic pressure. Emulsified oil droplets and suspended particles are retained, concentrated, and removed continuously as a fluid concentrate. The pore structure of the membrane acts as a filter, passing small solutes such as salts, while retaining larger emulsified and suspended matter. The pores of ultrafiltration membranes are much smaller than the particles rejected, and particles cannot enter the membrane structure. As a result, the pores cannot become plugged. Pore structure and size (less than 0.005 microns) of ultrafiltration membranes are quite different from those of ordinary filters in which pore plugging results in drastically reduced filtration rates and requires frequent backflushing or some other regeneration step. In addition to pore size, another important consideration is the membrane capacity. This is termed *flux* and it is the volume of water permeated per unit membrane area per unit time. The standard units are gallons per day per square foot (gpd/ft^2) or cubic meters per day per square meter (m^3/day/m^2).

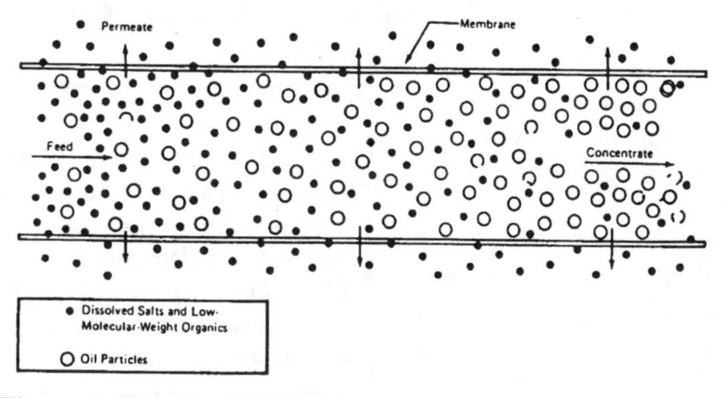

Figure 5. Ultrafiltration basics.

Because membrane equipment, capital costs, and operating costs increase with the membrane area required, it is highly desirable to maximize membrane flux.

Ultrafiltration utilizes membrane filters with small pore sizes ranging from 0.015μ to 8μ in order to collect small particles, to separate small particle sizes, or to obtain particle-free solutions for a variety of applications. Membrane filters are characterized by a smallness and uniformity of pore size difficult to achieve with cellulosic filters. They are further characterized by thinness, strength, flexibility, low absorption and adsorption, and a flat surface texture. These properties are useful for a variety of analytical procedures. In the analytical laboratory, ultrafiltration is especially useful for gravimetric analysis, optical microscopy, and X-ray fluorescence studies.

All particles larger than the actual pore size of a membrane filter are captured by filtration on the membrane surface. This absolute surface retention makes it possible to determine the amount and type of particles in either liquids or gases-quantitatively by weight or qualitatively by analysis. Since there are no tortuous paths in the membrane to entrap particle sizes smaller than the pore size, particles can be separated into various size ranges by serial filtration through membranes with successively smaller pore sizes. Figure 6 shows pore size in relation to commonly known particle sizes. Fluids and gases may be cleaned by passing them through a membrane filter with a pore size small enough to prevent passage of contaminants. This capability is especially useful in a variety of process industries which require cleaning or sterilization of fluids and gases.

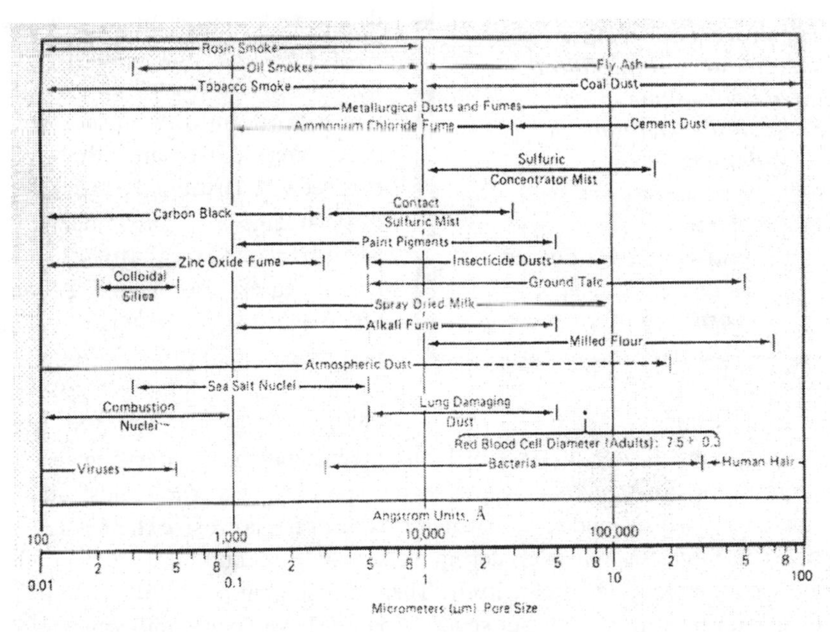

Figure 6. Range of common particle sizes (diameter) over range of UF pore size.

The retention efficiency of membranes is dependent on particle size and concentration, pore size and length, porosity, and flow rate. Large particles that are smaller than the pore size have sufficient inertial mass to be captured by inertial impaction. In liquids the same mechanisms are at work. Increased velocity, however, diminishes the effects of inertial impaction and diffusion. With interception being the primary retention mechanism, conditions are more favorable for fractionating particles in liquid suspension.

In contrast to reverse osmosis, where cellulose acetate has occupied a dominant position, a variety of synthetic polymers has been employed for ultrafiltration membranes. Many of these membranes can be handled dry, have superior organic solvent resistance, and are less sensitive to temperature and pH than cellulose acetate. Polycarbonate resins, substituted olefins, and polyelectrolyte complexes have been employed among other polymers to form ultrafiltration membranes. Preparation details for most of the membranes are proprietary. As noted earlier, molecular weight cutoff is used as a measure of rejection, however, shape, size, and flexibility are also important parameters. For a given molecular weight, more rigid molecules are better rejected than flexible ones. Ionic strength and pH often help determine the shape and rigidness of large molecules. Membrane lifetimes are usually two years or more for treating clean streams (water processing), but are drastically reduced when treating comparatively dirty streams (e.g., oily emulsions). Membrane guarantees by manufacturers are determined only after pilot work is done on the particular stream in question. In some cases as little as a 90-day guarantee may be given for oil/water waste applications. There are in fact both current and many emerging uses of ultrafilters in the areas of biological research, processing sterile fluids, air and water pollution analysis, and recovery of corrosive or noncorrosive chemicals. The technology is applicable to dewatering some sludges, but this use is highly dependent on the particular sludge itself. There are no commercial uses of UF for sludge dewatering at this time, but several sources have been found which claim that this represents a possible near-future application. Pollution of water supplies within the food industry is a significant problem, since many food wastes possess

SOME KEY POINT FOR YOU TO REMEMBER ABOUT ULTRAFILTRATION

☞ ***It is a moderately tight membrane.***
☞ ***It can remove some of the colloidal particles including bacteria.***
☞ ***The process can replace sedimentation, normal filtration and disinfection, to some extent.***
☞ ***Some of the operating information are:***

- ***Process time per cycle: 16-20 hrs***
- ***System size: 1-1,000 m^2***
- ***Cleaning: By acid or caustic cycle***

extremely high biochemical oxygen demand (BOD) requirement. In the potato starch industry, for example, waste effluent containing valuable proteins, free amino acids, organic acids, and sugars can be processed by ultrafiltration. Reclamation of these materials, which are highly resistant to biodegradation, are providing an economic solution to this waste removal problem. For the concentration of juices and beverages, RO is preferable to evaporation due to lower operating costs and no degradation of the product. Processing by RO retains more flavor components than does heated, vacuum pan concentration. Since ultrafiltration does not retain the low molecular weight flavor components and some sugar, UF is employed as a complement to RO. A two-stage process may be used in which the first stage, UF, allows the passage of sugars and other low molecular weight compounds. This permeate is then dewatered by RO and recycled back to the main stream. High juice concentrations are possible in this manner because the UF removes the colloidal and suspended solids which would foul the RO, and helps relieve some of the high hydraulic pressure due to high osmotic pressure of the juice. In a process as shown in Figure 7, a citrus press liquor, or multiphase suspension, is ultrafiltered following a coarse prefiltration. The resultant clear permeate is processed through ion exchange and granular activated carbon adsorption units to remove low molecular weight contaminants and inorganic salts. The product is a natural citrus sugar solution suitable for reuse. The concentrated suspended solids are used in making animal feed. Pectins are a family of complex carbohydrates which are used to form gels with sugar and acid in the production of jellies, preserves, and other confections from fruit juices. The recovery of starch and other high molecular weight compounds from waste effluents is an important application for UF. The output from a 30-ton-per-day starch plant would be about 432,000 gpd with a solids content of 0.5 percent to 1.0 percent, and 9,000 to 14,000 mg/Liter COD.

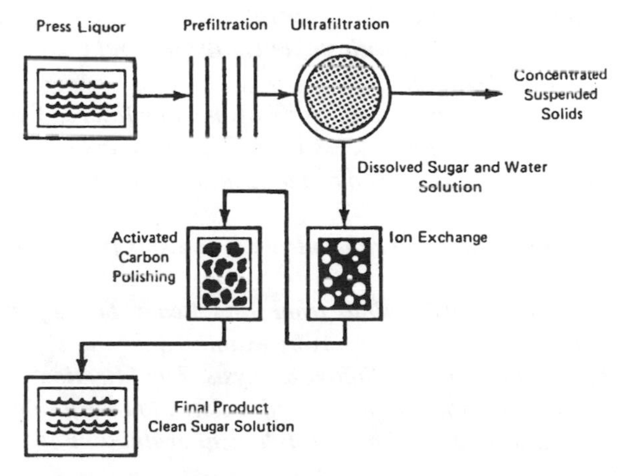

Figure 7. Process flow scheme for sugar recovery from citrus press liquors.

Treatment systems which can both reduce the strength of this waste and recover valuable by-products, such as proteins, are an obvious advantage to this industry. Heat and acid coagulation, distillation, and freezing techniques are more costly and less efficient than UF in protein recovery. Reverse osmosis is also a competing process, but protein recovery by UF would lead to a somewhat higher purity. In the production of single-cell proteins as a food source, UF has several applications. For harvesting cells, UF can replace centrifugation in some applications since the efficiency of centrifugation decreases rapidly with particle size. Ultrafiltration is also well suited for recovering and concentrating the metabolic products of fermentation (enzymes, for example). In a related application, UF is also able to concentrate and desalt protein products, being more efficient than dialysis for this purpose. Moreover, a UF membrane module may be coupled with a fermenter so that toxic metabolites can be continuously removed from the system as fresh substrate is introduced. This permits the growth limitations of a batch fermenter to be relieved and permits a substantial increase in productivity. A membrane

IMPORTANT APPLICATIONS

☞ ***Recovery of paint from watersoluble coat bases (primers) applied by the wet electrodeposition process (electrocoating) in auto and appliance factories.***

☞ ***Recovery of proteins in cheese whey (a waste from cheese processing) for dairy applications. This is done if a market for protein can be found, in particular for feeding cattle and farm animals. In cheese whey processing, a typical unit might process 500,000 pounds a day of whey for 300 days a year.***

☞ ***The concentration of waste-oil emulsions from machine shops, which are produced in association with cooling, lubricating, machining, and heavy metal rolling operations. The separation of the oil from the water works well with stable emulsions, but with unstable emulsions the oil will clog the filter.***

☞ ***Biologically active particles and fractions may be filtered from fluids using ultrafilters. This process is used extensively by beer and wine manufacturers to provide cold stabilization and sterilization of their products.***

☞ ***Water pollution analysis to concentrate organisms from water samples.***

☞ ***Filtering cells and cell fractions from fluid media. These particles, after concentration by filtration, may be examined through subsequent quantitative or qualitative analysis. The filtration techniques also have applications in fields related to immunology and implantation of tissues as well as in cytological evaluation of cerebrospinal. fluid.***

enzymatic reactor is similar to a membrane fermenter with the exception that no microorganisms are present. Instead, enzyme-catalyzed reactions take place and reuse of the enzyme is simplified. That is, purification problems and enzyme removal from end products can be eliminated.

FOULING CONSIDERATIONS

A critical consideration with UF technology is the problem of fouling. Foulants interfere with UF by reducing product rates-- sometimes drastically--and altering membrane selectivity. The story of a successful UF application is in many respects the story of how fouling was successfully controlled. Fouling must be considered at every step of UF process development in order to achieve success.

When we talk about this subject, the term *foulant* or *foulant layer* comes to the forefront. Foulant, or fouling layer, are general terms for deposits on or in the membrane that adversely affect filtration. The term "fouling" is often used indiscriminately in reference to any phenomenon that results in reduced product rates. "Fouling" in this casual sense can involve several distinct phenomena. These phenomena can be desirable or undesirable, reversible or irreversible. Different technical terms apply to each of these possibilities.

You may be surprised, but fouling is not always detrimental. The term *dynamic membrane* describes deposits that benefit the separation process by reducing the membrane's effective MWCO (*Molecular Weight cut-off*) so that a solute of interest is better retained. Concentration polarization refers to the reversible build-up of solutes near the membrane surface. Concentration polarization can lead to irreversible fouling by altering interactions between the solvent, solutes and membrane.

UF fits between nanofiltration and microfiltration in the filtration spectrum and involves separations of constituents ranging from about 1-100 nanometers in size, or about 500 to 500,000 daltons in molecular weight. UF separations involve proteins, polysaccharides and other macromolecules important to the food industry. separation is primarily according to size, but surface forces are important in determining the separation as well. UF is different from conventional filtration, also called normal or dead-end filtration, in that it operates in the crossflow mode; that is, the feed stream flows parallel to the filtration media (membrane). Crossflow acts as a sweep stream to continuously cleanse the surface of the membrane from accumulated retentate. There are two products of UF: the permeate, containing components small enough to pass through the membrane, and the concentrate, containing the retentates.

Cake layer formation builds on the membrane surface and extends outward into the feed channel. The constituents of the foulant layer may be smaller than the pores of the membrane. A gel layer can result from denaturation of some proteins. Internal pore fouling occurs inside the membrane. The size of the pore is reduced and pore flow is constricted. Internal pore fouling is usually difficult to clean.

Fouling can be characterized by mechanism and location. Membranes can foul in three places: on, above or within the membranes (refer to the sidebar on the next page). The term *agglomeration* in the general sense, describes colloidal precipitates resulting from solute-solute attractions. Agglomerates can deposit on the membrane surface, reducing permeability. On the other hand, controlled aggregation of solutes can facilitate ultrafiltration.

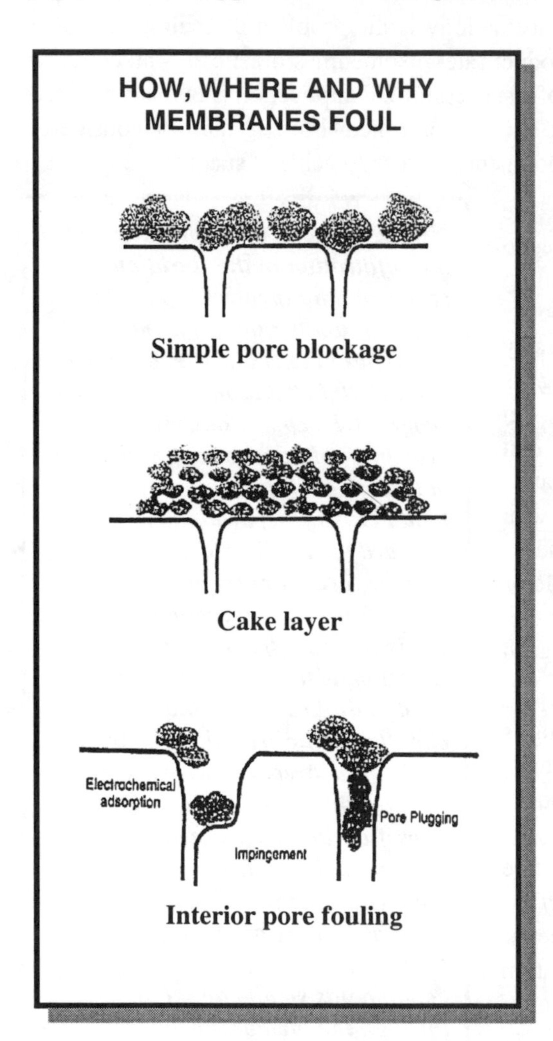

Sorption or adsorption refers to deposition of foulants on the membrane surface resulting from electrochemical attractions. These attractions arise from non-covalent, intermolecular forces such as Van der Waals forces and hydrogen bonding. Adsorption is associated with internal pore fouling, since most of the surface area of the membrane occurs internally. The high internal surface area of UF membranes is readily apparent from photomicrographs of cross-sections of UF membranes. The photomicrographs show sponge-like structures that suggest convoluted, tortuous pore pathways. Adsorption can lead to more extensive fouling. For instance, a protein might denature upon adsorbing to the surface of an ultrafilter. The denatured protein attracts other proteins, the process repeats, and a deposit builds on the membrane surface.

UF membranes are often rated by molecular weight cut-off (MWCO); solutes above the MWCO are retained and those below the MWCO permeate through the membrane. MWCO can be determined by challenging a UF membrane with a polydisperse solute, such as dextran, in a crossflow filtration experiment A retention profile or curve is determined by comparing the dextran molecular weight distribution in the feed to that in the permeate using size exclusion chromatography (SEC). MWCO is typically defined as the 90 percent retention level, or the molecular weight value on the ordinate where the retention curve crosses 90 percent on the axis. MWCO ratings are relative. Membrane retentivity depends upon many factors, including the shape of the solute used, the fluid mechanics and the various interactions possible between the solvent, solute and membrane.

MWCO curves are useful for identifying membranes with appropriate selectivity for an intended separation. Predicting the best membrane MWCO is not always straightforward, however. A common assumption is that the membrane MWCO should closely match the molecular weight of the solute of interest. Remarkably, better UF performances are sometimes achieved with membranes having MWCO's significantly higher or lower than the molecular weight of the solute to be retained. For example, lower protein adsorption and flux loss are reported in the literature in the filtration of albumin with polyethersulfone membranes when membrane MWCO's were much larger and smaller than the molecular weight of albumin. How is this possible? Better performances with the low and high MWCO membranes are explained by considering the effects of fouling on the membranes. Higher product rates are sometimes realized with lower MWCO membranes because they exclude more potential foulants and internal pore fouling is reduced. Membranes with higher MWCO's will sometimes effectively separate smaller solutes because solutes aggregate into larger entities or because foulant forms an effective dynamic membrane. The dynamic membrane reduces the effective MWCO of the ultrafilter so that the solute is retained. The larger pores suffer less flow restriction due to adsorption, and the greater hydraulic permeability of the larger pores yields high product rates.

A useful analytical tool for predicting and diagnosing fouling s Fourier Transform Infrared Spectroscopy (FTIR). FTIR can reveal important information that is useful for predicting and measuring foulants. The FTIR is a standard laboratory instrument for chemical analysis, and has been applied for many years in the field of membrane science. It has been successfully applied in identifying which solutes in complex mixtures may cause fouling. The ability to distinguish foulants is advantageous in applications where complex process streams predominate. Fit with an attenuated total reflectance (ATR) accessory, FTIR allows us to look quickly and easily at the chemistry of the foulant layer and membrane surface. The ATR technique can also provide quantitative estimates of fouling. There are some references at the end of this chapter that will give you more details.

FTIR can be used to screen membranes for fouling tendencies prior to the first ultrafiltration experiment. Screening can be done by means of a simple static adsorption test. Membranes showing greater static adsorption are expected to foul more during ultrafiltration and are disfavored. Figure 8 illustrates the FTIR results of a static adsorption test using a polysulfone ultrafilter as the substrate and a water extract of soy flour as the source of potential adsorbates.

FTIR can be used to diagnose fouling as well as to predict it. The techniques are similar. Among the diagnostic possibilities, one can:

- chemically identify the foulant(s) by searching spectral libraries
- estimate the thickness of foulant layer by comparing the relative size of peaks due to the membrane and foulant
- evaluate the effectiveness of various cleaners by measuring the disappearance of foulant peaks
- surmise internal pore fouling if foulant peaks persist after the surface of the membrane has been thoroughly cleansed.

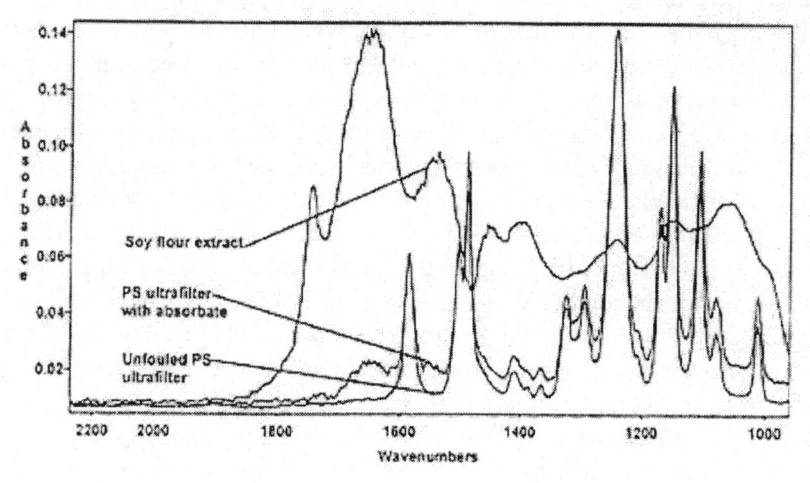

Figure 8. Example of FTIR analysis of Polysulfone (PS) ultrafilter static adsorption test.

WHAT MICROFILTRATION AND NANOFILTRATION ARE

In the case of microfiltration, a more porous membrane is used than in the other membrane separation technologies, thus yielding a relatively higher flux. It is mainly useful in removing turbid causing materials and can replace conventional granular filtration processes. The most significant design parameters are:

- **Transmembrane pressure**
- **Tangential velocity**
- **Size and geometry of modules**
- **Recirculation factor**

The clean water flux across a membrane without any material being deposited follows Darcy's Law:

$$J_w = \Delta P/\mu R_m \tag{2}$$

The net pressure differential across a membrane, taking into consideration the osmotic pressure is given by (ΔP - $\Delta\Pi$), and hence, the expression for the permeate flux is:

$$J_w = (\Delta P - \Delta\Pi)/\mu(R_m + R_c) \tag{3}$$

where J_w = permeate flux, m/s

ΔP = pressure difference, N/m^2

R_m = internal membrane resistance, i/m

R_c = resistance due to deposit on the surface, 1/m

μ = dynamic viscosity, $N\text{-}sec/m^2$

The resistance due to the deposit on the surface is given by the following relationship:

$$R_c = 180(1 - \epsilon)^2\, \delta/(d_p^{\,2}\epsilon^3) \tag{4}$$

where ϵ = porosity of deposit

δ = thickness of the deposit, m

d_p = average diameter of the particles, m

Commercial systems for wastewater treatment are designed to be submerged into built on-site rectangular concrete tanks. These are pre-engineered modular membrane systems that typically use a membrane with a 0.2-micron nominal pore size. A vacuum pump draws water through the membrane fibers of sub-modules submerged in the open top filter tanks. The fibers are the same polypropylene material as those used in the conventional filtration process. A typical system operates under vacuum, and utilizes improved filter cake characteristics at low pressures with a maximum driving pressure in the league of 85-100kPa.

There have been only a few studies have evaluated membrane microfiltration of secondary wastewater effluent. Microfiltration membranes might be used to achieve very low turbidy effluents with very little variance in treated water quality. Because bacteria and many other microorganisms are also removed, such membrane disinfection might avoid the need for chlorine and subsequent dechlorination. Metal salts of iron or aluminum may also be added to enhance membrane performance. For example, iron or aluminum coagulants may be added to precipitate otherwise soluble species such as phosphorus and arsenic as well as improving the removal of viral particles. Coagulation of colloidal materials may also increase the effective size of particles applied to membranes and increase permeate flux by 1) reducing foulant penetration into membrane pores, 2) forming a more porous cake on the membrane surface, 3) decreasing the accumulation of materials on the membrane due to particle size effects of particle transport, and 4) improving the backflushing characteristics of the membrane.

Membrane microfiltration at the pilot-scale has produced a permeate of similar or better quality than that produced by conventional filtration. Good removal of particulate contaminants, including coliform bacteria, have been observed. In this regard, the process appears to be as effective as chlorination for the removal of coliforms from secondary waste effluent. A key advantage is the ability to filter and disinfect in a single step without the need for subsequent dechlorination. Preliminary results indicate that coagulation pretreatment in conjunction with membrane microfiltration can be used to reduce phosphorus concentrations as well. There does not appear to be any advantage in running the microfiltration unit in a crossflow mode and there may even be some disadvantages. The permeate quality and evolution of pressure drop obtained from the membrane operated in the dead-end mode is found similar or superior to that obtained under crossflow conditions.

Although membrane processes have been used successfully for many years in desalting brackish water and seawater, new kinds of membrane processes are now capable of treating water for a wide range of other uses. Some of these new, robust processes promise to do a better job meeting our current water treatment goals than such conventional processes as granular media filtration, carbon filtration and disinfection with chlorine. Engineers classify membranes in many different ways, including describing them by the driving forces used for separating materials (i.e., pressure, temperature, concentration and electrical potential), the mechanism of separation, the structure and chemical composition, and the construction geometry. In water treatment, the membranes most widely used are broadly described as pressure driven. Each membrane process is best suited for a particular water treatment function. For example, microfiltration (MF) and ultrafiltration (UF), which are very low pressure processes, most effectively remove particles and microorganisms. The reverse osmosis (RO) process most effectively desalts brackish water and seawater and removes natural organic matter and synthetic organic and inorganic chemicals. The nanofiltration (NF) process softens water by removing calcium and magnesium ions. These so-called nanofilters are also effective in removing the precursors to disinfection by-products that result from such oxidants as chlorine.

Nearly a decade ago, the use of low-pressure membranes such as MF and UF for disinfection and particle removal was only a concept being studied or was used only on a limited basis. The water community foresaw the possibility of providing primary disinfection without the use of chemicals. Moreover, *Cryptosporidium*, a waterborne enteric pathogen responsible for several disease outbreaks, was gradually showing resistance to traditional disinfectants such as chlorine. Thus, researchers believed that greater emphasis should be placed on removing organisms through physical means as opposed to chemical means.

Membrane processes also offer other advantages over conventional treatments. They reduce the number of unit processes in treatment systems for clarification and disinfection and increase the potential for process automation and plant compactness. Designers also thought membrane plants could be much smaller than conventional plants of the same capacity and, given their modular configuration, could be easily expanded. Additionally, these plants would produce less sludge than conventional plants because they wouldn't use such chemicals as coagulants or polymers.

Today many of the projected benefits of MF and UF have been realized. These technologies provide effective disinfection for potable water supplies as they reduce the levels of *Giardia* and *Cryptosporidium*, as well as a variety of bacteria, below detectable levels. MF and UF plants are now in operation throughout the world. In Europe there are several large UF plants. In the U.S., the San Jose Water Co. in Saratoga, Calif., was the first to construct a major MF plant (17,000 m^3/day). A plant with 15,000 m^3/day capacity followed in Rancho Cucamonga, Calif., and a 68,000 m^3/day plant is under construction in Kenosha, Wisconsin. The largest plant for disinfection and particle removal, a 106,000 m^3/day UF installation, is being planned in Del Rio, Tex. Today, there is more than 400,000 m^3/day of MF and UF capacity in the U.S., either in operation or in the planning stage.

The anticipated U.S. EPA Disinfectant/Disinfection By-Product Rule will lower the maximum contaminant level (MCL) for trihalomethanes (THM) from 100 to 80 μg/L, and set an MCL for haloacetic acids (HAA) at 60 μg/L. Thus, NF and RO membrane processes are receiving considerably more attention since they are efficient at removing the precursors to these by-products. RO and NF membranes, in addition to desalting brackish water and seawater, are being used to remove inorganic chemicals such as nitrates, as well as synthetic organic chemicals. NF plants have been installed since the 1980s in Florida to remove color and hardness from groundwater. Today, there is 568,000 m^3/day of installed NF and RO capacity in the U.S.

Membrane plant design begins with the selection of the membrane, which can be organic or inorganic in composition. Membrane manufacturers strive to formulate membranes that provide a desired permeate quality, are durable and resistant to fouling, and can be produced at a competitive cost. Most commercial water treatment NF and RO membranes are made up of organic polymers and are asymmetric. The active layer responsible for the separation process is typically a few micrometers thick and is supported on a highly permeable layer that adds mechanical strength to the membrane. One type of asymmetric membrane for NF and RO systems, the thin-film composite (TFC), shows great promise for potable water treatment. These membranes generally consist of an ultrathin active layer coated onto a microporous layer that, in turn, is supported on a mechanically strong base. TFC membranes typically have higher water permeability and chemical resistance than symmetric membranes. MF and UF membranes are constructed in either asymmetric or symmetric configurations. A number of hydrophilic and hydrophobic polymeric materials are used in manufacturing these membranes. These include cellulosic polymers, polypropylene, polysulfones and polyamides. The choice of material will influence contaminant rejection characteristics, durability and fouling potential.

Membrane systems consist of membrane elements or modules. For potable water treatment, NF and RO membrane modules are commonly fabricated in a spiral configuration. An important consideration of spiral elements is the design of the feed spacer, which promotes turbulence to reduce fouling. MF and UF membranes often use a hollow fiber geometry. This geometry does not require extensive pretreatment because the fibers can be periodically backwashed. Flow in these hollow fiber systems can be either from the inner lumen of the membrane fiber to the outside (inside-out flow) or from the outside to the inside of the fibers (outside-in flow). Tubular NF membranes are now just entering the marketplace.

MF and UF systems can be designed to operate in various process configurations. A common configuration is one in which the feedwater is pumped with a cross-flow tangential to the membrane. The only pretreatment usually provided is a crude prescreening (usually 50 to 300 μm). The water that permeates the membrane is clean. The water that does not permeate is recirculated as concentrate and blended with additional feedwater just after the preliminary filter. To control the concentration of the solids in the recirculation loop, some of the concentrate is discharged at a specified rate.

MF and UF systems may also operate in a direct filtration configuration, with no cross-flow (or recirculation). This is often termed *dead-end filtration*. All of the prescreened feedwater passes through the membrane. Therefore, there is 100% recovery of this water, except for the small fraction of the water used to periodically backwash the system. MF and UF plants typically rely on either liquid or pneumatic backwashing systems. Most MF and UF water treatment plants use this direct flow configuration, since it saves considerably on energy by not requiring recirculation. There are also capital cost savings since there is no need to purchase recirculation pumps and associated piping.

RO and NF systems usually operate in a series of stages. In a three-stage system, the first stage consists of three pressure vessels, which usually contain four to eight membrane elements; the second stage has two pressure vessels; and the final stage has one. In full-scale plants, elements are approximately 1,000 mm long and have a diameter of 200 mm. Permeate is collected from each pressure vessel. The concentrate from the first stage serves as the feed to the second; concentrate from the second stage serves as the feed to the third. Consequently, each successive stage of the array increases the total system recovery.

For many groundwater applications, the pretreatment required for RO or NF

consists in adding acid or antiscalant and then passing the feed through a cartridge filter. However, for surface waters, more extensive pretreatment is necessary. This may involve conventional treatment, MF, UF, slow sand filtration or, in some cases, granular activated carbon adsorption.
One innovative process configuration for surface water and tertiary wastewater treatment involves the use of double-membrane systems, consisting of a low-pressure and a high-pressure membrane in series. This treatment is effective for both microbial and chemical contaminant control. The first membrane (MF or UF) is used to help prevent fouling of the second, higher-pressure membrane system (RO or NF). In the Netherlands, the Heemskirk water treatment plant, which will be completed late this year, will use a 53,000 m^3/day double-membrane process (UF and TFC-RO) to treat water from the Ijssel River. San Diego, California, will also use a double-membrane system integrated into its treatment train as part of an approach for purifying tertiary wastewater for potable reuse. Another system uses membranes designed to be immersed in a process tank and suction instead of pressure to draw water through the membrane hollow fiber lumen.

Pilot testing is often a key aspect of successful membrane plant design. One of the most important reasons for conducting a pilot study is to evaluate the influence of water quality on membrane fouling. It is critical to determine if the process is feasible for a specific water source, particularly for those that exhibit significant water quality changes on a seasonal basis. Other reasons for pilot tests include demonstrating regulatory compliance, identifying the most effective and appropriate processes, evaluating new membrane products and establishing design criteria for a full-scale plant. Designers can also evaluate pretreatment options by conducting experiments using parallel treatment trains with varying pretreatment processes but identical operating conditions. This avoids the confounding factors in data interpretation, such as changing source water quality. Designers can also verify and fine-tune chemical cleaning procedures for site-specific conditions. In pilot studies, designers are able to optimize the operating conditions of each individual stage of a multistage installation.

As membranes filter out the impurities from the water, the membranes themselves become fouled (or clogged) and less effective. The fouling of membranes has been one the primary impediments to their more widespread application in water treatment. Membrane systems operate in one of two modes: constant transmembrane water flux (flow rate per unit membrane area) with variable pressure; or constant pressure with variable transmembrane water flux. The former is the more common. Membrane fouling occurs during an increase in transmembrane pressure to maintain a particular water flux or during a decrease in water flux when the system is operated at constant pressure. In general, membranes can be fouled by an accumulation of inorganic particles (for example, clays, iron, manganese and silica) and organic compounds (such as humic and fulvic acids, hydrophilic and hydrophobic materials, and proteins). Bacteria can also adhere to the membranes and create a biofilm. Accurate tests to predict fouling still do not

exist. However, researchers can conduct "autopsies" on fouled membranes prior to chemical cleaning to analyze the nature and composition of the contamination. They can then adjust pretreatment and chemical cleaning procedures. Fouling can be controlled by hydrodynamic and chemical methods, periodic backwashing, and chemical cleaning. Other methods include improving pretreatment and changing operating conditions. Membrane fouling rates are functions of the operating conditions such as water flux and recovery. Typically, reducing the operating flux and recovery will reduce fouling. Research shows that, for MF and UF, increasing the frequency of backwashing also decreases the rate of fouling. Because filtered or raw water is used for backwashing, the net recovery of direct-flow MF and UF systems decreases as the frequency of backwashing increases. Improved pretreatment also reduces membrane fouling. Water treatment operators have decreased MF and UF fouling rates by using coagulation/flocculation/sedimentation and dissolved air flotation pretreatment.
Chemical scaling is another form of fouling that occurs in NF and RO plants. The thermodynamic solubility of salts such as calcium carbonate and calcium and barium sulfate imposes an upper boundary on the system recovery. Thus, it is essential to operate systems at recoveries lower than this critical value to avoid chemical scaling, unless the water chemistry is adjusted to prevent precipitation. It is possible to increase system recovery by either adjusting the pH or adding an antiscalant, or both.

WHAT REVERSE OSMOSIS IS

When pure water and a salt solution are introduced on opposite sides of a semipermeable membrane in a vented container, the pure water diffuses through the membrane and dilutes the salt solution, At equilibrium, the liquid level on the saline water side of the membrane will be above that on the freshwater side; this process is known as osmosis and is depicted in Figure 9. The view on the left illustrates the commencement of osmosis and the center view presents conditions at equilibrium. The effective driving force responsible for the flow is osmotic pressure. This pressure has a magnitude dependent on membrane characteristics, water temperature, and salt solution properties and concentration. By applying pressure to the saline water, the flow process through the membrane can be reversed. When the applied pressure on the salt solution is greater than the osmotic pressure, fresh water diffuses in the opposite direction through the membrane and pure solvent is extracted from the mixed solution; this process is termed reverse osmosis (RO). The fundamental difference between reverse osmosis and electrodialysis is that in reverse osmosis the solvent permeates the membrane, while in electrodialysis the solute moves through the membrane.

Reverse osmosis is a means for separating dissolved solids from water molecules in aqueous solutions as a result of the membranes being composed of special polymers which allow water molecules to pass through while holding back most other types of molecules; since true "pores" do not exist in the membrane, suspended solids are also retained by *superfiltration*. In an actual reverse osmosis system, operating in a continuous-flow process, feed water to be treated or desalinated is circulated through an input passage of the cell, separated from the output product water passageway by the membrane.

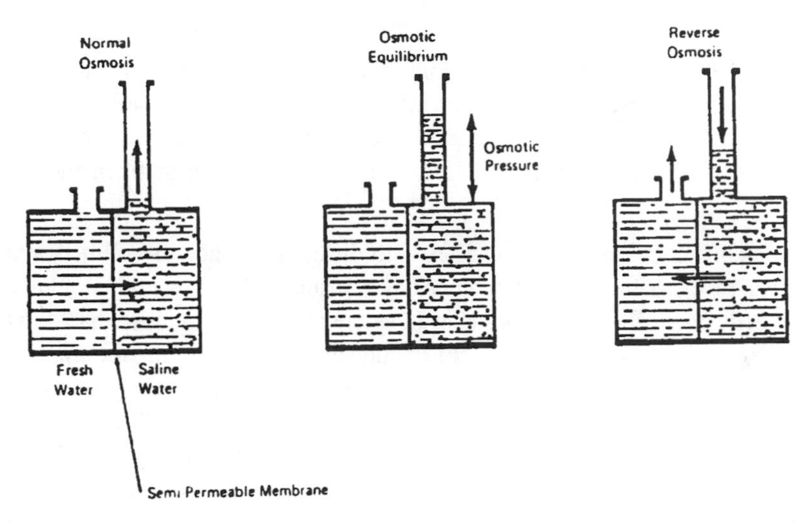

Figure 9. Principle of Reverse Osmosis.

The feed stream is split into two fractions - a purified portion called the product water (or permeate) and a smaller portion called the concentrate' containing most of the impurities in the feed stream. At the far end of the feed-water passage, the concentration (dewatered) reject stream exits from the cell. After permeating the membrane, the product (fresh-water) flow is collected. The percentage of product water obtained from the feed stream is termed the *recovery,* typically around 75 percent.

The ratio *(F-P)/F,* or the concentration of a solute species in the feed *(F)* minus that in the product *(P)* over the concentration in the feed, is called the *rejection* of that species. Rejections may be stated for particular ions, molecules, or conglomerates such as TDS or hardness. Solids rejection depends on factors such as types and forms of solids, membrane types, recovery, pressure, and pH. Suspended solids (typically defined as particles larger than 0.5 micron mean diameter, and including colloids, bacteria, and algae) are rejected 100 percent; that is, none can pass through the membrane. Weakly ionized dissolved solids (usually organics, but may include other materials such as silicates) undergo about a 90 percent rejection at normal recoveries for certain membranes. Although pH can strongly influence the rejection, when the molecular weight of these solids is less than 100, rejection decreases appreciably. Ionized solids, or salts, are rejected independent of molecular weight and at molecular weights considerably below 100. At 75 percent recovery and pressures greater than 250 pounds per square inch, overall rejection of total dissolved solids (TDS) is about 90 percent. Rejections vary with pressure because the actual salt flow through the membrane remains fairly constant, but the water permeation depends nearly linearly on pressure, affecting the ratio of concentrations. For example, rejection of sodium chloride can fall from 90 percent at 300 pounds per square inch to 20 percent at 50 pounds per square inch, indicating the need to operate at the highest pressures possible.

Cellulose acetate is a common membrane material, but others include nylon and aromatic polyamides. The mechanism at the membrane surface involves the influent water and impurities attempting to pass through the pressurized side, but only pure water and certain impurities soluble in the membrane emerge from the opposite side.

Various configurations of membranes with different surface-to-volume ratios and different flux capabilities (gallons per day per square foot, or gpd/ ft') have been developed. Each type of membrane is a flexible plastic filmno more than 4 to 6 mils thick, firmly supported. Basic designs include the plate and frame, the spiral-wound module (jellyroll configuration), the tubular, and the newest of the process designs, the hollow-fine fiber. Fibers range from 25 to 250 microns (0.001 to 0.01 in.) in diameter, can withstand enormous pressure, are self-supporting, and can be bundled very compactly within a containment pipe. While product flow per square foot of fiber surface is less than that for an equivalent area of flat membrane, the difference in surface area more than compensates for the reduced unit flux.

Major problems inherent in general applications of RO systems have to do with (1) the presence of particulate and colloidal matter in feed water, (2) precipitation of soluble salts, and (3) physical and chemical makeup of the feed water. All RO membranes can become clogged, some more readily than others. This problem is most severe for spiral-wound and hollow-fiber modules, especially when submicron and colloidal particles enter the unit (larger particulate matter can be easily removed by standard filtration methods). A similar problem is the occurrence of concentration-polarization, previously discussed for ED processes. Concentration-polarization is caused by an accumulation of solute on or near the membrane surface and results in lower flux and reduced salt rejection.

The degree of concentration that can be achieved by RO may be limited by the precipitation of soluble salts and the resultant scaling of membranes. The most troublesome precipitate is calcium sulfate. The addition of polyphosphates to the influent will inhibit calcium sulfate scale formation, however, and precipitation of many of the other salts, such as calcium carbonate, can be prevented by pretreating the feed either with acid or zeolite softeners, depending on the membrane material.

Hydrolysis of cellulose acetate membranes is another operational problem and occurs whenever the feed is too acid or alkaline; that is, the pH deviates beyond designed range limits. As may readily happen, whenever C02 passes through the membrane, the resultant permeate has a low pH. The operational solution is to remove the gas from the permeate by deaerators, by strong-base anion resins or a complementary system-for example, RO and ion exchange, in series. Aromatic polyamide or nylon membranes are much less sensitive to pH than cellulose acetate. Compounds such as phenols and free chlorine that are either soluble in the membrane or vice versa will be poorly rejected and may damage the membrane. Procedures to improve feed-water makeup and thus reduce such membrane damage include acid pretreatment of the feed water, dechlorination, periodic cleaning or replacement of the membrane, sequestration of cations, coagulation and filtration of organics, and use of alternative, more durable membrane materials.

Reverse osmosis process is applied-or undergoing evaluation for imminent application-to a number of water-upgrading needs including high-purity rinse water production for the electronics industry (semiconductor manufacturing), potable municipal water supplied for newly-developed communities (for example, large coastal plants to upgrade brackish well water contaminated by seawater intrusion), boiler feed-water supplies, spent liquor processing for pulp and paper mills, and treatment of acid mine drainage.

In desalting operations, distillation plants have provided the major portion of the world's capacity. As the world's requirements for treated water increase, however, and water quality standards become more stringent, the membrane treatment processes in general and commercial RO processes in particular have been undergoing appreciable development. Important factors in the expansion of commercial RO applications are their favorably low power requirements and the realization of continuous technical improvements in membranes which are used in RO systems. A general guideline in water benefication is that RO is most frequently considered for cases in which the TDS is greater than 2,000 to 3,000 ppm; ED generally applies when the TDS is less than 2,000 to 3,000 ppm. However, many exceptions exist, based on feed-water species and product requirements.

Glossary of Salt Water

The general term for all water over 1,000 ppm (mg/L) total dissolved solids.

- **Fresh Water : <1,000 TDS**
  >
- **Brackish : 1,000-5,000 TDS**
- **Highly Brackish : 5,000-15,000 TDS**
- **Saline : 15,000-30,000 TDS**
- **Sea Water : 30,000-40,000 TDS**

One of the most important applications of RO is in the reclamation of large volumes of municipal and industrial wastewaters and the concentration of the solids for simplified disposal. The value of the reclaimed water offsets the cost of RO, and dilute wastewater concentration leads to economies in any further required liquid waste treatment.

Unrestricted use of reclaimed wastewater for drinking water, however, requires careful examination. While practically a complete barrier to viruses, bacteria, and other toxic entities that must be kept out of a potable supply, RO membranes could pose serious problems should any defect develop in their separation mechanism. Given the purity and clarity of RO-treated wastewaters, however, it might be advantageous to use RO and then subject the product to well-established disinfection procedures.

You should remember that RO uses a semi-permeable membrane. As such, the membrane is permeable to only very light molecules like water. Under atmospheric condirtions the fresh water flows into the solution which is called osmotic flow. But for purification purposes, this is no use, and hence we employ the reverse of osmotic flow. For this to happen, we need to apply external pressure in excess of osmotic pressure. The osmotic pressure is given by:

$$\Delta p = nRT \tag{3}$$

Of course, you should be familiar with this equation (the *Ideal Gas Law*), where 'n' is the molar concentration of solute, R is the universal gas law constant, and T is absolute temperature in °K. The permeate flow can be calculated from:

$$J_w = A_m (\Delta P - \Delta p) \tag{4}$$

In this expression, A_m is the membrane permeability coefficient.

It is useful to compare the merits of the various processes for seawater desalination. Although the comparison will be primarily qualitative, it should be helpful in providing a deeper insight into the strengths and weaknesses of process. Foremost among the aspects of comparison is the energy consumption of each process you consider. With the known process specification, it is theoretically possible to calculate the minimum work or energy needed for separation of pure water from brine. For the real process, however, the actual work required is likely to be many times the theoretically possible minimum. This is because the bulk of the work is required to keep the process going at a finite rate rather than to achieve the separation.

The minimum work needed is equal to the difference in free energy between the incoming feed (i.e. seawater or brackish water) and outgoing streams (i.e. product water and discharge brine). For the normal seawater (3.45 per cent salt) at a temperature of 25° C, for usual recoveries the minimum work has been calculated as equal to about 0.86 kWh/m^{-3}. Table 5 makes the desired comparison.

Table 5. Energy requirements of four industrial desalination processes. (*Source: International Atomic Energy Agency 1992.*)

	MSF	**MEB**	**MEB/VC**	**RO**
Possible unit size	60,000	60,000	24,000	24,000
Energy consumption (kWh/m^3) - Electrical/mechanical	4-6	2-2.5	7-9	5-7
Energy consumption (kWh/m^3) - Thermal	55-120	30-120	None	None
Electrical equivalent for thermal energy (kWh/m^{-3})	8-18	2.5-10	None	None
Total equivalent energy (kWh/m^{-3})	12-24	4.5-12.5	7-9	5-7

There are no major technical obstacles to desalination as a means of providing an unlimited supply of fresh water, but the high energy requirements of this process pose a major challenge. Theoretically, about 0.86 kWh of energy is needed to desalinate 1 m^3 of salt water (34,500 ppm). This is equivalent to 3 kJ kg^{-1}. The present day desalination plants use 5 to 26 times as much as this theoretical minimum depending on the type of process used. Clearly, it is necessary to make desalination processes as energy-efficient as possible through improvements in technology and economies of scale.

Desalination as currently practiced is driven almost entirely by the combustion of fossil fuels. These fuels are in finite supply; they also pollute the air and contribute to global climate change. The whole character of human society in the 20th century in terms of its history, economics and politics has been shaped by energy obtained mostly from oil. Almost all oil produced to date is what is called conventional oil, which can be made to flow freely from wells (i.e. excluding oil from tar sands and shale). Of this vast resource, about 1600 billion barrels have so far been discovered, and just over 800 billion barrels had been used by the end of 1997. It is estimated that there may be a further 400 billion barrels of conventional oil yet to be found. With current annual global consumption of oil being approximately 25 billion barrels, and rising at 2 per cent per annum, the "business as usual" scenario would suggest that the remaining oil will be exhausted by 2050. The supply of oil will undoubtedly be boosted by an increase of supplies from unconventional sources, notably the tar sands and shale of Canada and the "Orinoco sludge" of Venezuela. This oil can only be extracted using high energy inputs, and at very high environmental costs. There will be strong political and international pressure against development of these resources, but, when world oil prices are high enough, production will inevitably increase. In theory, unconventional oil could stretch the world's oil supply by another 30 years. In practice, of course, the rate of consumption of oil will be heavily influenced by economic and many other factors, so that prediction in this area is very difficult. The political situation of two of the world's largest potential producers, Iran and Iraq, could be highly relevant to supplies as well as to the global political economy. It is clear, however, that one of the most important of the influencing factors will be the relative cost of renewable energy and how quickly the world can switch to sustainable technologies. There is nothing to gain by deferring investment in this area, and everything to lose by postponing it any longer.

While salinity or salty water, is generally used to describe and measure seawater or certain industrial wastes, we use the term total dissolved solids ("TDS") to describe water high in various salt compounds and dissolved minerals. While one could have very high total dissolved solids, and very low salinity from a chemistry standpoint, here we are talking about high TDS. Total Dissolved Solids (TDS) refers to the amount of dissolved solids (typically various compounds of salts, minerals and metals) in a given volume of water. It is expressed in parts per million (also known as milligrams per liter) and is determined by evaporating a small amount of amount of water in the lab, and weighing the remaining solids. Another way to approximately determine TDS is by measuring the conductivity of a water sample and converting the resistance in micromhos to TDS. TDS in municipally-treated waters in our area range from 90 ppm to over 1000 ppm. The most common range on city water is 200 - 400 ppm. The maximum contaminant level set by USEPA is 500 ppm. California sets its standard as 1000 ppm, probably due to the high number of ground water sources in the state. The MCL is known as a Secondary Standard and in one sense, refers to the aesthetic quality of a a given water. The higher the TDS, the less palatable the water is thought to be. Sea water ranges from 30,000 to 40,000 ppm. Many brackish ground water supplies are used around California and we have many clients whose private well water has a TDS of 1500 - 2000 ppm. In some cases the levels exceed 7000 ppm. Generally, one wants a TDS of less than 500 for household use. In our experience, it appears that folks can tolerate for general household use, soft clean water with a TDS of up to 1500 ppm. When the levels start to exceed 1500 ppm, most people start to complain of dry skin, stiff laundry, and corrosion of fixtures. White spotting and films on surfaces and fixtures is also common at these levels and can be very difficult or impossible to remove.

TDS affects taste also, and waters over 500 - 600 ppm can taste poor. When the levels top 1500 ppm, most people will report the water tastes very similar to weak alka-seltzer. TDS is removed by distillation, reverse-osmosis or electrodialysis. In our area, most desalination projects, both large and small are accomplished with reverse-osmosis. Depending on the water chemistry, reverse osmosis systems are the most popular, given their low cost and ease of use. Distillers work very well also, and produce very high quality water, but require electricity and higher maintenance than reverse osmosis systems. For whole house treatment, commercial-sized reverse osmosis systems are usually the best approach. You will find a compilation of research and review articles at the end of this chapter that will provide you more in-depth information on each of the technologies covered.

RECOMMENDED RESOURCES FOR THE READER

1. Vigneswaran, S. and Ben Aim, R., Water, Wastewater and Sludge Filtration, CRC Press, Boca Raton, FL, pp 139-224, (1989).
2. Paulson, David J.; Wilson, Richard L.; and Spatz, D. Dean, "Crossflow Membrane Technology and Its Applications," *Food Technology 38 (12)* 77-87,&111(1984).
3. Mohr, C.M.; Leeper, S.A.; Engelgau, D.E.; and Charboneau, B.L., *Membrane Applications and Research in Food Processing,* (Noyes Data Corporation, Park Ridge, NJ) p. 305 (1989).
4. Morr, C.V., "Current Status of Soy Protein Functionality in Food Systems," Journal of thc American Oil Chemists' Society 67 (5) 265-27 (1990).
5. Lawhorn, J.T. and Lusas, E.W., "New Techniques in Membrane Processing of Oilseeds," *Food Technology* 38 (12) 95-106 (1984).
6. Lorchirachoonkul, S., *Diss. Abst Intem* B45 (1) 117 (1984).
7. Cheryan, Munir, *Ultrafiltration Handbook* (Technomic Publishing Company, Inc., Lancaster, PA) p. 374 (1986).
8. Paulson, David J., "An Overview of and Definitions for Membrane Fouling," Osmonics, Inc. Presented at 5th Annual Membrane Technology/Planning Confcrence, October 1987, Cambridge, MA.
9. Van den Berg, G.B.; Hanemajer, J.H. and Smolders, C.A., "Ultrafiltration of Protein Solutions; the Role of Protein Association in Rejection and Osmotic Pressure," Journal of Membrane Science, 31 (1987) 307-320.
10. Porter, Mark C., *Handbook of Industrial Membrane Technology,* 1990 (Noyes Publications, Park Ridge, NJ) p. 174.
11. Bailey, A.I., Mohammed, R.A., Luckham, P.F., Taylor, S.E., Dewatering of Crude Oil Emulsions. III - Emulsion Resolution by Chemical Means, *Colloids and Surfaces*, v.83, pp.261-271, 1994.
12. Bailey, A.I., Chen, T.Y., Mohammed, R.A., Luckham, P.F., Taylor, S.E. , Dewatering of Crude Oil Emulsions. IV - Emulsion Resolution by Electrical Means, *Colloids and Surfaces*, v.83, pp.273-284, 1994.
13. Clay, P.G., Adeleye, S.A., Olapido, M.O.A. , Sorption of Caesium,

Strontium and Europium Ions on Clay Minerals, *J. Materials Science*, v.29, pp.954-958, 1994.

14. Livingston, A.G., Brookes, P.R., Biological Detoxification of a 3-chloronitrobenzene Manufacture Wastewater in an Extractive Membrane Bioreactor, *Water Research*, v.28, pp.1347-1354, 1994.
15. Livingston, A.G., Extractive Membrane Bioreactors : A New Process Technology for Detoxifying Industrial Waste waters, *J. Chem. Tech. Biotech.*, v.60, pp.117-124, 1994.
16. Livingston, A.G., Freitas dos Santos, L.M., Extraction and Biodegradation of a Toxic Volatile Organic Compound (1,2 dichloroethane) from Wastewater in a Membrane Bioreactor, *Applied Microbiology and Biotechnology*, v.42, pp.421-431, 1994.
17. White, D.A., Rautiu, R., Adeleye, S.A., Adkins, L., The use of zeta potential measurements in inorganic ion exchange studies, *Hydrometallurgy*, v.35, pp.361-374, 1994.
18. White, D.A., Fathurrachman, Extraction of uranium (VI) and uranium (IV) from hydrochloric acid using tri-n-octylamine in a benzene diluent, *Hydrometallurgy*, v.36, pp.161-168, 1994.
19. Livingston, A.G., Brookes, P.R., Detoxification of chemical industry wastewaters using an extractive membrane bioreactor, *49th Annual Purdue Industrial Waste Conference*, May 9-11, 1994.
20. Livingston, A.G., Freitas dos Santos, L., Membrane attached biofilms for desctruction of volatile oragnic compounds (VOC), *Proc. 3rd Asia Pacific Biochemical Eng. Cong*, Singapore, 1994.
21. Stuckey, D.C., Caridis, K.A., Leak, D.J., Design of a novel bioreactor with cell recycle for continuous biotransformation and product extraction, *Proc. 3rd Asia Pacific Biochemical Eng. Conf.*, Singapore, pp.315-317, 1994.
22. Stuckey, D.C., Caridis, K.A., Leak, D.J., Kinetics of Mycobacterium M156 for chiral biotransformations, *Biotechnology '94, 2nd Int. Symp. on Applied Biocatalysis*, Brighton, pp.37-39, 1994.
23. Stuckey, D.C., Rosjidi, M., Leak, D.J., The downstream separation of chiral epoxides using colloidal liquid aphrons (CLAs), *"Separations for Biotechnology III"* ed. D.L. Pyle, pp.440-446, SCI Publishing, 1994.
24. Bailey, A.I., Cardenas-Valera, A.E., Doroszkowsi, A., Graft copolymers as stabilizers for oil-in-water emulsions. Part 1. Synthesis of the copolymers and their behaviour as monolayers spread at the air-water and oil-water interfaces, *Colloids and Surfaces*, v.96, pp.53-67, 1995.
25. Bailey, A.I., Cardenas-Valera, A.E., Graft copolymers as stabilizers for oil-in-water emulsions. Part 2. Preparation of the emulsions and the factors affecting their stability, *Colloids and Surfaces*, v.97, pp.1-12, 1995.
26. Freitas dos Santos, L.M., Biological treatment of VOC containing

wastewaters: novel extractive membrane bioreactor vs. conventional aerated bioreactor, *Process Safety and Environmental Protection*, v.3, part B, pp.227-234, 1995.

27. Livingston, A.G., Freitas dos Santos, L.M., Novel membrane bioreactor for detoxification of VOC wastewaters : biodegradation of 1,2 dichloroethane, *Water Research*, v.29, pp.179-194, 1995.

28. White, D.A., Assabumrungrat, S., Puttick, S., Sizing of inorganic microfiltration monoliths for water and air purification, *Trans.of the IChemE*, v.73B, pp.108-114, 1995.

29. Freitas dos Santos, L.M., Livingston, A.G., Membrane attached biofilms: novel technique for measurement of biofilm thickness, density and diffusivity, *Proc. IAWQ Conf. Workshop on Biofilm Structure, Growth and Dynamics - Need for New Concept?,* Noordwijkerjout, The Netherlands, August 1995.

30. Freitas dos Santos, L.M., Pavasant, P., Pistikopoulos, E.N., Livingston, A.G., Growth of immobilised cells: results and predictions for membrane-attached biofilms using a novel in situ biofilm thickness measurement technique, *Immobilised Cells: Basics and Applications*, Ed. Wijfells, R.H., Buitelaar, R.M., Bucke, C., Tramper, J., Progess in Biotech. 11, pp.290-297, 1996.

31. Livingston, A.G., Gikas, P., Viability of immobilised cells: use of specific ATP levels and oxygen uptake rates., *Immobilised Cells: Basics and Applications*, Ed. Wijfells, R.H., Buitelaar, R.M., Bucke, C., Tramper, J., Progess in Biotech. 11, pp.264-271, 1996.

32. Mendes-Tatsis, M.A., Agble, D., Mass transfer with interfacial convection and added surfactants, *Int. Solvent Extraction Conf. (ISEC '96) Value adding through solvent extraction*, Ed. Shallcross, D.C., Paimin, R. Prvcic, L.M., Melbourne, Australia, pp.267-272, 1996.

33. Perez de Ortiz, E.S., Dias Lay, M. De L., Gruentges, K., Aluminium and iron extraction by DNNSA and DNNSA-DEHPA reverse micelles, *Int. Solvent Extraction Conf. (ISEC '96) Value adding through solvent extraction*, Ed. Shallcross, D.C., Paimin, R. Prvcic, L.M., Melbourne, Australia, pp.409-411, 1996.

34. Perez de Ortiz, E.S., Henman, T.E., Hall, T., Le Febvre, W., Coagulant recycle from water treatment residuals: technical and economic feasibility of processes for purifying recovered coagulants, *Proc. of the 10th Annual Residuals and Biosolids Management Conference: 10 Years of Progress and Look Toward the Future*, Water Environment Federation, 10-1 to 10-8, 1996.

35. White, D.A., Helbig, J., Production of flocs for water treatment by electrodialysis in a cell fitted with an inorganic membrane, *Progress in Membrane Science & Technology Conf*, pp.35-36, Univ. of Twente, The Netherlands, 1996.

QUESTIONS FOR THINKING AND DISCUSSING

1. What should be the pressure across an ultrafiltration module, in order to achieve a permeate flux of 400 Liter/m^2-day. The clean membrane resistance is 2.4×10^{10} 1/m.

2. If the pressure drop across a tubular membrane is 2.8 bars, determine the permeat velocity across the membrane module. The thickness and the porosity of the deposit are 2 mm and 40 %, respectively. The average diameter of the partices is 5 microns. The initial membrane resistance is estimated to be 1.7×10^{10} 1/m.

3. The energy requirements for the desalination of brackish waster based on RO is typcially around 9 kWh/m^3. Develop a simplified calculation basis for multistage distillation and compare the energy requirements for the two techniques. All things being considered equal, which of these approaches might best be described as pollution prevention technology. The following process flow scheme applies to a distillation operation.

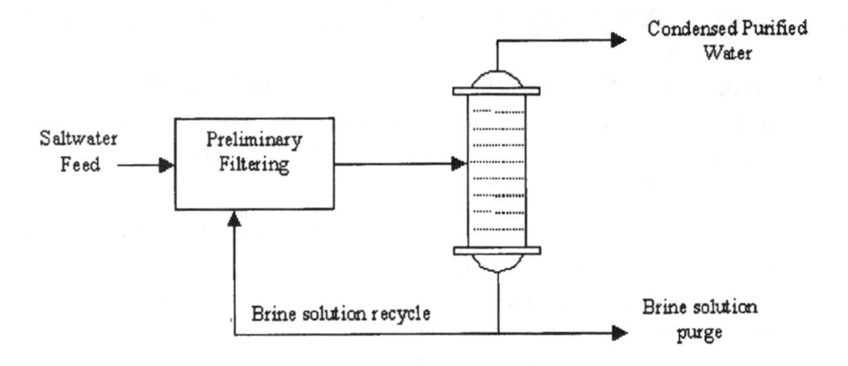

4. Explain the various mechanisms that result in fouling membrane surfaces.

5. Consider a typical electrodialysis process. What will happen if the electrodes on either side are interchanged? Explain the performance of the modified system.

6. Compute the osmotic pressure for a typical sea water.

7. An electrodialysis cell has the following dimensions (110 cm $\times$ 60 cm $\times$ 0.04 cm (thickness), and is used to treat water with a throughput velocity of 10 cm/sec. The product concentration is 0.0092 eq/Liter. The cell current efficiency is 0.892. Resistance across the cell is 0.205 ohm. The influent concentration is 125 mg/Liter of NaCl. Calculate the following: (a) cell current, (b) cell power output, (c) the cell voltage, and (d) the energy consumption per equivalent of product transferred.

8. The following data was obtained from a series of pilot tests for a microfiltration process, in which the incoming particles have an average size of 3 microns. Determine the following: (a) Prepare a plot of time versus permeate flow and deposit resistance. Examine these results and establish a basis for recommending when the process should be backwashed. Discuss these recommendations and your criteria; (b) Calculate the time averaged permeate flux between the start of filtration till the time of backwashing. Assume that this time averaged permeate flux will remain constant for the remainder of the runs, and compute the membrane area required for treating a surface water source of capacity 1 Mliter/day (Hint: apply Simpson's rule to performing a curve fitting of the data); (c) If the average particle size of particles in the influent is 8 microns, compute the flux at 35 minutes. Compare this flux with the corresponding flux obtained with the 3 micron particles. Discuss the implication of these results in a commercial water treatment facility.

Time, min.	Deposit porosity	Deposit thickness, cm
5	0.5	0.10
10	0.45	0.15
15	0.41	0.20
20	0.34	0.21
25	0.32	0.23
30	0.30	0.24
35	0.29	0.25

9. Develop some pollution prevention options for the disposal of brine or salt solution derived from a desalting operation.

Chapter 10

ION EXCHANGE AND CARBON ADSORPTION

INTRODUCTION

Ion exchange and carbon adsorption are unrelated technologies, and often have different objectives. They are however oftentimes used in compliment to achieve high water quality attributes.

Ion exchange is a reversible chemical reaction wherein an ion (an atom or molecule that has lost or gained an electron and thus acquired an electrical charge) from solution is exchanged for a similarly charged ion attached to an immobile solid particle. These solid ion exchange particles are either naturally occurring inorganic zeolites or synthetically produced organic resins. The synthetic organic resins are the predominant type used today because their characteristics can be tailored to specific applications. An organic ion exchange resin is composed of high-molecular-weight polyelectrolytes that can exchange their mobile ions for ions of similar charge from the surrounding medium. Each resin has a distinct number of mobile ion sites that set the maximum quantity of exchanges per unit of resin. The industry application most familiar with ion exchange technology is metal plating. Most plating process water is used to cleanse the surface of the parts after each process bath. To maintain quality standards, the level of dissolved solids in the rinse water must be regulated. Fresh water added to the rinse tank accomplishes this purpose, and the overflow water is treated to remove pollutants and then discharged. As the metal salts, acids, and bases used in metal finishing are primarily inorganic compounds, they are ionized in water and could be removed by contact with ion exchange resins. In a water deionization process, the resins exchange hydrogen ions ($H+$) for the positively charged ions (such as nickel. copper, and sodium). and hydroxyl ions ($OH-$) for negatively charged sulfates, chromates. and chlorides. Because the quantity of $H+$ and OH ions is balanced, the result of the ion exchange treatment is relatively pure, neutral water. Ion exchange technology is applied in many other industry sectors, including the petroleum and chemical industries, as well as general wastewater treatment applications. The technology is most often compared to reverse osmosis, since both technologies are often aimed at similar objectives. In this regard, in addition to discussing ion exchange as a technology, we will also review some of the operational tradeoffs and economics of the two processes in this chapter.

The history of carbon adsoprtion in the pruification of water dates back to ancient times. Adsorption on porous carbons was described as early as 1550 B.C. in an ancient Egyptian papyrus and later by Hippocrates and Pliny the Elder, mainly for medicinal purposes. In the 18th century, carbons made from blood, wood and animals were used for the purification of liquids. All of these materials, which can be considered as precursors of activated carbons, were only available as powders. The typical technology of application was the so-called batch contact treatment, where a measured quantity of carbon and the liquid to be treated were mixed and, after a certain contact time, separated by filtration or sedimentation. At the beginning of the 19th century the decolourisation power of bone char was detected and used in the sugar industry in England. Bone char was available as a granular material which allowed the use of percolation technology, where the liquid to be treated was continuously passed through a column. Bone char, however, consists mainly of calcium phosphate and a small percentage of carbon; this material, therefore was only used for sugar purification. At the beginning of the 20th century the first processes were developed to produce activated carbons with defined properties on an industrial scale. However, the steam activation and chemical activation processes could only produce powder activated carbon. During the First World War, steam activation of coconut char was developed in the United States for use in gas masks. This activated carbon type contains mainly fine adsorption pore structures suited for gas phase applications.

After World War II technology advances were made in developing coal based granular activated carbons with a substantial content of transport pore structure and good mechanical hardness. This combination allowed the use of activated carbon in continuous decolourisation processes resulting superior performance. In addition optimization of granular carbon reactivation was achieved. Today many users are switching from the traditional use of powdered activated carbon as a disposable chemical to continuous adsorption processes using granular activated carbon combined with reactivation. By this change they are following the modern tendency towards recycling and waste minimization, thereby reducing the use of the world's resources. In this chapter we will explore the use of activated carbon in standard water treatment applications.

This overview will familiarize with the technology, which can be used as standalone, or in conjunction with other technologies such as RO, ion exchange and others. Although not discussed per se, this technology has found historical use in more recent times in many groundwater remediation projects. It is most often thought of as the workhorse in groundwater pump-and-treat applications.

Remember to refer to the ***Glossary*** at the end of the book if you run across any terms that are unfamiliar to you.

THEORY AND PRACTICE OF ION EXCHANGE

Water can contain varying concentrations of dissolved salts which dissociate to form charged particles called ions. These ions are the positively charged cations and negatively charged anions that permit the water or solution to conduct electrical currents and are therefore called electrolytes. Electrical conductivity is thus a measure of water purity, with low conductivity corresponding to a state of high purity. The process of ion exchange is uniquely suited to the removal of ionic species from water supplies for several reasons. First, ionic impurities may be present in rather low concentrations. Second, modern ion-exchange resins have high capacities and can remove unwanted ions preferentially. Third, modern ion-exchange resins are stable and readily regenerated, thereby allowing their reuse. Other advantages ion exchange offers are: (1) the process and equipment are a proven technology. Designs are well developed into pre-engineered units that are rugged and reliable, with well-established applications; (2) fully manual to completely automatic units are available; (3) there are many models of ion-exchange systems on the market which keep costs competitive; (4) temperature effects over a fairly wide range (from 0° to 35° C) are negligible; (5) the technology is excellent for both small and large installations, from home water softeners to large utility/industrial applications.

Ion exchange is a well-known method for softening or for demineralizing water. Although softening could be useful in some instances, the most likely application for ion exchange in wastewater treatment is for demineralization. Many ion-exchange materials are subject to fouling by organic matter. It is possible that treatment of secondary effluent for suspended-solids removal and possibly soluble organic removal will be required before carrying out ion exchange. Many natural materials and, more importantly, certain synthetic materials have the ability to exchange ions from an aqueous solution for ions in the material itself. Cation-exchange resins can, for example, replace cations in solution with hydrogen ions. Similarly, anion-exchange resins can either replace anions in solution with hydroxyl ions or absorb the acids produced from the cation-exchange treatment. A combination of these cation-exchange and anion-exchange treatments results in a high degree of demineralization.

Since the exchange capacity of ion-exchange materials is limited, they eventually become exhausted and must be regenerated. The cation resin is regenerated with an acid; the anion resin is regenerated with a base. Important considerations in the economics of ion exchange are the type and amounts of chemicals needed for regeneration. Often, water to be demineralized is first passed through a cation-exchange material requiring a strong acid, usually sulfuric, for regeneration. The exchange material is referred to as *strong acid resin.* The amount of acid regenerant is somewhat more than the stoichiometric amount, possibly 100 percent excess or more. If sulfuric acid is the regenerating acid, a waste brine is produced consisting of sulfates of the various actions in the water being treated. Because the partially treated water contains mineral acids, it is common to pass it next through an acid-absorbing resin or *weak base resin.* This resin can be regenerated with either a weak or strong base. The efficiency of regenerant use is quite high with these resins. If sodium hydroxide is the regenerating base, a waste brine is produced consisting of the sodium salts of the various anions in the water being treated. Certain anionic materials are not removed by the weak-base resin and must be further treated with strong-base resin if thorough demineralization is desired. Regenerant usage by the strong-base resins is poorer than for the weak base resins. The reasons for applying this technology in the removal of mineral species should be quite apparent to those of you who work with applications involving heat exchange. Water problems in cooling, heating, steam generation, and manufacturing are caused in large measure from the kinds and concentrations of dissolved solids, dissolved gases, and suspended matter in the makeup water supplied. Table 1 lists the major objectionable ionic constituents present in many water supplies that can be removed by demineralization. Prevention of scale and other deposits in cooling and boiling waters is best accomplished by removal of dissolved solids. Whereas in municipal water purification such removal is limited to the partial reduction of hardness and the removal of iron and manganese, in industrial water treatment it is often carried much further and may include the complete removal of hardness, the reduction or removal of alkalinity, the removal of silica, or even the complete removal of all dissolved solids.

Table 1. Common Ionic Constituents Contained in Water.

Constituent of concern	Chemical designation	Resultant problems
Hardness	Calcium and magnesium salts in the forms of $CaCO_2$, Ca, Mg.	This is the primary source of scaling in heat exchange equipment, boilers, pipelines/transfer lines, etc. Tends to form curds with soap and interferes with dyeing applications as well.
Alkalinity	Bicarbonate (HCO_3), carbonate (CO_3), and hydrate (OH), expressed as $CaCO_3$.	Causes foaming and carryover of solids with steam. Can cause embrittlement of boiler steel. Biocarbonate and carbonate generate CO_2 in steam, a source of corrosion.

Constituent of concern	Chemical designation	Resultant problems
Free mineral acidity	H_2SO_4, HCl, and other acids, expressed as $CaCO_3$.	Causes rapid corrosion and deterioration of surfaces.
Chloride	Cl^-	Interferes with silvering processes and increase TDS.
Sulfates	$(SO_4)^=$	Results in the formation of calcium sulfate scale.
Iron and manganese	Fe^{+2} (ferrous) Fe^{+3} (ferric) Mn^{+2}	Discolors water, and results in the formation of deposits in water lines, boilers and other heat exchangers. Can interfere with dying, tanning, paper manufacture and various process works.
Carbon dioxide	CO_2	Results in the corrosion of water lines, especially steam and condensate lines.
Silica	SiO_2	Results in the formation of scale in boilers and cooling water systems. can produce insoluble scale on turbine blades due to silica vaporization in high pressure boilers (usuallu over 600 psi).

The two most frequently encountered water problems-scale formation and corrosion-are common to cooling, heating, and steam-generating systems. Hardness (calcium and magnesium), alkalinity, sulfate, and silica all form the main source of scaling in heat-exchange equipment, boilers, and pipes. Scales or deposits formed in boilers and other exchange equipment act as insulation, preventing efficient heat transfer and causing boiler tube failures through overheating of the metal. Free mineral acids (sulfates and chlorides) cause rapid corrosion of boilers, heaters, and other metal containers and piping. Alkalinity causes embrittlement of boiler steel, and carbon dioxide and oxygen cause corrosion, primarily in steam and condensate lines. Low-quality steam can produce undesirable deposits of salts and alkali on the blades of steam turbines; much more difficult to remove are silica deposits which can form on turbine blades even when steam is satisfactory by ordinary standards. At steam pressures above 600 psi, silica from the boiler water actually dissolves in the gaseous steam and then reprecipitates on the turbine blades at their lower-pressure end.

In the operation of every cooling, heating, and steam-generating plant, the water changes temperature. Higher temperatures, of course, increase both corrosion rates and scale-forming tendency. Evaporation in process steam boilers and in evaporative cooling equipment increases the dissolved-solids concentration of the water, compounding the problem.

In addition to the formation of scale or corrosion of metal within boilers, auxiliary equipment is also susceptible to similar damage. Attempts to prevent scale formation within a boiler can lead to makeup line deposits if the treatment chemicals are improperly chosen. Thus, the addition of normal phosphates to an unsoftened feed water can cause a dangerous condition by clogging the makeup line with precipitated calcium phosphate. Deposits in the form of calcium or magnesium stearate deposits, otherwise known as "bathtub ring" can be readily seen, and are caused by the combination of calcium or magnesium with negative ions of soap stearates.

HOW ION EXCHANGE WORKS

Ion exchangers are materials that can exchange one ion for another, hold it temporarily, and then release it to a regenerant solution. In a typical demineralizer, this is accomplished in the following manner: The influent water is passed through a hydrogen cation-exchange resin which converts the influent salt (e.g., sodium sulfate) to the corresponding acid (e.g., sulfuric acid) by exchanging an equivalent number of hydrogen (H^+) ions for the metallic cations (Ca^{+2}, Mg^{+2}, Na^+). These acids are then removed by passing the effluent through an alkali regenerated anion-exchange resin which replaces the anions in solution (Cl^-, $SO_4^=$, NO_3^-) with an equivalent number of hydroxide ions. The hydrogen ions and hydroxide ions neutralize each other to form an equivalent amount of pure water. During regeneration, the reverse reaction takes place. The cation resin is regenerated with either sulfuric or hydrochloric acid and the anion resin is regenerated with sodium hydroxide. Figure 1 illustrates a basic scheme for ion exchange demineralization.

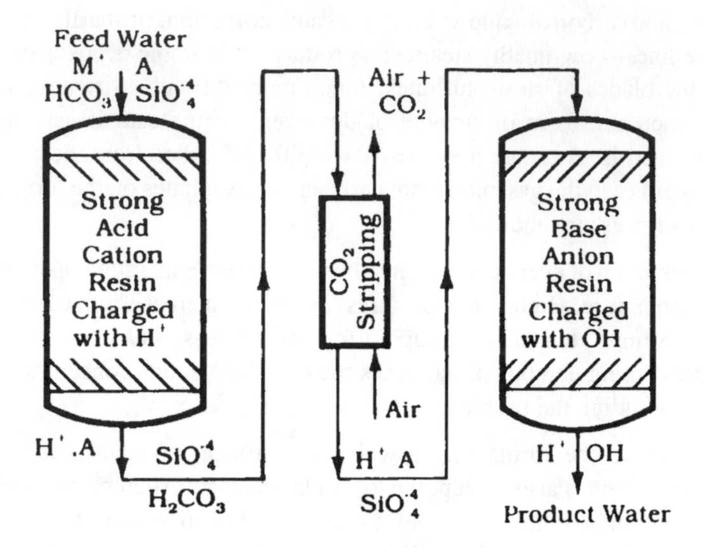

Figure 1. Shows ion exchange demineralization scheme.

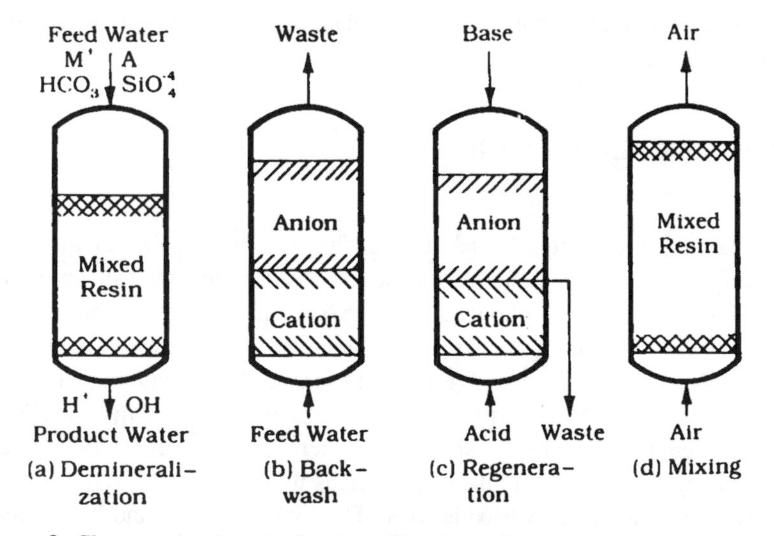

Figure 2. Shows mixed resin demineralization scheme.

There are various arrangements or equipment possible but in all cases, except in mixed-bed demineralization, the water should first pass through a cation exchanger. In mixed demineralization, the two exchange materials (that is, the cation-exchange resin and the anion-exchange resin) are placed in one shell instead of two separate shells. In operation, the two types of exchange materials are thoroughly mixed so that we have, in effect, a number of multiple demineralizers in series. Higher-quality water is obtained from a mixed-bed unit than from a two-bed system. (see Figure 2 for an example). Operation of cation and anion exchanges is shown in Figure 3 (for fundamental processes) and Figure 4 (operation modes for both cation/anion exchanges).

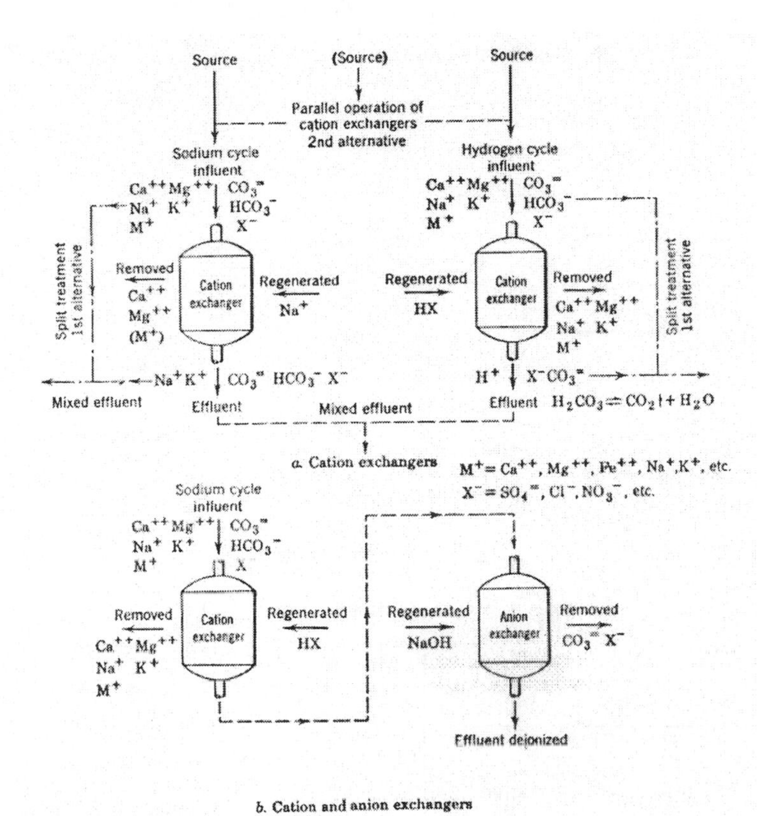

Figure 3. Operational schemes of ion exchange.

To be suitable for industrial use, an ion-exchange resin must exhibit durable physical and chemical characteristics which are summarized by the following properties.

Functional Groups - The molecular structure of the resin is such that it must contain a macroreticular tissue with acid or basic radicals. These radicals are the basis of classifying ion exchangers into two general groups: (1) Cation exchangers, in which the molecule contains acid radicals of the HSO_3 or HCO type able to fix mineral or organic cations and exchange with the hydrogen ion H^+; (2) Anion exchangers, containing basic radicals (for example, amine functions of the type NH_2) able to fix mineral or organic anions and exchange them with the hydroxyl ion OH^- coordinate to their dative bonds. The presence of these radicals enable a cation exchanger to be assimilated to an acid of form H-R and an anion exchanger to be a base of form OH-R when regenerated.

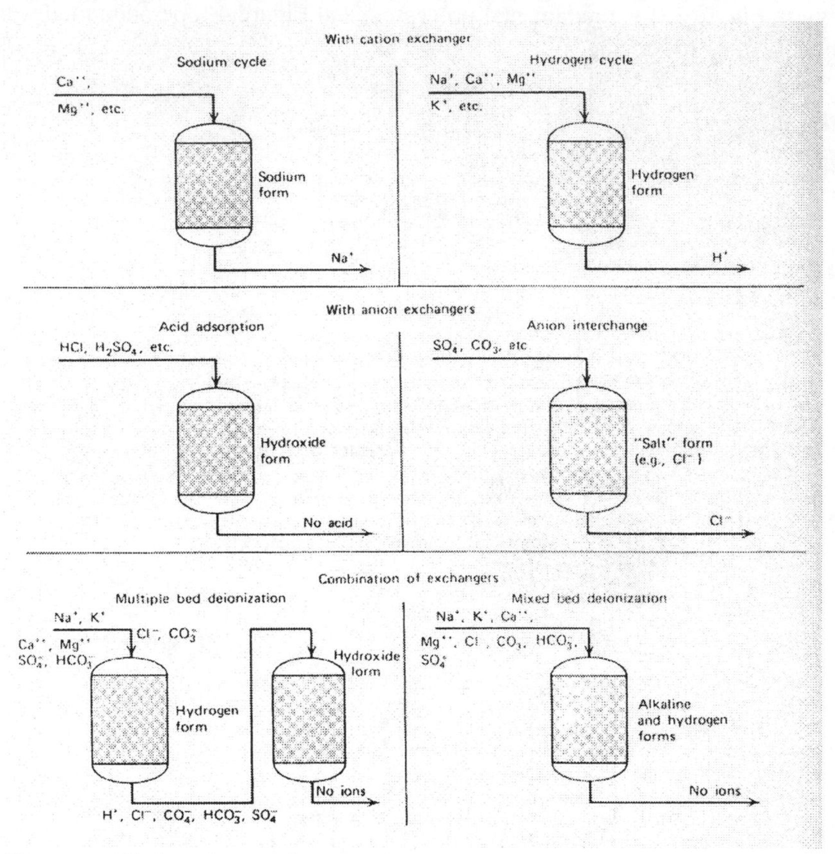

Figure 4. Various ion exchange schemes.

These radicals act as immobile ion-exchange sites to which are attached the mobile cations or anions. For example, a typical sulfonic acid cation exchanger has immobile ion-exchange sites consisting of the anionic radicals SO to which are attached the mobile cations, such as H^+ or Na^+. An anion exchanger similarly has immobile cationic sites to which are attached mobile (exchangeable) hydroxide anions OR The radicals attached to the molecular nucleus further determine the nature of the acid or base, whether it will be weak or strong. Exchangers are divided into four specific classifications depending on the kind of radical, or functional group, attached; strong acid, strong base, weak acid, or weak base. Each of these four types of ion exchangers is described in detail below.

Solubility - The ion-exchange substance must be insoluble under normal conditions of use. Most ion-exchange resins in current use are high molecular weight polyacids or polybases which are virtually insoluble in most aqueous and nonaqueous media. This is no longer true of some resins once a certain temperature has been reached. For example, some anion-exchange resins are limited to a maximum temperature of 105° F. Liquid ion-exchange resins exist also, yet we do not consider their applicability here and they also exhibit very limited solubility in aqueous solutions.

Bead Size - The resins must be in the form of spherical granules of maximum homogeneity and dimensions so that they do not pack too much, the void volume among their interstices is constant for a given type, and the liquid head loss in percolation remains acceptable. Most ion-exchange resins occur as small beads or granules usually between 16 and 50 mesh in size.

Resistance to Fracture - The ion or ionized complexes that the resins are required to fix are of varied dimensions and weights. The swelling and contraction of the resin bead that this causes must obviously not cause the grains to burst.

Another important factor is the bead resistance to osmotic shock which will inevitably occur across its boundary surface, as there will be a salinity gradient of different magnitude during the cycle of the exchange material. The design of ion-exchange apparatus also must take into consideration the safe operation of the ion-exchange resin and avoid excessive stresses or mechanical abrasion in the bed, which could lead to breakage of the beads.

TYPES OF RESINS

As noted earlier, ion-exchange materials are grouped into four specific classifications depending on the functional group attached; strong-acid cation, strong-base anion, weak-acid cation, or weak-base anion.. In addition to these, we also have inert resins that do not have chemical properties.

Strongly Acidic Cation Resins

Strongly acidic cation resins derive their exchange activity from sulfonic functional groups (HSO). The major cations in water are calcium, magnesium, sodium, and potassium and they are exchanged for hydrogen in the strong acid cation exchanger when operated in the hydrogen cycle. The following stoichiometric equation represents the exhaustion phase, and is written in the molecular form (as if the salts present were undissociated). It shows the cations in combination with the major anions, the bicarbonate, sulfate, and chloride anions:

$$\begin{matrix} Ca & 2HCO_2 & & Ca & 2H_2CO_3 \\ Mg & \cdot SO4 + 2RSO_3H \leftrightarrow & 2RSO_3 & Mg + & H_2SO_4 \\ Na & 2Cl & & Na & 2HCl \end{matrix}$$

Note that R represents the complex resin matrix. Because these equilibrium reactions are reversible, when the resin capacity has been exhausted it can be recovered through regeneration with a mineral acid. The strong-acid exchangers operate at any pH, split strong or weak salts, require excess strong-acid regenerant (typical regeneration efficiency varies from 25 percent to 45 percent in concurrent regeneration), and they permit low leakage. In addition, they have rapid exchange rates, are stable, exhibit swelling less than 7 percent going from Na^+ to H^+ form, and may last 20 years or more with little loss of capacity.

These resins have found a wide range of application, being used on the sodium cycle for softening, and on the hydrogen cycle for softening, dealkalization, and demineralization.

Weakly Acidic Cation-exchange Resins

Weakly acidic cation-exchange resins have carboxylic groups (COOH) as the exchange sites. When operated on the hydrogen cycle, the weakly acidic resins are capable of removing only those cations equivalent to the amount of alkalinity present in the water, and most efficiently the hardness (calcium and magnesium) associated with alkalinity, according to these reactions:

$$\left.\begin{matrix} Ca \\ Mg \\ 2Na \end{matrix}\right\} (HCO_3) + RCOOH \leftrightarrow 2RCOO \begin{matrix} Ca \\ Mg \\ 2Na \end{matrix} + H_2CO_4$$

These reactions are also reversible and permit acid regeneration to return the exhausted resin to the hydrogen form. The resin is highly efficient, for it is regenerated with 110 percent of the stoichiometric amount of acid as compared to 200 to 300 percent for strong-acid cation-exchange resins. It can be regenerated with the waste acid from a strong-acid cation exchanger and there is little waste problem during the regeneration cycle. In order to prevent calcium sulfate precipitation when regenerated with H_2SO_4, it is usually regenerated stepwise with initial H_2SO_4 at 0.5 percent. The resins are subject to reduced capacity from increasing flow rate (above 2 gpm/ft), low temperatures, and/or a hardness-alkalinity ratio especially below 1.0.

Weakly acidic resins are used primarily for softening and dealkalization, frequently in conjunction with a strongly acidic polishing resin. Systems which use both resins profit from the regeneration economy of the weakly acidic resin and produce treated water of quality comparable to that available with a strongly acidic resin.

Strongly Basic Anion-exchange Resins

Strongly basic anion-exchange resins derive their functionality from the quaternary ammonium exchange sites. All the strongly basic resins used for demineralization purposes belong to two main groups commonly known as type I and type 11. The principal difference between the two resins, operationally, is that type I has a greater chemical stability, and type II has a slightly greater regeneration efficiency and capacity. Physically, the two types differ by the species of quaternary ammonium exchange sites they exhibit. Type I sites have three methyl groups, while in type 11, an ethanol group replaces one of the methyl groups. In the hydroxide form, the strongly basic anion will remove all the commonly encountered inorganic acids according to these reactions:

$$\left.\begin{matrix} H_2SO_4 \\ 2HCl \\ H_2SiO_3 \\ H_2CO_3 \end{matrix}\right\} + 2ZOH \rightarrow \begin{matrix} SO_4 \\ 2Cl \\ 2HSiO3 \\ 2HCO_3 \end{matrix} + H_2O$$

Like the cation resin reactions, the anion-exchange reactions are also reversible and regeneration with a strong alkali, such as caustic soda, will return the resin to the hydroxide form.

The strong-base exchangers operate at any pH, can split strong or weak salts, require excess high-grade NaOH for regeneration (with the typical efficiency varying from 18 percent to 33 percent), are subject to organic fouling from such compounds when present in the raw water and to resin degradation due to oxidation and chemical breakdown. The strong-base anion resins suffer from capacity decrease and silica leakage increase at flow rates above 2 gpm/ft^3 of resin, and cannot operate over 130° to 150° F depending on resin type. The normal maximum continuous operating temperature is 120° F, and to minimize silica leakage, warm caustic (up to 120° F) should be used. Type I exchangers are for maximum silica removal. They are more difficult to regenerate and swell more (from Cl to OH form) than type Il. The major case for selecting a type I resin is where high operating temperatures and/or very high silica levels are present in the influent water or superior resistance to oxidation or organics is required.

Type II exchangers remove silica (but less efficiently than type I) and other weak anions, regenerate more easily, are less subject to fouling, are freer from the odor of amine, and are cheaper to operate than type I. Where free mineral acids are the main constituent to be removed and very high silica removal is not required, type II anion resin should be chosen.

Weakly Basic Anion Resins

Weakly basic anion resins derive their functionality from primary (R-NH), secondary (R-NHR'), tertiary (R-N-R'2), and sometimes quaternary amine groups. The weakly basic resin readily absorbs such free mineral acids as hydrochloric and sulfuric, and the reactions may be represented according to the following:

$$H_2SO_4 + 2ZOH = 2ZSO_4 + 2HO$$

$$2HCl \qquad\qquad 2Cl$$

Because the preceding reactions are also reversible, the weakly basic resins can be regenerated by applying caustic soda, soda ash, or ammonia. The weak-base exchanger regenerates with a nearly stoichiometric amount of base (with the regeneration efficiency possibly exceeding 90 percent) and can utilize waste caustic following strong-base anion-exchange resins. Weakly basic resins are used for high strong-acid waters (Cl, SO_4, NO_3), and low alkalinity, do not remove anions satisfactorily above pH 6, do not remove CO or silica, but have capacities about twice as great as for strong-base exchangers. Weak-base resins can be used to precede a strong-base anion resin to provide the maximum protection of the latter against organic fouling and to reduce regenerant costs. Make note of the sidebar discussions. These summarize for you the important sodium cation-exchanger reactions. Note in the sidebar discussions that Z denotes the sodium cation exchanger. An important term we should make note of in reviewing these stocihiometric relations is *Compensated Hardness*. What this refers to is that the hardness of the water for softening by the zeolite process should be compensated when: (1) the *total hardness* (TH) is greater than 400 ppm as $CaCO_3$, or (2) the sodium salts (Na) are over 100 ppm as $CaCO_3$. Compensated hardness can be calculated from the following formula:

Sodium Cation-Exchanger Reactions for Regeneration

[Ca, Mg] Z + 2NaCl = Na_2Z + [Ca, Mg] Cl_2

Calcium and/or Magnesium with Cation Exchanger (Insoluble) + Sodium Chloride (Soluble) = Sodium Cation Exchanger (Insoluble) + Calcium and/or Magnesium Chlorides (Soluble).

Compensated Hardness (ppm) =

TH (ppm) × {9,000/(9,000 - Total Cations (ppm))}

Express compensated hardness according to the following:

- Next higher tenth of a grain up to 0.5 grains per gallon;
- Next higher half of grain from 5.0 to 10.0 grains per gallon;
- Next higher grain above 10.0 grains per gallon.

The salt consumption with a sodium cation exchange water softener ranges between 0.275 and 0.533 Lbs of salt per 1,000 grains of hardness, expressed as calcium carbonate, removed. This range is attributed to two factors: (1) the water composition, and (2) the operating exchange value at which the exchange resin is to be worked. The lower salt consumption may be attained with waters that are not excessively hard nor high in sodium salts, and where the exchange resin is not worked at its maximum capacity.

Sodium Cation-Exchanger Reactions for Softening

$$Na_2Z + [Ca, Mg][(HCO_3)_2, SO_4, Cl_2] = [Ca, Mg]Z + [2NaSO_4, 2NaCl]$$

Sodium Cation Exchanger (Insoluble) + Calcium and/or Magnesium with Bicarbonates, Sulfates and/or Chlorides (Soluble) = Calcium and/or Magnesium Cation Exchnager (Insoluble) + (Sodium Bicarbonate, Sodium Sulfate and/or Sodium Chloride (Soluble).

The next group of sidebar discussion summarize for you the important reactions for hydrogen cation exchange resins. The symbol Z denotes the hydrogen cation-exchanger resin. Three groups of reactions are summarized. these are the reactions with bicarbonates, the reactions with sulfates or chlorides, and finally regeneration reactions. The reactions are fairly straightforward, and if you work regularly with water softening systems, it's best to try and memorize these reactions.

Reactions for Hydrogen Cation-Exchanger Resins - Regeneration Reactions

$$[Ca, Mg, Na_2]Z + H_2SO_4 = H2Z + [Ca, Mg, Na_2]SO_4$$

Calcium, Magnesium and/or Sodium-Cation Exchanger (Insoluble) + Sulfuric Acid (Soluble) = Hydrogen Cation Exchanger (Insoluble) + Calcium, Magnesium and/or Sodium Sulfates (Soluble).

Inert Resins

There also exists a type of resin with no functional groups attached. This resin offers no capacity to the system but increases regeneration efficiency in mixed-bed exchangers. These inert resins are of a density between cation and anion resins and when present in

Reactions for Hydrogen Cation-Exchanger Resins - Reactions with Sulfates or Chlorides

[Ca, Mg, Na_2] [SO_4, Cl_2] + H_2Z = [Ca, Mg, Na_2]Z + H_2SO_4 or HCl

Calcium, Magnesium and/or Sodium as Sulfates and/or Chlorides (Soluble) + Hydrogen Cation Exchanger (Insoluble) = Calcium, Magnesium and/or Sodium-Cation Exchanger (Insoluble) + Sulfuric Acid and/or Hydrochloric Acid (Soluble).

mixed-bed vessels help to separate cation and anion resins during backwash. Advantages of inert resins include: (1) Classify cation and anion resins so that little or no mixing of cation or anion resin occurs before regeneration, and a buffering mid-bed collection zone exists; (2) Improve regeneration efficiency, thereby reducing resin quantities needed; (3) Protect against osmotic shock since the inert layer effectively prevents the exposure of cation resin to the caustic regenerant solution and the exposure of anion resin to the acid regenerant solution.

ION EXCHANGE SOFTENING (SODIUM ZEOLITE SOFTENING)

This is one of the ion-exchange processes used in water purification. In this process, sodium ions from the solid phase are exchanged with the hardness ions from the aqueous phase. Consider a bed of ion-exchange resin having sodium as the exchangeable ion, with water containing calcium and magnesium hardness allowed to percolate through this bed. Let us denote the ion-exchange resinous material as RNa, where R stands for resin matrix and Na is its mobile exchange ion. The hard water will exchange Ca. and Mg ions rapidly, so that water at the effluent will be almost completely softened. Calcium and magnesium salts will be converted into corresponding sodium salts.

The reaction will proceed toward the right-hand side to its completion until the bed gets completely exhausted or saturated with Ca. and Mg ions. In order to reverse the equilibrium so that the reaction proceeds toward the left-hand side, the concentration of sodium ions has to be increased. This increase in sodium ions is accomplished by using a brine solution of sufficient strength so that the total sodium ions present in the brine are more than the total equivalent of Ca and Mg in the exhausted bed. This reverse reaction is carried out in order to bring the exhausted resin back to its -sodium form. This process is known as *regeneration*. When the softener with the fresh resin in sodium form is put in service, the sodium ions in the surface layer of the bed are immediately exchanged with calcium and magnesium, thereby producing soft water with very little residual hardness in the effluent. As the process continues, the resin bed keeps exchanging its sodium ions with calcium and magnesium ions until the hardness concentration increases rapidly and the softening run is ended.

This softening process can be extended to a point where the hardness coming in and going out is the same. When this condition is reached, the bed is completely exhausted and does not have any further capacity to exchange ions. This capacity is called the *total breakthrough capacity*. In practice, the softening process is never extended to reach this stage as it is ended at some predetermined effluent hardness, much lower than the influent hardness. This capacity is called the *operating exchange capacity*. After the resin bed has reached this capacity, the resin bed is regenerated with a brine solution.

The regeneration of the resin bed is never complete. Some traces of calcium and magnesium remain in the bed and are present in the lower-bed level. In the service run, sodium ions exchanged from the top layers of the bed form a very dilute regenerant solution which passes through the resin bed to the lower portion of the bed. This solution tends to leach some of the hardness ions not removed by previous regeneration. These hardness ions appear in the effluent water as *leakage*. Hardness leakage is also dependent on the raw water characteristics. If the Na/Ca ratio and calcium hardness are very high in the raw water, leakage of the hardness ions will be higher.

Reactions for Hydrogen Cation-Exchanger Resins - Reactions with Bicarbonates

$$[Ca, Mg, Na_2] + H_2Z = [Ca, Mg, Na_2]\,Z + 2H_2O + 2CO_2$$

Calcium, Magnesium and/or Sodium Bicarbonates (Soluble) + Hydrogen Cation Exchanger (Insoluble) = Calcium, Magnesium and/or Sodium Cation Exchanger (Insoluble) + Water + Carbon Dioxide (Soluble Gas).

RESIN PERFORMANCE

Variances in resin performance and capacities can be expected from normal annual attrition rates of ion-exchange resins. Typical attrition losses that can be expected include: (1) Strong cation resin: 3 percent per year for three years or 1,000,000 gals/ cu.ft; (2) Strong anion resin: 25 percent per year for two years or 1,000,000 gals/ cu.ft; (3) Weak cation/anion: 10 percent per year for two years or 750,000 gals/ cu. ft. A steady falloff of resin-exchange capacity is a matter of concern to the operator and is due to several conditions:

Improper backwash. Blowoff of resin from the vessel during the backwash step can occur if too high a backwash flow rate is used. This flow rate is temperature dependent and must be regulated accordingly. Also, adequate time must be allotted for backwashing to insure a clean bed prior to chemical injection.

Channeling. Cleavage and furrowing of the resin bed can be caused by faulty operational procedures or a clogged bed or underdrain. This can mean that the solution being treated follows the path of least resistance, runs through these furrows, and fails to contact active groups in other parts of the bed.

Incorrect chemical application. Resin capacities can suffer when the regenerant is applied in a concentration that is too high or too low. Another important parameter to be considered during chemical application is the location of the regenerant distributor. Excessive dilution of the regenerant chemical can occur in the vessel if the distributor is located too high above the resin bed. A recommended height is 3 inches above the bed level.

Mechanical strain. When broken beads and fines migrate to the top of the resin bed during service, mechanical strain is caused which results in channeling, increased pressure drop, or premature breakthrough. The combination of these resulting conditions leads to a drop in capacity.

Resin Fouling. In addition to the physical causes of capacity losses listed previously, there are a number of chemically caused problems that merit attention, specifically the several forms of resin fouling that may be found.

Organic fouling occurs on anion resins when organics precipitate onto basic exchange sites. Regeneration efficiency is then lowered, thereby reducing the exchange capacity of the resin. Causes of organic fouling are fulvic, humic, or tannic acids or degradation products of DVB (divinylbenzene) cross linkage material of cation resins. The DVB is degraded through oxidation and causes irreversible fouling of downstream anion resins.

Iron fouling is caused by both forms of iron ions; the insoluble form will coat the resin bead surface and the soluble form can exchange and attach to exchange sites on the resin bead. These exchanged ions can be oxidized by subsequent cycles and precipitate ferric oxide within the bead interior.

Silica fouling is the accumulation of insoluble silica on anion resins. It is caused by improper regeneration which allows the silicate (ionic form) to hydrolyze to soluble silicic acid which in turn polymerizes to form colloidal silicic acid with the beads. Silica fouling occurs in weak-base anion resins when they are regenerated with silica-laden waste caustic from the strongbase anion resin unless intermediate partial dumping is done.

Microbiological fouling (MB) becomes a potential problem when microbic growth is supported by organic compounds, ammonia, nitrates, and so on which are concentrated on the resin. Signs of MB fouling are increased pressure drops, plugged distributor laterals, and highly contaminated treated water.

Calcium sulfate fouling occurs when sulfuric acid is used to regenerate a cation exchanger after exhaustion by a water high in calcium. The precipitate of calcium sulfate (gypsum) that forms can cause calcium and sulfate leakage during subsequent service runs. Given a sufficient calcium input in the water to treat, calcium sulfate fouling is especially prevalent when the percent solution of regenerant is greater than 5 percent, or the temperature is greater than 100*F, or when the flow rate is less than 1 gpm/cu ft. Stepwise injection of sulfuric acid during regeneration can help prevent fouling.

Aluminum fouling of resins can appear when aluminum floc from alum or other coagulants in pretreatment are encountered by the resin bead. This floc coats the resin bead and in the ionic form will be exchanged. However, these ions are not efficiently removed during regeneration so the available exchange sites continuously decrease in number.

Copper fouling is found primarily in condensate polishing applications. Capacity loss is due to copper oxides coating the resin beads.

Oil fouling does not cause chemical degradation but gives loss of capacity due to filming on the resin beads and the reduction of their active surface. Agglomeration of beads also occurs causing increased pressure drop, channeling, and premature breakthrough. The oil-fouling problem can be alleviated by the use of surfactants.

OPERATIONAL SEQUENCING CONSIDERATIONS

The mode of operation for ion-exchange units can vary greatly from one system to the next, depending on the user's requirements. Service and regeneration cycles can be fully manual to totally automatic, with the method of regeneration being cocurrent, countercurrent, or external. The exhaustion phase is called the service run. This is followed by the regeneration phase which is necessary to bring the bed back to initial conditions to cycle. The regeneration phase includes four steps: backwashing to clean the bed, introduction of the excess regenerant, a slow rinse or displacement step to push the regenerant slowly through the bed, and finally a fast rinse to remove the excess regenerant from the resin and elute the unwanted ions to waste.

Service Cycle - The service cycle is normally terminated by one or a combination of the following criteria:

- High effluent conductivity.
- Total gallons throughput.
- High-pressure drop.
- High silica.
- High sodium.
- Variations in pH.
- Termination of the service cycle can be manually or automatically initiated.

Backwash Cycle - Normally, the first step in the regeneration sequence is designed to reverse flow from the service cycle using sufficient volume and flow rate to develop proper bed expansion for the purpose of removing suspended material (crud) trapped in the ion-exchange bed during the service cycle. The backwash waste water is collected by the raw water inlet distributor and diverted to waste via value sequencing. Backwash rate and internal design should avoid potential loss of whole bead resin during the backwash step. (Lower water temperature means more viscous force and more expansion.)

Regenerant Introduction - This introduction of regenerant chemicals can be cocurrent or countercurrent depending on effluent requirements, operating cost, and so on. Regenerant dosages (pounds per cubic foot), concentrations, flow rate, and contact time are determined for each application. The regenerant distribution and collection system must provide uniform contact throughout the bed and should avoid regenerant hideout. Additional effluent purity is obtained with countercurrent systems since the final resin contact in the service will be the most highly regenerated resin in the bed, creating a polishing effect.

Displacement Slow Rinse Cycle - The final steps in the regeneration sequence are generally terminated on acceptable quality. Displacement, which precedes the rinse step, is generally an extension of the regenerant introduction step. The displacement step is designed to give final contact with the resin, removing the bulk of the spent regenerant from the resin bed.

Fast Rinse Cycle - The fast rinse step is essentially the service cycle except that the effluent is diverted to waste until quality is proven. This final rinse is always in the same direction as the service flow. Therefore, in countercurrent systems the displacement flow and rinse flows will be in opposite directions.

SEQUENCE OF OPERATION: MIXED BED UNITS

In mixed-bed units, both the cation and the anion resins are mixed together thoroughly in the same vessel by compressed air. The cation and the anion resins being next to each other constitute an infinite number of cation and anion exchangers. The effluent quality obtainable from a well-designed and operated mixed-bed exchanger will readily produce demineralized water of conductivity less than 0.5 mmho and silica less than 10 ppb.

Service Cycle - As far as the mode of operation is concerned, the service cycle of a mixedbed unit is very similar to a conventional two-bed system, in that water flows into the top of the vessel, down through the bed, and the purified effluent comes out the bottom. It is in the regeneration and the preparation of it that the mixed-bed differs from the two-bed equipment. The resins must be separated, regenerated separately, and remixed for the next service cycle.

Backwash Cycle - Prior to regeneration, the cation and the anion resins are separated by backwashing at a flow rate of 3.0 to 3.5 gpm/ft. The separation occurs because of the difference in the density of the two types of resin. The cation resin, being heavier, settles on the bottom, while the anion resin, being lighter, settles on top of the cation resin. After backwashing, the bed is allowed to settle down for 5 to 10 minutes and two clearly distinct layers are formed. After separation, the two resins are independently regenerated.

Regenerant Introduction - The anion resin is regenerated with caustic flowing downward from the distributor placed just above the bed, while the cation resin is regenerated with either hydrochloric or sulfuric acid, usually flowing upward. The spent acid and caustic are collected in the interface collector, situated at the interface of the two resins. The regenerant injection can be carried out simultaneously as described or sequentially. In sequential regeneration, the cation-resin regeneration should precede the anion-resin regeneration to prevent the possibility of calcium carbonate and magnesium hydroxide precipitation, which may occur because of the anion-regeneration waste coming in contact with the exhausted cation resin. If this precipitation occurs, it can foul the resins at the interface. This becomes very critical when only the mixed-bed exchanger is installed to demineralize the incoming raw water.

In the case of sequential regeneration, during the caustic and acid injection period, a blocking flow of the demineralized water is provided in the opposite direction of the regenerant injection. This is required to prevent the caustic from entering the cation resin and acid from entering the anion resin. When regeneration is carried out simultaneously, acid and caustic injection flows act like blocking flows to each other and no additional blocking flow with water is needed. In a few sequential-type regeneration systems, acid is injected to flow downward through the central interface collector which now also acts as an acid distributor.

Rinsing and Air Mix Cycles - After completion of the acid and caustic injection, both the cation and anion resins are rinsed slowly to remove the majority of the regenerant, without attempting to eliminate it completely. After the use of 7 to 10 gallons of slow rinse volume per cubic foot of each type of resin, the unit is drained to lower the water to a few inches above the resin bed. The resins are now remixed with an upflow of air. After remixing, the unit is filled completely with water flowing slowly from the top, to prevent anion-resin separation in the upper layers. The mixed-bed exchanger is then rinsed at fast flow rates. The conductivity of the effluent water may be very high for a few minutes and will then drop suddenly to the value usually observed in the service cycle. This phenomenon is characteristic of mixed beds and is due to the absorption of the remaining acid or caustic in different parts of the bed, by one or the other resin. This, no doubt, results in the loss of resin capacity, but this loss is negligible as compared to the length of the service cycle and the savings in the overall time required for regeneration.

SEQUENCE OF OPERATION: SOFTENER UNITS

Following are the basic steps involved in a regeneration of a water softener.

Backwashing - After exhaustion, the bed is backwashed to effect a 50 percent minimum bed expansion to release any trapped air from the air pockets, minimize the compactness of the bed, reclassify the resin particles, and purge the bed of any suspended insoluble material. Backwashing is normally carried out at 5-6 gpm/ft. However, the backwash flow rates are directly proportional to the temperature of water.

Brine Injection - After backwashing, a 5 percent to 10 percent brine solution is injected during a 30-minute period. The maximum exchange capacity of the resin is restored with 10 percent strength of brine solution. The brine is injected through a separate distributor placed slightly above the resin bed.

Displacement or Slow Rinse - After brine injection, the salt solution remaining inside the vessel is displaced slowly, at the same rate as the brine injection rate. The slow rinsing should be continued for at least 15 minutes and the slow rinse volume should not be less than 10 gallons/cu ft of the resin. The actual duration of the slow rinse should be based on the greater of these two parameters.

Fast Rinse - Rinsing is carried out to remove excessive brine from the resin. The rinsing operation is generally stopped when the effluent chloride concentration is less than 5-10 ppm in excess of the influent chloride concentration and the hardness is equal to or less than 1 ppm as CaCO.

Each arrangement will vary substantially in both operating and installed costs. Important factors for selection are:

1. Influent water analysis.
2. Flow rate.
3. Effluent quality.
4. Waste requirements.
5. Operating cost.

SELECTIVITY AND GENERAL CONSIDERATIONS

As noted. ion exchange reactions are stoichiometric and reversible, and in that way they are similar to other solution phase reactions. For example:

$$NiSO_4 + Ca(OH)_2 = Ni(OH)_2 + CaSO_4$$

In this reaction, the nickel ions of the nickel sulfate ($NiSO_4$) are exchanged for the calcium ions of the calcium hydroxide [$Ca(OH)_2$] molecule. Similarly, a resin with hydrogen ions available for exchange will exchange those ions for nickel ions from solution. The reaction can be written as follows:

$$2(R\text{-}SO_3H) + NiSO_4 = (R\text{-}SO_3)2Ni + H_2SO_4$$

R indicates the organic portion of the resin and SO_3 is the immobile portion of the ion active group. Two resin sites are needed for nickel ions with a plus 2 valence (Ni^{+2}). Trivalent ferric ions would require three resin sites. As shown, the ion exchange reaction is reversible. The degree the reaction proceeds to the right will depend on the resins preference. or selectivity, for nickel ions compared with its preference for hydrogen ions. The selectivity of a resin for a given ion is measured by the selectivity coefficient. K. which in its simplest form for the reaction

$$R^{-}A^{+} + B^{+} = R^{-}B^{+} + A^{+}$$

is expressed as: K = (concentration of B^+ in resin/concentration of A^+ in resin) × (concentration of A^+ in solution/concentration of B+ in solution).

The selectivity coefficient expresses the relative distribution of the ions when a resin in the A^+ form is placed in a solution containing B^+ ions. Table 2 gives the selectivity's of strong acid and strong base ion exchange resins for various ionic compounds. It should be pointed out that the selectivity coefficient is not constant but varies with changes in solution conditions. It does provide a means of determining what to expect when various ions are involved. As indicated in Table 2, strong acid resins have a preference for nickel over hydrogen. Despite this preference, the resin can be converted back to the hydrogen form by contact with a concentrated solution of sulfuric acid (H_2SO_4):

$$(R\text{--}SO_4)_2Ni + H_2SO_4 \rightarrow 2(R\text{-}SO_3H) + NiSO_4$$

As we noted above, but a little differently, this step is known as regeneration. In general terms, the higher the preference a resin exhibits for a particular ion, the greater the exchange efficiency in terms of resin capacity for removal of that ion from solution. Greater preference for a particular ion, however, will result in increased consumption of chemicals for regeneration.

Resins currently available exhibit a range of selectivity's and thus have broad application. As an example. for a strong acid resin. the relative preference for divalent calcium ions (Ca^{+2}) over divalent copper ions (Cu^{+2}) is approximately 1.5

to 1. For a heavy-metal-selective resin. the preference is reversed and favors copper by a ratio of 2.300 to 1.

Table 2. Selectivity of ion Exchange Resins in Order of Decreasing Preference.

Strong acid cation exchanger	Strong base anion exchanger
Barium	Iodide
Lead	Nitrate
Calcium	Bisulfate
Nickel	Chloride
Cadmium	Cyanide
Copper	Bicarbonate
Zinc	Hydroxide
Magnesium	Fluoride
Potassium	Sulfate
Ammonia Sodium	
Hydrogen	

Ion exchange resins are classified as cation exchangers, which have positively charged mobile ions available for exchange, and anion exchangers, whose exchangeable ions are negatively charged. Both anion and cation resins are produced from the same basic organic polymers. They differ in the ionizable group attached to the hydrocarbon network. It is this functional group that determines the chemical behavior of the resin. Resins can be broadly classified as strong or weak acid cation exchangers or strong or weak base anion exchangers.

Strong acid resins are so named because their chemical behavior is similar to that of a strong acid. The resins are highly ionized in both the acid ($R\text{-}SO_3H$) and salt ($R\text{-}SO_3Na$) form. They can convert a metal salt to the corresponding acid by the reaction:

$$2(R\text{-}SO_3H) + NiCl_2 \rightarrow (R\text{-}SO_4),Ni + 2HCl$$

The hydrogen and sodium forms of strong acid resins are highly dissociated and the exchangeable Na^+ and H^+ are readily available for exchange over the entire pH range. Consequently, the exchange capacity of strong acid resins is independent of solution pH. These resins would be used in the hydrogen form for complete deionization; they are used in the sodium form for water softening (calcium and magnesium removal). After exhaustion, the resin is converted back to the hydrogen form (regenerated) by contact with a strong acid solution, or the resin can be convened to the sodium form with a sodium chloride solution. In the above, the hydrochloric acid (HCl) regeneration would result in a concentrated nickel chloride (NiCl,) solution.

In a weak acid resin. the ionizable group is a carboxylic acid (COOH) as opposed to the sulfonic acid group (SO_3H) used in strong acid resins. These resins behave similarly to weak organic acids that are weakly dissociated. Weak acid resins exhibit a much higher affinity for hydrogen ions than do strong acid resins. This characteristic allows for regeneration to the hydrogen form with significantly less acid than is required for strong acid resins. Almost complete regeneration can be accomplished with stoichiometric amounts of acid. The degree of dissociation of a weak acid resin is strongly influenced by the solution pH. Consequently, resin capacity depends in part on solution pH. Figure 1 shows that a typical weak acid resin has limited capacity below a pH of 6.0. making it unsuitable for deionizing acidic metal finishing wastewater.

Like strong acid resins. strong base resins are highly ionized and can be used over the entire pH range. These resins are used in the hydroxide (OH) form for water deionization. They will react with anions in solution and can convert an acid solution to pure water:

$$R\text{--}NH_3OH + HCl \rightarrow R\text{-}NH_3Cl + HOH$$

Regeneration with concentrated sodium hydroxide (NaOH) converts the exhausted resin to the hydroxide form.

Weak base resins are like weak acid resins. in that the degree of ionization is strongly influenced by pH. Consequently, weak base resins exhibit minimum exchange capacity above a pH of 7.0. These resins merely sorb strong acids: they cannot split salts.

In an ion exchange wastewater deionization unit. the wastewater would pass first through a bed of strong acid resin. Replacement of the metal cations (Ni^{+2}. Cu^{+2}) With hydrogen ions would lower the solution pH. The anions (SO_4^{-2}. Cl^-) can then be removed with a weak base resin because the entering wastewater will normally be acidic and weak base resins sorb acids. Weak base resins are preferred over strong base resins because they require less regenerant chemical. A reaction between the resin in the free base form and HCl would proceed as follows:

$R\text{-}NH_2 + HCl \rightarrow R\text{-}NH_3Cl$

The weak base resin does not have a hydroxide ion form as does the strong base resin. Consequently, regeneration needs only to neutralize the absorbed acid: it need not provide hydroxide ions. Less expensive weakly basic reagents such as ammonia (NH_3) or sodium carbonate can be employed. Chelating resins behave similarly to weak acid cation resins but exhibit a high degree of selectivity for heavy metal cations. Chelating resins are analogous to chelating compounds found in metal finishing wastewater; that is, they tend to form stable complexes with the heavy metals. In fact. the functional group used in these resins is an EDTAa compound. The resin structure in the sodium form is expressed as R-EDTA-Na. The high degree of selectivity for heavy metals permits separation of these ionic compounds from solutions containing high background levels of calcium, magnesium, and sodium ions. A chelating resin exhibits greater selectivity for heavy metals in its sodium form than in its hydrogen form. Regeneration properties are similar to those of a weak acid resin; the chelating resin can be converted to the hydrogen form with slightly greater than stoichiometric doses of acid because of the fortunate tendency of the heavy metal complex to become less stable under low pH conditions. Potential applications of the chelating resin include polishing to lower the heavy metal concentration in the effluent from a hydroxide treatment process or directly removing toxic heavy metal cations from wastewaters containing a high concentration of nontoxic, multivalent cations. Table 3 shows the preference of a commercially available chelating resin for heavy metal cations over calcium ions. (The chelating resins exhibit a similar magnitude of selectivity for heavy metals over sodium or magnesium ions.) The selectivity coefficient defines the relative preference the resin exhibits for different ions. The preference for copper (shown in Table 3) is 2300 times that for calcium. Therefore, when a solution is treated that contains equal molar concentrations of copper and calcium ions, at equilibrium. the molar concentration of copper ions on the resin will be 2300 times the concentration of calcium ions. Or, when solution is treated that contains a calcium ion molar concentration 2300 times that of the copper ion concentration, at equilibrium. the resin would hold an equal concentration of copper and calcium.

Table 3. Chelating Cation Resin Selectivities for Metal Ions.

Metal	**KM/Ca**[a]
Hg^{+2}	2800
Cu^{+2}	2300
Pb^{+2}	1200
Ni^{+2}	57

Metal	KM/Ca[a]
Zn^{+2}	17
Cd^{+2}	15
Co^{+2}	6.7
Fe^{+2}	4.7
Mn^{+2}	1.2
Ca^{+2}	1

a -- *Selectivity coefficient for the metal over calcium ions at a pH of 4.*

EQUIPMENT AND OPERATION

Ion exchange processing can be accomplished by either a batch method or a column method. In the first method, the resin and solution are mixed in a batch tank, the exchange is allowed to come to equilibrium, then the resin is separated from solution. The degree to which the exchange takes place is limited by the preference the resin exhibits for the ion in solution. Consequently, the use of the resins exchange capacity will be limited unless the selectivity for the ion in solution is far greater than for the exchangeable ion attached to the resin. Because batch regeneration of the resin is chemically inefficient, batch processing by ion exchange has limited potential for application.

Passing a solution through a column containing a bed of exchange resin is analogous to treating the solution in an infinite series of batch tanks. Consider a series of tanks each containing 1 equivalent (eq) of resin in the X ion form . A volume of solution containing 1 eq of Y ions is charged into the first tank. Assuming the resin to have an equal preference for ions X and Y, when equilibrium is reached the solution phase will contain 0.5 eq of X and Y. Similarly. the resin phase will contain 0.5 eq of X and Y. This separation is the equivalent of that achieved in a batch process.

If the solution were removed from Tank 1 and added to Tank 2, which also contained 1 eq of resin in the X ion form, the solution and resin phase would both contain 0.25 eq of Y ion and 0.75 eq of X ion. Repeating the procedure in a third and fourth tank would reduce the solution content of Y ions to 0.125 and 0.0625 eq. respectively. Despite an unfavorable resin preference. using a sufficient number of stages could reduce the concentration of Y ions in solution to any level desired. This analysis simplifies the column technique, but it does provide insights into the process dynamics. Separations are possible despite poor selectivity for the ion being removed. Most industrial applications of ion exchange use fixed-bed column systems, the basic component of which is the resin column (Figure 5). The column design must:

- Contain and support the ion exchange resin
- Uniformly distribute the service and regeneration flow through the resin bed
- Provide space to fluidize the resin during backwash
- Include the piping, valves, and instruments needed to regulate flow of feed, regenerant. and backwash solutions

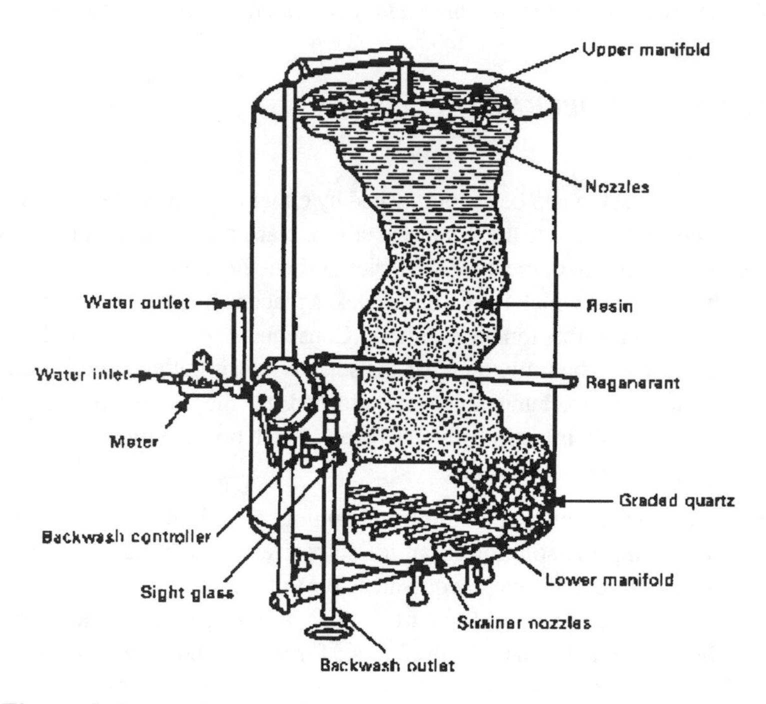

Figure 5. Ion exchange unit.

After the feed solution is processed to the extent that the resin becomes exhausted and cannot accomplish any further ion exchange, the resin must be regenerated. In normal column operation, for a cation system being converted first to the hydrogen then to the sodium form, regeneration employs the following basic steps:

- The column is backwashed to remove suspended solids collected by the bed during the service cycle and to eliminate channels that may have formed during this cycle. The back- wash flow fluidizes the bed, releases trapped particles. and reorients the resin particles according to size.

During backwash the larger, denser panicles will accumulate at the base and the particle size will decrease moving up the column. This distribution yields a good hydraulic flow pattern and resistance to fouling by suspended solids.

- The resin bed is brought in con- tact with the regenerant solution. In the case of the cation resin. acid elutes the collected ions and converts the bed to the hydrogen form. A slow water rinse then removes any residual acid.
- The bed is brought in contact with a sodium hydroxide solution to convert the resin to the sodium form. Again, a slow water rinse is used to remove residual caustic. The slow rinse pushes the last of the regenerant through the column.
- The resin bed is subjected to a fast rinse that removes the last traces of the regenerant solution and ensures good flow characteristics.
- The column is returned to service.

For resins that experience significant swelling or shrinkage during regeneration, a second backwash should be performed after regeneration to eliminate channeling or resin compression. Regeneration of a fixed-bed column usually requires between 1 and 2 hr. Frequency depends on the volume of resin in the exchange columns and the quantity of heavy metals and other ionized compounds in the wastewater. Resin capacity is usually expressed in terms of equivalents per liter (eq/L) of resin. An equivalent is the molecular weight in grams of the compound divided by its electrical charge. or valence. For example. a resin with an exchange capacity of 1 eq/L could remove 37.5 g of divalent zinc (Zn^{+2}, molecular weight of 65) from solution. Much of the experience with ion exchange has been in the field of water softening: therefore, capacities will frequently be expressed in terms of kilograins of calcium carbonate per cubic foot of resin. This unit can be converted to equivalents per liter by multiplying by 0.0458. Typical capacities for commercially available cation and anion resins are shown in Figure 4. The capacities are strongly influenced by the quantity of acid or base used to regenerate the resin. Weak acid and weak base systems are more efficiently regenerated; their capacity increases almost linearly with regenerant dose. Columns are designed to use either cocurrent or countercurrent regeneration. In cocurrent units, both feed and regenerant solutions make contact with the resin in a downflow mode. These units are the less expensive of the two in terms of initial equipment cost. On the other hand, cocurrent flow uses regenerant chemicals less efficiently than countercurrent flow: it has higher leakage concentrations (the concentration of the feed solution ion being removed in the column effluent), and cannot achieve as high a product concentration in the regenerant. Efficient use of regenerant chemicals is primarily a concern with strong acid or strong base resins. The weakly ionized resins require only slightly greater than stoichiometric chemical doses for complete regeneration regardless of whether cocurrent or countercurrent flow is used. With strong acid or strong base resin systems. improved chemical efficiency can be achieved by reusing a part of the spent regenerants. In strongly ionized resin systems, the degreeof column regeneration is the major factor in determining the chemical efficiency of the regeneration process. To realize 42 percent of the resin's theoretical exchange capacity requires 1.4 times the stoichiometric amount of reagent [2 lb HCl/ft^3 (32 g HCI/L)]. To increase the exchange capacity available to 60 percent of theoretical increases consumption to 2.45 times the stoichiometric dose [5 Ib HCl/ft^3 (80 g HCI/Liter)]. The need for acid doses considerably higher than stoichiometric means that there is a significant concentration of acid in the spent regenerant. Further. as the acid dose is increased incrementally, the concentration of acid in the spent regenerant increases. By discarding only the first part of the spent regenerant and saving and reusing the rest, greater exchange capacity can be realized with equal levels of regenerant consumption. For example, if a regenerant dose of 5 Ib HCl/ft^3 (80 g HCI/Liter) were used in the resin system, the first 50 percent of spent regenerant would contain only 29 percent of the original acid concentration. The rest of the acid regenerant would contain 78 percent of the original acid concentration. If this second part of the regenerant is reused in the next regeneration cycle before the resin bed makes contact with 5 Ib/ft3 (80 g/Liter) of fresh HCI, the exchange capacity would increase to 67 percent of theoretical capacity. The available capacity would then increase from 60 to 67 percent at equal chemical doses. Figure 5 shows the improved reagent utilization achieved by this manner of reuse over a range of regenerant doses. Regenerant reuse has disadvantages in that it is higher in initial cost for chemical storage and feed systems and regeneration procedure is more complicated. Still. where the chemical savings have provided justification, systems have been designed to reuse parts of the spent regenerant as many as five times before discarding them.

COST CONSIDERATIONS AND COMPARISONS TO RO

The cost of resins is a major consideration. Heavy-metal-selective chelating resins are the most expensive. Table 4 provides some cost ranges for commercially available resins.

Table 4. Cost Ranges of Commercial Resins.

Resin Type	Cost ($/f^{t3})
Strong acid cation	70-120
Weak acid cation	150-200
Strong base anion	180-250

Resin Type	Cost ($/f^{t3})
Weak base anion	180-200
Chelating cation	330-600

The technology that competes with ion exchange in wastewater application is reverse osmosis (RO), therefor it is appropriate to make some comparisons. Direct cost comparisons are not straightforward, and requires comparison of some of the hidden cost parameters. Since there appear to be few detailed comparisons in the open literature, there exists the general impression that RO is more economical than ion exchange. Whereas this may be true in a number of applications, as a general rule this is not the case.

The following are factors that should be taken into consideration when making case-specific cost comparisons between the two technologies. First, ion exchange is generally run batch, whereas RO is essentially continuous. This implies a higher degree of operator attention for ion exchange. However, one must remember that RO membranes must be cleaned, and this can be frequently depending upon the treatment application. Furthermore, although ion exchange systems are discontinuous, commercial systems are fully automated, and hence operator attention can be brought to a minimum.

A second consideration is that RO tends to be sensitive to incoming suspended matter. Comprehensive and sometimes expensive pre-treatment technologies are generally needed with RO, whereas ion exchange is less sensitive to the suspended matter. Further, RO systems are sensitive to hardness, so that softening is usually required as a pre-treatment. As a rule, RO membranes cannot handle high silica waters.

RO systems are quite sensitive to certain temperature ranges. Most critical is the range from 25 to 15° C, where RO systems have been reported to loose up to 30 % performance. RO has increased salt passage if the temperature increases, whereas ion exchange is insensitive to temperature variations. With the new generation of high performance resins, ion exchange can be kept fairly small using short operating cycles, and regeneration utilization approaches stoichiometric theoretical values. This translates into lower running costs.

Between the two technologies, ion exchange can be thought of as more within the arena of a pollution prevention technology. We can state this because of the relative recoveries or yields. If we take a boiler application as an example, with ion exchange, the difference between "net throughput" (the water you produce for the boiler) and "gross throughput" (the water you consume is minimal. You simply need a few cubic meters for the dilution of regenerants and for rinse. Typically, for a medium TDS water, the wasted water is about 5 % or less. With TDS and older co-flow regeneration systems, it can reach or exceed 10 %. In comparison, with

RO, only about 70 to 75 % of the water that is pumped into the system can be recovered. RO rejects large volumes of concentrate. Ion exchange removes all ions down to extremely low residuals. It does not remove non-ionic species, however (only partially). In contrast, RO removes all compounds based on their size. Very small ions or molecules, such as Na, Cl, CO_2 are only partially removed. Other ions (Ca, SO_4) are harmful to the membrane. Globally, RO is a partial demineralization process, wehreas complete demineralization can be achieved with a simple ion exchange plant. To achieve the same salt residual as obtained with a simple ion exchange plant, a double-pass RO system is needed, and is considerably more expensive. The company Rhom and Hass (go to www.rohmhaas.com for details) has done a cost comparison between several design cases for the two technologies. Although Rhom and Haas favors ion exchange because they are marketing their technology, the comparisons are well done, and clearly support a cost advantage ion exchange. The following is one cost comparison summary. In this case the plant specs are as follows:

- Daily water production = 91 m^3/hr
- Operating basis = 300 days per year
- Treated water specifications: Conductivity = 1 μS/cm; Residula silica = 30 μg/Liter as SiO_2;
- Feed water salinity = 6.54 meq/Liter (327 mg/Liter as $CaCO_3$)

The comparison is made against a simple ion exchange plant versus a single pass RO system. The following is the overall cost estimate comparison, followed by a breakdown of the individual cost item details, including operating and energy costs. In reviewing these comparative costs, you see that there most definitely a significant cost savings in favor of ion exchange over RO, and given the water conservation advantage of ion exchange over RO, clearly it is more within the arena of pollution prevention technologies. This is not necessarily the case for every situation, and a detailed cost comparison for any investment needs to be made on a case-specific basis. When you visit Chapter 12, some of the principles for cost estimating and making technology comparisons are discussed.

Table 5. Overall Investment and Operating Cost for Design Case Study.

Investment Cost	Ion Exchange	Reverse Osmosis
U.S. $	840,000	1,020,000
Operating Costs (Refer to Table 6 for detailed breakdown)		
U.S. $/$m^3$ treated water	0.442	0.690
U.S. $ per annum (330 day basis)	241,330	376,500

Table 6. Unit Cost Details for Design Case Study. (*Note: Usage is expressed per m^3 produced water, and Cost is in U.S. \$ per m^3 produced*).

Detailed Cost Estimate	**Unit Cost**	**Ion Exchange**		**Com-ments**	**RO**		**Comments**
Component	**U.S. \$**	**Usage**	**Cost**		**Usage**	**Cost**	
Energy, kWh	0.06	0.25	0.015		0.99	0.059	
H_2SO_4, kg	0.11	0.40	0.044		0.02	0.002	for MB
NaOH, kg	0.56	0.15	0.084		0.06	0.032	for MB
Neutralizat-ion, kg	0.11	0.00	0.000	Self neutral	0.15	0.017	for RO
Anti-scale, kg	3.89	0.00	0.000		0.01	0.019	5 ppm
Extra Water, m^3	0.26	0.03	0.008	Dilut-ion and rinse	0.45	0.117	Rejected
Wastewater, m^3	0.33	0.03	0.011	Spent reg.	0.45	0.149	Concen-trate
Resin Replace-ment			0.020			0.003	for MB
Membrane Replace-ment			0.000			0.031	
Subtotal			0.182			0.430	
Net feed water, m^3	0.26	1.00	0.260		1.00	0.260	
Total			0.442	\$/m^3		0.690	\$/m^3

CARBON ADSORPTION IN WATER TREATMENT

HOW ADSORPTION WORKS

Activated carbon is a crude form of graphite, the substance used for pencil leads. It differs from graphite by having a random imperfect structure which is highly porous over a broad range of pore sizes from visible cracks and crevices to molecular dimensions. The graphite structure gives the carbon it's very large surface area which allows the carbon to adsorb a wide range of compounds. Activated carbon can have a surface of greater than 1000 m^2/g. This means 5g of activated carbon can have the surface area of a football field.
Adsorption is the process by which liquid or gaseous molecules are concentrated on a solid surface, in this case activated carbon. This is different from absorption, where molecules are taken up by a liquid or gas. Activated carbon can made from many substances containing a high carbon content such as coal, wood and coconut shells. The raw material has a very large influence on the characteristics and performance activated carbon.
The term *activation* refers to the development of the adsorption properties of carbon. Raw materials such as coal and charcoal do have some adsorption capacity, but this is greatly enhanced by the activation process. There are three main forms of activated carbon.

- **Granular Activated Carbon (GAC)** - irregular shaped particles with sizes ranging from 0.2 to 5 mm. This type is used in both liquid and gas phase applications.
- **Powder Activated Carbon (PAC)** - pulverized carbon with a size predominantly less than 0.18mm (US Mesh 80). These are mainly used in liquid phase applications and for flue gas treatment.
- **Pelleted Activated Carbon** - extruded and cylindrical shaped with diameters from 0.8 to 5 mm. These are mainly used for gas phase applications because of their low pressure drop, high mechanical strength and low dust content.

Activated carbon has the strongest physical adsorption forces or the highest volume of adsorbing porosity of any material known to mankind.

Activated carbon is also available in special forms such as a cloth and fibres. Activated Charcoal Cloth (ACC) represents a family of activated carbons in cloth form. These products are fundamentally unique in several important ways compared with the traditional forms of activated carbon and with other filtration media that incorporate small particles of activated carbon. Developed in the early 1970's ACC products are similar to the traditional activated carbon products in that they are 100 % activated carbon. This gives the products the same high capacity for adsorption of organic compounds and other odorous gases as the more traditional, pelletised, granular and powder forms of activated carbon. As with the traditional forms of activated carbons, ACC products can be impregnated with a range of chemicals to enhance the chemisorption capacity for selected gases. By being constructed of bundles of activated carbon filaments and fibres in a textile form, several important advantages are imparted to ACC. The diameter of these fibres is approximately 20 mm, so the kinetics for ACC products are similar to that of a very tine carbon particle. Gases and liquids can flow through the fabric and the accelerated adsorption kinetics mean that the ACC can retain the advantages of mass transfer zones associated with deeper filter beds. Faster adsorption rates mean smaller adsorption equipment and up to twenty times less carbon on line.

> **Gas Phase Adsorption** - ***This is a condensation process where the adsorption forces condense the molecules from the bulk phase within the pores of the activated carbon. The driving force for adsorption is the ratio of the partial pressure and the vapour pressure of the compound.***
> **Liquid Phase Adsorption** - ***The molecules go from the bulk phase to being adsorbed in the pores in a semi-liquid state. The driving force for adsorption is the ratio of the concentration to the solubility of the compound.***

Adsorption is the process where molecules are concentrated on the surface of the activated carbon. Adsorption is caused by London Dispersion Forces, a type of Van der Waals Force which exists between molecules. The force acts in a similar way to gravitational forces between planets. London Dispersion Forces are extremely short ranged and therefore sensitive to the distance between the carbon surface and the adsorbate molecule They are also additive, meaning the adsorption force is the sum of all interactions between all the atoms. The short range and additive nature of these forces results in activated carbon having the strongest physical adsorption forces of any material known to mankind. All compounds are adsorbable to some extent. In practice, activated carbon is used for the adsorption of mainly organic compounds along with some larger molecular weight inorganic compounds such as iodine and mercury. In general, the adsorbability of a compound increases with : (1) increasing molecular weight, (2) a higher number of functional groups such as double bonds or halogen compounds, and (3) increasing polarisability of the molecule. This is related to electron clouds of the molecule. Refer to Figure 6. The most common manufacturing process is high temperature steam activation though activated carbon can also be manufactured with chemicals. Along with the raw material, the activation process has a very large influence on the characteristics and performance of activated carbon. Figure 7 illustrates the production of granular activated carbons by steam activation of selected grades of pulverized and then reaglommerated bituminous coal.

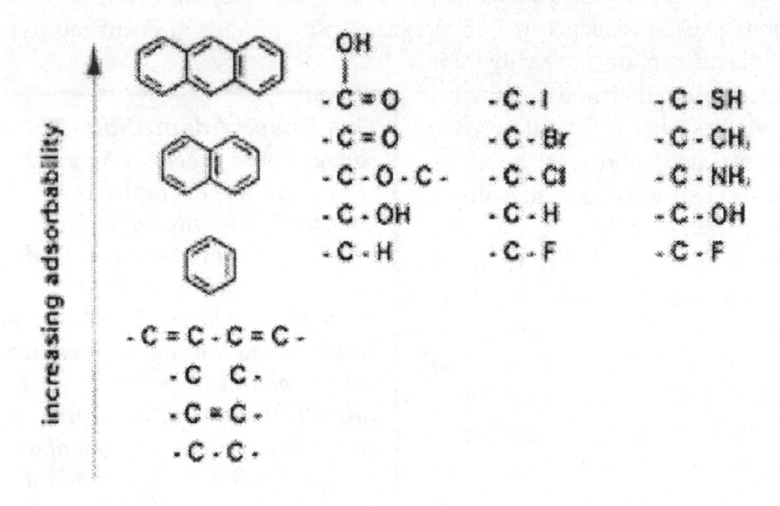

Figure 6. Relative adsorptivity of organic materials.

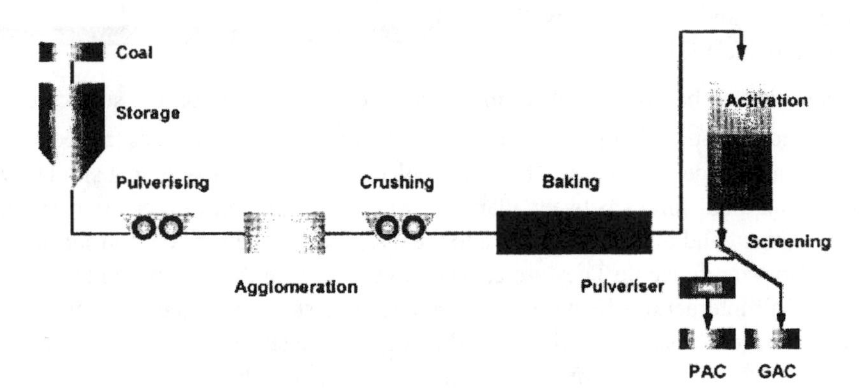

Figure 7. Manufacture process for activated carbon.

APPLICATION

The way activated carbon is used is normally determined by the form of activated carbon. With granular and pelleted activated carbon, in most cases the carbon is installed in a fixed bed with the liquid or gas passing through the bed. The compounds to be removed are retained on the activated carbon. The carbon is used until exhaustion. It can then be :

Reactivated, normally off-site

- **Functional Groups -** The molecular structure of the resin is such that it must contain a macroreticular tissue with acid or basic radicals. These radicals are the basis of classifying ion exchangers into two general groups: (1) Cation exchangers, in which the molecule contains acid radicals of the HSO_3 or HCO type able to fix mineral or organic cations and exchange with the hydrogen ion H^+; (2) Anion exchangers, containing basic radicals (for example, amine functions of the type NH_2) able to fix mineral or organic anions and exchange them with the hydroxyl ion OH^- coordinate to their dative bonds. The presence of these radicals enable a cation exchanger to be assimilated to an acid of form H-R and an anion exchanger to be a base of form OH-R when regenerated. In-situ regenerated. This is possible for most gas phase and some liquid phase applications.

It may also be replaced with new carbon and disposal of the exhausted carbon Most adsorbers are pressure vessels constructed in carbon steel, stainless steel or plastic. Large systems for drinking water are often constructed in concrete. In some cases, a moving or pulsed bed adsorber is employed to optimize the use of the granular activated carbon.

The main factors in the design of an adsorption system are the: (1) **Carbon consumption** - The amount of carbon required to treat the liquid or gas, normally expressed per unit of the fluid treated; and (2) **Contact time** - For a fixed flow rate, the contact time is directly proportional to the volume of carbon and is the main factor influencing the size of the adsorption system and capital cost.

With powder activated carbon, in most cases, the carbon is dosed into the liquid, mixed and then removed by a filtration process. In some cases, two or more mixing steps are used to optimise the use of powder carbon. Powder activated carbon is used in a wide range of liquid phase applications and some specific gas phase applications such as Incinerator flue gas treatment and where it is bonded into filters such as fabrics for personnel protection.

THE CONTAMINANTS ACTIVATED CARBON CAN AND CANNOT REMOVE

Activated carbon (AC) filtration is most effective in removing organic contaminants from water. Organic substances are composed of two basic elements, carbon and hydrogen. Because organic chemicals are often responsible for taste, odor, and color problems, AC filtration can generally be used to improve aesthetically objectional water. AC filtration will also remove chlorine. AC filtration is recognized by the Water Quality Association as an acceptable method to maintain certain drinking water contaminants within the limits of the EPA National Drinking Water Standards (refer to Table 7).

Table 7. Water contaminants that can be reduced to acceptable standards by activated carbon filtration. (*Source: Water Quality Association, 1989*)

Primary Drinking Water Standards

Contaminant	**MCL, mg/L*
Inorganic Contaminants	
Organic Arsenic Complexes	0.05
Organic Chromium Complexes	0.05
Mercury (Hg+2) Inorganic	0.05
Organic Mercury Complexes	0.002
Organic Contaminants	
Benzene	0.005
Endrin	0.0002
Lindane	0.004
Methoxychlor	0.1
1,2-dichloroethane	0.005
1,1-dichloroethylene	0.007
1,1,1-trichloroethane	0.200
Total Trihalomethanes (TTHMs)	0.10
Toxaphene	0.005
Trichloroethylene	0.005
2,4-D	0.1
2,4,5-TP (Silvex)	0.01
Para-dichlorobenzene	0.075

Secondary Drinking Water Standards

Contaminant	***SMCL*
Color	15 color units
Foaming Agents (MBAS)	0.5 mg/L
Odor	3 threshold odor number

**Maximum Contaminant Level*
***Secondary Maximum Contaminant Level*

AC filtration does remove some organic chemicals that can be harmful if present in quantities above the EPA Health Advisory Level (HAL). Included in this category are trihalomethanes (THM), pesticides, industrial solvents (halogenated hydrocarbons), polychlorinated biphenyls (PCBs), and polycyclic aromatic hydrocarbons (PAHs).

THMs are a byproduct of the chlorination process that most public drinking water systems use for disinfection. Chloroform is the primary THM of concern. EPA does not allow public systems to have more than 100 parts per billion (ppb) of THMs in their treated water. Some municipal systems have had difficulty in meeting this standard.

The Safe Drinking Water Act mandates EPA to strictly regulate contaminants in community drinking water systems. As a result, organic chemical contamination of municipal drinking water is not likely to be a health problem. Contamination is more likely to go undetected and untreated in unregulated private water systems. AC filtration is a viable alternative to protect private drinking water systems from organic chemical contamination. Note that eadon gas can also be removed from water by AC filtration, but actual removal rates of radon for different types of AC filtration equipment have not been established. This makes AC ideal for use both in industrial applications as well as in residential.. Similar to other types of water treatment, AC filtration is effective for some contaminants and not effective for others. AC filtration does not remove microbes, sodium, nitrates, fluoride, and hardness. Lead and other heavy metals are removed only by a very specific type of AC filter. Unless the manufacturer states that its product will remove heavy metals, one should assume that the AC filter is not effective in removing them.

Regular water testing is recommended to reduce the risk of consuming contaminated water. Many contaminants are not detected by the senses. Even if contamination can be detected by color, smell, or taste, only a laboratory test can tell you the quantity of contaminant actually present. Testing should always be done by a reputable or certified laboratory. Prior to sending in your water sample, determine what you want your water tested for. Contact the laboratory to find out how to take a proper water sample. Remember, there are thousands of substances that can contaminate your water, and they all have slightly different chemical behavior. Proper sampling and handling for one type of contaminant may cause erroneous results for other types of contaminants.

AC works by attracting and holding certain chemicals as water passes through it. AC is a highly porous material; therefore, it has an extremely high surface area for contaminant adsorption. The equivalent surface area of 1 pound of AC ranges from 60 to 150 acres. AC is made of tiny clusters of carbon atoms stacked upon one another. The carbon source is a variety of materials, such as peanut shells or coal. The raw carbon source is slowly heated in the absence of air to produce a high carbon material. The carbon is activated by passing oxidizing gases through the material at extremely high temperatures. The activation process produces the pores that result in such high adsorptive properties. The adsorption process depends on the following factors: 1) physical properties of the AC, such as pore size distribution and surface area; 2) the chemical nature of the carbon source, or the amount of oxygen and hydrogen associated with it; 3) chemical composition and concentration of the contaminant; 4) the temperature and pH of the water; and 5) the flow rate or time exposure of water to AC.

MAKE NOTE OF PHYSICAL AND CHEMICAL PROPERTIES

Forces of physical attraction or adsorption of contaminants to the pore walls is the most important AC filtration process. The amount and distribution of pores play key roles in determining how well contaminants are filtered. The best filtration occurs when pores are barely large enough to admit the contaminant molecule. Because contaminants come in all different sizes, they are attracted differently depending on pore size of the filter. In general AC filters are most effective in removing contaminants that have relatively large molecules (most organic chemicals). Type of raw carbon material and its method of activation will affect types of contaminants that are adsorbed. This is largely due to the influence that raw material and activation have on pore size and distribution.

Processes other than physical attraction also affect AC filtration. The filter surface may actually interact chemically with organic molecules. Also electrical forces between the AC surface and some contaminants may result in adsorption or ion exchange. Adsorption, then, is also affected by the chemical nature of the adsorbing surface. The chemical properties of the adsorbing surface are determined to a large extent by the activation process. AC materials formed from different activation processes will have chemical properties that make them more or less attractive to various contaminants. For example chloroform is adsorbed best by AC that has the least amount of oxygen associated with the pore surfaces. You can't possibly determine the chemical nature of an AC filter. However, this does point out the fact that different types of AC filters will have varying levels of effectiveness in treating different chemicals.

CONTAMINANT PROPERTIES

As already noted, large organic molecules are most effectively adsorbed by AC. A general rule of thumb is that similar materials tend to associate. Organic molecules and activated carbon are similar materials; therefore there is a stronger tendency for most organic chemicals to associate with the activated carbon in the filter rather than staying dissolved in a dissimilar material like water. Generally, the least soluble organic molecules are most strongly adsorbed. Often the smaller organic molecules are held the tightest, because they fit into the smaller pores.

Concentration of organic contaminants can affect the adsorption process. A given AC filter may be more effective than another type of AC filter at low contaminant concentrations, but may be less effective than the other filter at high concentrations. This type of behavior has been observed with chloroform removal. The filter manufacturer should be consulted to determine how the filter will perform for specific chemicals at different levels of contamination.

CONDITIONS THAT IMPACT ON PERFORMANCE

Adsorption usually increases as pH and temperature decrease. Chemical reactions and forms of chemicals are closely related to pH and temperature. When pH and temperature are lowered many organic chemicals are in a more adsorbable form. The adsorption process is also influenced by the length of time that the AC is in contact with the contaminant in the water. Increasing contact time allows greater amounts of contaminant to be removed from the water. Contact is improved by increasing the amount of AC in the filter and reducing the flow rate of water through the filter.

AC filters can be a breeding ground for microorganisms. The organic chemicals that are adsorbed to the AC are a source of food for various types of bacteria. Pathogenic bacteria are those that cause human diseases such as typhoid, cholera, and dysentery. AC filtration should only be used on water that has been tested and found to be bacteria free or effectively treated for pathogenic bacteria. Other types of non-pathogenic bacteria that do not cause diseases have been regularly found in AC filters. There are times when high amounts of bacteria (non-pathogenic) are found in water filtered through an AC unit. Research shows little risk to healthy people that consume high amounts of non-pathogenic bacteria. We regularly take in millions of bacteria every day from other sources. However, there is some concern for certain segments of the population, such as the very young or old and people weakened by illness. Some types of non-pathogenic bacteria can cause illness in those whose natural defenses are weak. Flushing out bacteria that have built up in the filter can be accomplished by backflushing the AC filter prior to use. Water filtered after the initial flushing will have much lower levels of bacteria and ingestion of a high concentration of bacteria will have been avoided. Some compounds of silver have been used as disinfectants, especially in European operations. Silver has been added to certain AC filters as a solution to the bacteria problem. Unfortunately, product testing has not shown silver impregnated AC to be much more effective in controlling bacteria than normal AC filters. The areas that require definition when specifying and szing a carbon adsorption system include:

Processing conditions:

- Concentration of adsorbate
- Temperature of liquid stream
- pH of liquid stream
- Flow rates and operating frequency
- Pressure drop in system

Characteristics of the adsorbate:

- Relative molecular mass
- Solubility of the adsorbate
- Concentration relative to solubility limits
- Polarity of adsorbate
- Temperature of solution

Selection of adsorbent for optimum efficiency:

- Specific adsorption isotherm
- Selection of optimum activity level
- Cost sensitivity analysis
- Consideration of thermal reactivation

We will now touch upon some of these factors. First, let's look at what we mean by system isotherm. Freundlich liquid phase isotherm studies can be used to establish the adsorptive capacity of activated carbon over a range of different concentrations. Under standard conditions, the adsorptive capacity of activated carbon increases as the concentration increases, until we reach a point of maximum saturation capacity. An example of an isotherm for phenol is shown in Figure 8.

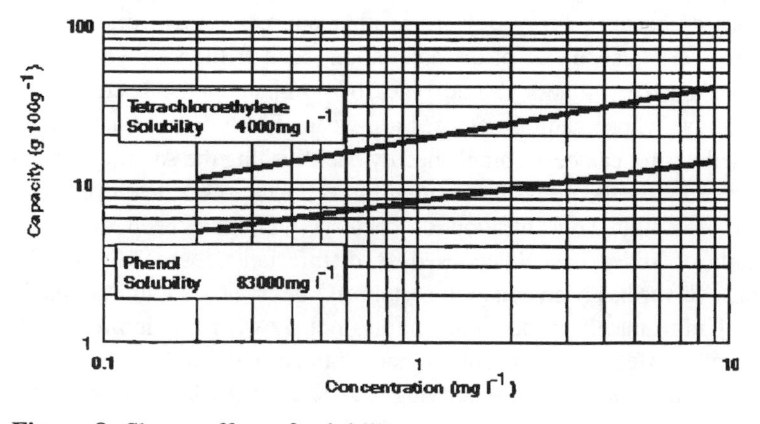

Figure 8. Shows effect of solubility.

The Freundlich liquid phase isotherm can be used to determine the effect of solubility on the adsorptive capacity of activated carbon over a range of different concentrations. Phenol is highly soluble due to its polar nature whilst, in comparison, tetrachloroethylene (PCE) has a low solubility due to being non-polar. In the isotherms illustrated, the concentration of phenol is low relative to its solubility limit and consequently, the adsorptive capacity peaks at 18% maximum (see Figure 9). In comparison the concentration of tetrachloroethylene is relatively close to its solubility limit and, accordingly, the adsorptive capacity is exceptionally good.

The effect of linear flow rate through an activated carbon adsorption unit can be illustrated by the mass transfer zone (MTZ diagram, Figure 10). After a period of operation, the top portion of the bed becomes fully loaded with the adsorbed organics. This is known as the saturation zone. Removal of dissolved organic matter occurs in the section of the bed referred to as the MTZ. The length of the MTX can be optimized at lower flow rates. Other considerations such as concentration and particle size influence the MTZ. The equilibrium zoneremains unaffected until the MTZ proceeds through the bed.

Initially, upon start-up of an adsorption unit, a slight increase in the effluent water's pH is observed. This is attributed to trace leaching of the soluble matter from the matrix of the activated carbon.

Leaching can be controlled by application of efficient backwashing, which will readily remove any soluble materials. The system will then reach equilibrium with the pH of the feedwater.

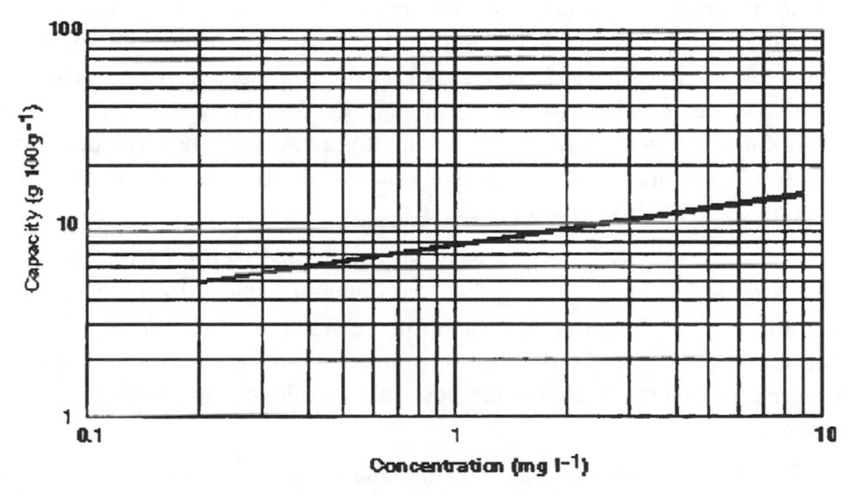

Figure 9. Example of isotherm for phenol adsorption.

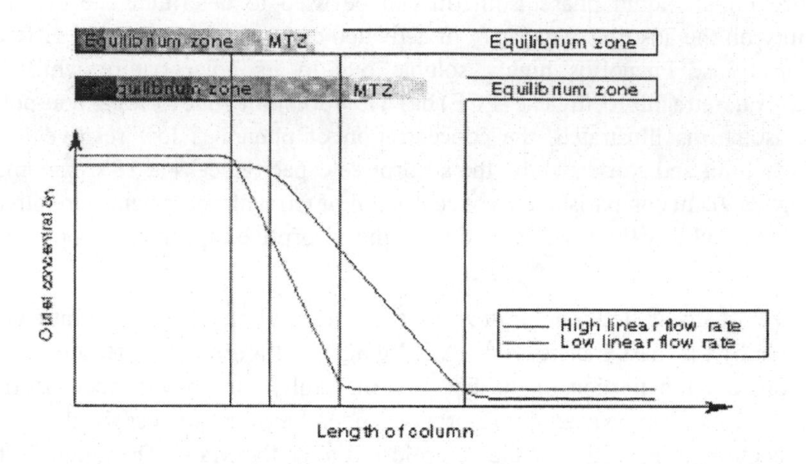

Figure 10. MTZ plot showing equilibrium zones.

Adsorption efficiency can be optimized by using finer particle size products which will improve the diffusion rate to the surface of the activated carbon. However, there is a tradeoff in using finer particles with pressure drop and, hence energy use. Note that during start-up of an activated carbon filter bed, a bed expansion of 25 to 35 % is recommended in order to remove soluble matter and to stratify particles in order to ensure that the MTZ is maintained when future backwashing is performed.

To best understand adsorptive solvent recovery we have to consider some fundamentals of adsorption and desorption. In a very general sense, adsorption is the term for the enrichment of gaseous or dissolved substances (the adsorbate) on the boundary surface of a solid (the adsorbent). On their surfaces adsorbents have what we call *active centers* where the binding forces between the individual atoms of the solid structure are not completely saturated. At these active centers an adsorption of foreign molecules takes place.

The adsorption process generally is of an exothermal nature. With increasing temperature and decreasing adsorbate concentration the adsorption capacity decreases. For the design of adsorption processes it is important to know the adsorption capacity at constant temperature in relation to the adsorbate concentration. Figure 11 shows the adsorption isotherms for several common solvents.

Remember that this technology is versatile, and is applied equally well to solvent recovery and pollution control applications in gas as well as liquid systems. Let's now focus attention on the applications in water treatment.

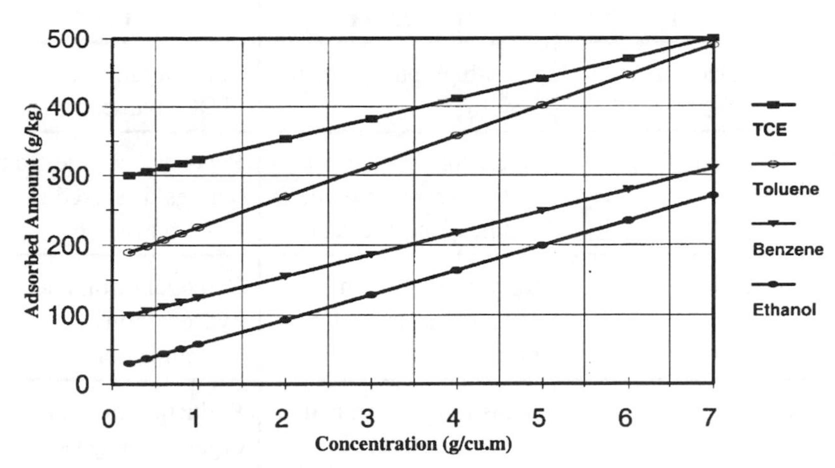

Figure 11. Adsorption isotherms of commons solvents in vapor state.

APPLICATIONS

Applications of carbon adsorption go far beyond conventional water treatment applications which we will discuss in a general sense shortly. Table 8 provides a summary of the key applications of carbon adsorption systems for liquid phase applications.

Table 8. Liquid Phase Applications of Carbon Adsorption.

Industry	Description	Use
Potable water treatment	Granular activated carbons (GAC) installed in rapid gravity filters	Removal of dissolved organic contaminants, control of taste and odor problems
Soft drinks	Potable water treatment, sterilization with chlorine	Chlorine removal and adsorption of dissolved organic materials
Brewing	Potable water treatment	Removal of trihalomethanes (THM) and phenolics

Industry	Description	Use
Semi-conductors	Ultra-high purity water	Total organic carbon (TOC) reduction
Gold recovery	Operation of carbon in leach, carbon in pulp, and heap leach circuits	Recovery of gold from tailings dissolved in sodium cyanide
Petrochemical	Recycling of steam condensate for boiler feed water	Removal of oil and hydrocarbon contamination
Groundwater	Industrial contamination of ground water reserves	Reduction of total organic halogens (TOX) and adsorbable organic halogens (AOX) including chloroform, tetrachloroethylene, and trichloroethylene
Industrial wastewater	Process effluent treatment to meet environmental discharge standards	Reduction of total organic halogens (TOX), biological oxygen demand (BOD), and chemical oxygen demand (COD)
Swimming pools	Ozone injection for removal of organic contaminants	Removal of residual ozone and control of chloramine levels

The most common application of carbon adsorption in municipal water treatment is in the removal of taste and odor compounds. Figure 12 provides an example of a process flow diagram for a municipal water treatment plant. In this example water is pumped from the river into a flotation unit, which is used for the removal of suspended solids such as algae and particulate matter. Dissolved air is the injected under pressure into the basin. This action creates microbubbles which become attached to the suspended solids, causing them to float. This results in a layer of suspended solids on the surface of the water, which is removed using a mechanical skimming technique. Go back to Chapter 8 if you need to refresh your memory on air flotation systems.

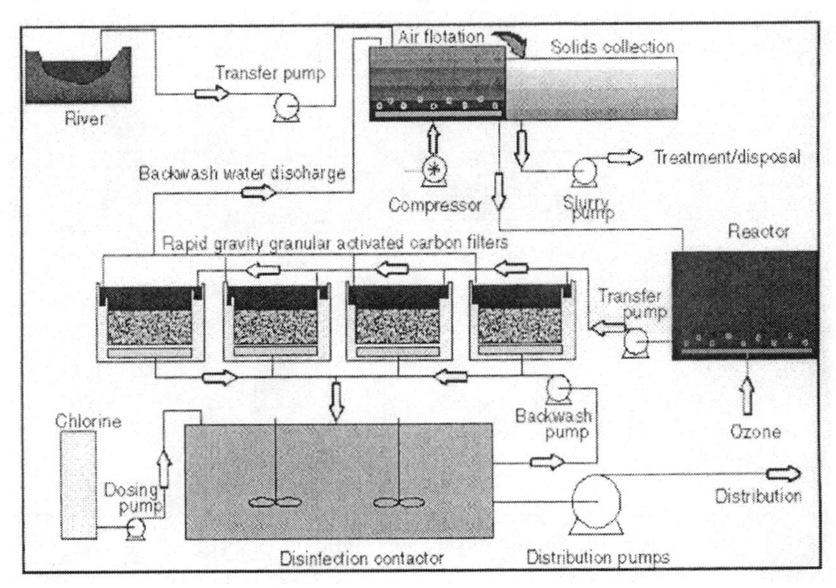

Figure 12. Process flow sheet for municipal water plant in Europe.

The next step in the process involves the production of ozone bypassing high tension, high frequency electrical discharges through air in specially designed units. Ozone is injected into the water to provide bactericidal action and to break down the natural humic compounds that are the cause of taste and odor problems. The water then passes through a rapid gravity filtration system filled with activated carbon (GAC), which adsorbs the compounds resulting from the ozone treatment. Following adsorption, the water is disinfected for supply to the distribution network. Understand that treatment plants are unique, in many ways like oil refineries - i.e., design basis can be substantially different depending on the nature of the water being treated. Figure 13 provides another example of a municipal water treatment facility using PAC. Again the plant is used for the removal of taste and odor compounds.

There are regions where the treatment of water is intended for potable purposes is not necessary at all times during the year. The presense of taste, odor and naturally occurring toxins largely depends on the biological action in areas where lake or reservoir water supply is common. In these situations it is more cost effective to use intermittent dosing of activated carbon into the water during those times of the year where it is needed. The use of PAC is preferred in these case, mainly because no costly fixed bed filtration equipment is required. The PAC can be dosed directly to existing flocculant tanks at a prescribed rate to achieve the level of pollutant removal required.. Shown in Figure 13, following the dosing of PAC the activated carbon is removed as part of the flocculation process, or it can be filtered out by mechanical means. The final stage of water treatment is disinfection, whereupon the water is pumped to the distribution network.

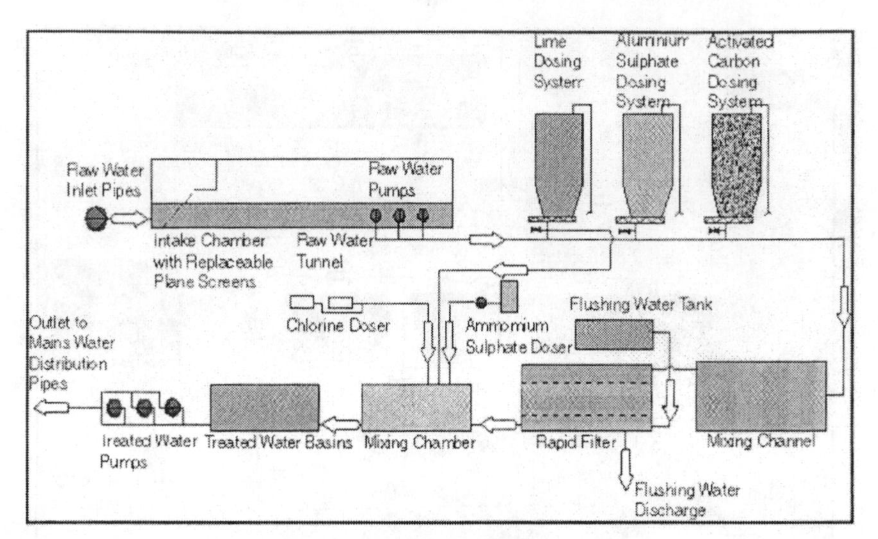

Figure 13.Example of a municipal water treatment plant for taste and odor control.

Nonpotable water treatment is also well within the economical applications of liquid phase adsorption systems. There in fact are so many unique examples of process water treatment throughout the chemical industry that we could go on for days discussing speicif systems. One example of process water treatment is shown in Figure 14. This system (designed by CPL Carbon Link, *go to www.activated-carbon.com on the Web; a good site with lots of technical information from this supplier!*) Shown in Figure 14 shows a process diagram for the removal of creosote and pesticides from the liquid phase in a timber treatment facility. A storage dosing tank is used for smoothing the flow, from where the water is pumped into a chemical dosing system for pH adjustment. Then, ferric sulphate is added to form a precipitate with suspended solids, which is subsequently flocculated by the addition of polyelectolyte.

The water is then pumped through series operated sand filters, which provide the final stage of suspended solids removal and protect the garnualr activated carbon (GAC) filters from particulate contamination. Series operated GAC filters are then used to remove the dissolved creosote and pesticides from the water. To achieve compliance with specifications levels, water should be sampled and analyzed after leaving the first GAC filter. The second GAC filter normally serves as a guard bed.

A final example of application and process layout is shown in Figure 15. In this example the process relies on activated carbon to remove color bodies from a recycled glucose intermediary prior to use in the production of confectionary. The glucose containing the color taint must be mildly heated (to about 70° C), so that the normally solid product becomes less viscous and easier to pump. The syrup is pumped through a series of high efficiency filters (mechanical type) that remove entrained particulate matter and crystallized sugar formed during the heating process.

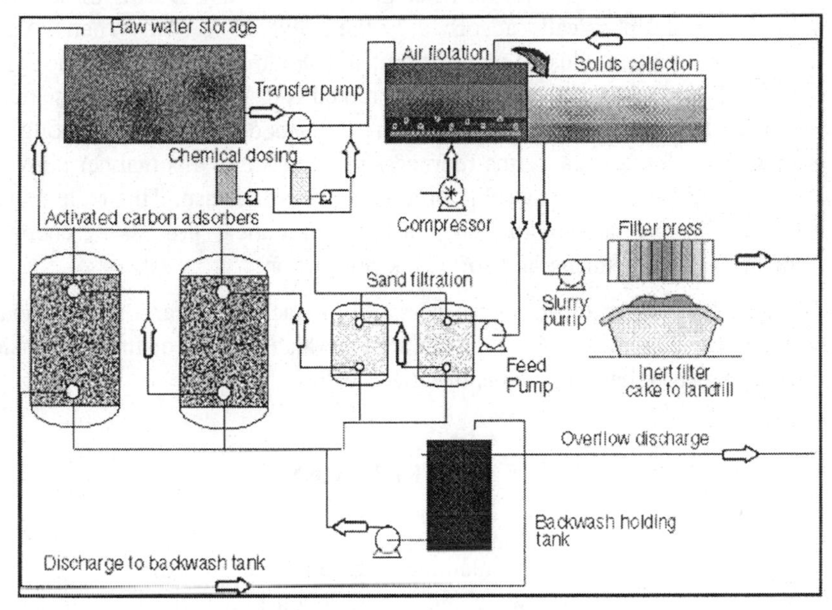

Figure 14. Example of a process water treatment facility.

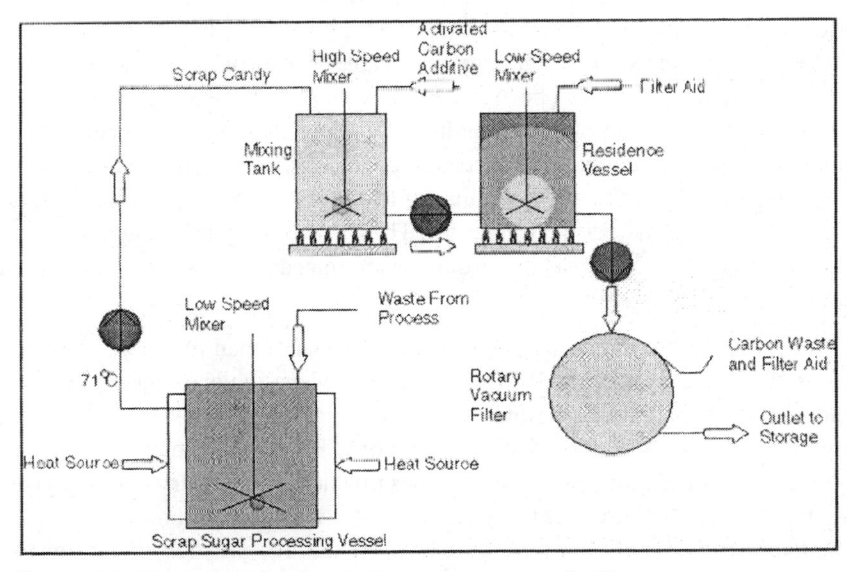

Figure 15. Example of a decolorization treatment facility.

Filtered syrup is then passed through columns containing GAC using a high residence time (a variation is simply the addition of PAC on an as-needed basis – this obviously has cost advantages for batch operated systems). During these stages the color bodies are physically adsorbed by the activated carbon. When PAC is added to the process, the heated syrup is agitated. Following agitation, the syrup undergoes mechanical filtration to remove entrained PAC prior the glucose being used to manufacture the confectionary. This is a good example of a pollution prevention technology, because the reprocessing of waste in this manner allows it to be suitable for re-use as a saleable product after further use. This technique is adaptable in diverse applications such as pharmaceutical processes, chemical intermediaries manufacturing and soft drink production.

These examples help to illustrate the versatility of activated carbon in standard water treatment applications. Another application which merits a distinct discussion is groundwater remediation. This is discussed below.

APPLICATION TO SUBSURFACE TREATMENT

The application of adsorption to contaminated groundwater remediation is not only an important subject, but one we could expand upon into several volumes unto itself. At best, all we can do is try to provide a concise overview in this volume.

When we discuss this subject, we cannot separate groundwater treatment from subsurface soil treatment, as the two often go hand in hand. Furthermore, carbon adsorption is not the only technology applicable to subsurface (soil and water) remediation, but it has indeed been successfully applied over the last twenty or so years to countless site remediation activities. Te selection of the proper remediation technology depends largely upon local site conditions such as the hydrogeology, properties of the groundwater, the nature of the contaminants, soil properties, and a range of other parameters and properties. The standard jargon among remediation specialists that focus only on the groundwater remediation aspect, refer to the treatment technology as *pump and treat*.

Applications have traditionally focused on the removal of man-made contaminants. These include members of the BETX (benzene-ethylene-toluene-xylene) family of components associated with gasoline, and industrial solvents, of which the most notorious ones are the chlorinated hydrocarbons. BETX constituents entered the groundwater from a combination of activities practiced by industry over the years. Perhaps the most common activity in the U.S. whereby groundwater became contaminated by gasoline constituents is from the old practice of storing and disepnsing of this products from single-wall, bare steel underground tanks. Having worked for many years with contractors in site remmediations of auto service stations and industrial facilities, one begins to appreciate how poor this old technology was, and that better than 90 % of these vessels leaked after say 15 years of service (see Figure 16 as an example). It wasn't until the Underground Storage tank regulations (Subtitle C) of RCRA (Resource Conservation and Recovery Act) that the magnitude of this problem was really understood, and groundwater contamination issues began to be addressed on an aggressive basis. Substandard underground storage tank systems led to discharges of gasoline from pitting and corrosion of the vessel walls and in transfer lines and dispensers, and from the overfilling of these vessels. In many cases these were not catastrophic leaks, but rather slow steady losses that cumulatively saturated the soil with contaminants slowly permeating down to the groundwater.

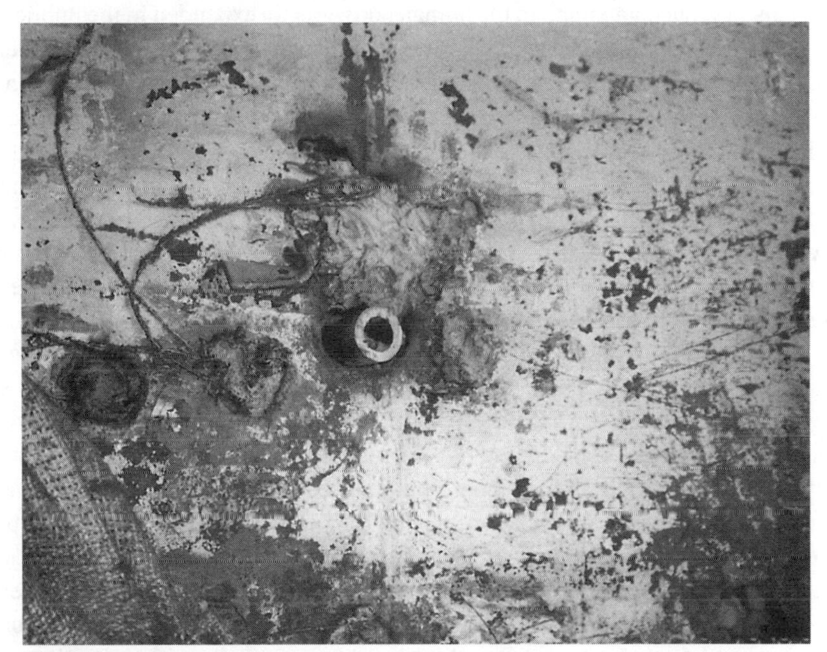

Figure 16. Shows severe pitting and corrosion on the wall of a steel underground storage tank.

Other industrial solvents have entered into the groundwater simply by poor industry practices, and in some cases, downright negligence and disregard for public safety. This may seam like a harsh statement, but we must recognize that environmental legislation as we know it today, did not exist 30 some odd years ago. As such there was no real driving force for industry to protect the environment, other than say their own sense of public safety. Since pollution controls cost money, from a business standpoint, it makes little sense to invest in technologies aimed at cleaning up pollution, and three or more decades ago, the concepts of pollution prevention were simply not within the mainstream of industry thinking. Although a gross over-generalization, certainly there certainly were companies that simply dumped spent solvents directly on bare ground, and in some situations, were well aware of the fact that these discharges could potentially reach groundwater sources that supplied drinking water to the public. In other situations, companies acted out of pure ignorance. Up until and throughout the 1950s, and even later, many industry people believed that because many of the solvents discussed below evaporate quickly, a simply means of disposal was to spread waste or spent solvents directly onto the ground or in unlined lagoons.

Today we have laws that protect our water resources, along with widespread public awareness of the health effects of toxic pollutants. At the same time there are rapidly expanding demands on the potable water supplies by the ever growing population of the United States. Consequently, research into the behavior of chemicals in groundwater and in the human body greatly expanded in the public and private sectors of many developed countries. Many industrial areas have discovered contaminated water supplies, which should not at all be surprising, given the large scale of the petrochemical industries, and their rapid and spectacular growth over the last century in the developed world. Government agencies responded by establishing Maximum Contaminant Level Goals (MCLGs) and maximum contaminant levels (MCLs) for many toxic chemical compounds. The MCLGs are EPA's non-enforceable levels, based solely on possible health risks and exposure. Based on the MCLGs, EPA has established enforceable MCLs, which are set as close to the MCLGs as possible, considering the ability of public water systems to detect and remove contaminants using suitable treatment technologies. Today, once a chemical regulated under current federal, state, or local law is identified in soil or water above the MCL, facility operators or property owners are required to initiate assessment and remediation of the contamination within a specified amount of time. If a responsible party cannot be identified or does not have the financial means to clean up the contamination, the government (namely the taxpayers) pays for the required cleanup. Unfortunately, the cost of cleaning up discharges of certain chemicals can greatly exceed the value of the contaminated property. Although the remediation of contaminated soil is a simple process, usually involving the excavation and disposal of the impacted media, or in-situ treatment (thermal methods or vapor extraction), if the contaminant has reached the groundwater (the source of potable water used by most municipal water suppliers), the risk to the public welfare, remedial cost, and amount of time required to remove the contaminants can increase substantially.

The most commonly used remediation technique for the recovery of organic contaminants from ground water has been pump- and-treat, which recovers contaminants dissolved in the aqueous phase. In this regard, the application of carbon adsorption has found extensive, but not exclusive use. Vacuum extraction (also called soil venting) has also become popular for removal of volatile organic contaminants from the unsaturated zone in the gaseous phase. Both of these techniques can, in the initial remediation phase, rapidly recover contaminants at concentrations approximately equal to the solubility limit (pump-and-treat), or the maximum gas phase concentration of the contaminant (vacuum extraction). The maximum gas phase concentration will depend on whether the contaminant is present as a free phase or as a solute in the aqueous phase. During this initial phase, large amounts of the contaminants may be removed. The second phase of the remediation, however, is characterized by rapidly declining contaminant concentration in the effluent as the rate of mass transfer into the flowing phase controls the rate of removal. The third phase of the remediation is characterized by a *tailing* in the effluent of low contaminant concentrations. However, low effluent concentrations may not be a reliable indication of low contaminant levels remaining in the subsurface. Diffusion of contaminants from less-permeable areas into the regions where flow is occurring or the slow desorption of contaminants from the soil surface may control contaminant removal during this phase, and termination of the extraction process before these processes are complete may lead to significant rebounding of the ground water and/or soil air concentrations. This means that the rate-limiting properties of the systems are different in each of the three phases of the remediation: in the first phase, the solubility of the contaminant in the aqueous phase (pump-and-treat) or its maximum gas phase concentration (vacuum extraction); in the second phase, it is the mass transfer step, i.e., dissolution into the aqueous phase (pump-and-treat), or vaporization (vacuum extraction); and, during the third phase, it is diffusion from low permeability areas or the desorption rate.

CONTAMINANT PROPERTIES

Organic contaminants in the subsurface can be present as a separate nonaqueous phase liquid, dissolved in the aqueous phase, in the vapor phase in the soil gas, partitioned into the soil organic matter, or adsorbed onto the solid mineral phase. The relative amount of the contaminant in each of these phases is determined largely by the properties of the contaminant. Generally, the most important property of the soil in determining the distribution of contaminants is the soil organic matter, which normally controls the absorption of hydrophobic compounds. Table 9 is a compendium of common chemicals found to have contaminated groundwaters. The table lists some of the relevant physical properties of volatile and semi-volatile organic chemicals that have been, or have the potential to be, recovered using carbon adsorption and other techniques. Many of these are commonly found at Superfund sites and other sites where ground-water contamination has occurred. Most of these organic chemicals are essentially immiscible with water; acetone and methanol are the exceptions. Many of these compounds have low viscosities and, thus, have the potential to flow readily in the liquid phase. Approximately half of these compounds are less dense than water; the other half are more dense than water. The density of an organic liquid relative to that of water is important in determining the vertical mobility of the contaminant. Those that are less dense than water will tend to float on the groundwater table, while those that are more dense than water may move downward through the aquifer if the pressure in the organic liquid is greater than the displacement pressure of the aquifer materials. Low permeability clay layers in the aquifer may restrict the vertical movement and allow the liquids to accumulate on top of the layer. The properties listed in Table 9 are taken from various literature sources and should be considered average values.

In tackling a groundwater remediation problem, careful attention should be given to the properties of the contaminant relative to water. Let's take for example the two chlorinated hydrocarbons - trichloroethylene (TCE) and perchloroethylene (PCE, tetrachloroethylene). Both of these solvents were often used interchangeably in industrial applications involving the degreasing of metal surfaces, and significant numbers of sites throughout the United Sates have been found to have groundwaters contaminated by these solvents. They both represent genuine health risks (TCE is a suspected carinogen, and both TCEand PCE have low MCLs).

Table 9. Properties of Organic Chemicals Found at Contaminated Sites.

Chemical	Boiling Point °C	Desnity gm/cm^3	Viscosity cP @ 25° C	Water Solubility mg/L	Vapor Pressure mm Hg	Diffusion Coefficient, cm^2/day	
						in H_2O	in Air
Methylene Chloride (Dichloromethane)	40	1.3182 @ 25° C	0.413	20,000 @ 20° C	> 760 @ 50° C		28.82
1,2-Dichloroethylene (trans)	49	1.2444	0.317	600 @ 20° C	> 760 @ 50° C		
Acetone	56.3	0.7899	0.306	∞	622.4 @ 50° C	9417. 6 0° C	30.99
1,1-Dichloroethane	57.4	1.17	0.464	5,500 @ 20° C	608.6 @ 50° C		30.62
1,1-Dichloroethylene (cis)	60	1.2649 @ 25° C	0.445	800 @ 20° C	580.0 @ 50° C		
Trichloromethane (Chloroform)	61.2	1.49	0.537	8,000 @ 20° C	541.3 @ 50° C	7862 0° C	31.28

Chemical	Boiling Point °C	Desnity gm/cm³	Viscosity cP @ 25° C	Water Solubility mg/L	Vapor Pressure mm Hg	Diffusion Coefficient, cm²/day	
						in H_2O	in Air
1-Hexene	63.5	0.675	0.252	50 @ 20° C	485.3 @ 50° C	6212 20° C	30.61
Methanol	64.6	0.791	0.544	∞	400 @ 50° C	14,688	37.43
n-Hexane	68.7	0.659	0.300	9.5 @ 20° C	407.5 @ 50° C	6143 20° C	31.6
1,1,1-Trichloroethane	74.1	1.3303 @ 25° C	0.793	4,400 @ 20° C	360.1 @ 50° C		32.50
Carbon Tetrachloride	76.8	1.5833 @ 25° C	0.908	800 @ 20° C	58.3 @ 10° C		32.43
2-Butanone (Methyl Ethyl Ketone)	79.6	0.7994 @ 25° C	0.405	26,800	52.6 @ 10° C		34.76
Benzene	80.1	0.88	0.604	1770 @ 25° C	47.8 @ 10° C	7460 20° C	33.83
Cyclohexane	80.7	0.7731 @ 25° C	0.894	58 @ 25° C	50.5 @ 10° C	7430. 420° C	33.01
1,2-Dichloroethane	84	1.257	0.779	8700 @ 25° C	40.0 @ 10° C		35.61
Trichloroethylene	87.3	1.4578 @ 25° C	0.545	1100 @ 25° C	37.6 @ 10° C	7030	34.54
Toluene	110.6	0.8647 @ 25° C	0.56	515-540 @ 25° C	14.3 @ 10° C	6570 20° C	38.01
4-Methyl 2-Pentanone	116.6	0.802	0.545	19,000	381.0 @ 100° C		40.61

Chemical	Boiling Point °C	Desnity gm/cm³	Viscosity cP @ 25° C	Water Solubi-lity mg/L	Vapor Pressure mm Hg	Diffusion Coefficient, cm²/day	
						in H_2O	in Air
Tetrachloro-ethylene	121.3	1.613 @ 25° C	0.844	150 @ 25° C	400 @ 100° C		39.68
n-Octane	126	0.6986 @ 25° C	0.508	0.7 @ 20° C	368.7 @ 100° C	4910 20° C	41.49
Chloro-benzene	131.7	1.1007 @ 25° C	0.753	490 @ 25° C	323.7 @ 100° C	6394 26° C	40.97
Ethyl-benzene	136.2	0.8654 @ 25° C	0.631	160 @ 25° C	295.7 @ 100° C	6333 20° C	42.24
Xylenes	138.4-144.4	0.8764 @ 25° C	0.608–0.802	160-180 @ 25° C	238.9-280.8 @ 100° C	5980 20° C	42.40-43.43
n-Decane	174.2	0.730	0.838	0.052	77.7 @ 25° C		51.38
Dichloro-Benzene (3 isomers)	173-180	1.2988 25° C (ortho isomer)	1.044-1.324	80-150 @ 25° C	67.1 @ 100° C		36.18-49.00
Dodecane	216.5	0.75	1.383	0.0034	19.5 @ 100° C		61.51
Naphthalene	218	0.97		32 @ 25° C	22.7 @ 100° C	4432 0° C	
1-Methyl naphthalene	244.8	1.020		28.5	0.043 kg/m³		
Hexadecane	286.9	0.773	3.032	0.0063	<1		81.38
Phenan-threne	340	0.98		1.18			72.50

Chemical	Boiling Point °C	Desnity gm/cm³	Viscosity cP @ 25° C	Water Solubi-lity mg/L	Vapor Pressure mm Hg	Diffusion Coefficient, cm²/day	
						in H_2O	in Air
Gasoline		0.73 @ 20° C	32 @ 20° C	30-300		6273 20° C	

Both TCE and PCE are organic molecules composed mainly of carbon and hydrogen atoms; they are unsaturated aliphatic chlorinated hydrocarbon molecules because of the C=C double bonds, and both have the absence of the aromatic ring. Examining the properties of these molecules in comparison with the characteristics of water, shows that both solvents are colorless liquids at room temperature, are less viscous than water, will sink in water, and will dissolve slightly in water. Once discharged onto the ground surface, these chemicals will percolate very quickly into the unsaturated zone (the soil zone between the surface and the soil/water interface, also called the vadose zone). Rainwater and moisture in the soil facilitates rapid transportation of the chemicals to the groundwater. Once the chemicals reach the groundwater, they sink through the water, deeper underground. The liquids will continue to sink until they reach a less permeable layer of earthen material (called a confining layer). There, they will spread out under the influence of gravity or sink through holes in the confining layer.

About 15 gallons of TCE can impact an area 1,000 meters in length, 100 meters in width, and 20 meters in depth with an average concentration of 100 ppb, or roughly 528 million gallons

Remediation of groundwater impacted by dense phase chlorinated solvents is more difficult than spills of chemicals such as gasoline or diesel fuel. Gasoline and diesel fuel are less dense than water and tend to float near the surface of the watertable. Recovery of contaminants dissolved in the water or floating on the water is undertaken through the installation of one or more recovery wells. Such wells are installed to a depth that will facilitate the creation of a cone of depression in the watertable that will capture the contaminant plume. By pumping the groundwater out to a reservoir, and then passing them through a combination activated carbon filter and vapor extraction unit, the solvents can be recovered as waste, and clean water returned to ground. However, this is not as easy as it sounds, because the nature of these solvents can make recovery very difficult at times. Since these solvents will sink though the water, sometimes hundreds of feet, the volume of contaminated water can be much greater. Just one typical industrial drum (55 gallons in size) can impact a very large area and volume of groundwater. These factors make TCE, TCA, and PCE very difficult, and expensive, to clean up.

A further complication is the biotransformations of the hydrocarbons in soil and groundwater. In the early 1980s, chloroethene, cis- and trans-1,2-dichloroethene were detected in groundwater beneath industrial sites that traditionally only used TCE and PCE as degreasers or solvents. Because microorganisms present within the sources of recharge for the area aquifers were found to be active in the decomposition of organic matter, researchers speculated that the organisms could also act on the hydrocarbons, resulting in exotic organic byproducts that are poisonous themselves. It was concluded that some organisms can promote successive transformation of PCE to TCE, and of PCE and TCE to other compounds. Some of the chemical transformations are illustrated in Figure 17. As a result, there are several concerns when dealing with groundwater remediation. First, because of daughter compounds or by-products of environmental transformations, the identification of the source of contamination becomes much more difficult. As a result, the abatement of the source of discharge takes more time, during which additional contaminants may be introduced into the environment. Second, treatment methods developed to remove one compound may prove ineffective in removing the breakdown compounds.

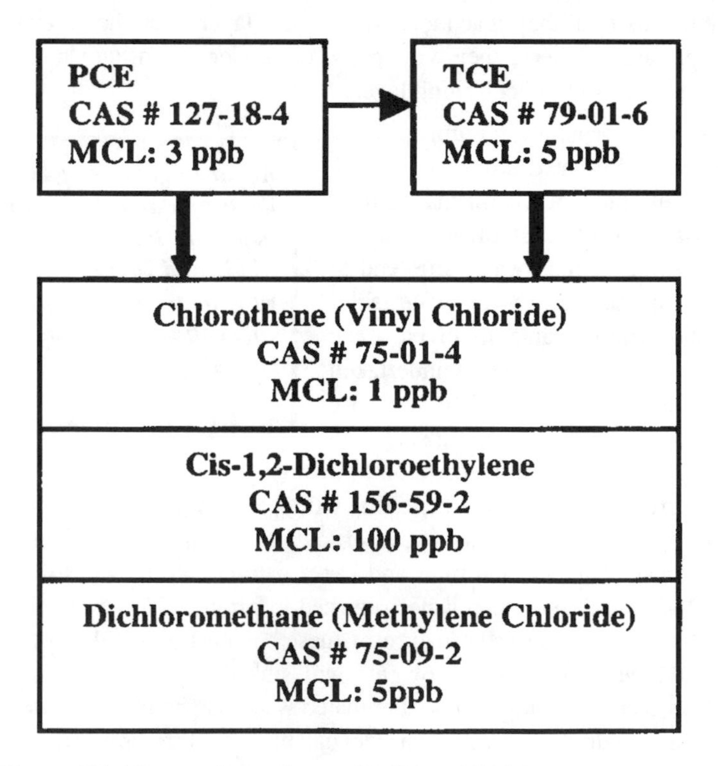

Figure 17. Biotransformations of PCE and TCE.

Third, the chlorinated hydrocarbons may be transformed into contaminants that have lower MCLs or higher MCLs, making remedial activities much more difficult to accomplish. This can mean then that activated carbon, and other standard technologies such as thermal treatment and extraction may have to be used in combination to achieve proper levels of cleanup. Unfortunately many groundwater contamination sites are never fully characterized to the extent of defining satisfactory levels of cleanup. This in fact becomes an argument that once groundwater contamination has occurred, there is essentially permanent impact (or at least many, many years) to the environment and to property values. Table 10 provides you with a summary and comparison of common treatment technologies that can separate organic chemicals from liquid wastestreams.

Table 10. Compares Treatment Technologies that Separate Organics from Groundwater.

Treatment Technology	Feed Stream Properties	Output Stream Characteristics
Carbon adsorption	Aqueous solutions; typical concentration < 1 %; SS < 50 ppm	Adsorbate on carbon; usually regenerated thermally or chemically
Resin adsorption	Aqueous solutions; typical concentration < 8 %; SS < 50 ppm; no oxidants	Adsorbate on resin; always chemically regenerated
Ultrafiltration	Solution or colloidal suspension of high molecular weight organics	One stream concentrated in high molecular weight organics; one containing dissolved ions
Air stripping	Solution containing ammonia; high pH	Ammonia vapor in air
Steam stripping	Aqueous solutions of volatile organics	Concentrated aqueous streams with volatile organics and dilute stream with residuals
Solvent extraction	Aqueous or non-aqueous solutions; concentrations < 10 %	Concentrated solution of organics in extraction solvent
Distillation	Aqueous or non-aqueous solutions; high organic concentrations	Recovered solvent; still bottom liquids, sludge, and tars
Steam distillation	Volatile organics, nonreactive with water or steam	Recovered volatiles plus condensed steam with traces of volatiles

EVALUATING THE MERITS OF CARBON ADSORPTION

Where activated carbon is a potential treatment technology, the first evaluation step is generally to run simple isotherms to determine feasibility. Isotherms are based on batch treatment where impurities reach equilibrium on available carbon surface. While such tests provide an indication of the maximum amount of impurity a GAC can adsorb, it cannot give definite scale up data for a GAC operation due to several factors:

- In a GAC column, dynamic adsorption occurs along an adsorption wave front where the impurity concentration changes.
- GAC rarely becomes totally exhausted in a column.
- Ground GAC exhibits a significantly greater rate of adsorption than a normal GAC.
- Effects of recycling with regeneration cannot be studied.

Because of these factors, pilot column tests should be conducted using the most promising carbons as indicated by the isotherms in order to give an accurate comparison of the carbons. In addition to this, by utilizing reliable scaling-up calculations, pilot tests can be used for system sizing. Pilot column tests can be operated with columns arranged in parallel or in series. The choice of arrangement of the columns is dependent on the scope of the test; such as (1) comparison of GAC grades and/or the effect of regeneration, or (2) scale up to full plant design. In the first case, the columns should be arranged in parallel. In the secind case, the columns arranged in series loaded with one GAC grade. Mobile pilot column systems can be leased from some vendors or assembled using guidelines described below. The columns are mounted vertically and arranged as shown in Figure 18 for the example of scale up tests and downflow operation. It is often good practice to operate columns in upflow as this reduces the opportunity for channeling.

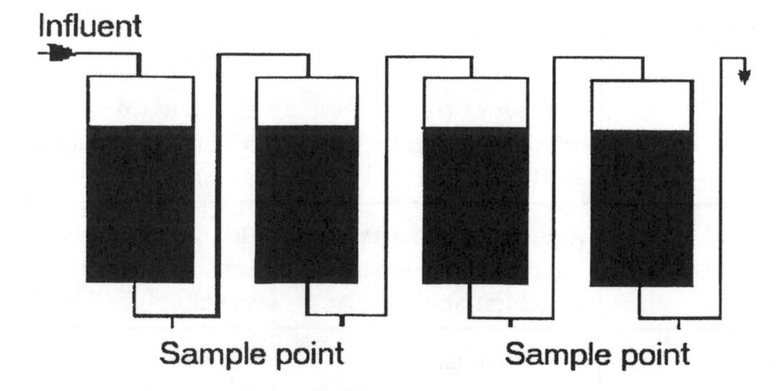

Figure 18. Configuration of column scale-up tests.

It is also preferred where suspended solids create a high pressure drop, or dissolved gases create bubbles in the carbon bed. For a downflow or percolation system, an influent line should be installed at the top of the column, with an effluent at the bottom. To prevent the column from draining during operation, the effluent line from the last column should extend from the bottom of the column to above the top of the column. This will keep the column filled with liquid at all times during operation and prevent siphoning from occurring.

It is recommended that suspended solids be removed from the feed stream to a GAC column. If this is not possible in the scale-up design, then the effect of suspended solids should be included in the pilot run. In the upflow operation, most of the suspended solids work their way up through the GAC bed without a significant increase in pressure drop.

The carbon bed should be at least 60 cm deep with a 4 cm internal diameter. A smaller column is not recommended as the wall effect becomes significant. The carbon bed can be supported by glass wool, wire cloth, etc. Columns and fixtures can be constructed from glass, plastic, reinforced fiberglass, or metal. Borosilicate glass is commonly used. It is essential that all columns used in the pilot system have at least the same internal diameter.

NOTE: To develop reasonably good data for scale-up to full plant design, it is important to have the operation of the pilot column system as near as possible to the anticipated plant conditions. The most critical factors, flow rate and feed impurity concentration, must be constant for the entire test run.

Contact flow rates, Hourly Space Velocity (HSV), or the quantity of feed liquor, are expressed in the number of carbon bed volumes passing through the column per hour. A bed volume is the volume occupied by the carbon bed, including carbon volume and void volume. The recommended HSV range is between 0. 1 and 3.0 depending upon the degree of purification required, the type and concentration of impurity, the nature of the process liquor and the pressure drop. Generally, high levels of purification, high impurity concentrations, and/or high viscosities will require a lower HSV As an example - a typical HSV. for decolourisation of starch based sweeteners is 0.25. The carbon will perform more efficiently, but with a tradeoff in the maount of liquid that can be processed through a column in a given period of time. On the other hand, for the removal of traces of organics in drinking water and wastewater, a HSV range of 2 to 3 produces good results.

As the flow rate and quantity of liquor are the most important controllable variables in developing design data, a feed pump suitable for accurate and continuous flow is required. Depending on the size of the pilot column system, the use of peristaltic, diaphragm, piston-type or centrifugal pumps are recommended. The feed pump should be used in combination with a volumetric or gravimetric flow control device. Before process liquor is delivered to the carbon, suspended matter should be removed, preferably by the same method planned for the plant system.

Adsorption to activated carbon is a function of diffusion rate. This means that the columns should be operated at a temperature at which the liquor approaches the viscosity of water but will not decompose or form too much color. If the process requires operation at elevated temperature, a jacketed tube should be used. The column temperature is controlled by circulating water of the required temperature through the jacket. This requires a thermostatic bath enabling accurate temperature control. When loading the column, care should be taken to avoid entrapping air in the carbon bed. Entrapped air can cause channeling during column operation, preventing complete contact of the process liquor with the carbon particles. In small columns, entrapped air can be avoided by pouring out the carbon in boiling water just before loading. Most of the excess water can be poured off, along with most of the fine carbon particles. About one quarter of the column should be filled with water before loading the carbon. As the carbon is added to the column it should be submerged in the water at all times. Sometimes it is useful to backwash the carbon prior to operation in order to remove remaining dust and entrapped air. It is recommended that a T-junction is used between dosing pump and the columns to allow any entrapped air to be released. Before starting the test, the complete system should be checked by running on water for several hours. After setting the appropriate flowrate, the liquor to be treated can be fed to the columns and this will displace the water. Several samples of the feed liquor should be taken over the duration of the test to highlight any drift in impurity concentration. Samples of effluent liquor after each column should be taken at regular time intervals or after a fixed number of processed bed volumes. If the accuracy of the analysis is affected by undissolved solids, all samples should be filtered prior to examination. For examination of purity levels of taken samples, your standard purity test can be used. When the effluent from the parallel columns or last column in series exceeds the purity requirement, the test should be stopped. It is not recommended to stop the liquor feed during the test. A procedure with intermittent daytime runs will give deviating test results in most cases. When you collect your data, tabulate the information on a spreadsheet. Effluent impurity concentrations for all columns should be plotted against elapsed time or processed bed volumes, generating "Breakthrough curves". The points at which the purity requirement is exceeded are defined as "Breakthrough points".

SOME FINAL COMMENTS ON BOTH TECHNOLOGIES

Both technologies are extremely important to achieving high quality water characteristics, and both are complex – each posing a different set of challenges in scaling up to commercial size operations. You will find that most equipment suppliers have the expertise to tailor their equipment and processes to specific applications, but that in many situations, pilot scale testing will be required.

The overview of these technologies presented in this chapter should give you a flavor for the applications and general principles. For carbon adsorption, we can add one more helpful piece of information. Table 11 provides you with some representative organic chemicals that can be removed from water using activated carbon systems. Following this table is a short glossary of important terms for ion exchange and carbon adsorption.

Table 11. Representative Organic Chemicals and Typical Retentivities on Activated Carbons

Chemical	Formula	Molecular Weight	Boiling Point @ 760 mm Hg, °C	Avg. Retentivity in % at 20° C and 760 mm Hg
Methane Series	C_nH_{2n+2}			
Methane	CH_4	16.04	-184	1
Ethane	C_2H_6	30.07	-86	1
Propane	C_3H_8	44.09	-12	5
Butane	C_4H_{10}	58.12	1	8
Pentane	C_5H_{12}	72.15	37	12
Hexane	C_6H_{14}	86.17	69	16
Heptane	C_7H_{16}	100.20	98.4	23
Octane	C_8H_{18}	114.23	125.5	25
Nonane	C_9H_{20}	128.25	150.0	25
Decane	$C_{10}H_{22}$	142.28	231.0	25
Acetylene Series	C_nH_{2n-2}			
Acetylene	C_2H_2	26.04	-88.5	2
Propyne	C_3H_4	40.06	-23.0	5
Butyne	C_4H_6	54.09	27.0	8
Pentyne	C_5H_8	68.11	56.0	12

Chemical	Formula	Molecular Weight	Boiling Point @ 760 mm Hg, ºC	Avg. Retentivity in % at 20º C and 760 mm Hg
Hexyne	C_6H_{10}	82.14	71.5	16
Ethylene Series	C_nH_{2n}			
Ethylene	C_2H_4	28.05	-103.9	3
Propylene	C_3H_6	42.08	-17.0	5
Butylene	C_4H_8	56.10	-5.0	8
Pentylene	C_5H_{10}	70.13	40.0	12
Hexylene	C_6H_{12}	84.16	64.0	-
Heptylene	C_7H_{14}	98.18	94.9	25
Octalene	C_8H_{16}	112.21	123.0	25
Benzene Series	C_nH_{2n-6}			
Benzene	C_6H_6	78.11	80.1	24
Toluene	C_7H_8	92.13	110.8	29
Xylene	C_8H_{10}	106.16	144.0	34
Isoprene	C_5H_8	68.11	34.0	15
Turpentine	$C_{10}H_{16}$	136.23	180.0	32
Naphthalene	$C_{10}H_8$	128.16	217.9	30
Phenol	C_6H_5OH	94.11	182.0	30
Methyl Alcohol	CH_3OH	32.04	64.7	15
Ethyl Alcohol	C_2H_5OH	46.07	78.5	21
Propyl Alcohol	C_3H_7OH	60.09	97.19	26
Butyl Alcohol	C_4H_9OH	74.12	117.71	30
Amyl Alcohol	$C_5H_{11}OH$	88.15	138.0	35
Cresol	C_7H_7OH	108.13	202.5	30
Methanol	$C_{10}H_{19}OH$	156.26	215	20

Chemical	Formula	Molecular Weight	Boiling Point @ 760 mm Hg, °C	Avg. Retentivity in % at 20° C and 760 mm Hg
Formaldehyde	H_3CHO	30.03	-21.9	3
Acetaldehyde	CH_3CHO	44.05	21.0	7

SHORT GLOSSARY

Acidity: An expression of the concentration of hydrogen ions present in a solution.

Adsorbent: A synthetic resin possessing the ability to attract and to hold charged particles.

Adsorption: The attachment of charged particles to the chemically active group on the surface and in the pores of an ion exchanger.

Alkalinity: An expression of the total basic anions (hydroxyl groups) present in a solution. It also represents, particularly in water analysis, the bicarbonate, carbonate, and occasionally, the borate, silicate, and phosphate salts which will react with water to produce the hydroxyl groups.

Anion: A negatively charged particle or ion. Anion interchange: The displacement of one negatively charged particle by another on an anion-exchange material.

Attrition: The rubbing of one particle against another in a resin bed; frictional wear that will affect the site of resin particles.

Backwash: The countercurrent flow of water through a resin bed (that is, in at the bottom of the exchange unit, out at the top) to clean and regenerate the bed after exhaustion.

Base exchange: The property of the trading of cations shown by certain in- soluble naturally occurring materials (zeolites) and developed to a high degree of specificity and efficiency in synthetic resin adsorbents.

Batch operation: The utilization of ion-exchange resins to treat a solution in a container wherein the removal of ions is accomplished by agitation of the solution and subsequent decanting of the treated liquid.

Bed: A mass of ion-exchange resin particles contained in a column. Bed depth: The height of the resinous material in the column after the exchanger has been properly conditioned for effective operation. Bed expansion: The effect produced during backwashing when the resin particles become separated and rise in the column. The expansion of the. bed due to the increase in the space between resin particles may be controlled by regulating backwash flow.

Bicarbonate alkalinity: The presence in a solution of hydroxyl (OH-) ions resulting from the hydrolysis of carbonates or bicarbonates. When these salts react with water, a strong base and a weak acid are produced, and the solution is alkaline.

Breakthrough: The first appearance in the solution flowing from an ion-ex- change unit of unabsorbed ions similar to those which are depleting the activity of the resin bed. Breakthrough is an indication that regeneration of the resin is necessary.

Capacity: The adsorption activity possessed in varying degrees by ion-ex- change materials. This quality may be expressed as kilograins per cubic foot, gram-milliequivalents per gram, pound-equivalents per pound, gram-milliequivalents per milliliter, and so on, where the numerators of these ratios represent the weight of the ions adsorbed and the de- nominators represent the weight or volume of the adsorbent.

Carbonaceous exchangers: Ion-exchange materials of limited capacity pre- pared by the sulfonation of coal, lignite, peat, and so on.

Carboxylic: A term describing a specific acidic group (COOH) that contrib- utes cation-exchange ability to some resins.

Cation: A positively charged particle or ion. Channeling: Cleavage and furrowing of the bed due to faulty operational procedure, in which the solution being treated follows the path of least resistance, runs through these furrows, and fails to contact active groups in other parts of the bed.

Chemical stability: Resistance to chemical change which ion-exchange resins must possess despite contact with aggressive solutions.

Color-throw: Discoloration of the liquid passing through an ion-exchange material; the flushing from the resin interstices of traces of colored organic reaction intermediates.

Column operation: Conventional utilization of ion-exchange resins in columns through which pass, either upflow or downflow, the solution to be treated.

Cycle: A complete course of ion-exchange operation. For instance, a complete cycle of cation exchange would involve regeneration of the resin with acid, rinse to remove excess acid, exhaustion, backwash, and finally regeneration.

Deashing: The removal from solution of inorganic salts by means of adsorption by ion-exchange resins of both the cations and the anions that comprise the salts. See *deionization.*

Deionization: Deionization, a more general term than *deashing,* embraces the removal of all charged constituents or ionizable salts (both inorganic and organic) from solution.

Demineralizing: See *deashing*.

Density: The weight of a given volume of exchange material, backwashed and in place in the column.

Dissociation: Ionization.

Downflow: Conventional direction of solutions to be processed in ion-ex- change column operation, that is, in at the top, out at the bottom of the column.

Dynamic system: An ion-exchange operation wherein a flow of the solution to be treated is involved.

Efficiency: The effectiveness of the operational performance of an ion ex- changer. Efficiency in the adsorption of ions is expressed as the quantity of regenerant required to effect the removal of a specified unit weight of adsorbed material, for example, pounds of acid per kilogram of salt removed.

Effluent: The solution which emerges from an ion-exchange column. Electrolyte: A chemical compound which dissociates or ionizes in water to

produce a solution which will conduct an electric current; an acid, base, or salt.

Elution: The stripping of adsorbed ions from an ion-exchange material by the use of solutions containing other ions in concentrations higher than those of the ions to be stripped.

Equilibrium reactions: The interaction of ionizable compounds in which the products obtained tend to revert to the substance from which they were formed until a balance is reached in which both reactants and pacts are present in definite ratios.

Equivalent weight: The molecular weight of any element or radical expressed as grams, pounds, and so on divided by the valence.

Exchange velocity: The rate with which one ion is displaced from an ex- changer in favor of another.

Exhaustion: The state in which the adsorbent is no longer capable of useful ion exchange; the depletion of the exchanger's supply of available ions. The exhaustion point is determined arbitrarily in terms of: (1) a value in parts per million of ions in the effluent solution; and (2) the reduction in quality of the effluent water determined by a conductivity bridge which measures the resistance of the water to the flow of an electric current.

Fines: Extremely small particles of ion-exchange materials.

Flow rate: The volume of solution which passes through a given quantity of resin within a given time. Flow rate is usually expressed in terms of feet per minute per cubic foot of resin or as millimeters per minute per millimeter of resin.

Freeboard: The space provided above the resin bed in an ion-exchange column to allow for expansion of the bed during backwashing.

Grain: A unit of weight; 0.0648 grams; 0.000143 pounds.

Grains per gallon: An expression of concentration of material in solution. One grain per gallon is equivalent to 17.1 parts per million.

Gram: A unit of weight; 15.432 grains; 0.0022 pounds.

Gram-milliquivalents: The equivalent weight in grams, divided by 1,000.

Greensands: Naturally occurring materials, composed primarily of complex silicates, which possess ion-exchange properties.

Hardness: The scale-forming and lather-inhibiting qualities which water, high in calcium and magnesium ions, possesses.

Hardness as calcium carbonate: The expression ascribed to the value obtained when the hardness-forming salts are calculated in terms of equivalent quantities of calcium carbonate; a convenient method of reducing all salts to a common basic for comparison.

Head loss: The reduction in liquid pressure associated with the passage of a solution through a bed of exchange material; a measure of the resistance of a resin bed to the flow of the liquid passing through it.

Hydraulic classification: The rearrangement of resin particles in an ion-ex- change unit. As the backwash water flows up through the resin bed, the particles are placed in a mobile condition wherein the larger particles settle and the smaller particles rise to the top of the bed.

Hydrogen cycle: A complete course of cation-exchange operation in which the adsorbent is employed in the hydrogen or free acid form.

Hydroxyl: The term used to describe the anionic radical (OH-) which is responsible for the alkalinity of a solution.

Influent: The solution which enters an ion-exchange unit.

Ion: Any particle of less than colloidal size possessing either a positive or a negative electric charge.

Ionization: The dissociation of molecules into charged particles.

Ionization constant: An expression in absolute units of the extent of dissociation into ions of a chemical compound in solution.

Ion exchange: See fundamental description beginning page 273. Kilograin: A unit of weight; 1,000 grains.

Leakage: The phenomenon in which some of the influent ions are not ad- sorbed and appear in the effluent when a solution is passed through an underregenerated exchange resin bed.

Negative charge: The electrical potential which an atom acquires when it gains one or more electrons; a characteristic of an anion.

pH: An expression of the acidity of a solution; the negative logarithm of the hydrogen-ion concentration (pH 1 very acidic; pH 14, very basic; pH 7, neutral).

pOH: An expression of the alkalinity of a solution; the negative logarithm of the hydroxyl-ion concentration.

pK: An expression of the extent of dissociation of an electrolyte; the negative logarithm of the ionization constant of a compound.

Physical stability: The quality which an ion-exchange resin must possess to resist changes that might be caused by attrition, high temperatures, and other physical conditions.

Positive charge: The electrical potential acquired by an atom which has lost one or more electrons; a characteristic of a cation.

Raw water: Untreated water from wells or from surface sources.

Regenerant: The solution used to restore the activity of an ion exchanger. Acids are employed to restore a cation exchanger to its hydrogen form; brine solutions may be used to convert the cation exchanger to the sodium form. The anion exchanger may be rejuvenated by treatment with an alkaline solution.

Regeneration: Restoration of the activity of an ion exchanger by replacing the ions adsorbed from the treated solution by ions that were adsorbed initially on the resin.

Rejuvenation: See *regeneration.*

Reverse deionization: The use of an anion-exchange unit and a cation-ex- change unit-in that order-to remove all ions from solution.

Rinse: The operation which follows regeneration; a flushing out of excess regenerant solution.

Siliceous gel zeolite: A synthetic, inorganic exchanger produced by the aqueous reaction of alkali with aluminum salts.

Static system: The batch-wise employment of ion-exchange resins, wherein (since ion exchange is an equilibrium reaction) a definite endpoint is reached in which a finite quantity of all the ions involved is present. Opposed to a dynamic, column-type operation.

Sulfonic: A specific acidic group (S03H) on which depends the exchange activity of certain cation adsorbents.

Swelling: The expansion of an ion-exchange W which occurs when the re- active groups on the resin are converted from one form to another.

Throughput volume: The amount of solution passed through an exchange W before exhaustion of the resin is reached.

Upflow: The operation of an ion-exchange unit in which solutions are passed in at the bottom and out at the top of the container.

Voids: The space between the resinous particles in an ion-exchange bed.

Zeolite: Naturally occurring hydrous silicates exhibiting limited base exchange.

RECOMMENDED RESOURCES FOR THE READER

Recommend that you surf the following Web sites for detailed equipment information on ion exchangers and carbon adsorption.

1. **Ion Exchange Chromatography** - Basic principles of ion exchange chromatography and studies conducted from Texas A&M University. http://ntri.tamuk.edu/fplc/ion.html.
2. **Ion Exchange** - Encyclopedia article - good general introduction. (Encarta® Concise Encyclopedia Article)...http://encarta.msn.com/index/conciseinde.
3. **Indian Ion Exchange and Chemical Industries** - Produces reverse osmosis and demineralization systems, base exchange softeners, clarifiers and filters, degassers and de-aerators, filtration and micro filtration systems, effluent treatment plant...http://www.indianionexchange.com .
4. **Dionex Corporation** - Manufacturers of liquid chromatography systems (IC and HPLC), chromatography software data systems, reversed-phase and ion-exchange columns, and accelerated solvent extraction systems...http://www.dionex.com.
5. **Remco Engineering** - Manufacturer of systems for water treatment, waste water recycling, heavy metal recovery and filtration. Provides some good general information plus vendor specific. http://www.remco.com .
6. **Purolite Corporation** - London company offers ion exchange products and polymeric absorbents. Find corporate info, offices worldwide, products, services, and contacts...http://www.purolite.com..
7. **Ion Exchange for Introduction to Biochemical Engineering** - General and company specific information – "Ion exchange can be defined as a reversible exchange of ions between a solid and a liquid in which there is no substantial change in the structure of the solid (Dowex Dow Chemical)". The solids are the ion exchange resin., etc. http://www.rpi.edu/dept/chem-eng/Biotech.
8. **Ion Exchange Resins** - Spectra/Gel Ion Exchange resins are ion exchange media for use in low-pressure liquid chromatography. They are based on a polystyrene/divinylbenzene support and are available for both anion and cation exchange applications. This site will give you a reasonable

background on resin selection criteria and some economic factors for comparisons to other technologies. http://www.lplc.com/misc/ionex.htm..

9. **Ion exchange for Metal Recovery** - A discussion of the trade-offs Author Karrs, Stanley R. Buckley, Deborah Morey Document Type Proceedings article Source 7th AESF/EPA Conference on Pollution Control in the Metal Finishing Industry Subject Resource. Although specific to the metal industry, still some very good technical data that will assist you on resin selection for ion exchange applications. Also, some good schematics of process operations. http://es.epa.gov/techpubs/0/7250.html.

10. **Application of Ion Exchange to Materials Recovery** - A compilation of links to internet recycling resources provided by Allan Barton of Murdoch University, author of Resources Recycling and Recovery. Visit this at the following URL... http://wwwscience.murdoch.edu.au/teachin.

11. **Ion Exchange Resins** - Very good site for you to visit! Contains cost comparisons between RO and ion exchange systems for several design cases. http://www.rohmhaas.com/ionexchange.

12. **Comparison with Reverse Osmosis and Ion Exchange** - Another great site to visit for comparison with reverse osmosis and ion exchange The BioDentmProcessReverse OsmosisIon Exchange THE BioDentm PROCESS The BioDentm process has several important advantages over other nitrate treatment systems: http://www.nitrateremoval.com..

13. **Ion Exchange Simulation** - Uses the Java applet that simulates an ion exchange experiment. Select from topics for information on using the applet. Setting up the experiment Running the simulation. http://www.rit.edu/~pac8612/webionex/web.

14. **JCE 1996 (73) 639 [Jul] Visible Ion Exchange Demonstration for Large or Small Lecture Halls** - Visible Ion Exchange Demonstration for Large or Small Lecture Halls Jerry A. Driscoll Department of Chemistry, University of Utah, Salt Lake City, UT 84112 This demonstration is a colorful illustration of how an ion exchange column works. Some great visual graphics! http://jchemed.chem.wisc.edu.

15. **Ion Exchange Resin Cross Reference Guide** - American L.B. Science & Technology Group Co., Ltd. is a group of manufacturer & manufacturer representative specializing in electronic and mechanical and water treatment products. http://www.americanlb.com/resin/res.html.

16. **Care & Use of Vydac VHP Ion-Exchange Columns** - Guide to Column Care and Use Vydac VHP Protein Ion-Exchange Columns (300VHP, 301VHP, 400VHP) Vydac VHP-Series Protein Ion-Exchange columns consist of a polystyrene-divinylbenzene copolymer bead with a chemically attached hydrophilic surface. A thorough treatment of the theory and some experimental data. http://www.vydac.com/vydacpubs/CMGuides/.

17. **IEXTOOLS (TM) Ion Exchange Software** - IexTools Ion exchange is extensively used in ultrapure (high-purity) water manufacturing. Ion exchange is used in the following industries: power, microelectronics, food, water treatment (potable, wastewater), and hydrometallurgy. http://www.ultrapureh2o.com/product1.htm.

18. **The Zeolite Researchers' List: Catalysis, ion-exchange and separation** - List of researchers in the field of zeolites (including URLs and e-mail-adresses). Subfield: Catalysis, ion-exchange and separation. http://www.tn.utwente.nl/cdr/Staff/Haral.

19. **Adsorption and ion exchange group - Loughborough University** - adsorption and ion exchange group loughborough university department of chemical engineering The adsorption and ion exchange group is concerned with research into environmental pollution control, especially the removal of trace toxic metals and organochemicals. You will find abstracts of research and links to research articles. http://www-staff.lboro.ac.uk/~cgbs2.

20. **Home Water Treatment Using Activated Carbon** - Discussion and guidelines from a 1997 article from the Michigan State University Extension. http://www.msue.msu.edu/msue/imp/modwq/w.

21. **Water Treatment FAQ** - By Patton Turner. Excellent overview of all water treatment methods and associated problems. http://www.providenceco-op.com/waterfaq.

22. **Treatment Systems for Household Water Supplies - Activated Carbon Filtration** - 1992 article from the North Dakota State University Extension Service explaining in detail what activated charcoal systems can and cannot do. http://www.ext.nodak.edu/extpubs/h2oqual.

23. **Activated Carbon for Process Water Treatment: Activated Carbon from CPL Carbon Link** - Activated carbon from CPL Carbon Link for liquid and gas phase purification by adsorption. Activated carbons for all applications including chemical, water, air, solvent recovery, gold recovery, food, automotive, industrial, catalysis.. http://www.activated-carbon.com.

24. **Carbochem** - Supplies carbon and other chemical products based in copper, cerium, nickel, and cobalt. http://www.carbochem.com.

25. **Water Quality and Treatment Handbook, 5th Edition** - State-of-the-art handbook of community water supplies The leading source of information on water quality, water treatment, and quality control for 60 years is now available in an up-to-the-minute new edition. Go to this site for detailed contents of this important publication. http://preview.mcgraw-hill.com/info/com.

26. **Activated Carbon Treatment of Drinking Water** - Cornell Cooperative Extension, New York State College of Human Ecology Activated Carbon Treatment of Drinking Water Linda Wagenet and Ann Lemley Fact Sheet 3, December 1995 Activated carbon filtration (AC) is effective in reducing certain organic chemicals... http://www.cce.cornell.edu/factsheets/wq.

27. **Granular Activated Carbon (GAC) Adsorption (Liquid Phase) page** - Description of granular activated carbon (GAC) adsorption (liquid phase) remediation technology used to clean up pumped ground water contaminated with volatile/semi-volatile organics and PCBs. http://erb.nfesc.navy.mil/restoration/te.

28. **Tailoring Activated Carbon Surfaces for Water, Wastewater and Hazardous Waste Treatment Operations** - Tailoring Activated Carbon Surfaces for Water, Wastewater and Hazardous Waste Treatment Operations EPA Grant Number: R828157 Title: Tailoring Activated Carbon Surfaces for Water, Wastewater and Hazardous Waste Treatment Operations Investigators: Tan. http://es.epa.gov/ncerqa_abstracts/grant.

29. **Ground Water Pumping and Treatment** - Last updated: 17-Jul-96 Ground Water Pumping and Treatment: extracts contaminated ground water and separates the contamination from the water, then destroys the contaminants. Techniques used to treat ground water including carbon adsorption are discussed. http://www.deq.state.la.us/remediation.

30. **Home Water Treatment Using Activated Carbon** - Michigan State University Extension MSU Extension Water Quality Bulletins - WQ239201 07/14/97 Home Water Treatment Using Activated Carbon Introduction Activated carbon (AC) filters. Useful site for residential applications in water purification. http://www.msue.msu.edu/imp/modwq/wq2392...

Recommend you check these hardcopy references out and add them to your library.

31. McPeak, John F. and Harold L. Aronovitch, *Iron in Water and Processes for its Removal*, Hungerford and Terry Inc., Clayton, N.J., 1983.
32. *Guidelines for Canadian Drinking Water Quality*, 4th ed., Minister of National Health and Welfare, Ottawa, 1989.
33. *Nalco Water Treatment Handbook (The)*, 2nd ed., edited by Frank N. Kemmer, McGraw-Hill Inc., New York, 1988.
34. Owens, Dean L., *Practical Principles of Ion Exchange Water Treatment*, Tall Oaks Publishing, Inc., Voorhees NJ, 1985.
35. Sybron Chemicals Inc.,*A Look at the Synthesis of Ion-Exchange Resins*, McGraw-Hill Publishing Co. Inc., New York, 1963.
36. Meltzer, Theodore H., *High Purity Water Preparation*, Tall Oaks Publishing, Inc., Littleton, CO, 1993.

QUESTIONS FOR THINKING AND DISCUSSING

1. Go to the proper U.S. environmental legislation and obtain the MCLs for the chemicals listed in Table 9.
2. Complete the following reactions that take place with hydrogen cation-exchange resins:
   $Na_2(HCO_3)_2 + H_2 Z =$
   $CaSO_4 + H_2Z =$
   $MgZ + H_2SO_4 =$
3. Complete the following reactions that take place for anion exchange reactions (weakly basis and strongly basis):
   $R_3N + H_2SO_4 =$
   $R_3N + 2HCl =$
   $R_3N{\cdot}H_2SO_4 + Na_2CO_3 =$
   $R_4NOH + H_2SiO_3 =$
   $(R_4N)_2SO_4 + 2NaOH =$
   The symbol R_3N represents the complex weakly basic anion-exchanger radical. The symbol R_4N represents the complex strongly basic anion-exchanger radical.
4. The company Rhom and Hass has done a cost comparison between several design cases of RO versus ion exchange. Go to www.rohmhaas.com and review the three design cases presented. Then develop criteria and an analysis which shows a point where the two technologies are equivalent in terms of cost. Be specific in identifying the conditions under which the two technologies become nearly identical in cost in achieving the same degree of water treatment.
5. Although we did not discuss this, you should be able to readily identify commercial adsorbents that can compete with activated carbon in water treatment applications. What are they, what are their properties, and how do unit costs compare? In performing the cost analysis, take into consideration the volumes of adsorbents needed to achieve comparable degrees of water treatment. To do this, you should develop a base case scenario.
6. For a period of ten years, a company was regularly dumping about 50 gallons per month of spent TCE from their degreaser operation on the back lot. It is estimated that about 2 % of this solvent, as free phase, reached the groundwater table. Assuming an average groundwater contamination level of 100 ppb, what volume of groundwater may have been impacted by the illegal dumping?
7. Soil samples have been found to contain 500 ppm of TCE and 750 ppm of benzene. Two technologies are being considered to remediate the soil. One is based on vapor extraction combined with carbon adsorption. The other technique is simply to inject steam into the ground, vaporize the solvents and then withdraw them as a vapor extraction technique, discharging the

vapors directly to the air. It is estimated that the area and soil depth contaminated are 500 m^2 and 10 m, respectively. Address the following: (a) what MCLs will need to be achieved in soil?; (b) instead of steam injection, what other thermal techniques could be applied to treat the soil in-situ?; (c) Of the two solvents, which one is likely to be more effectively removed by carbon adsorption and why?; (d) What possible biotransformations could occur for either of these slvents, and how would you go about removing these daughter compounds?

8. Develop a detailed process scheme for decolorizing recycle water used in a fabric dyeing operation. Assume the operation to be continuous.
9. Develop a detailed list of all parameters that impact on the operation of a carbon filter.
10. The following data were obtained from a pilot adsorption test for refinery wastewater, where the concern is for COD removal. Develop the breakthrough curves for this process.

Time, Hours	**COD Remaining, %**			
	Column 1	**Column 2**	**Column 3**	**Column 4**
25	20	17	13	12
50	25	17	13	11
100	45	22	16	13
150	67	27	17	13
200	80	33	19	14
250	88	44	23	16
300	96	47	28	19
350	97	60	34	24
400	99	68	37	26
450	100	70	41	29
475	97	75	45	-

Chapter 11

WATER STERILIZATION TECHNOLOGIES

INTRODUCTION

This chapter will provide you with a very basic overview of the principles and technologies associated with water purification, or more specifically, sterilization. In very simplistic terms, there are two general classes of technologies, namely those based on chemical methods, and those based on non-chemical technologies. The major application is in purifying water for human consumption purposes. The subject is much broader and complex than presented in this chapter, but our intent is simply to gain a general working knowledge of the techniques that are available to us. Remember to refer to the ***Glossary*** at the end of the book if you run across any terms that are unfamiliar to you.

WHAT WATERBORNE DISEASES ARE

Untreated waters contain a number of harmful pollutants which give the water color, taste, and odor. These pollutants include viruses, bacteria, organic materials, and soluble inorganic compounds, and these must be removed or rendered harmless before the water can be used again. A breakdown of the documented outbreaks identifies acute gastroenteritis, hepatitis shigellosis, ciardiasis, chemical poisoning, typhoid fever, and salmonellosis. Sources of contaminated water can be traced to sernipublic water systems, municipal water systems, and to individual water systems.

In cell culture, it has been shown that one virion can produce infection. In the human host, because of acquired resistance and a variety of other factors, the one virion/one infection possibility does not exist.

Very little is known of the epidemiology of waterborne diseases. The current database is insufficient to determine the scope and intensity of the problem. The devastating effect of epidemics is sufficient to rank water-associated epidemics as a most important public health problem.

Viruses and bacteria may be eliminated by chemical methods or by irradiation, and organic poisons may also be controlled. Inorganic matter must be removed by other means.

WHAT VIRUSES ARE

Viruses are ultramicroscopic organisms. They are parasites; they need to infest a host in order to duplicate themselves. Viruses excreted with human and animal feces are called enteric viruses, and more than 100 such organisms have been identified. As many as one million viruses can be found in one gram of excrement. The concentration in raw sewage varies over a wide range; as many as 463,500 infectious particles per liter of raw sewage have been detected. Viruses found in surface waters are introduced from three major sources. Viruses of human origin can be traced to untreated or inadequately treated domestic sewage. Runoffs from agricultural land, feedlots, and forests introduce viruses from domestic and wild animals and birds. Plant viruses, insect viruses, and other forms of life associated with the aquatic environment may also infect the waters.

WHAT BACTERIA IS

In addition to viruses, bacteria (microscopic organisms that can reproduce without a host in the proper conditions) are also found in water. In general, damage to the human body from bacterial infection is due to the action of the toxins they produce. Bacteria found in water are derived from contact with air, soil, living and decaying plants and animals, and animal excrements. Many of these bacteria are aerobic and anaerobic spore-forming organisms associated with varying densities of coliforms, fecal coliforms, fecal streptococci, staphylococci, chromogenic forms, fluorescent strains, nitrifying and denitrifying groups, iron and sulfur bacteria, proteus species, and pathogenic bacteria. Many bacteria are of little sanitary significance and die rapidly in water. Fecal pollution adds a variety of intestinal pathogens. The most common genera found in water are salmonella, shigella, vibrio, mycobacterium, pasteurella, and leptospira.

HOW WATER BECOMES CONTAMINATED

The circumstances under which water becomes contaminated are as varied as the ways water is taken internally. It is then conceivable that almost any virus could be transmitted through the water route. The increased use of water for recreational purposes increases the incidence of human contact with bodies of water and, consequently, with waterborne viruses and bacteria. The major waterborne viruses among pathogens, and the most likely candidates for water transmission, are the picornaviruses (from pico, meaning very small, and RNA, referring to the presence of nucleic acid). The characteristics of picornaviruses are shown in Table 1. Among the picornaviruses are the enteroviruses (polioviruses, coxsackieviruses, and echoviruses) and the rhinoviruses of human origin. Also included are enteroviruses from excrements of cattle, swine, and other domesticated animals; and rhinoviruses of nonhuman origin, viruses of foot and mouth disease, teschen disease, encephalomyocarditis, mouse encephalomyelitis, avian encephalomyelitis, and vesicular exantherm. of pigs. Additionally, certain viruses can be transported by the water route because their vectors, water molds, and nematodes live in the soil and move with the movement of water. Plant pathogenic viruses also enter the water route and contribute to the problem, though this area has been given little attention in the past. Viruses associated with industrial abattoirs, meat packing, food processing, pharmaceutical, and chemical operations are also a potential problem. All enteric viruses occur in sewage in considerable numbers, and recent detection techniques make it possible to find these viruses in almost all streams that receive sewage effluents. Enteric viruses have been isolated from surface waters around the world. Samples collected from tidal rivers in the United States contained viruses in 27 percent to 52 percent of the cases. The contamination of surface water by enteric viruses appears to be ubiquitous.

Table 1. Picornavirus Characteristics (Very Small RNA Viruses)

1. Small spheres about 20-30 It in diameter
2. RNA core, icosahedral form of cubic symmetry
3. Resistant to ether, chloroform, and bile salts, indicating lack of essential lipids
4. Heat stabilized in presence of divalent cations (Molar MgC12)
5. Enteroviruses separated from rhinoviruses by acid lability of the latter viruses (inactivated at pH 3.0-5.0)

WHY THEY SURVIVE

A variety of factors is responsible for the survival of viruses in water bodies. Some of the more significant ones are listed for you in Figure 1. The survival of enteric viruses under laboratory conditions and in estuaries varies from a few hours to up to 200 days. Survival in winter is superior to that at summer temperatures. It is not known exactly what happens to these multitudes of viruses introduced in water bodies. The inability of rhinoviruses to withstand pH changes, temperature fluctuations, and the lack of protective covering

Survival of the fittest

Sea or estuary water	**2-130 days**
River water	**2->188 days**
Tap water	**5-168 days**
Soil	**25-175 days**
Oysters	**6-90 days**
Landfill leachates	**7->90 days**
Sediments	**?**

offered by feces and other organic materials probably makes the water route of minor importance in their transmission. These factors do not affect the enteroviruses, which are stable and persist in water for long periods. Coxsackieviruses, it has been found, are relatively resistant to concentrations of chlorine normally used for disinfection of bacteria in water. Studies have shown that enteric viruses easily survive present sewage treatment methods and may survive in waters for a considerable time. It has been shown that oysters incorporate poliovirus into their tissues even when grown in sea water with only small amounts of virus. The traditional processes and techniques currently in use for the removal of viruses from water and wastewater include methods effecting physical removal of the particles and those causing the inactivation or destruction of the organism. Among the first are sedimentation, adsorption, coagulation and precipitation, and filtration. The seconncompasses high pH and chemical oxidation by disinfectants such as halogens. Let's take a look at these technologies.

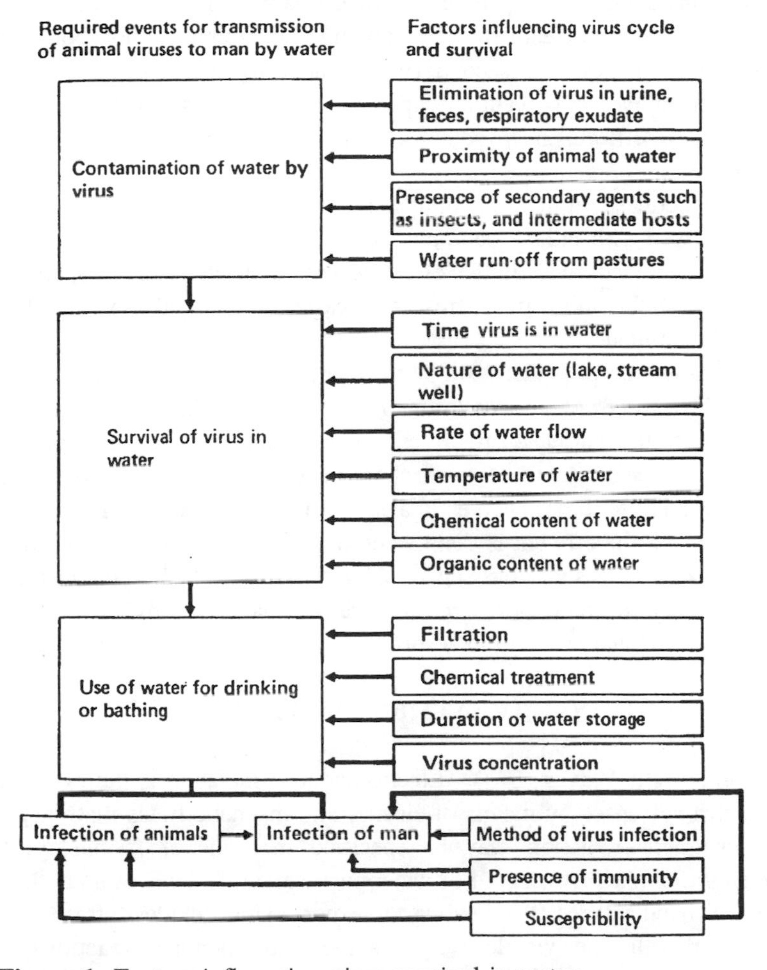

Figure 1. Factors influencing virus survival in water.

TREATMENT OPTIONS AVAILABLE TO US

Primary treatment of municipal waste involving settling and retention removes very few viruses. Sedimentation effects some removal. Virus removal of up to 90 percent (which is a minimal removal efficiency) has been observed after the activated sludge step. Further physical-chemical treatment can result in large reductions of virus titer, coagulation being one of the most effective treatments achieving as much as 99.99 percent removal of virus suspended in water. If high pH (above 11) is maintained for long periods of time, 99.9 percent of the viruses can be removed.

Of all the halogens, chlorine at high doses (40 mg/l for 10 min) is very effective, achieving 99.9 percent reduction. Lower doses (for example, 8 mg/ 1) result in no decrease in virus.

As a result of several studies, the following conclusions regarding viruses in sewage warrant consideration: (1) primary sewage treatment has little effect on enteric viruses; (2) secondary treatment with trickling filters removes only about 40 percent of the enteroviruses; (3) secondary treatment by activated sludge treatment effectively removes 90 percent to 98 percent of the viruses; and (4) chlorination of treated sewage effluents may reduce, but may not eliminate, the number of viruses present.

The current concept of disinfection is that the treatment must destroy or inactivate viruses as well as bacillary pathogens. Under this concept, the use of coliform counting as an indicator of the effectiveness of disinfection is open to severe criticism given that coliform organisms are easier to destroy than viruses by several orders of magnitude.

An important concept is that a single disinfectant may not be capable of purifying water to the desired degree. Also, it might not be practicable or cost effective. This has given rise to a variety of treatment combinations in series or in parallel. The analysis further indicates that the search for the perfect disinfectant for all situations is a sterile exercise. It has been estimated that in the United States only 60 percent of municipal waste effluent is disinfected prior to discharge and, in a number of cases, only on a seasonal basis. Coupling this fact with the demonstration that various sewage treatment processes achieve only partial removal of viruses leaves us with a substantial problem to resolve.

NONCONVENTIONAL TREATMENT METHODS

Electromagnetic Waves (EM) - Electromagnetic radiation is the propagation of energy through space by means of electric and magnetic fields that vary in time. Electromagnetic radiation may be specified in terms of frequency, vacuum wavelength, or photon energy. For water purification, EM waves up to the low end of the UV band will result in heating the water. (This includes infrared as well as most lasers.) In the visible range, some photochemical reactions such as dissociation and increased ionization may take place. At the higher frequencies, it will be necessary to have thin layers of water because the radiation will be absorbed in a relatively short distance. It should be noted that the conductivity and dielectric constant of materials are, in general, frequency dependent. In case of the dielectric constant, it decreases as 1/wavelength. Hence, the electromagnetic absorption will vary with the frequency of the applied field. There may be some anomalies in the absorption spectra in the vicinity of frequencies that could excite molecules. At those frequencies, the absorption could be unusually large.

Ultraviolet radiation in the region between 0.2 μ to 0.3 μ has germicidal properties. The peak germicidal wavelength is around 0.26 μ. This short UV is attenuated in air and, hence, the source must be very near the medium to be treated. The medium must be very thin as the UV will be attenuated in the medium as well.

X-rays and gamma rays are high-energy photons and will tend to ionize most anything with which they collide. They could generate UV in air. At higher energies it is possible for the gamma rays to induce nuclear reactions by stripping protons or neutrons from nuclei. This could result in the production of isotopes and/or the production of new atoms.

Sound - A sound wave is an alteration in pressure, stress, particle displacement, particle velocity, or a combination of these that is propagated in an elastic medium. Sound waves, therefore, require a medium for transmission; that is, they may not be transmitted in a vacuum. The sound spectrum covers all possible frequencies. The average human ear responds to frequencies between 16 Hz and 16 kHz. Frequencies above 20 kHz are called ultrasonic frequencies. Sound waves in the 50-200 kHz range are used for cleaning and degreasing. In water purification applications, ultrasonic waves have been used to effect disintegration by cavitation and mixing of organic materials. The waves themselves have no germicidal effect but, when used with other treatment methods, can provide the necessary mixing and agitation for effective purification.

ELECTRON CONSTANTS

Mass	**0. 109 × 10^{-31} kg**
Charge	**1.602 × 10^{-14} Coulomb**
Spin	**0.5**

Electron Beams - The electron is the lightest stable elementary particle of matter known and carries a unit of negative charge. It is a constituent of all matter and can be found free in space. Under normal conditions, each chemical element has a nucleus consisting of a number of neutrons and protons, the latter equal in number to the atomic number of the element. Electrons are located in various orbits around the nucleus. The number of electrons is equal to the number of protons, and the atom is electrically neutral when viewed from a distance. The number of electrons that can occupy each orbit is governed by quantum mechanical selection rules. The binding energy between an electron and its nucleus varies with the orbit number, and in general the electrons with the shortest orbit are the most tightly bound. An electron can be made to jump from one orbit into another by giving it a quantum of energy. This energy quantum is fixed for any given transition and whether a transition will occur is again governed by selection rules. In other words, although an electron is given a quantum of energy sufficient to raise it to an adjacent higher state, it will not go up to that state if the transition is not permitted. In that case, it is theorized that if the electron absorbs the quantum, it will most probably go up to the excited state, remain there for a time allowed by the uncertainty principle, reradiate the quantum, and return to its original state. If an electron is given a sufficiently large quantum of energy, it will completely leave the atom. The electron will carry off as kinetic energy the difference between the input quantum and the energy required to ionize. The remaining atom will now become a positively charged ion, and the stripped electron will become a free electron. This electron may have sufficient energy when it leaves the atom (or it may acquire sufficient energy from an external field) to collide with another atom and strip it of an electron. This is the basis for electric discharge where free electrons are accelerated by an applied field and, as they collide with neutral atoms, generate additional free electrons. This process avalanches as the electrons approach the positive electrode. At the same time, the positively charged ions are accelerated toward the negative electrode. In a vacuum, when a voltage is applied between two electrodes, electrons will move from the cathode to the anode. Of course, in a vacuum there will be no avalanching effects. Electrons are emitted from the cathode by a number of mechanisms:

- **Thermionic Emission -** Because of. the nonzero temperature of the cathode, free electrons are continuously bouncing inside. Some of these have sufficient energy to overcome the work function of the material and can be found in the vicinity of the surface. The cathode may be heated to increase this emission. Also to enhance this effect, cathodes are usually made of, or coated with, a low work-function material such as thorium.
- **Shottky Emission -** This is also a therinionic type of emission except that in this case, the applied electric field effectively decreases the work function of the material, and more electrons can then escape.
- **High Field Emission -** In this case, the electric field is high enough to narrow the work-function barrier and allow electrons to escape by tunneling through the barrier.
- **Photoemission -** Electromagnetic radiation of energy can cause photoemission of electrons whose maximum energy is equal to or larger than the difference between the photon energy and the work function of the material.
- **Secondary Emission -** Electrons striking the surface of a cathode could cause the release of some electrons and, hence, a net amplification in the number of electrons. This principle is used in the construction of photomultipliers where light photons strike a photoemitting cathode releasing photoelectrons. These electrons are subsequently amplified striking a number of electrodes (called dynodes) before they are finally collected by the anode.

Electromagnetism - In a high-gradient magnetic separator, the force on a magnetized particle depends on the intensity of the magnetizing field and on the gradient of the field. When a particle is magnetized by an applied magnetic field, the particle develops an equal number of north and south poles. Hence, in a uniform field, a dipolar particle experiences a torque, but not a net tractive force. In order to develop a net tractive force, a field gradient is required; that is, the induced poles at the opposite ends of the particle must view different magnetic fields. In a simplified, one-dimensional case, the magnetomotive force on a particle is given by:

$$F_m = \mu(\delta H/\delta x) = MV(\delta H/\delta x) = \chi HV(\delta H/\delta x)$$

where μ *is* the magnetic moment of the particle under field intensity, $H \delta H/\delta x$ *is* the field gradient. The magnetic moment μ is the product of the magnetization of the particle and its volume ($\mu u = MV$). And magnetization is the product of the particle susceptibility, χ, and the field intensity, H. In water purification, this magnetic force may be used to separate magnetizable particles.

Direct and Alternating Currents - Electrolytic treatment is achieved when two different metal strips are dipped in water and a direct current is applied from a rectifier. The higher the voltage, the greater the force pushing electrons across the gap between the electrodes. If the water is pure, very few electrons cross the path between the electrodes. Impurities increase conductivity, hence decreasing the required voltage. Additionally, chemical reactions occur at both the cathode and the anode. The major reaction taking place at the cathode is the decomposition of water with the evolution of hydrogen gas. The anode reactions are oxidations by four major means: (1) oxidation of chloride to chlorine and hypochlorite, (2) formation of highly oxidative species such as ozone and peroxides, (3) direct oxidation by the anode, and (4) electrolysis of water to produce oxygen gas.

APPLICATIONS

Electrolytic Treatment - A great deal of interest was generated in the United States prior to 1930 in electrolytic treatment of wastewater, but all plans were abandoned because of high cost and doubtful efficiency. Such systems were based on the production of hypochlorite from existing or added chloride in the wastewater system. A great deal of effort has been made in reevaluating such techniques.

Reduction in Number of Viable Microorganisms by Adsorption onto the Electrodes - Protein and microorganism adsorption on electrodes with anodic potential has been documented. Microorganism adsorption on passive electrodes (in the absence of current) has been observed with subsequent electrochemical oxidation. This does not appear to be a major route for inactivation.

Electrochemical Oxidation of the Microorganism Components at the Anode - Oxidation of various viruses due to oxidation at the surface of the working electrode has been indicated, although the peak voltage used in many experiments would not be sufficient for the generation of molecular or gaseous oxygen.

Destruction of the Microorganisms by Production of a Biocidal Chemical Species - It has been shown that NaCl is not needed for effective operation in the destruction of microorganisms. Biocidal species such as Cl, HO -, 0, CIO, and HOCI occur but have very low diffusion coefficients. Hence, if this phenomenon occurs, the probability is that organisms are destroyed at the electrode surface rather than in the bulk solution.
Destruction by Electric Field Effects - It has been observed that some organisms are killed in midstream without contact with the electrodes. The organisms were observed to oscillate in phase with the electric field. Hence, microorganism kill can also be ascribed to changes caused by changing electromotive forces resulting from the impressed AC.
Electromagnetic Separation - In the typical operation, a magnetized fine-particle seed (typically iron oxide) and a flocculent (typically aluminum sulfate) are added to the wastewater, prompting the formation of magnetic microflocs. The stream then flows through a canister packed with stainless steel wire and a magnetic field is applied. The stainless steel wool captures the flocs by magnetic forces.

OZONATION

Ozone has been used continuously for nearly 90 years in municipal water treatment and the disinfection of water supplies. This practice began in France, then extended to Germany, Holland, Switzerland, and other European countries, and in recent years to Canada. Ozone is a strong oxidizing substance with bactericidal properties similar to those of chlorine. In test conditions it was shown that the destruction of bacteria was between 600 and 3,000 times more rapid by ozone than by chlorine. Further, the bactericidal action of ozone is relatively unaffected by changes in pH while chlorine efficacy is strongly dependent on the pH of the water.
Ozone's high reactivity and instability as well as serious obstacles in producing concentrations in excess of 6 percent preclude central production and distribution with its associated economies of scale.
In the electric discharge (or corona) method of generating ozone, an alternating current is imposed across a discharge gap with voltages between 5 and 25 kV and a portion of the oxygen is converted to ozone. A pair of large-area electrodes are separated by a dielectric (1-3 mm in thickness) and an air gap (approximately 3 mm). Although standard frequencies of 50 or 60 cycles are adequate, frequencies as high as 1,000 cycles are also employed.
The mechanism for ozone generation is the excitation and acceleration of stray electrons within the high-voltage field. The alternating current causes the electron to be attracted first to one electrode and then to the other. As the electrons attain sufficient velocity, they become capable of splitting some oxygen molecules into free radical oxygen atoms. These atoms may then combine with O_2 molecules to form O_3.
Besides the disinfection of sewage effluent, ozone is used for sterilizing industrial containers such as plastic bottles, where heat treatment is inappropriate. Breweries

use ozone as an antiseptic in destroying pathogenic ferments without affecting the yeast. It is also used in swimming pools and aquariums. It is sometimes used in the purification and washing of shellfish and in controlling slimes in cooling towers. Ozone has also been shown to be quite effective in destroying a variety of refractory organic compounds.

ULTRAVIOLET RADIATION

It has been shown that:

- Ultraviolet radiation around 254 mm renders bacteria incapable of reproduction by photochemically altering the DNA of the cells
- A fairly low dose of ultraviolet light can kill 99 percent of the fecal coliform and fecal streptococcus.
- Bacterial kill is independent of the intensity of the light but depends on the total dose.
- Simultaneous treatment of water with UV and ozone results in higher microorganism kill than independent treatment with both UV and ozone.
- When ultrasonic treatment was applied before treating with the UV light, a higher bacteria kill was obtained.
- The UV dose required to reduce the survival fraction of total coliform and fecal streptococcus to 102 (99 percent removal) is approximately 4 x 10 ff Einsteins/ml.

Some limitations are associated with UV radiation for disinfection. These include: (1) The process performance is highly dependent on the efficacy of upstream devices that remove suspended solids; (2) Another key factor is that the UV lamps must be kept clean in order to maintain their peak radiation output; (3) A further drawback is associated with the fact that a thin layer of water (<0.5 cm) must pass within 5 cm of the lamps.

One way of implementing the UV disinfection process at existing activated sludge plants involves suspending the UV lights (in the form of low-pressure mercury arc UV lamps with associated reflectors) above the secondary clarifiers. The effluent is exposed to the UV radiation as it rises over the wire in a thin film.

ELECTRON BEAM

The idea of using ionizing radiation to disinfect water is not new. Ionizing radiations can be produced by various radioactive sources (radioisotopes), by X-ray and particle emissions from accelerators, and by high-energy electrons. The advances in reliable, relatively low-cost devices for producing high-energy electrons are more significant.

Unlike X-rays and gamma rays, electrons are rapidly attenuated. The maximum range of a 1-million-volt electron is about 4 m in air and about 5 cm in water. In transit in matter, an electron loses energy through collisions that ionize atoms and molecules along its path. Bacteria and viruses are destroyed by the secondary ionization products produced by the primary traversing electron. The energetic electrons dissociate water into free radicals H^+ and OH^-. These may combine to form active molecules-hydrogen, peroxides, and ozone. These highly active fragments and molecules attack living structures to promote their oxidation, reduction, dissociation, and degradation. Studies have indicated that 400,000 rads would be adequate for sewage disinfection. At 100 ergs per gram rad, 400,000 rads would raise the temperature of the water or sludge by 1° C. At this dose, each cm^2 of moving sludge would receive about 12×10^{12} electrons, each electron producing some 30,000 secondary ionizations.

BIOLOGY OF AQUATIC SYSTEMS

Before examining the various techniques for purifying water, an understanding of the key biological organisms is necessary. These key organisms include bacteria, algae, protozoa, crustaceans, and fish. Bacteria and protozoa are the major groups of microorganisms. There are a number of waterborne diseases of man caused by bacteria. Some of these organisms are used in evaluating the sanitary quality of water for drinking and recreational purposes.

WATERBORNE DISEASES

There are a number of infectious, enteric (that is, intestinal) diseases of man which are transmitted through fecal wastes. Pathogens (disease-producing agents) include bacteria, viruses, protozoa, and parasitic worms. Widespread diseases generally occur in regions where sanitary disposal of human feces is not practiced. The most common waterborne bacterial diseases are typhoid fever *(Salmonella typhosa), Asiatic cholera* (vibrio comma), and bacillary dysentery (Shi*gella dysenteriae)*. The first of these is an acute infectious disease. Symptoms of typhoid fever are high fever and infection of the spleen, gastrointestinal tract, and blood. For cholera, symptoms include diarrhea, vomiting, and severe dehydration. Dysentery produces diarrhea, bloody stools, and high fever. These diseases can cause death and are still prevalent in many underdeveloped nations. However, in this country, proper environmental control has virtually eliminated these problems. Waterborne outbreaks of infectious hepatitis have occurred. However, the main transmission mechanism is by person-to-person contact. The probability of outbreaks from municipally treated water supplies is low. Symptoms include loss of appetite, nausea, fatigue, and pain. Also, a yellowish color appears in the white of eyes and skin (yellow jaundice is an older term for the disease). It is generally not fatal except to individuals with weaker or older metabolisms. Amoebic dysentery is the most common enteric protozoal infection. It is caused by Endamoeba histolytica and is transmitted by direct contact, food, and through the water in tropical climates. It is not transmittable via water in temperate climates. Disinfection, as with all these diseases, is the safest means of prevention. Bilharziasis (or Schistosomiasis) is a parasitic disease generated by a small, flat worm that can infest the internal organs, such as the heart, lungs, and liver, and even the veins. Eggs of these worms existing in human abdominal organs can be transmitted to water via fecal discharges. Once in water, they hatch into miracida and enter into snails. They then develop into sporocysts that produce fork-tailed cercariae which eventually abandon their shells and attach onto humans. They bore through the skin, enter the bloodstream, and eventually find their way to the internal organs to establish their homes. There is no immunization for this disease. Many feel it is one of the world's worst health problems, particularly in agricultural regions of Africa and South America. Fortunately, this disease does not occur in the United States (the intermediate snail host just happens to be one of several specific species not found on the continental United States).

Bacteria consists of simple, colorless, one-celled plants that utilize soluble food. They are capable of self-reproduction without the aid of sunlight. As decomposers, they represent decaying organic matter in nature. They typically range in size from 0.5-5 and as such are only visible through a microscope. Individual bacteria cells take on various geometries. Typical configurations include spheres, rods, or spirals. They may be single, in pairs, packets, or chains.

Reproduction is by binary fission, meaning a cell divides into two new cells, each of which matures and divides again. Fission takes place every 1,530 mill under ideal conditions. Ideal conditions mean that the growth environment has abundant food, oxygen, and essential nutrients.

Bacteria are named according to a binomial system. The first word is the genus and the second is the species name. The most frequently referred to bacterium in the sanitary field is Escherichia coli. E. coli is a common coliform that can be used as an indicator of water's bacteriological quality. Under a microscope and magnified 1,000 times, cells appear as individual short rods.

There are two major classifications of bacteria called heterotrophic and autotrophic. Heterotrophs, also called saprohytes, utilize organic substances both as a source of energy and carbon. Heterotrophs are further subclassified into three groups. Subclassifications are based on the bacteria's action toward free oxygen. Aerobes need free dissolved oxygen to decompose organics to derive energy for growth and reproduction. This can be described by:

$$\text{Organics} + \text{Oxygen} \rightarrow CO_2 + H_2O + \text{Energy}$$

The second subgroup are anaerobes, which oxidize organics in the absence of dissolved oxygen. This is accomplished by using the oxygen which is found in other compounds (such as nitrate and sulfate). Anaerobic behavior can be described by the following reactions:

$$\text{Organics} + NO_3 \rightarrow CO_2 + N_2 + \text{Energy}$$
$$\text{Organics} + SO_4^{=} \rightarrow CO_2 + H_2S + \text{Energy}$$

Note also that:

Organics → Organic + S + CO_2 + H_2O + Energy

and that the organic acids undergo further reaction:

Organic Acids → CH_4 + CO_2 + Energy

Facultative bacteria comprise the last group and use free dissolved oxygen when available. However, they can also survive in its absence (that is, they also gain energy from the anaerobic reaction). Heterotrophic bacteria decompose organics to obtain energy for the synthesis of new cells, respiration, and motility. Some energy is lost in the process as heat. Autotrophic bacteria oxidize inorganic constituents for energy and utilize carbon dioxide as a source of carbon. The major bacteria types in this class are nitrifying, sulfur, and iron bacteria. Nitrifying bacteria will oxidize ammonium nitrogen to nitrate. Sulfur bacteria perform a reaction given which causes crown corrosion in sewers. Water in sewers quite frequently turns septic and generates hydrogen sulfide gas by generating hydrogen sulfide. The H_2S generated absorbs in the condensation moisture on the sewer side walls and the crown of the pipe. Those sulfur bacteria able to survive at very low pH (pH < 1) oxidize weak H_2S acid to strong sulfuric acid. This oxidation reaction depletes the oxygen from the sewer air. Crown corrosion of concrete-lined systems can greatly reduce the structural integrity of piping and eventually cause walls to collapse. Iron bacteria oxidize soluble inorganic ferrous iron to insoluble ferric. Certain types of filamentous bacteria (Leptothrix and Crenothrix) deposit oxidized iron in the form of $Fe(OH)_3$ in their sheath. This produces yellow or reddish-colored slimes. Water pipes are ideal environments for these type bacteria as they have an abundance of highly dissolved iron content to provide energy and bicarbonates to serve as a carbon source. As these microorganisms mature and die, they decompose, generating obnoxious odors and foul tastes.

PRODUCTS OF CELL SYNTHESIS

ENERGY + ORGANICS →

- **New Cell Growth**
- **Respiration and Motility**
- **Lost Heat**

FUNGI AND MOLDS

Fungi are microscopic nonphotosynthetic plants which include in their classification yeast and molds. Yeasts have a commercial value as they are used for fermentation operations in distilling and brewing. When anaerobic conditions exist, yeasts metabolize sugar, manufacturing alcohol from the synthesis of new cells. Alcohol is not manufactured under aerobic conditions and the yield of new yeast cells is greater. Filamentous forms of fungi are molds. These best resemble higher orders of plant life, having branched or threadlike growths. They grow best in environments consisting of acid solutions with high sugar concentrations. Molds are nonphotosynthetic, multicellular, heterotrophic, and aerobic. The growth of molds can be suppressed by increasing the pH.

ALGAE, PROTOZOA, AND MULTICELLULAR ANIMALS

Algae are microscopic photosynthetic plants. They are among the simplest plant forms, having neither roots, stems, nor leaves. Algae typically range from single-cell entities (which impart a green color to surface waters) to branched forms that can be seen by the naked eye. The latter often appear as attached green slime on surface bodies of water. Diatoms refers to singlecelled algae which are housed in silica shells. The blue-green algae generally associated with water pollution are Anacystis, Anabaena, and Aphanizomellon. Green algae are Oocystis and Pediastrum. Algae are autotrophic; that is, they use carbon dioxide or bicarbonates as sources of carbon. Inorganic nutrients of phosphate and nitrogen as ammonia or nitrate are also used. Some trace nutrients are also necessary (magnesium, boron, cobalt, calcium). The reaction or process by which algae propagate is known as photosynthesis.The products of photosynthesis are new plant growth and oxygen. The energy supplied to the reaction is derived from sunlight. Pigments biochemically convert solar energy into useful energy for plant reproduction and survival. In prolonged absence of sunlight, plant matter performs a dark reaction to exist. In this case, algae absorb oxygen and degrade stored food to produce yield energy for respiratory functions. The reaction rate for the dark reaction is much slower than photosynthesis. Macrophytes are aquatic photosynthetic plants (excluding algae). They often appear on surface bodies of water as floating, submerged, and immersed aggregates. Floating plants are not anchored or rooted. In the animal kingdom, one of the simplest forms is the protozoan. Protozoa are single-celled aquatic animals that have relatively complex digestive systems. They use solid organic material as food and multiply by binary fission. They are aerobic organisms and digest bacteria and algae and, consequently, play an essential role in the aquatic food chain. The smallest type are the flagellated protozoa which range in size from 10 μ to 50 μ. These have long hairlike strands which provide motility by a whiplike action. The amoeba is a member of the protozoa family. Rotifiers are simple, multicelled, aerobic animals. These metabolize solid food. Rotifiers are found in natural waters, stabilization ponds, and extended aeration basins in municipal treatment plants. Crustaceans are multicellular animals (about 2 mm in size). They are herbivores which ingest algae and are in turn eaten by fish.

BIOLOGICAL GROWTH FACTORS

Major factors affecting biological growth are temperature, nutrient availability, oxygen supply, pH, degree of sunlight, and the presence of toxins. Bacteria are classified by their optimum temperature range for growth. For example, mesophilic bacteria grow best in an optimum temperature range for growth. For example, mesophilic bacteria grow best in a temperature range of 10 - 40° C (optimum at 37° C In general, the rate of biological activity almost doubles for every 10-15° C rise in temperature within the range of 5 - 35° C. Beyond 40° C, mesophilic activity drops dramatically and thermophilic growth is initiated (thermophilic bacteria have a range between 45 - 75° C, with an optimum of about 55° C). Thermophilic bacteria are typically more sensitive to temperature variations.

WATER QUALITY TEST METHODS

Determination of the bacteriological quality of water is not a straightforward analysis. The testing for a specific pathogenic bacteria can often lead to erroneous conclusions. Analyses for pathogenic bacteria are difficult to perform. In general, data are not quantitatively reproducible. As an example, if Salmonella was found to be absent from a water sample, this does not exclude the possible presence of Shigella, Vibrio, or disease-producing viruses. The bacteriological quality of water is based on test procedures for nonpathogenic indicator organisms (principally the coliform group).

Coliform bacteria, typified by Escherichia coli and fecal streptococci (enterococci), reside in the intestinal tract of man. These are excreted in large numbers in the feces of humans and other warm-blooded animals. Typical concentrations average about 50,000,000 coliforms per gram. Untreated domestic wastewater generally contains more than 3,000,000 coliforms per 100 ml. Pathogenic bacteria and viruses causing enteric diseases originate from the same source (that is, fecal discharges of diseased persons). Consequently, water contaminated by fecal pollution is identified as being potentially dangerous by the presence of coliform bacteria.

Standards for drinking water specify that a water is safe provided that the test method does not reveal more than an average of one coliform organism per 100 ml. The number of pathogenic bacteria, such as Salmonella typhosa, in domestic wastewater is generally less than 1 per mil coliforms, and the average density of enteric viruses has been measured as a virus-tocoliform ratio of 1:100,000. The die-off rate of pathogenic bacteria is greater than the death rate of coliforms outside of the intestinal tract of animals. Consequently, upon exposure to treatment, a reduction in the number of pathogens relative to coliforms will occur. Water quality based on a standard of less than one coliform per 100 ml is statistically safe for human consumption. That is, there is a high improbability of ingesting any pathogens. This is an Environmental Protection Agency (EPA) standard applicable only to processed water where treatment includes chlorination.

Coliform criteria for body-contact water use and recreational use have been established by most states. Upper limits of 200 fecal coliforms per 100 ml and 2,000 total coliforms per 100 ml have been established. These values are only guidelines since there is no positive epidemiological evidence that bathing beaches with higher coliform counts are associated with transmission of enteric diseases. Some experts feel that these standards may be too conservative from a standpoint of realistic public health risk. In recent years, no cases of enteric disease have been linked directly to recreational water use in this country. Coliform standards applied to water used for swimming are linked to water-associated diseases of the skin and respiratory passages rather than enteric diseases. This, naturally, is entirely different than the purpose of the coliform standard for drinking water, which is related to enteric disease transmission. Here tighter restrictions are imperative, since a water distribution system has the potential of mass transmission of pathogens in epidemic proportions.

Water sample collection techniques differ depending on the source being tested. The minimum number of water samples collected from a distribution system which are examined each month for coliforms is a function of the population. For example, the minimum number required for populations of 1,000 and 100,000 are 2 and 100, respectively. To ascertain compliance with the bacteriological requirements of drinking water standards, a certain number of positive tests must not be exceeded. When 10-ml standard portions are examined, not more than 10 percent in any month should be positive (that is, the upper limit of coliform density is an average of one per 100 ml).

Coliforms are defined as all aerobic and facultative anaerobic, nonsporeforming species. Gram-stain negative rods ferment lactose and produce gas within 48 h of incubation (at 35° C). The initial coliform analysis is the presumptive test which is based on gas production from lactose. In this test, 10ml portions of water samples are transferred into prepared fermentation tubes using sterile pipettes. The tubes contain lactose or lauryl tryptose, broth, and inverted vials. Inoculated tubes are placed in a warm-air incubator (at 35° C $\pm$ 0.5° C). Growth with the production of gas (the gas is identified by the presence of bubbles in the inverted vial) means a positive test. That is, it indicates that coliform bacteria may be present. A negative reaction, either no growth or growth without gas, excludes coliforms.

Such tests are employed to substantiate or refute the presence of coliforms in a positive presumptive test. In normal, potable water coliform testing, the test is confirmed using brilliant green bile broth. Occasionally, one may desire to run a completed test. This involves transferring a colony from an Endo (or EMB plate) to nutrient agar and into lactose broth. If gas is not produced in the lactose fermentation tube, the colony transferred did not contain coliforms and the test is negative. If gas is generated, a portion of growth on the nutrient agar is smeared onto a glass slide and prepared for observation under a microscope using the Gram-stain technique. If the bacteria are short rods, with no spores present, and the Gramstain is negative, the coliform group is present and the test is completed. If the culture Gram-stains positive (purple color), the completed test is negative.

In examining surface water quality, an elevated-temperature coliform test is used to separate microorganisms of the coliform group into those of fecal and nonfecal sources. This approach is applicable to studies of stream pollution, raw-water sources, wastewater treatment systems, bathing waters, and general water quality monitoring. It is not recommended as a substitute for the coliform tests used in examination of potable waters.

The water analysis is incomplete unless the number of coliform bacteria present is determined as well. A multiple-tube fermentation technique can be used to enumerate positive presumptive, confirmed, and fecal coliform tests. Results of the tests are expressed in terms of the most probable number (MPN). That is, the count is based on a statistical analysis of sets of tubes in a series of serial dilutions. MPN is related to a sample volume of 100 ml. Thus, an MPN of 10 means 10 coliforms per 100 ml of water.

For MPN determination, sterile pipettes calibrated in 0.1-ml increments are used. Other equipment includes sterile screw-top dilution bottles containing 99 ml of water and a rack containing six sets of five lactose broth fermentation tubes. A sterile pipette is used to transfer 1.0-ml portions of the sample into each of five fermentation tubes. This is followed by dispensing 0.1 ml into a second set of five. For the next higher dilution (the third), only 0.01 ml of sample water is required. This small quantity is very difficult to pipette accurately, so 1.0 ml of sample is placed in a dilution bottle containing 99 ml of sterile water and mixed. The 1.0-ml portions containing 0.01 ml of the surface water sample are then pipetted into the third set of five tubes. The fourth set receives 0.1 ml from this same dilution bottle. The process is then carried one more step by transferring 1.0 ml from the first dilution bottle into 99 ml of water in the second for another hundredfold dilution. Portions from this dilution bottle are pipetted into the fifth and sixth tube sets. After incubation (48 h at 35'C), the tubes are examined for gas production and the number of positive reactions for each of the serial dilutions is recorded.

A final testing technique worth noting is the membrane filter method for coliform testing. This procedure involves passing a measured water sample through a membrane filter to remove the bacteria. The filter is then placed on a growth medium in a petri dish. The bacteria retained by the filter pad grow and establish a small colony. The number of coliforms present is established by counting the number of colonies and expressing this value in terms of number per 100 ml of water. This technique has been widely adopted for use in water quality monitoring studies, especially since it requires considerably less laboratory apparatus than the standard multiple-tubes technique. Also, this technique can be adapted to field studies.

Equipment needed to perform the membrane filter coliform test includes filtration units, filter membranes, absorbent pads, forceps, and culture dishes. The common laboratory filtration unit consists of a funnel that fastens to a receptacle bearing a porous plate to support the filter membrane. The filterholding assembly can be constructed of glass, porcelain, or stainless steel. It is sterilized by boiling, autoclaving, or ultraviolet radiation. For filtration, the assembly is mounted on a side-arm filtering flask which is evacuated to draw the sample through the filter. For field use, a small hand-sized plunger pump or syringe is used to draw a sample of water through the small assembly holding the filter membrane.

Commercial filter membranes are normally 2-in diameter disks with pore openings of 0.45 (±0.02) R. This is small enough to retain microbial cells. Filters used in determining bacterial counts have a grid printed on the surface. To facilitate counting colonies, the filter membranes must be sterilized prior to use, either in a glass petri dish or wrapped in heavy paper. After sterilization, the pads are placed in culture dishes to absorb the nutrient media on which the membrane filter is placed. During the testing, filters are handled on the outer edges with forceps that are also sterilized before use.

Glass or disposable plastic culture dishes are used. If glass petri dishes are employed, a humid environment must be maintained during incubation. This prevents losses of media by evaporation (the dishes have loose-fitting covers). Disposable plastic dishes have tight-fitting lids which minimize the problem of dehydration.

The size of the filtered sample is established by the anticipated bacterial density. An ideal quantity results in the growth of about 50 coliform colonies and not more than 200 colonies of all types. Often it may be difficult to anticipate the number of bacteria in a sample. Two or three volumes of the same sample must be tested. When the portion being filtered is less than 20 ml, a small amount of sterile dilution water is added to the funnel before filtration. This uniformly disperses the bacterial suspension over the entire surface of the filter. The filter-holding assembly is placed on a suction flask. A sterile filter is placed grid side up over the porous plate of the apparatus using sterile forceps. The funnel is then locked in place holding the membrane. Filtration is performed by passing the sample through the filter under partial vacuum. A culture dish is prepared by placing a sterile absorbent pad in the upper half of the dish and pipetting enough enrichment media on top to saturate the pad. M-Endo medium is used for the coliform group and M-FC for fecal coliforms. The filter is then removed from the filtration apparatus and placed directly on the pad in the dish. The cover is replaced and the culture is incubated (for 24 h at 35° C). For fecal coliforms, incubation is performed by placing the culture dishes in watertight plastic bags and submerging them in a water bath at 44.5° C. Coliform density is calculated in terms of coliforms per 100 ml by multiplying the colonies counted by 100 and dividing this value by the milliliters of the sample filtered.

DISINFECTION BY CHLORINATION

Disinfection has received increased attention over the past several years from regulatory agencies through the establishment and enforcement of rigid bacteriological effluent standards. In upgrading existing wastewater treatment facilities, the need for improved disinfection as well as the elimination of odor problems are frequently encountered. Adequate and reliable disinfection is essential in ensuring that wastewater treatment plants are both environmentally safe and aesthetically acceptable to the public. Chlorine is the most widely used disinfectant in water and wastewater treatment. It is used to destroy pathogens, control nuisance microorganisms, and for oxidation. As an oxidant, chlorine is used in iron and manganese removal, for destruction of taste and odor compounds, and in the elimination of ammonia nitrogen. It is, however, a highly toxic substance and recently concerns have been raised over handling practices and possible residual effects of chlorination.

PROPERTIES OF CHLORINE AND ITS CHEMISTRY

Chlorine (Cl_2) is a greenish-yellow-colored gas having a specific gravity of 2.48 as compared to air under standard conditions of temperature and pressure. It was discovered in 1774 from the chemical reaction of manganese dioxide ($MnNO_2$) and hydrochloric acid (HCl) by the Swedish chemist, Scheele, who believed it to be a compound containing oxygen. In 1810, it was named by Sir Humphrey Davy, who insisted it was an element (from the Greek work chloros, meaning greenish-yellow). In nature, it is found in the combined state only, usually with sodium as salt (NaCl), carnallite ($KMgCl_36H_2O$), and sylvite (Kcl).

Chlorine is a member of the halogen (salt-forming) group of elements and is derived from chlorides by the action of oxidizing agents and, most frequently, by electrolysis. As a gas, it combines directly with nearly all elements. At 10° C, 1 volume of water dissolves about 3.10 volumes of chlorine; at 30° C, only 1.77 volumes of Cl_2 are dissolved in 1 volume of water.

In addition to being the most widely used disinfectant for water treatment, chlorine is extensively used in a variety of products, including paper products, dyestuffs, textiles, petroleum products, pharmaceuticals, antiseptics, insecticides, foodstuffs, solvents, paints, and other consumer products. Most chlorine produced is used in the manufacture of chlorinated compounds for sanitation, pulp bleaching, disinfectants, and textile processing. It is also used in the manufacture of chlorates, chloroform, and carbon tetrachloride and in the extraction of bromine. Among other past uses, chlorine served as a war gas during World War I.

As a liquid, chlorine is amber colored and is 1.44 times heavier than water. In solid form, it exists as rhombic crystals. Various properties of chlorine are given in Table 2.

Chlorine gas is a highly toxic substance, capable of causing death or permanent injury due to prolonged exposures via inhalation. It is extremely irritating to the mucous membranes of the eyes and the respiratory tract. It will combine with moisture to liberate nascent oxygen to form hydrochloric acid. If both these substances are present in quantity, they can cause inflammation of the tissues with which they come in contact. Pulmonary edema may result if lung tissues are attacked. Chlorine gas has an odor detectable at a concentration as low as 3.55 ppm. Irritation of the throat occurs at 15 ppm. A concentration of 50 ppm is considered dangerous for even short exposures. At or above concentrations of 1,000 ppm, exposure may be fatal. Chlorine can also cause fires or explosions upon contact with various materials. Table 2.5 lists various substances chlorine can react with to create fire hazards. It emits highly toxic fumes when heated and reacts with water or steam to generate toxic and corrosive hydrogen chloride fumes.

Table 2. General properties of chlorine.

Symbol	Cl
(as gas)	CL_2
Atomic Number	17
Atomic Weight	35.453
Melting Point (°C)	-101
Boiling Point (°C)	-34.5
Liquid Density (0° C and 3.65 atm; g/1)	1.47
Vapor Pressure (mmHg @ 20° C)	4800
Vapor Density (@ STP: g/1)	2.49
Viscosity (micropoises) at	
Temperature = 12.7° C	129.7
= 20°	132.7
= 50°	146.9
= 100°	167.9
= 150°	187.5
= 200°	208.5

In the United States, chlorine was first used as a disinfectant for municipal wastewater treatment in the Jersey City, New Jersey, Boonton reservoir in 1908. This also marked the first legal recognition of chlorine as a disinfectant for public health protection. Chlorine is a strong oxidizing agent and can be used to modify the chemical character of water. For example, it is used to control bacteria, algae, and macroscopic biological-fouling organisms in condenser cooling towers. It is also used to alter the chemical character of some industrial process waters, such as the destruction of sulfur dioxide and ammonia, the reduction of iron and manganese, and the reduction of color (examples include bleaching operations in the pulp and paper industry and oxidation of organic constituents). In water chlorine hydrolyzes to form hypochlorous acid (HOCL), as shown by the following reactions:

$$Cl_2 + H_2O = HOCl + H^+Cl^-$$

The hypochlorous acid undergoes further ionization to form hypochlorite ions (OCl^-):

$$HOCl = H^+ + OCl^-$$

Equilibrium concentrations of HOlI and OCl depend on the pH of the wastewater. Increasing the pH shifts the preceding equilibrium relationships to the right, causing the formation of higher concentrations of HOCl.

Chlorine may also be applied as calcium hypochlorite and sodium hypochlorite. Hypochlorites are salts of hypochlotous acid. Calcium hypochlorite ($Ca(OCl)_2$)

represents the predominant dry form used in the United States. Calcium hypochlorite is commercially available in granular powdered or tablet forms. Either of these forms readily dissolves in water and contains approximately 70 percent available chlorine. Sodium hypochlorite (NaOCl) is commercially available in liquid form at concentrations typically between 5 percent to 15 percent available chlorine. Hypochlorites react in water as follows:

$NaOCl \rightarrow Na^{+}OCl^{-}$

$Ca(OCl)_2 \rightarrow Ca^{+2} + 2OCl^{-}$

$H^{+} + OCL^{-} \rightarrow HOCl$

The amount of HOCl plus OCl in wastewater is referred to as the *free available chlorine*. Chlorine is a very active oxidizing agent and is therefore highly reactive with readily oxidized compounds such as ammonia. Chlorine readily reacts with ammonia in water to form chloramines.

$HOCl + NH_3 \rightarrow H_2O + NH_2Cl$ (monochloramine)

$HOCl + NH_2Cl \rightarrow H_2O + NHCl_2$ (dichloramine)

$HOCl + NHC1_2 \rightarrow H_2O + NC1_3$ (trichloramine)

The specific reaction products formed depend on the pH of the water, temperature, time, and the initial chlorine-to-ammonia concentration ratio. In general, monochloramine and dichloramine are generated in the pH range of 4.5 to 8.5. Above pH 8.5, monochloramine usually exists alone. However, below pH 4.4, trichloramine is produced.When chlorine is mixed with water containing ammonia, the residuals developed produce a curve similar to the one shown in Figure 2. The positive sloped line from the origin represents the concentration of chlorine applied or the residual chlorine if all of that applied appears as residual. The solid curve represents chlorine residuals corresponding to various dosages that remain after some specified contact time. The chlorine demand at a specified dosage is obtained from the vertical distance between the applied and residual curves. Chlorine demand represents the amount of chlorine reduced in chemical reactions (that is, it is the amount that is no longer available). For molar chlorine to ammonia-nitrogen ratios below 1, monochloramine and dichloramine are formed with their relative amounts dependent on pH and other factors. When higher dosages of chlorine are added, the chlorine-to- nitrogen ratio increases, resulting in an oxidation of the ammonia and a reduction of the chlorine. Three moles of chlorine react with two moles of ammonia, generating nitrogen gas and reducing chlorine to the chloride ion:

$2\ NH_3 + 3Cl_2 \rightarrow N_2 + 6HCl$

Residuals of chloramine decline to a minimum value that is referred to as the *breakpoint*. When dosages exceed the breakpoint, free chloride residuals result. Breakpoint curves are unique for different water samples since the chlorine demand is a function of the concentration of ammonia, the presence of other reducing agents, and the contact time between chlorine application and residual testing.

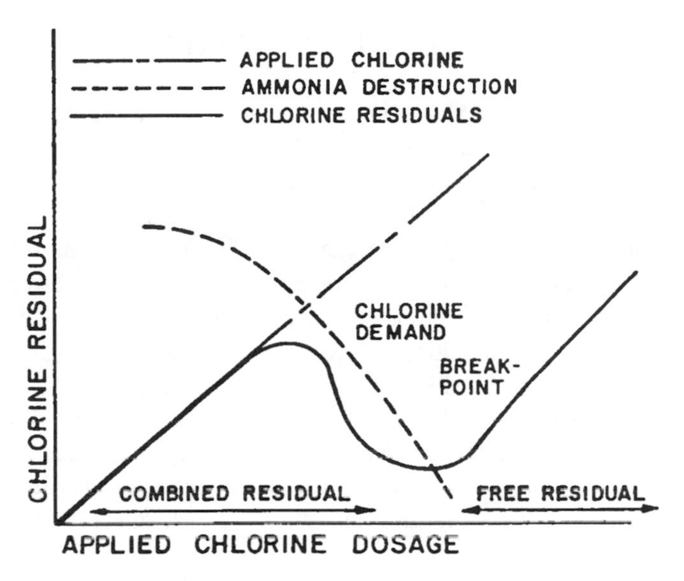

Figure 2. Breakpoint chlorination curve.

GERMICIDAL DESTRUCTION

Chlorine's ability to destroy bacteria and various microorganisms results from chemical interference in the functioning of the organism. Specifically, it is the chemical reaction between HOCl and the bacterial or viral cell structure which inactivates the required life processes. The high germicidal efficiency of HOCl is attributed to the ease by which it is able to penetrate cell walls. This penetration is comparable to that of water and is due both to its low molecular weight (that is, it's a small molecule) and its electrical neutrality. Organism fatalities result from a chemical reaction of HOCI with an enzyme system in the cell which is essential to the metabolic functioning of the organism. The enzyme attacked is triosephosphate dehydrogenase, found in most cells and essential for digesting glucose. Other enzymes also undergo attack. However, triosephosphate dehydrogenase is particularly sensitive to oxidizing agents. The OCl^- ion resulting from the dissociation is a relatively poor disinfectant because of its inability to diffuse through a microorganism's cell walls. This is because of its negative charge. The sensitivity of bacteria to chlorination is well known. However, the effect on protozoans and viruses has not been entirely delineated. Protozoal cysts and enteric viruses are more resistant to chlorine than are coliforms and other enteric bacteria.

However, very little evidence exists to indicate that current water treatment practices are inadequate (no outbreaks of viral or protozoal infections have been reported and waterborne diseases attributed to these pathogens are rare in this country).

CONTACT TIME, pH, AND TEMPERATURE EFFECTS

Hypochlorous acid and hypochlorite ion are known as free available chlorine. The chloramines are known as combined available chlorine and are slower than free chlorine in killing microorganisms. For identical conditions of contact time, temperature, and pH in the range of 6 to 8, it takes at least 25 times more combined available chlorine to produce the same germicidal efficiency. The difference in potency between chloramines and HOCl can be explained by the difference in their oxidation potentials, assuming the action of chloramine is of an electrochemical nature rather than one of diffusion, as seems to be the case for HOCl.

The effect of pH alone on chlorine efficiency is shown in Figure 3. Chlorine exists predominantly as HOCI at low PH levels. Between pH of 6.0 and 8.5, a dramatic change from undissociated to completely dissociated hypochlorous acid occurs. Above pH 7.5, hypochlorite ions prevail; while above 9.5, chlorine exists almost entirely as OCl. Increased pH also diminishes the disinfecting efficiency of monochloramine.

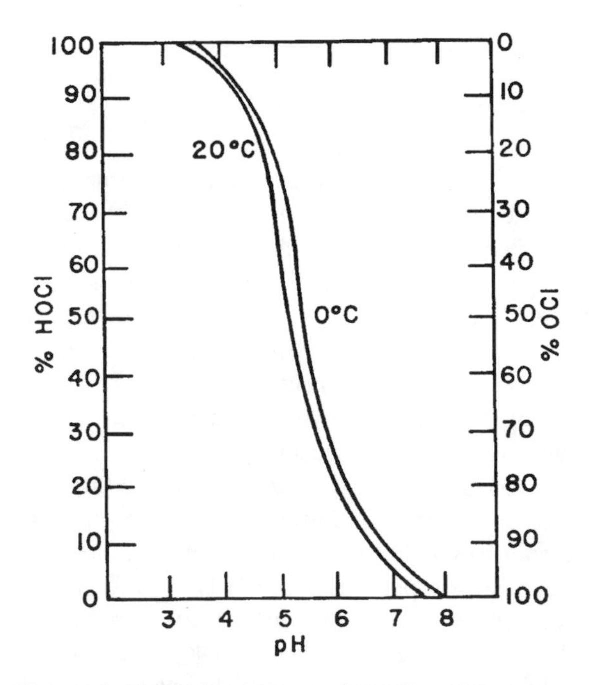

Figure 3. Relative amounts of HOCl and Ocl formed as a function of pH.

It has also been demonstrated that the germicidal effectiveness of free and combined chlorine is markedly diminished with decreasing water temperature. In any situation in which the effects of lowered temperature and high pH value are combined, reduced efficiency of free chlorine and chloramines is marked. These factors directly affect the exposure time needed to achieve satisfactory disinfection.
Under the most ideal conditions, the contact time needed with free available chlorine may only be on the order of a few minutes; combined available chlorine under the same conditions might require hours.

CHLORINE DOSAGE RATES AND RESIDUALS

Table 3 gives recommended ranges of chlorine dosages for disinfection of various wastewaters. Recommended minimum bactericidal chlorine residuals are given in Table 4. Data in Table 4 are based on water temperatures between 20° C to 25° C after a 10-minute contact for free chlorine and a 60 minute contact for combined available chlorine.
The minimum residuals required for cyst destruction and inactivation of viruses are much greater. Although chlorine residuals in Table 4 are generally adequate, surface waters from polluted waterways are usually treated with much heavier chlorine dosages. Ordinary chlorination will destroy all strains of coli, aerogenes, pyocyaneae, typhsa, and dysenteria.
In addition to these microorganisms, three other types are readily destroyed: Enteric vegetative bacteria (Eberthella, Shigella, Salmonella and Vibrio species); Worms such as the block flukes (Schistosoma. species); Viruses (for example, the virus of infectious hepatitis). Each of these groups of organisms differs in its reaction with chlorine.
There is evidence that the comparative reaction of different organisms to one form of chlorine is not necessarily maintained relative to other forms.

Table 3. Recommended Chlorine Dosage Ranges.

Wastewater Type	Chlorine Dosage (mg/1)
Raw Sewage	6-12
(Septic) Raw Sewage	12-25
Settled Sewage	5-10
Chemical Precipitation Effluent	3-10
Trickling Filter Effluent	3-10
Activated Sludge Effluent	2-8
Sand Filter Effluent	1-5

Table 4. Minimum Bactericidal Chlorine Residuals (mg/l).

pH Value	Free Available Chlorine Residual After 10-min Contact	Combined Available Chlorine Residual After 60-min Contact
6.0	0.2	1.0
7.0	0.2	1.5
8.0	0.4	1.8
9.0	0.8	>3.0

CHLORINATION SYSTEMS

Water chlorination is carried out by using both free and combined residuals. The latter involves chlorine application to produce chloramine with natural or added ammonia. Anhydrous ammonia is used if insufficient natural ammonia is present in the wastewater. Although the combined residual is less effective than free chlorine as a disinfectant, its most common application is as a post-treatment following free residual chlorination to provide initial disinfection.

Free residual chlorination establishes a free residual through the destruction of naturally present ammonia. High dosages of chlorine applied during treatment may result in residuals that are esthetically objectionable or undesirable for industrial water uses. Dechlorination is sometimes performed to reduce the chlorine residual by adding a reducing agent (called a dechlor). Sulfur dioxide is often used as the dechlor in municipal plants. Aeration by submerged or spray aerators also diminishes the residual chlorine concentration.

The chlorine used for disinfection is available in three forms: liquified compressed gas, calcium hypochlorite or sodium hypochlorite, and chlorine bleach solutions. Liquid chlorine is shipped in pressurized steel cylinders with sizes typically 100 and 500 lb; one-ton containers are used in large installations. There are two types of chlorine dispensing systems: direct feed and solution feed. The first involves metering dry chlorine gas and conducting it under pressure to the water. Solution-feed systems meter chlorine gas under vacuum and dissolve it in a small amount of water, forming a concentrated solution which is then applied to the water being treated. At 20° C, 1 volume of water dissolves 2.3 volumes of chlorine gas (about 7,000 mg/1). At concentrations of total chlorine below 1,000 mg/l, none of the gas exists in solutions as $C1_2$; all of it is present as HOCl or dissociated ions. Calcium hypochlorite is a dry bleach which is available in granular and tablet forms. Calcium hypochlorite is relatively stable under normal conditions; however, it can undergo reactions with organic materials. It should be stored in an isolated area. Sodium hypochlorite is available in liquid form. It is marketed in carboys and rubber-lined drums for small quantities. Sodium hypochlorite solutions are highly corrosive, unstable, and require storage at temperatures below 85° F. Sodium hypochlorite can either be delivered to the site in liquid form in 500 - 5,000-gallon tank cars or trucks, or manufactured on site. It is normally sold at a concentration of 12 percent to 15 percent by weight of available chlorine. It can be manufactured on site from salt or from sea water.

The main component in a chlorine gas feed is the variable orifice inserted in the feed line to control the rate of flow out of the cylinder. The orifice basically consists of a grooved plug sliding in a fitted ring. Feed rate is adjusted by varying the V-shaped opening. Since a chlorine cylinder pressure varies with temperature, the discharge through such a throttling valve does not remain constant without frequent adjustments of the valve setting. Also, conditions on the outlet side vary with pressure changes at the point of application. Therefore, a pressure-regulating valve is used between the cylinder and the orifice, with a vacuum-compensating valve on the discharge side. A safety pressure-relief valve is held closed by vacuum.

Chlorine feeders can be controlled either manually or automatically based on flow or chlorine residual, or both. In manual mode, a continuous feed rate is established. This is satisfactory when chlorine demand and flow are relatively constant and where operators are available to make adjustments. Automatic proportional control equipment is used to adjust the feed rate to provide a constant preestablished dosage for all rates of flow. This is accomplished by metering the main flow and using a transmitter to signal a chlorine feeder. An analyzer located downstream from the point of application is used to monitor the chlorinator. Combined automatic flow and residual control maintain a present chlorine residual in the water that is independent of the demand and flow variations. The feeder is designed to respond to signals from both the flow meter transmitter and the chlorine residual analyzer.

For hypochlorite solutions, positive-displacement diaphragm pumps (either mechanically or hydraulically actuated) are used. The hypochlorinator consists of a water-powered pump paced by a positive-displacement water meter. The meter register shaft rotates proportionately to the main line flow and controls a cam-operated pilot valve. This in turn regulates water now discharged of hypochlorite that is proportional to the main flow. Admitting main pressure behind the pumping diaphragm balances the water pressure in the pumping head. The advantage of this system is that the pump does not need electrical power. The hypochlorite dosage can be manually adjusted by changing the stroke length setting of the pump.

CHLORINE CONTACT TANKS

The configuration of contact tanks can result in appreciable differences between actual and theoretical contact times. Contact times and germicidal efficiency depend on a number of parameters, the most important being the mixing characteristics of the basin. Proper designs must account for possible flow pattern elimination via short circuiting, acceptable dosage rates, optimum pH range, and upstream removal of ammonia nitrogen.

Rapid dispersement of chlorine at the addition point increases chlorine contact and improves disinfection efficiency. Baffles can be designed to generate turbulence at the chlorine addition point and improve mixing. Baffled systems have the advantage of not requiring mechanical equipment. Mechanical mixing or air agitation can be used where plant hydraulics will not allow the use of baffles, or where a portion of the existing basin can be converted to a mixing chamber and the remainder of the basin and/or a long outfall sewer can be used to provide the needed contact time.

TOXIC EFFECTS OF CHLORINE

The toxicity of chlorine residuals to aquatic life has been well documented. Studies indicate that at chlorine concentrations in excess of 0.01 mg/l, serious hazard to marine and estuarine life exists. This has led to the dechlorination of wastewaters before they are discharged into surface water bodies. In addition to being toxic to aquatic life, residuals of chlorine can produce halogenated organic compounds that are potentially toxic to man. Trihalomethanes (chloroform and bromoform), which are carcinogens, are produced by chlorination.

CHLORINE DIOXIDE

Chlorine dioxide, discovered in 1811 by Davy, was prepared from the reaction of potassium chlorate with hydrochloric acid. Early experimentation showed that chlorine dioxide exhibited strong oxidizing and bleaching properties. In the 1930s, the Mathieson Alkali Works developed the first commercial process for preparing chlorine dioxide from sodium chlorate. By 1939, sodium chlorite was established as a commercial product for the generation of chlorine dioxide.

Chlorine dioxide uses expanded rapidly in the industrial sector. In 1944, chlorine dioxide was first applied for taste and odor control at a water treatment plant in Niagara Falls, New York. Other water plants recognized the uses and benefits of chlorine dioxide. In 1958, a national survey determined that 56 U.S. water utilities were using chlorine dioxide. The number of plants using chlorine dioxide has grown more slowly since that time.

At present, chlorine dioxide is primarily used as a bleaching chemical in the pulp and paper industry. It is also used in large amounts by the textile industry, as well as for the &aching of flour, fats, oils, and waxes. In treating drinking water, chlorine dioxide is used in this country for taste and odor control, decolorization, disinfection, provision of residual disinfectant in water distribution systems, and oxidation of iron, manganese, and organics. The principal use of chlorine dioxide in the United States is for the removal of taste and odor caused by phenolic compounds in raw water supplies.

Chlorine dioxide is a yellow-green gas and soluble in water at room temperature to about 2.9 g/l chlorine dioxide (at 30 mm mercury partial pressure) or more than 10 g/l in chilled water. The boiling point of liquid chlorine dioxide is 11° C; the melting point is - 59° C. Chlorine dioxide gas has a specific gravity of 2.4. The oxidant is used in a water solution and is five times more soluble in water than chlorine gas. In addition, chlorine dioxide does not react with water in the same manner that chlorine does. Chlorine dioxide is volatile; consequently, it can be stripped easily from a water solution by aeration.

Chlorine dioxide has a disagreeable odor, similar to that of chlorine gas, and is detectable at 17 ppm. It is distinctly irritating to the respiratory tract at a concentration of 45 ppm in air. Concentrations above 11 percent can be mildly explosive in air. As a gas or liquid, it readily decomposes upon exposure to ultraviolet light. It is also sensitive to temperature and pressure, two reasons why chlorine dioxide is generally not shipped in bulk concentrated quantities. Chlorine dioxide has a much greater oxidative capacity than chlorine and is therefore a more effective oxidant in lower concentrations. Chlorine dioxide also maintains an active residual in potable water longer than chlorine does. It does not react with ammonia or with trihalomethane precursors when prepared with no free residual chlorine.

Chlorine dioxide is prepared from feedstock chemicals by several methods. The specific method depends on the quantity needed and the safety limitations in handling the various feedstock chemicals. The most common processes are:

From sodium chlorite ($NaClO_2$):

- Acid and sodium chlorite
- Gaseous chlorine and sodium chlorite
- Sodium hypochlorite, acid, and sodium chlorite.

From sodium chlorate ($NaClO_3$):

- The sulfur dioxide process
- The methanol process.

The first group of processes is more commonly used. The second group of processes is frequently used by industry where the quantities produced are much greater than in water utilities.

Oxidation of phenols with chlorine dioxide or chlorine produces chlorinated aromatic intermediates before ring rupture. Oxidation of phenols with either chlorine dioxide or ozone produces oxidized aromatic compounds as intermediates which undergo ring rupture upon treatment with more oxidant and/or longer reaction times. In many cases, the same nonchlorinated, ringruptured aliphatic products are produced using ozone or chlorine dioxide.

In oxidizing organic materials, chlorine dioxide can revert back to the chlorite ion. In the presence of excess chlorine (or other strong oxidant), chlorite can be preoxidized to chlorine dioxide. Using large excesses of chlorine dioxide over the organic materials appears to favor oxidation reactions (without chlorination), but slight excesses appear to favor chlorination. When excess free chlorine is present with the chlorine dioxide, chlorinated organics usually are produced, but in lower yields, depending on the concentration of chlorine and its reactivity with the particular organic(s) involved. Treatment of organic compounds with pure chlorine dioxide containing no excess free chlorine produces oxidation products containing no chlorine in some cases, but products containing chlorine in others.

Under drinking water plant treatment conditions, humic materials and/ or resorcinol do not produce trihalomethanes with chlorine dioxide even when a slight excess of chlorine (1 percent to 2 percent) is present. Also, saturated aliphatic compounds are not reactive with chlorine dioxide. Alcohols are oxidized to the corresponding acids.

The gaseous chlorine-sodium chlorite process for producing chlorine dioxide uses aqueous chlorine and aqueous sodium chlorite to produce a mixture of chlorine dioxide and chlorine (commonly as HOCl). Figure 4 shows such a system, consisting of a chlorine dioxide generator, a gas chlorinator, a storage reservoir for liquid sodium chlorite, and a chemical metering pump. (Sodium chlorite solution can be prepared from commercially available dry chemical by adding it to water.) The recommended feed ratio of chlorine to sodium chlorite is 1:1 by weight. Additional chlorine can be injected into the reactor vessel without changing the overall production of chlorine dioxide.

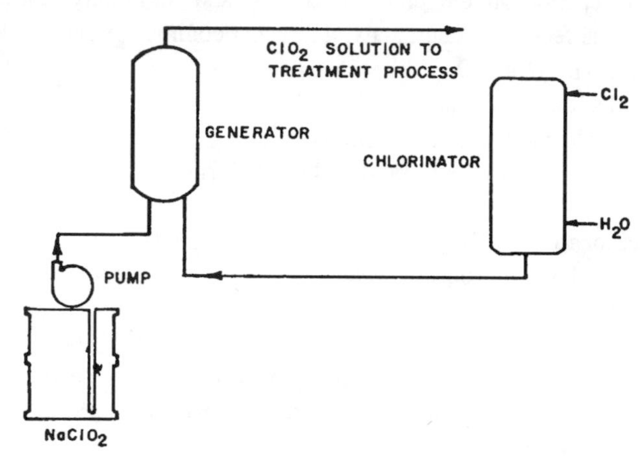

Figure 4. Components of a gaseous sodium chlorite-chlorine dioxide generation system.

A major disadvantage of this system is the limitation of the single-pass gas-chlorination phase. Unless increased pressure is used, this equipment is unable to achieve higher concentrations of chlorine as an aid to a more complete and controllable reaction with the chlorite ion. The French have developed a variation of this process using a multiple-pass enrichment loop on the chlorinator to achieve a much higher concentration of chlorine and thereby quickly attain the optimum pH for maximum conversion to chlorine dioxide. By using a multiple-pass recirculation system, the chlorine solution concentrates to a level of 5-6 g/l. At this concentration, the pH of the solution reduces to 3.0 and thereby provides the low pH level necessary for efficient chlorine dioxide production. A single pass results in a chlorine concentration in water of about 1 g/l, which produces a pH of 4 to 5. If sodium chlorite solution is added at this pH, only about 60 percent yield of chlorine dioxide is achieved. The remainder is unreacted chlorine (in solution) and chlorite ion. When upwards of 100 percent yield of chlorine dioxide is achieved, there is virtually no free chlorite or free chlorine carrying over into the product water. The French system can be designed for variable-feed rates with automatic control by an analytical monitor. This has the advantages of eliminating the chlorine dioxide storage reservoir. Production can be varied by 20 equal increments. A 10 kg/h (530 lb/day) reactor can be varied in 0.5 kg/h (26.5 lb/ day) steps over the range of 0-10 kg/h, and this can be accomplished by automatic control with the monitor located in the main plant control panel.

Another approach to chlorine dioxide production is the acid-sodium chlorite system. The combination of acid and sodium chlorite produces an aqueous solution of chlorine dioxide without production of significant amounts of free chlorine. The acid-based process avoids the problem of differentiating between chlorine and chlorine dioxide for establishing an oxidant residual. This system uses liquid chemicals as the feedstock. Each tank has a level sensor to avoid overfilling. The tanks are installed below ground in concrete bunkers which are capable of withstanding an explosion. There are no floor drains in these bunkers. Any spillage must be pumped with corrosion-resistant pumps. Primary and backup sensors with alarms warn of any spillage. Because of the potential explosiveness, chemicals are diluted prior to the production of chlorine dioxide. The dilution is carried out on a batch basis controlled by level monitors. Proportionate quantities of softened dilution water along with the chemical reagents are pumped to mixing vessels by means of calibrated double-metering pumps. After the reactor is properly filled, an agitator within the container mixes the solution. Dilutions of 9 percent HCl and 7.5 percent sodium chlorite are produced in the chemical preparation process. The chlorine dioxide is subsequently manufactured on a batch basis. The final strength of the solution is about 20 percent, 90 percent to 95 percent of this is chlorine dioxide and 4 percent to 7 percent is chlorine.

Chlorine is the most widely used disinfectant in water treatment. It appears, however, that it may not be the best disinfectant to use for drinking water where poor-quality raw water or completely recycled water is used. Other reasons for considering alternative disinfection techniques include the possibility that disinfection by chloramines will allow viruses to remain viable or that the inactivated virus particles have viable nucleic acids that may be released within humans, the reduction of germicidal efficiency with elevated pH, and the formation of persistent chlorinated organic compounds. Chlorine dioxide has proven to be a strong oxidizing agent. When free of chlorine, it does not form trihalomethane compounds in drinking water. It is less likely than chlorine to form chlorinated compounds with most organics commonly encountered in raw water supplies. Chlorine dioxide is effective in oxidizing organic complexes of iron and manganese, imparts no taste and odor to treated water, and provides a highly stable, long-lasting oxidant residual.

DISINFECTION WITH INTERHALOGENS AND HALOGEN MIXTURES

The interhalogen compounds are the bromine- and iodine-base materials. It is the larger, more positive halogen that is the reactive portion of the interhalogen molecule during the disinfection process. Although only used on a limited basis at present, there are members of this class that show great promise as environmentally safe disinfectants.

PROPERTIES OF BROMINE AND BROMIDES

Bromine (from the Greek word bromos, meaning stench) has an atomic weight 79.909, atomic number 35, melting point - 7.2° C, and boiling point 58.78° C. As a gas it has a density of 7.59 g/l and as a liquid 3.12 g/l (20° C). The element was discovered by Balard in 1826 but not prepared in quantity until 1860. It is a member of the halogen group of elements. Bromine is found mainly in the bromide form, widely distributed and in relatively small proportions. Extractable bromides occur in the ocean and salt lakes, brines, or saline deposits left after these waters evaporated during earlier geological periods. The average bromide content of ocean water is 65 ppm by weight (about 308,000 tons of bromine per cubic mile of sea water). The Dead Sea is one of the richest commercial sources of bromine in the world (containing nearly 0.4 percent at the surface and up to 0.6 percent at deeper levels). In the United States, major sources of bromine are the brine wells in Arkansas, Ohio, and Michigan (bromide contents range from 0.2 percent to 0.4 percent).

Bromine is considered a moderate fire hazard. As liquid or vapor, it can enter spontaneous chemical reactions with reducing materials. It is a very powerful oxidizer. Bromine is considered a highly dangerous material. Upon being heated, it emits highly toxic fumes. It will react with water or steam to produce toxic and corrosive fumes.

Bromine is the only liquid nonmetallic element. It is a heavy, mobile, reddish-brown liquid that readily volatilizes at room temperature to a red vapor having a strong pungent odor. Its disagreeable odor strongly resembles chlorine and has a very irritating effect on the eyes and throat. Bromine is readily soluble in water or carbon disulfide, forming a red solution. It is less active than chlorine but more so than iodine. Bromine unites readily with many elements and has a bleaching action.

The toxic action of bromine is similar to that of chlorine and can cause physiological damage to humans through inhalation and oral routes. It is an irritant to the mucous membranes of the eyes and upper respiratory tract. Severe exposures may result in pulmonary edema. Chronic exposure is similar to therapeutic ingestion of excessive bromides.

The most common inorganic bromides are sodium, potassium, ammonium, calcium, and magnesium bromides. Methyl and ethyl bromides are among the most common organic bromides. The inorganic bromides produce a number of toxic effects in humans: depression, emaciation, and in severe cases, psychoses and mental deterioration. Bromide rashes (called bromoderma) can occur especially on the facial area and resemble acne and furunculosis. This often occurs when bromide inhalation or administration is prolonged. Organic bromides such as methyl bromide and ethyl bromide are volatile liquids of relatively high toxicity. When any of the bromides are strongly heated, they emit highly toxic fumes.

INTERHALOGEN COMPOUNDS AND THEIR PROPERTIES

Interhalogen compounds are formed from two different halogens. These compounds resemble the halogens themselves in both their physical and chemical properties. Principal differences show up in their electronegativities. This is clearly shown by the polar compound ICl, which has a boiling point almost 40° C above that of bromine, although both have the same molecular weights. Interhalogens have bond energies that are lower than halogens and therefore in most cases they are more reactive. These properties impart special germicidal characteristics to these compounds. The principal germicidal compound of this group is bromine chloride. At equilibrium, BrCl is a fuming dark red liquid below 5° C. It exists as a solid only at relatively low temperatures. Liquid BrCl can be vaporized and metered as a vapor in equipment similar to that used for chlorine.
BrCl is prepared by the addition of equivalent amounts of chlorine to bromine until the solution has increased in weight by 44.3 percent: The reaction is as follows:

$$Br_2 + C1_2 \rightarrow 2BrCl$$

BrCl can be prepared by the reaction in the gas phase or in aqueous hydrochloric acid solution. In the laboratory, BrCl is prepared by oxidizing bromide salt in a solution containing hydrochloric acid.

$$kBrO_3 + 2kBr + 6HCI \rightarrow 3BrCI + 3kCl + 3H_2O$$

BrCl exists in equilibrium with bromine and chlorine in both gas and liquid phases. Table 5 lists various physical properties of BrCl. Due to the polarity of BrCl, it shows greater solubility than bromine in polar solvents. In water, it has a solubility of 8.5 gms per 100 gms of water at 20_0 C (that is, 2.5 times the solubility of bromine; 11 times that of chlorine). Bromine chloride's solubility in water is increased greatly by adding chloride ions to form the complex chlorobromate ion, $BrC1_2$.

Table 5. Physical Properties of BrCl.

Molecular weight	115.37
Melting point (°C)	-66
Boiling point (°C)	5
Density (g/cc), 20 ° C	2.34
Heat of fusion (cal/g)	17.6
Heat of vaporization (cal/g)	53.2
Heat formation (kcal/mole)	0.233
Heat capacity (cal./deg. mole, 298° K)	8.38
Entropy (cal./deg. mole, 298° K)	57.34
Dipole moment	0.56
Electrical conductivity (dm^{-1} cm^{-1})	-
Degree of dissociation (%, vapor 25° C)	21

CHEMISTRY OF BROMINE CHLORIDE

Various organic and inorganic species that act as reducing agents react with and destroy free halogen residuals during interaction with microorganisms (see Figure 2.13 for examples of competitive reactions). Competitive reactions depend on the reactivity of the chemical species, temperature, contact time, and pH. The quality of the effluent and the method of adding the disinfectant also help determine the specific reaction pathways. Bromine chloride is about 40 percent dissociated into bromine and chlorine in most solvents. Because of its high reactivity and fast equilibrium, BrCl often generates products that result almost entirely from it. This is illustrated by the disinfectant products shown in Figure 5. The major portion of the BrCl is eventually reduced to inorganic bromides and chlorides, with the exception of addition and substitution reactions with organic constituents.

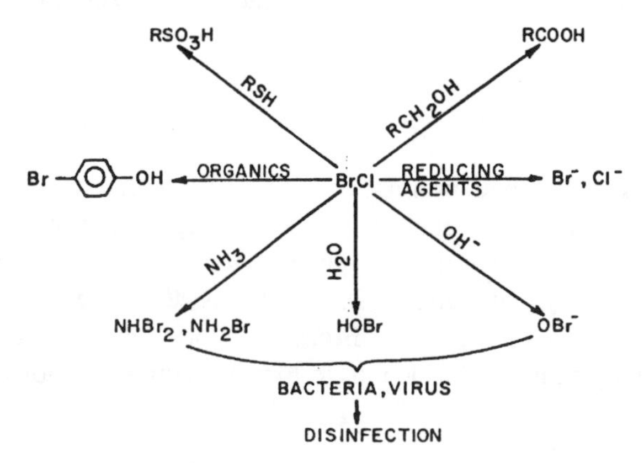

Figure 5. Reactions in wastewater disinfection.

It should be noted that although BrCl is mainly a brominating agent that is competitive with bromine, its chemical reactivity makes its action similar to that of chlorine (that is, disinfection, oxidation, and a bleaching agent). BrCl hydrolyzes exclusively to hypobromous acid, and if any hydrobromic acid (HBr) is formed by hydrolysis of the dissociated bromine, it quickly oxidizes to hydrobromous acid via hypochlorous acid.

Since hypohalous acid is a much more active disinfectant than the hypohalite ion, the effect of pH on ionization becomes important. Hypobromous acid has a lower ionization value than hypochlorous acid and this contributes to the higher disinfectant activity of BrCl compared with chlorine.

Bromine chloride also undergoes very specific reactions with ammonia and with organics. Monobromamine and dibromamine are the major products formed from reactions between BrCl and ammonia. These are unstable compounds in most conventional wastewater treatment plant effluents. In comparing the activities of bromarnine versus chloramines, the effects of ammonia and high pH tend to improve the bromarnine performance whereas the chloramine activity is reduced significantly. The reaction of ammonia with either BrCl or chlorine to form the halamine is very fast and generally goes to completion. As such, the presence of ammonia is essential to the disinfectant properties. Most sewage effluents typically have high ammonia concentrations in the range of 5 - 20 ppm. For such samples, the predominant bromine species (pH at 7 to 8) monobromamine and dibromamine are approximately equally distributed.

There are a large number of organics that undergo disinfection during the purification process. There are unfortunately a number of undesirable byproducts and side reactions which occur with some of them. One is the reaction between chlorine and phenol, producing chlorophenols, which are suspected carcinogens. Chlorophenols have obnoxious tastes and are toxic to aquatic life even at very low concentrations. Brominated phenolic products which are formed in the chlorobromination of wastewater are generally more readily degraded and often less offensive than their chlorinated counterparts.

The major organic reactions of BrCl consist of electrophilic brominations of aromatic compounds. Many aromatic compounds do not react in aqueous solution unless the reaction involves activated aromatic compounds (an example being phenol). Bromine chloride undergoes free-radical reactions more readily than bromine.

Metal ions in their reduced state also undergo reactions with BrCl. Examples include iron and manganese.

$$Fe^{+2} + BrCl \rightarrow Fe^{+3} + Br^{-} + Cl^{-}$$

$$Mn + BrCl \rightarrow Mn^{+2} + Br^{-} + Cl^{-}$$

Wastewater occasionally contains hydrogen sulfide and nitrites. These contribute to higher halogen demands. Many of these reactions reduce halogens to halide salts.

Bromine chloride's reactivity with metals is not as great as that of bromine; however, it is comparable to chlorine. Dry BrCl is typically two orders of magnitude less reactive with metals than dry bromine. Most BrCI is less corrosive than bromine. Like chlorine, BrCl is stored and shipped in steel containers. Also, Kynar and Viton plastics and Teflon@ are preferred over polyvinyl chloride (PVC) when BrCl is in the liquid or vapor states.

DISINFECTION WITH BROMINE CHLORINE

In chlorination, chlorine's reaction with ammonia forms chloramines, greatly reducing its bactericidal and virucidal effectiveness. The biocidal activity of monochloramine is only 0.02 - 0.01 times as great as that of free chlorine. Typical ammonia concentrations found in secondary sewage range from 5 - 20 ppm, which is about an order of magnitude greater than the amount needed to form monochloramine from normal chlorination dosages (which requires about 5 - 10 ppm). Therefore, monochloramine is the major active chlorine constituent in chlorinated sewage plant effluents. In contrast, BrCl ammonia reactions produce the major product bromamines. Bromamines have disinfectant characteristics which are significantly different than chloramines.

TOXICITY OF AQUATIC LIFE

Bromamines are considerably less stable than chloramines in receiving waters. Bromamines tend to break down into relatively harmless constituents typically in under 60 minutes. Consequently, BrCl is less damaging to marine life than chlorine. Chloramines at concentrations below 0.1 ppm have resulted in fish kills. There are also indirect effects from chloramine contamination. For example, fish populations tend to avoid toxic regions, even at very low levels of concentrations. Consequently, large areas of receiving waters can become unavailable to many species of fish and even cause blockage of upstream migrations during the spawning season. It should be noted that although chlorine efficiency is increased by nitrification, BrCl performance is not. Because of the high biocidal activity of bromamines, it is not necessary to utilize high concentrations and breakpoint conditions to achieve active halogen residuals, as is the case in chlorination. The breakpoint reaction with BrCl is achieved almost immediately in the presence of even slight excess amounts of bromine at pH levels of 7 to 8. There is, however, no need to reach the breakpoint to achieve good disinfectant properties with BrCl. In contrast, with chlorine it is necessary to add amounts in excess of the breakpoint to obtain sterilizing characteristics,

PROPERTIES OF IODINE

Iodine (from the Greek, *iodines,* meaning violet) has an atomic weight of 126.9044, atomic number 53, melting point 113.5° C, and boiling point 184.35° C. As a gas, its density is 11.27 g/l and as a solid its specific gravity is 4.93 (20° C). This halogen was discovered by Courtois in 1811. It occurs sparingly in the form of iodides in sea water from which it is assimilated by seaweeds, in Chilean saltpeter and nitrate-bearing soil, in brines from ancient sea deposits, and in brackish waters derived from oil and salt wells. Pure grades of iodine can be obtained from the reaction of potassium iodide with copper sulfate. Iodine is a grayish-black, lustrous solid that volatilizes at ordinary temperatures to a blue-violet gas. It forms compounds with many elements. However, it is less active than many of the other halogens which displace it from iodides. Iodine dissolves readily in chloroform, carbon tetrachloride, and carbon disulfide. It is only slightly soluble in water. Iodine is highly irritating to the skin, eyes, and mucous membranes. Its effect on the human body is similar to that of bromine and chlorine. However, it is more irritating to the lungs.

DISINFECTION WITH IODINE COMPOUNDS

Two interhalogens having strong disinfecting properties are iodine monochloride (ICl) and iodine bromide (IBr). Iodine monochloride has found use as a topical antiseptic. It may be complexed with nonionic or anionic detergents to yield bactericides and fungicides that can be used in cleansing or sanitizing formulations. These generally have a polymer structure which establishes its great stability, increased solubility, and lower volatility. By reducing the free halogen concentration in solution, polymers reduce both the chemical and bactericidal activity. Complexes of ICl are useful disinfectants which compromise lower bactericidal activity with increased stability. Iodine monochloride is itself a highly reactive compound, reacting with many metals to produce metal chlorides. Under normal conditions it will not react with tantalum, chromium, molybdenum, zirconium, tungsten, or platinum. With organic compounds, reactions cause iodination, chlorination, decomposition, or the generation of halogen addition compounds. In water, ICI hydrolyzes to hypoiodous and hydrochloric acids. In the absence of excess chloride ions, hypoiodous acid will disproportionate into iodic acid and iodine. Iodine bromide has a chemistry similar to ICl. Iodine bromide reacts with aromatic compounds to produce iodination in polar solvents and bromination in nonpolar solvents. It has complex chemical properties, as its solubility is increased more effectively by bromide than by chlorided ions. Primary hydrolysis takes place in the presence of hydrobromic acid. As a disinfectant, IBr is used in its complexed or stabilized forms. Unfortunately, it undergoes hydrolysis and dissociation reactions in aqueous solutions, both reactions being major limitations. Its disinfecting properties are similar to ICI and as in the case of ICl, germicidal activity should not be reduced by haloamine formation since bromamines are highly reactive and iodoamines are not generated. Upon application of prepared solutions to control microorganisms, the complex releases IBr gradually. This process forms free iodine during the decomposition of IBr (the decomposition takes place as fast as the IBr is released).

DISINFECTION WITH HALOGEN MIXTURES

Two approaches that have been investigated recently for disinfection are mixtures of bromine and chlorine, and mixtures containing bromide or iodide salts. Some evidence exists that mixtures of bromine and chlorine have superior germicidal properties than either halogen alone. It is believed that the increased bacterial activity of these mixtures can be attributed to the attacks by bromine on sites other than those affected by chlorine. The oxidation of bromide or iodide salts can be used to prepare interhalogen compounds or the hypollalous acid in accordance with the following reaction:

$$HOCl + NaBr \rightarrow HOBr + HCl$$

It has been reported that the rate of bacterial sterilization by chlorine in the presence of ammonia is accelerated with small amounts of bromides. As little as 0.25 ppm of bromamines can be significant under some conditions. However, if chloramines are produced prior to contact with bromide ions, the reaction and subsequent effect are reduced. Improved germicidal activity has also been shown for mixtures containing bromides and iodides with various chlorine releasing compounds. Bromide improves the disinfecting properties of dichloroisocyanuric acid and hypochlorite against several bacteria. Bromine-containing compounds are useful for their combined bleaching and disinfectant properties. There has been the concern that the use of interhalogen compounds in wastewater disinfection could produce unknown organic and inorganic halogen-containing substances. In the case of iodine, concern has been expressed over the physiological aspects in water supplies. Extensive studies have been reported on the role played by iodine and iodides in the thyroid glands of animals and man. Information on acute inhibition of hormone formation by excessive amounts of iodine is well known. Despite the fact that no strong evidence exists that iodine is harmful as a water disinfectant, only limited use has been attempted.Chronic bromide intoxication from continuous exposure to dosages above 3 - 5 g is called bromism. Typical symptoms are skin rash, glandular excretions, gastrointestinal disturbances, and neurological disturbances. Bromide can be absorbed from the intestinal tract and contaminate the body in a manner very similar to that for chloride. Brominated drinking water does not, however, significantly increase the amount of bromine admitted internally. The amount of additional bromine in chlorobrominated waters will not significantly increase human bromine concentrations nor result in bromism.

STERILIZATION USING OZONE

Ozone (O_3) is a powerful oxidant, and application to effluent treatment has developed slowly because of relatively high capital and energy costs compared to chlorine. Energy requirements for ozone are in the range of 10 to 13 kWh/lb generated from air, 4 kWh/lb from oxygen, and 5.5 kWh/lb from oxygen-recycling systems. Operating costs for air systems are essentially the electric power costs; for oxygen systems the cost of oxygen (2 to 30/lb) must be added to the electrical cost. Capital costs of large integrated ozone systems are $300 to $400 a pound per day of ozone generated and $100 a pound per day of ozone for the generator alone.

Actual uses of ozone include odor control, industrial chemicals synthesis, industrial water and wastewater treatment, and drinking water. Lesser applications appear in fields of combustion and propulsion, foods and pharmaceuticals, flue gas-sulfur removal, and mineral and metal refining. Potential markets include pulp and paper bleaching, power plant cooling water, and municipal wastewater treatment.

The odor control market is the largest and much of this market is in sewage treatment plants. Use of ozone for odor control is comparatively simple and efficient. The application is for preservation of environmental quality; in addition, alternative treatment schemes requiring either liquid chemical oxidants (like permanganate or hydrogen perioxide) or incineration can significantly increase capital and costs.

Ozone applications in the United States for drinking water are far fewer than in Europe. However, the potential market is large, if environmental or health needs ever conclude that an alternate disinfectant to chlorine should be required. Although energy costs of ozonation are higher than those for chlorination, they may be comparable to combined costs of chlorination dechlorination-reaeration, which is a more equivalent technique. One of ozone's greatest potential uses is for municipal wastewater disinfection.

Technical, economic, and environmental advantages exist for ozone bleaching of pulp in the paper industry as an alternate to hypochlorite or chlorine bleaching which yields deleterious compounds to the environment.

PRINCIPLES OF OZONE EFFLUENT TREATMENT

Ozone was first discovered by the Dutch philosopher Van Marun in 1785. In 1840, Schonbein reported and named ozone from the Greek word ozein, meaning to smell. The earliest use of ozone as a germicide occurred in France in 1886, when de Meritens demonstrated that diluted ozonized air could sterilize polluted water. In 1893, the first drinking water treatment plant to use ozone was constructed in Oudshorrn, Holland. Other plants quickly followed at Wiesbaden (1901) and Paderborn (1902) in Germany. In 1906, a plant in Nice, France, was constructed using ozone for disinfection. Today, there are over 1,000 drinking water treatment plants in Europe utilizing ozone for one or more purposes. In the United States, the first ozonation plant was constructed in Whiting, Indiana, in 1941 for taste and odor control.

Over 100 years ago it had been demonstrated that ozone (O_3), the unstable triatomic allotrope of oxygen, could destroy molds and bacteria and by 1892 several experimental ozone plants were in operation in Europe. In the 1920s, however, as a result of wartime research, during World War I, chlorine became readily available and inexpensive, and began to displace ozone as a purifier in municipalities throughout the United States. Most ozone studies and development were dropped at this time, leaving ozonation techniques, equipment, and research at a primitive stage. Ozone technology stagnated, and the development and acceptance of ozone for water and wastewater treatment was discontinued.

In addition to the popular use of chlorination as a wastewater disinfectant and the consequent technology lag in ozonation research, there was a third impediment to ozone commercialization: the comparatively high cost of ozonation in relation to chlorination. Ozone's instability requires on-site generation for each application, rather than centralized generation and distribution. This results in higher capital requirements, aggravated by a comparatively large electrical energy requirement. Ozone's low solubility in water and the generation of low concentrations, even under ideal conditions, also necessitates more elaborate and expensive contacting and recycling systems than chlorination.

In spite of such obstacles there is interest from time to time in the use of ozone, particularly for wastewater treatment. The technology for the destruction of organics and inorganics in water has not kept pace with the increasingly more sophisticated water pollution problems arising from greater loads, new products, and new sources of pollutant entry into the environment and increased regulation. The growing trend toward water reuse and the fact that some highly toxic pollutants may be refractory to conventional treatment methods has spurred investigation into new treatments, including ozonation.

A significant impetus from time to time for developing new methods is dissatisfaction with chlorination. Chlorine affects taste and odor and produces chloramines and a wide variety of other potentially hazardous chlorinated compounds in wastewaters. It seriously threatens the environment with an estimated 1,000 tons per year of chlorinated organic compounds discharged into U.S. waters (chloramines are not easily degradable and pose a hazard to the environment) and is questionable as a drinking water viricidal disinfectant. Ozone's development, on the other hand, could parallel a greater environmental awareness and a resulting demand for higher-quality effluents, as its potential for overcoming these problems is possible.

PROPERTIES OF OZONE

Ozone is an unstable gas, having a boiling point of -112° C at atmospheric conditions. Its molecular weight is 48. Ozone is partially soluble in water (approximately 20 times more soluble than oxygen), and has a characteristic penetrating odor which is readily detectable at concentrations as low as 0.01 - 0.05 ppm. Ozone is the most powerful oxidant currently available for use for wastewater treatment. Commercial generation equipment generates ozone at concentrations of 1 percent to 3 percent in air (that is, 2 percent to 6 percent in oxygen). Ozone is unstable in water, however, it is more stable in air, especially in cool, dry air.

As a strong oxidant, ozone reacts with a wide variety of organics. Ozone oxidizes phenol to oxalic and acetic acids. It oxidizes trihalomethane (THM) compounds to a limited extent within proper pH ranges and reduces their concentration by air stripping. Trihalomethanes are also oxidized by ozone in the presence of ultraviolet light. Oxidation by ozone does not result in the formation of THMs as does chlorination. A combination of ozone and ultraviolet light destroys DDT, malathion, and other pesticides. However, high dosages and extended contact times that are not normally encountered in drinking water treatment are needed. Ozonized organic substances are usually more biodegradable and absorbable than the starting, unoxidized substances. When ozonation is employed as the final treatment step for potable water systems in water containing significant concentrations of dissolved organics, bacterial regrowth in the distribution system can occur. Consequently, ozonation is not typically used as the final treatment step but rather followed by granular activated carbon filtration and sometimes by the addition of a residual disinfectant. Humic materials are the precursors of THMs. Humic substances can be oxidized by ozonation. Under proper conditions significant reduction in THM formation can be realized when ozone is applied prior to a chlorination step. Because of ozone's instability, it is able to produce a series of almost instantaneous reactions when in contact with oxidizable compounds. One example follows.

$$O_3 + 2KI + H_2O \rightarrow I_2 + O_2 + 2KOH$$

In this reaction, iodine is liberated from a solution of potassium iodide. This reaction can be used to assess the amount of ozone in either air or water. For determination in air or oxygen, a measured volume of gas is drawn through a wash bottle containing potassium iodide solution. Upon lowering the pH with acid, titration is effected with sodium thiosulfate, using a starch solution as an indicator. There is a similar procedure for determining ozone in water.

A typical ozone treatment plant consists of three basic subsystems: feedgas preparation; ozone generation; and ozone/water contacting. Commercially, ozone is generated by producing a high-voltage corona. discharge in a purified oxygen-containing feedgas. The ozone is then contacted with the water or wastewater; the treated effluent is discharged and the feedgas is recycled or discharged.

Ozone's high reactivity and instability, as well as serious obstacles in producing concentrations in excess of 6 percent, preclude central production and distribution with its associated economies of scale. The requirement for on-site generation and application of ozone must yield a cost-efficient, lowmaintenance operation in order to be useful. The feedgas employed in ozonation systems is either air, oxygen, or oxygen-enhanced air. The particular selection of feedgas for each application is based on economics and depends on several factors: total quantity of ozone required; desired concentration of ozone in the feedgas; and fate (recycle or discharge) of the feedgas. For a given ozone generator with a specified power input and gas flow, two to three times as much ozone may be generated from oxygen as from air. The maximum concentration economically produced from air is about 2 percent, while that generated from pure oxygen is approximately 6 percent.

The use of higher concentrations of ozone provides two advantages: capital and operating costs per pound of ozone produced are substantially reduced, and a greater concentration gradient for mass transfer of ozone is provided in the contacting step, yielding increased ozone-utilization efficiency. These advantages, however, must be weighed against the increased cost of oxygen production. Air is generally employed in those applications requiring less than 50 pounds/day of low-concentration ozone. If air is the feedgas, it must be dried and cooled to reduce accumulation of corrosive nitric acid and nitrogen oxides that occur as by-products when the dew point is above 40° C.

Ozone may be produced by electrical discharge in an oxygen-containing feedgas or by photochemical action using ultraviolet light. For large-scale applications, only the electric-discharge method is practical since the use of ultraviolet energy produces only low-volume, low-concentration ozone.

In the electric-discharge (or corona) method, an alternating current is imposed across a discharge gap with voltages between 5 and 25 kV and a portion of the oxygen is converted to ozone. A pair of large-area electrodes is separated by a dielectric (1-3 mm in thickness) and an air gap (approximately 3 mm.) as shown in Figure 6. Although standard frequencies of 50 or 60 cycles are adequate, frequencies as high as 1,000 cycles are also employed.

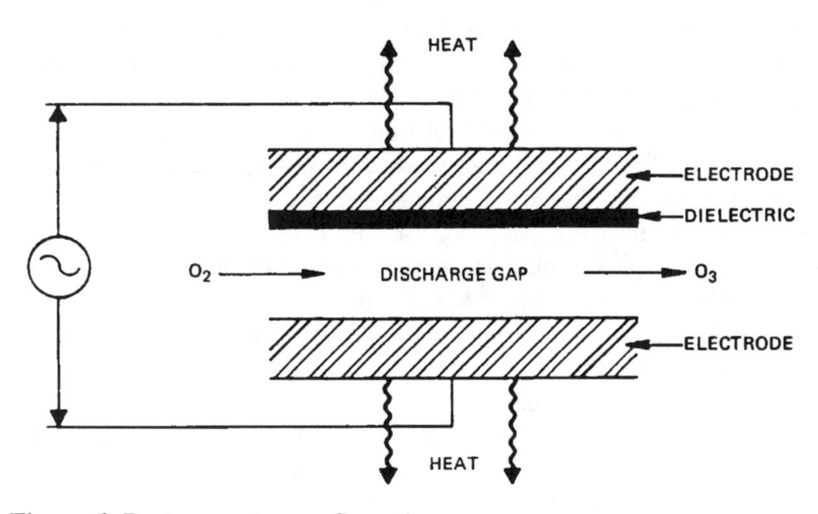

Figure 6. Basic ozonator configuration.

Only about 10 percent of the input energy is effectively used to produce the ozone. Inefficiencies arise primarily from heat production and, to a lesser extent, from light and sound. Since ozone decomposition is highly temperature dependent, efficient heat-removal techniques are essential to the proper operation of the generator.

The mechanism for ozone generation is the excitation and acceleration of stray electrons within the high-voltage field. The alternating current causes the electrons to be attracted first to one electrode and then the other. As the electrons attain sufficient velocity, they become capable of splitting some O_2 molecules into free radical oxygen atoms. These atoms may then combine with02molecules to form O_3. Under optimum operating conditions (efficient heat removal and proper feedgas flow), the production of ozone in corona-discharge generators is represented by the following relationships, showing the factors to be considered in the design of these generators:

$$V \propto pg$$

$$(Y/A) \propto f\epsilon V^2/d$$

where Y/A = ozone yield per unit area of electrode surface
V = applied voltage
p = gas pressure in the discharge gap
g = discharge-gap width
f = frequency of applied voltage
ϵ = dielectric constant
d = thickness of the dielectric

The following requirements will facilitate optimization of the ozone yield:

⇨ The pressure/gap combination should be constructed so the voltage can be kept relatively low while maintaining reasonable operating pressures. Low voltage protects the dielectric and electrode surfaces. Operating pressures of 10 - 15 pounds per square inch gauge (psig) are applicable to many waste treatment uses.

⇨ For high-yield efficiency, a thin dielectric with a high-dielectric constant should be used. Glass is the most practical material. High-dielectric strength is required to minimize puncture, while minimal thickness maximizes yield and facilitates heat removal.

⇨ For reduced maintenance problems and prolonged equipment life, highfrequency alternating current should be used. High frequency is less damaging to dielectric surfaces than high voltage.

⇨ Heat removal should be as efficient as possible.

Basic configurations of ozone generators are shown in Figures 6 and 7. The thre designs are the Otto plate, the tube, and the Lowther plate. The least efficient of these generators is the Otto plate, developed at the turn of the century. The tube and Lowther plate units include modern innovations in material and design. The Lowther plate generator is the most efficient configuration due in large measure to advantages in heat removal. In addition to ozone yield, the concentration of ozone is an important consideration. Ozone concentration from a generator is usually regulated by adjusting the flow rate of the feedgas and/or voltage across the electrodes.

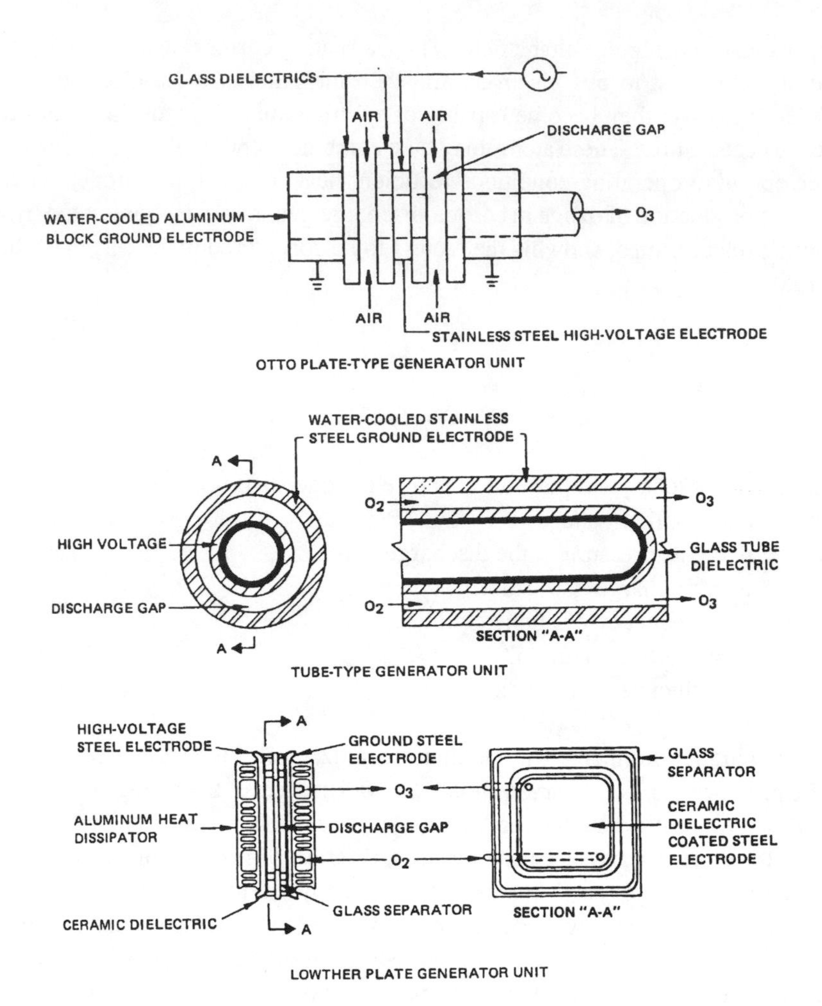

Figure 7. Types of ozone generators.

Contactor design is important in order to maximize the ozone-transfer efficiency and to minimize the net cost for treatment. The three major obstacles to efficient ozone utilization are ozone's relatively low solubility in water, the low concentrations and amounts of ozone produced from ozone generators, and the instability of ozone. Several contacting devices are currently in use including positive-pressure injectors, diffusers, and venturi units. Specific contact systems must be designed for each different application of ozone to wastewater. Further development in this area of gas-liquid contacting needs to be done despite its importance in waste treatment applications. In order to define the appropriate contactor, the following should be specified:

⇨ The objective: disinfection biochemical oxygen demand (BOD) or chemical

oxygen demand (COD) reduction to a particular level, trace refractory organics oxidation, and so on.
⇒ Relative rates of competitive reactions: chemical oxidation, lysing bacteria, decomposition of ozone in aqueous solutions, and so on.
⇒ Mass-transfer rate of ozone into solution.
⇒ Wastewater quality characteristics: total suspended solids, organic loading, and so on.
⇒ Operating pressure of the system.
⇒ Ozone concentration utilized.

Other considerations for the contacting system itself include contactor type (for example, packed bed, sparged column); number and configuration of contactor stages; points of gas-liquid contact, whether the mix is cocurrent or countercurrent; and the construction materials used. It is clear that designing an ozonation system for even a relatively simple application requires a thorough understanding of many factors in order to employ sound engineering methods and optimization techniques.

APPLICATIONS

Market areas of interest to manufacturers of ozone systems, actual uses defined as those which have been in operation for some time and not including "pilot" studies, arise in the following categories: odor control (sewage treatment and industrial), industrial chemicals synthesis, industrial water and wastewater treatment, and drinking water disinfection.
Ozone has proven to be effective against viruses. France has adopted a standard for the use of ozone to inactivate viruses. When an ozone residual of 0.4 mg/I can be measured 4 minutes after the initial ozone demand has been met, viral inactivation is satisfied. This property plus ozone's freedom from residual formation are important considerations in the public health aspects of ozonation. When ozonation is combined with activated carbon filtration, a high degree of organic removal can be achieved. Concerning the toxicity of oxidation products of ozone and the removal of specific compounds via ozonation, available evidence does not indicate any major health hazards associated with the use of ozone in wastewater treatment.

ODOR CONTROL

In the United States, the largest existing market for ozone systems is odor control. Much of this market is in sewage treatment plants. Industrial markets for ozone in odor control are smaller than for sewage treatment plants. Established applications include cooking odors at restaurants; pharmaceutical fermentations; fish, meat, and food processing; plastics and rubber processing; paint and varnish manufacture; and rendering plants. Nearly all of these industrial odor control applications use less than 100 lb/day of ozone and most use between 1 - 25 lb/day.

INDUSTRIAL CHEMICAL SYNTHESIS

There has been only one major use for ozone today in the field of chemical synthesis: the ozonation of oleic acid to produce azelaic acid. Oleic acid is obtained from either tallow, a by-product of meat-packing plants, or from tall oil, a by-product of making paper from wood. Oleic acid is dissolved in about half its weight of pelargonic acid and is ozonized continuously in a reactor with approximately 2 percent ozone in oxygen; it is oxidized for several hours. The pelargonic and azelaic acids are recovered by vacuum distillation. The acids are then esterified to yield a plasticizer for vinyl compounds or for the production of lubricants. Azelaic acid is also a starting material in the production of a nylon type of polymer.

INDUSTRIAL WATER AND WASTEWATER TREATMENT

The markets for ozone in industrial water and wastewater treatment are quite small. Industrial applications for ozone could grow. The use of ozone for treating photoprocessing solutions is a novel application that has been limited, but might grow. In this process, silver is recovered electrolytically; then the spent bleach baths of iron ferrocyanide complexes are ozonated. Iron cyanide complexes are stable to ozonation so that the ferrous iron is merely oxidized to ferric, which is its original form. Thus, the bleach is "regenerated" and is ready for recycling and reuse by the photoprocessor.

Ideally, 20.2 pounds of the ferrocyanide can be converted to 11.7 pounds of ferricyanide by one pound of ozone. Indeed, ozone oxidation efficiency is nearly 100 percent for ferrocyanide concentrations above 1.0 g/l.

A typical ozone system consists of 100 g/hr at a concentration of 1.0 percent to 1.5 percent in air fed to the bottom of bleach collection tanks through ceramic spargers (pore size of approximately 100 ~t). The system contains air compression and drying equipment, automatic control features, and a flat-plate, air-cooled ozone generator. Regeneration of bleach wastes totaling about 10,000 gallons a year, and recovery of other chemicals can also be cost effective.

MUNICIPAL DRINKING WATER

In the United States, Whiting, Indiana and Strasburg, Pennsylvania have used ozone in their drinking water treatment process. Other cities have run pilot studies. Ozone is used as a bleaching agent for miscellaneous items: petroleum, clays, wood products, and chemical baths. It has been proposed as a bleaching agent for hair and as a disinfectant for oils and emulsions. Ozone is used to modify tryptophan and indigo plant juice. It is an important factor in colorfastness. The desulfurization of flue gases by ozone has been considered an application where it promotes liquid-phase oxidation. The operations are carried out with vanadium catalysts, and the oxidation step is performed in gasfluidized beds. The desulfurizing effect of ozone on light petroleum distillates has also been reported.

The use of ozone has been proposed in special ore-flotation processes. Two widely different applications involve hydraulic cement and the fabrication of coating on insulators.

The metallurgical applications include steel refining, electrochemical processes, and gold recovery. The aggressive reactivity of ozone is evident in the corrosion of stainless steel and in chemical etching. The inhibition of ozone decomposition is accomplished in the presence of SF6, CC12172, or CF3Metal coatings, paints, and lacquers have been evaluated with respect to ozone resistivity. Ozone has been examined as a potential source of high-energy oxidation and for combustion and propulsion applications.

OZONATION EQUIPMENT AND PROCESSES

Ozonation systems are comprised of four main parts, including a gas-preparation unit, an electrical power unit, an ozone generator, and a contactor which includes an off-gas treatment stage. Ancillary equipment includes instruments and controls, safety equipment and equipment housing, and structural supports. The four major components of the ozonation process are illustrated in Figure 8.

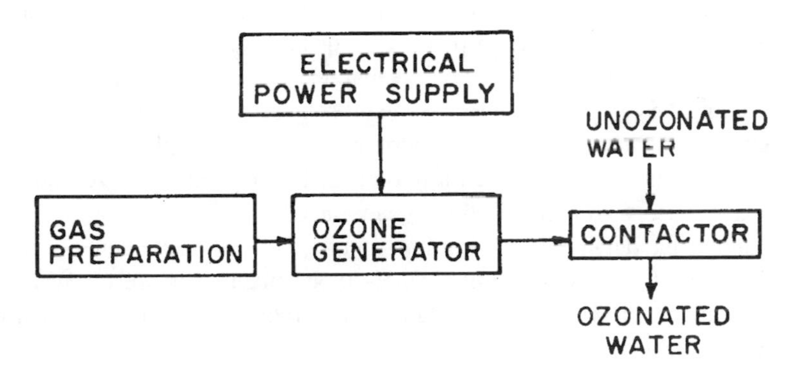

Figure 8. Components of ozonation.

A high level of gas preparation (usually air) is needed before ozone generation. The air must be dried to retard the formation of nitric acid and to increase the efficiency of the generation. Moisture accelerates the decomposition of ozone. Nitric acid is formed when nitrogen combines with moisture in the corona discharge. Since nitric acid will chemically attack the equipment, introduction of moist air into the unit must be avoided. Selection of the air-preparation system depends on the type of contact system chosen. The gas-preparation system will, however, normally include refrigerant gas cooling and desiccant drying to a minimum dew point of -40° C. A dew-point monitor or hygrometer is an essential part of any air preparation unit.

Conversion efficiencies can be greatly increased with the use of oxygen. However, the use of high-purity oxygen for ozone generation for disinfection is, cost effective. The Duisburg plant and the Tailfen plant of Brussels, Belgium, are the only operational municipal water treatment plants known which use high-purity oxygen instead of air as the ozone generator feedgas.

Electrical power supply units vary considerably among manufacturers. Power consumption and ozone-generation capacity are proportional to both voltage and frequency. There are two methods to control the output of an ozone generator: vary voltage or vary frequency. Three common electrical power supply configurations are used in commercial equipment:

- Low frequency (60 Hz), variable voltage.
- Medium frequency (600 Hz), variable voltage.
- Fixed voltage, variable frequency.

The most frequently used is the constant low-frequency, variable-voltage configuration. For larger systems, the 600-Hz fixed frequency is often employed as it provides double ozone production with no increase in ozone generator size.

The electrical (corona) discharge method is considered to be the only practical technique for generating ozone in plant-scale quantities. In principle, an ozone generator consists of a pair of electrodes separated by a gas space and a layer of glass insulator. An oxygen-rich gas is passed through the empty space and a high-voltage alternating current is applied. A corona discharge takes place across the gas space and ozone is generated when a portion of the oxygen is ionized and then becomes associated with nonionized oxygen molecules.

Figure 9 shows the details of a typical horizontal tube-type ozone generator. This unit is preferred for larger systems. Water-cooled plate units are often used in smaller operations. However, these require considerably more floor space per unit of output than the tube-type units. The air-cooled Lowther plate type is a relatively new design. It has the potential for simplifying the use of ozone-generating equipment. However, it has had only limited operating experience in water treatment facilities.

After the ozone has been generated, it is mixed with the water stream being treated in a device called a contactor. The objective of this operation is to maximize the dissolution of ozone into the water at the lowest power expenditure. There is a variety of ozone contactor designs. Principal ones employed in wastewater treatment facilities include:

- Multistage porous diffuser contactors, which involve a single application of an ozone-rich gas stream and application of fresh ozone gas to second and subsequent stages with off-gases recycled to the first stage.
- Eductor-induced, ozone vacuum injector contactors, which include total

or partial plant flow through the eductor; and subsequent stages with off-gases recycled to the first stage

- Turbine contactors, which involve positive or negative pressure to the turbine.
- Packed-bed contactors, which include concurrent or countercurrent water/ozone-rich gas flow.

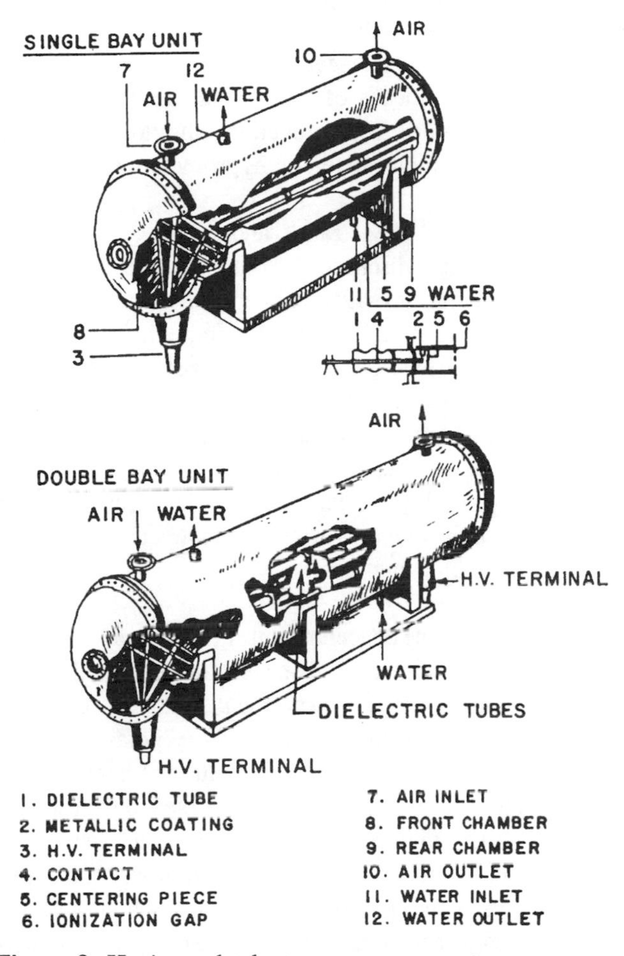

Figure 9. Horizontal tube-type ozone genrator.

Two-level diffuser contactors, which involve application of ozone-rich gas to the lower chamber. Lower chamber off-gases are applied to the upper chamber. Off-gas treatment from contactors is an important consideration. Methods employed for off-gas treatment include dilution, destruction via granular activated carbon, thermal or catalytic destruction, and recycling.

MEASUREMENT AND CONTROL

Favorable operational economics and good management practices require high levels of control of the ozonation system. Depending on the specific process of ozone applications, plant size, and design philosophy, the control system may be simple or complex. The trend in Europe is toward highly sophisticated and centralized control.

Several parameters should be measured to provide a fully operable ozonation system. There should be a means of providing full temperature and pressure profiles of the ozone generator feedgas from the initial pressurization (by fan, blower, or compressor) to the ozone generator inlet. Moisture content is also important. There should be a means of measuring the moisture content of the feedgas to the ozone generator. This procedure should be conducted with a continuously monitoring dew-point meter or hygrometer. Other parameters that require monitoring include:

- Temperature, pressure, flow rate, and ozone concentration of the ozonecontaining gas being discharged from all the ozone generators. This is the only effective method by which ozone dosage and the ozone production capacity of the ozone generator can be determined.
- Power supplied to the ozone generators. The parameters measured include amperage, voltage, power, and frequency, if this is a controllable variable.
- Flow rate and temperature of the cooling water to all water-cooled ozone generators. Reliable cooling is important to maintain constant ozone production and to protect the dielectrics in the generation equipment.
- There should be a means to monitor the several cycles of the desiccant drier, particularly the thermal-swing unit.

Analytical measurements of ozone concentrations must be made in the ozonized gas from the ozone generator, the contactor off-gases, and the residual ozone level in the ozonized water. Methods of ozone measurement commonly used are the: simple "sniff" test, Draeger-type detector tube, wet chemistry potassium iodide method, amperometric-type instruments, gas-phase chemiluminescence, and ultraviolet radiation adsorption. The use of control systems based on these measurements varies considerably. The key to successful operation is an accurate and reliable residual ozone analyzer. Continuous residual ozone monitoring equipment may be successfully applied to water that has already received a high level of treatment. However, a more cautious approach must be taken with the application of continuous residual ozone monitoring equipment for water that has only received chemical clarification because the ozone demand has not yet been satisfied and the residual is not as stable. Ozone production must be closely controlled because excess ozone cannot be stored. Changes in process demand must be responded to rapidly. Ozone production is costly; underozonation may produce undesired effects and overozonation may require additional costs where off-gas destruction is used.

OPERATION AND MAINTENANCE

Ozonation equipment typically has low maintenance requirements. The, airpreparation system requires frequent attention for air filter cleaning/changing and for assuring that the desiccant is drying the air properly. However, both are usually simple operations. Two factors which impact ozone generator operation and maintenance are the effectiveness of the air-preparation system and the amount of time that the generator is required to operate at maximum capacity. Maintenance of the ozone generators is commonly scheduled once a year. However, many plants perform this maintenance every six months. Typically, one man-week is necessary to service an individual ozone generation unit of the horizontaltube type. Dielectric replacement due to failure as well as breakage during maintenance may be as low as 1 percent to 2 percent. An average tube life of ten years can be expected if a feedgas dew point of - 60' is maintained and if the ozone generator is not required to operate for prolonged periods at its rated capacity. Plate-type ozone generators use window glass as dielectrics. However, the same attention to air preparation is taken as with the more expensive glass or ceramic tubes in order to avoid costly downtime. Operations and maintenance of the ozone contactor also requires attention. Turbines require electricity to power the drive motors, while porous diffusers require regular inspection and maintenance to insure a uniform distribution of ozone-rich gas in the contact chamber. It should be noted that serious safety problems exist with servicing some of these units. For example, even after purging the contact chambers with air, maintenance personnel entering the chambers should be equipped with a self-contained breathing apparatus, since the density of ozone is heavier than air and therefore is difficult to remove completely by air purging.

Chapter 12

TREATING THE SLUDGE

INTRODUCTION

Our discussions on water treatment began by making the analogy of a water treatment facility with any other manufacturing operation. The similarity between a water treatment facility and any manufacturing operation, whether it be a rubber producing plant or an auto-making facility or an iron and steel plant, are the reliance upon combinations of unit processes and unit operations that work in harmony to produce a high quality product. But that is where the similarity ends. A normal manufacturing operation aims not only to produce a high quality product, but efficient businesses strive to do so by eliminating or minimizing their wasteful by-products - simply because those by-products have little to no market value and add cost to production. If our cost of production is higher, then profit margins are lower. That in fact is the basis for pollution prevention and waste minimization practices of modern-day industry.

A water treatment facility differs in this regard because the primary objective is to produce high quality water by removing or destroying as much of the contaminants as possible. We cannot produce high quality water without generating the wasteful by-product, sludge, very often in large quantities. Water treatment plants are simply pollution control technologies, whether they are applied to industrial applications or municipal. That does not mean that pollution prevention practices are not appropriate for water treatment plants - they most certainly are and can minimize solid waste generation. But understand that we cannot eliminate the wasteful by-product of sludge as one might try and do if we had an manufacturing facility and we identified another technology to make our product and eliminate a wastestream generated by the older technology.

Sludge or solid waste is unavoidably produced in the treatment of water containing suspended solids. There are, however different technologies that we can select among that will indeed concentrate these solids, and thereby reduce the volumes that we ultimately must dispose of. In addition, some sludge can be stabilized and treated, which can impart a low, but none-the-less marketable value to this waste. These technologies and practices do indeed constitute pollution prevention and waste minimization programs within water treatment plant operations, and they can

have a very significant and positive impact on the overall costs of the operation.
This brings us to a collection of technologies that focus on: (1) sludge concentration, (2) sludge stabilization, (3) sludge handling and disposal. Some technologies fall into the category of pollution prevention, while others are within the normal arena of solid waste management and disposal.
Note that pollution prevention or P2 technologies, as in other industry sectors, are not necessarily the preferred choices. Specific technology selection quite often depends on localized conditions. By this, we mean the properties of the sludge, the volumes handled, and the comparative costs between technologies and or practices. In a very general sense, pollution prevention technologies are only appropriate when they are financially attractive for an operation. Like any other engineering project, the investment into a technology that falls within the pollution prevention arena must have financial attractiveness. An alternative way of stating this is that there are indeed situations where more conventional methods resulting in large volumes of sludge are more cost effective than a leading-edge technology that minimizes or reduces sludge volumes. The financial attractiveness of an investment needs to be assessed on a case by case basis.
This final chapter focuses on sludge processing and post-processing technologies. Where appropriate, we will point out which technologies may be considered P2. Recommended resources are given at the end of this chapter that will assist you in evaluating the relative investments required, as well as in obtaining more detailed technical information. There is also a section on ***Questions for Thinking and Discussing***, which will help you to generate ideas and approaches to selecting cost-effective methods. Remember to refer to the ***Glossary*** at the end of the book if you run across any terms that are unfamiliar to you.

WHAT SLUDGE IS

When we think of sludge, what automatically comes to mind is sewerage. Water carriage systems of sewerage provide a simple and economical means for removing offensive and potentially dangerous wastes from household and industry. The solution and suspension of solids in the transporting of water produces sewage. Thus, the role of solids and sludge removal at Sewage Treatment Plants is apparent. Sludge removal is complicated by the fact that some of the waste matters go into solution while others are colloidal or become finely divided in their flow through the sewage system. Ordinarily, less than half of such waste remains in suspension in a size or condition that can be separated by being strained out, skimmed off or settled out. The remainder must then be precipitated out by chemical means, filtered mechanically, or be subjected to biological treatment whereby they are either removed from the water or changed in character as to be rendered innocuous.
Sewage contains mineral and organic matter in suspension (coarse and fine suspended matter), in colloidal state (very finely dispersed matter) and in solution. Living organisms, notably bacteria and protozoa, find sewage to be an abundant source of food, and their lives' activities result in the decomposition of sewage.

Sewage becomes offensive due to its own instability together with the objectionable concentration of suspended materials. In addition, the potential presence of disease producing organisms makes sewage dangerous. Removal or stabilization of sewage matters may be accomplished in treatment works by a number of different methods or by a suitable combination of these methods.
While sewage sludge is rich in nutrients and organic matter, offering the potential for applications as a biosolid (discussed below) or it has a heating value making it suitable for incineration, many industrial sludges are often unsuitable for reuse. A more common practice with industrial sludge is to try and identify a reclaim value; i.e., if the sludge can be concentrated sufficiently there may be a portion of this waste which is reclaimable or may enter into a recycling market.

Figure 1. The wire is drawn through a cleaning solution bath, which removes the copper dust from the surface of the extruded wire. The wash liquor is slightly acidic and contains a detergent so that the wire surface is clean. From this point on the wire is spooled, and then sent to another part of the operation which manufactures multi-strand telecommunications cable.

An example is illustrated in Figures 1 through 4, which shows copper contaminated sludge from a cable manufacturing plant. In a concentrated enough form, and sufficiently dried (to at least 60 weight % solids) this waste can be sold to a smelter as opposed to landfilling. In the latter case we face disposal costs, which include storage, transportation and tipping fees at a landfill.

Figure 2. Photograph of copper wire being drawn (cold-extruded). The wire is cold-drawn by a series of extruders until the proper gauge is achieve. During this process, fine copper particulates are abraded and attrited off of the wire surface.

If instead we can sell the waste, then we not only have eliminated these costs, but in addition have generated a positive cash flow into our operation for the sale of the waste. This is what we mean by pollution prevention practice.

Of concern, however with any sludge management issue, are the costs for recovery of potentially valuable by-products as in the case of the cable plant and the costs associated with concentrating the sludge to a form that it can be handled in post processing operations.

There is an old saying among pollution control engineers that practiced pollution prevention long before it became fashionable. That saying is - 'If you see a dollar lying on the ground, it's worth your while to bend over and pick it up - but NOT if you have to break your back to do it.'

We will address some of these issues later on in this chapter. But for now, simply recognize the fact that the most basic part of waste water treatment is solids removal. Solids are removed in primary and secondary treatment tanks, but without such effective removal there is no treatment process efficiency.

Figure 3. The sludge is dried using a forced evaporation method to about 60 % solids, which makes it dry enough for bagging and transport.

Figure 4. At 60 % solids, the sludge has the consistency of dry dirt, and can be bagged in polyethylene bags and carted off to a smelter.

WHAT STABILIZATION AND CONDITIONING MEAN

PRE-STAGE BASICS

Before sludge undergoes treatment such as dewatering or thickening, it must be stored and pretreated. Sludge storage is an important, integral part of every wastewater sludge treatment and disposal system. Sludge storage provides many benefits including equalization of sludge flow to downstream processes, allowing sludge accumulation during times of non-operation of sludge-processing facilities, and allowing a uniform feed rate that enhances thickening, conditioning, and dewatering operations.

Sludge is stored within wastewater treatment process tankage, sludge treatment process systems, or separately in specially designed tanks. Sludge can be stored on a short-term or a long-term basis. Small treatment plants, where storage time may vary from several to 24 hours, may store sludge in wastewater clarification basins or sludge-thickening tanks. Larger plants often use aerobic digester, facultative lagoons, and other processes with long detention times to store sludge. The pretreatment of sludge is often necessary before dewatering or thickening can take place. It includes degritting and grinding. Sludge degritting involves the installation of grit removal and precessing facilities at the head works where raw wastewater first enters the treatment plant. As a result, there is reduced wear on influent pumping systems and primary sludge pumping, piping and thickening systems. Sludge grinding involves shearing of large sludge solids into smaller particles. This method is used to prevent problems with operation of downstream processes. In-line grinders reduce cleaning and maintenance down time of equipment. The grinders can shear sludge solids to 6 to13 mm, depending on design requirements.

Sludge-pumping systems play an important part in wastewater treatment plants, particularly those operations experiencing average flows of greater than 1 million gallons per day (mgd). There are different types of pumps within this process. Typical advantages of kinetic pumps for sludge transport include lower purchase cost, lower maintenance cost due to wear, less space used, and availability of both dry-well and submersible pumps. Advantages of positive displacement pumps include improved process control and pumping capability at high pressure and low flow.

Sludge cake storage (where a cake is the dewatered solid part of sludge) provides similar benefits for downstream disposal alternatives, like composting and incineration, to sludge storage which is used for thickening and dewatering. Storage of sludge cakes increases operational reliability, evens out flow fluctuations, and allows accumulation when downstream operations are not in service. Bins or hoppers are used to store sludge cakes. These can be made of any size form several cubic meters to 380 cubic meters capacity. Existing sludge dewatering operations can produce cakes that are 15 to 40% solids. These cakes range in consistency from pudding to damp cardboard. Since they will not flow by gravity in a pipe or channel, sludge cakes must be transported by one of the following methods: mechanical conveyors such as flat or troughed belt, corrugated belt, or Archimedes screw; gravity drop from dewatering equipment into storage hoppers directly below; and pumping by positive displacement pumps.

Thermal Stabilization

Thermal stabilization is a heat process by which the bound water (water associated with sludge) of the sludge solids is released by heating the sludge for short periods of time. Exposing the sludge to heat and pressure coagulates the solids, breaks down the cell structure, and reduces the hydration and hydrophilic (water loving) nature of the solids. The liquid portion of the sludge can then be separated from the solid by decanting and pressing.

Before any of the sludge can proceed to dewatering or thickening processes, it must be conditioned. Sludge conditioning involves *chemical* or *thermal treatment* to improve the efficiency of the downstream processes. Chemical conditioning involves use of inorganic chemicals or organic polyelectrolytes, or both. The most commonly used inorganic chemicals are ferric chloride and lime. Other chemicals are popular outside of the U.S.. Organic polymers, introduced during the 1960's, are used for both sludge-thickening and dewatering processes. Their advantage over inorganics is that polymers don't greatly increase the amount of sludge production: 1 kg of inorganic chemicals added will produce 1 kg of extra sludge. The disadvantage of polymers is their relatively high cost. There are several important factors that affect conditioning of sludge. They include: sludge characteristics, sludge handling, and sludge coagulation and flocculation. The fundamental purpose of sludge conditioning is to cause the aggregation of fine solids by coagulation with inorganic chemicals, flocculation with organic polymers, or both. A critical design parameter in conditioning is dosage. Selection of the right dosage of a chemical conditioner is critical for good performance. The dosage affects the solids content of sludge cakes as well as solids capture rate and solids disposal cost. Dosage is determined form pilot studies, bench tests, and on-line tests. In the following sections we will cover the basics of sludge stabilization and then conditioning. Our objective is to gain a working knowledge of these operations and to build our vocabulary.

CHEMICAL STABILIZATION

Chemical stabilization is a process whereby the sludge matrix is treated with chemicals in different ways to stabilize the sludge solids. Two common methods employed are lime stabilization, and the use of chlorine.

Chemical Stabilization

Chemical stabilization is a process whereby the sludge matrix is treated with chemicals in different ways to stabilize the sludge solids.

The lime stabilization process can be used to treat raw primary, waste activated, septage and anaerobically digested sludge. The process involves mixing a large enough quantity of lime with the sludge to increase the pH of the mixture to 12 or more. This normally reduces bacterial hazards and odor to a negligible value, improves vacuum filter performance and provides satisfactory means of stabilizing the sludge prior to ultimate disposal.

Stabilization by chlorine addition has been developed and is marketed under the registered trade name "Purifax". The chemical conditioning of sludge with chlorine varies greatly from the more traditional methods of biological digestion or heat conditioning. First, the reaction is almost instantaneous. Second, there is very little volatile solids reduction in the sludge. There is some breakdown of organic material and formation of carbon dioxide and nitrogen; however, most of the conditioning is by the substitution or addition of chlorine to the organic compound to form new compounds that are biologically inert.

STABILIZATION VIA AEROBIC DIGESTION

Aerobic digestion is an extension of the activated sludge aeration process whereby waste primary and secondary sludge are continually aerated for long periods of time. In aerobic digestion the microorganisms extend into the endogenous respiration phase. This is a phase where materials previously stored by the cell are oxidized, with a reduction in the biologically degradable organic matter. This organic matter, from the sludge cells is oxidized to carbon dioxide, water and ammonia. The ammonia is further converted to nitrates as the digestion process proceeds. Eventually, the oxygen uptake rate levels off and the sludge matter is reduced to inorganic matter and relatively stable volatile solids.

The primary advantage of aerobic digestion is that it produces a biologically stable end product suitable for subsequent treatment in a variety of processes. Volatile solids reductions similar to anaerobic digestion are possible. Some parameters affecting the aerobic digestion process are:

1. The rate of sludge oxidation,
2. sludge temperature,
3. system oxygen requirements,
4. sludge loading rate,

5. sludge age, and
6. sludge solids characteristics.

Aerobic digestion has been applied mostly to various forms of activated sludge treatment, usually "total oxidation" or contact stabilization plants. However, aerobic digestion is suitable for many types of municipal and industrial wastewater sludge, including trickling filter humus as well as waste activated sludge. Any design for an aerobic digestion system should include: an estimate of the quantity of sludge to be produced, the oxygen requirements, the unit detention time, the efficiency desired, and the solids loading rate. Aerobic digestion tanks are normally not covered or heated, therefore, they are much cheaper to construct than covered, insulated, and heated anaerobic digestion tanks. In fact, an aerobic digestion tank can be considered to be a large open aeration tank. Similar to conventional aeration tanks, the aerobic digesters may be designed for spiral roll or cross roll aeration using diffused air equipment. The system should have sufficient flexibility to allow sludge thickening by providing supernatant decanting facilities. The advantages most often claimed for aerobic digestion are:

- A humus-like, biologically stable end product is produced.
- The stable end product has no odors, therefore, simple land disposal, such as lagoons, is feasible.
- Capital costs for an aerobic system are low, when compared with anaerobic digestion and other schemes.
- Aerobically digested sludge usually has good dewatering characteristics. When applied to sand drying beds, it drains well and redries quickly if rained upon.
- The volatile solids reduction can be equal to those achieved by anaerobic digestion.

Supernatant liquors from aerobic digestion have a lower BOD than those from anaerobic digestion. Most tests indicated that BOD would be less than 100 ppm. This advantage is important because the efficiency of many treatment plants is reduced as a result of recycling high BOD supernatant liquors. There are fewer operational problems with aerobic digestion than with the more complex anaerobic form because the system is more stable. As a result, less skillful and costly labor can be used to operate the facility. In comparison with anaerobic digestion, more of the sludge basic fertilizer values are recovered.

The major disadvantage associated with aerobic digestion is high power costs. This factor is responsible for the high operating costs in comparison with anaerobic digestion. At small waste treatment plants, the power costs may not be significant but they certainly would be at large plants. Aerobically digested sludge does not always settle well in subsequent thickening processes. This situation leads to a thickening tank decant having a high solids concentration. Some sludge do not dewater easily by vacuum filtration after being digested aerobically. Two other minor disadvantages are the lack of methane gas production and the variable solids

> **Air Requirement**: ***15 - 20 cfm per 1,000 cubic feet of digester capacity is adequate. The air supplied must keep the solids in suspension; this requirement may exceed the sludge oxidation requirement. A dissolved oxygen concentration of 1 to 2 ppm should be maintained in the aerobic digestion tanks.***
>
> **Detention Time**: ***Waste activated sludge only, after sludge thickening. 10 - 15 days volumetric displacement time. If sludge temperatures are much less than 60°F, more capacity should be provided. Primary sludge mixed with waste activated or trickling filter humus. 20 days displacement time in moderate climates.***

reduction efficiency with varying temperature changes.

In a typical plant operation the pollutants dissolved in the wastewater or that would not settle in the primary clarifiers flow on in the wastewater to the Secondary treatment process. Secondary treatment further reduces organic matter (BOD_5) through the addition of oxygen to the wastewater which provides an aerobic environment for microorganisms to biologically break down this remaining organic matter. This process increases the percent removals of BOD and TSS to a minimum of 85 percent. A secondary treatment facility can be comprised of Oxygenation Tanks, Pure Oxygen Generating Plant, Liquid Oxygen Storage Tanks, Secondary Clarifiers, Return Sludge Pumping Station and Splitter Box, Sludge Thickeners and Pumping Station, Sludge Dewatering Building Addition and modifications to the existing Service Water Pumping Station. The Pure Oxygen Generation System often incorporates a pressure swing adsorption (PSA) system oxygen generating system A PSA system will provide a certain amount (as tons per day) of pure oxygen to the oxygenation system. As backup to the oxygen generating system, spare oxygen storage tanks containing liquid oxygen can be included in the design. Figure 5 illustrates what an aeration reactor looks like.

The oxygenation system is comprised of several covered oxygenation tanks, mechanical mixing system, and pressure-controlled oxygen feed and oxygen purity-controlled venting system. The primary effluent enters the head end of the tanks where it mixes with return activated sludge which consists of microorganisms

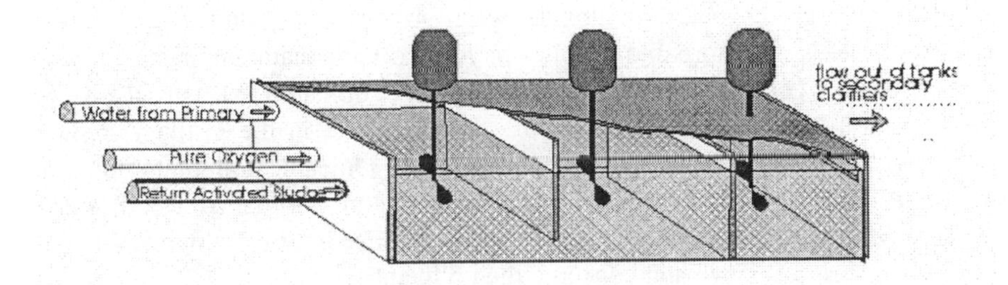

Figure 5. Example of aeration reactors.

"activated" by the organic matter and oxygen. This combination of primary effluent and return sludge forms a mixture known as "Mixed Liquor". This mixed liquor is continuously and thoroughly mixed by the mechanical mixer in each tank. The oxygen gas produced in the PSA system is introduced into the first stage of each tank and then remains in contact with the mixed liquor throughout the oxygenation system. Secondary clarifiers like the one illustrated in Figure 6 are used in this process.

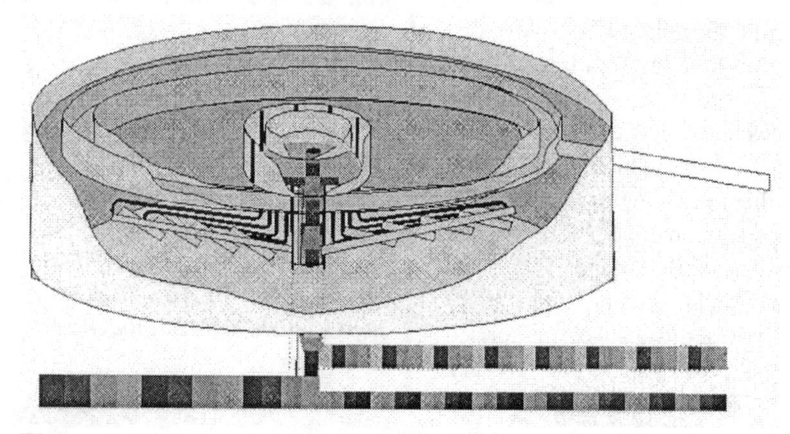

Figure 6. Diagram of a secondary clarifier.

Once the mixed liquor goes through the complete oxygenation process, it flows to four secondary clarifiers where the biological solids produced during the oxygenation process are allowed to settle and be pumped back to the head of the system. These settled solids being pumped, called return activated sludge, mix with the primary effluent to become mixed liquor. Since the population of microorganisms is growing some microorganisms in the return activated sludge are removed from the system. This solids waste stream is called waste activated sludge (WAS) and flows to the secondary gravity thickener for solids processing . The cleaned wastewater flows over the weir of the secondary clarifier and on to the disinfection (chlorination)-process. The activated sludge process describes is an aerobic, suspended growth, biological treatment method. It employs the metabolic reactions of microorganisms to produce a high quality effluent by oxidation and conversion of organics to carbon dioxide, water and biosolids (sludge). Basically the system speeds up nature and supplies oxygen so the aquatic environment will not have to. High concentrations of microorganisms (compared to a natural aquatic environment) in the activated sludge use the pollutants in the primary treated wastewater as food and remove the dissolved and non-settleable pollutants from the wastewater. These pollutants are incorporated into the microorganisms bodies and will then settle in the secondary clarifiers. Oxygen needs to be supplied for the microorganisms to survive and consume the pollutants.

STABILIZATION VIA ANAEROBIC DIGESTION

The purpose of digestion is to attain both of the objectives of sludge treatment -- a reduction in volume and the decomposition of highly putrescible organic matter to relatively stable or inert organic and inorganic compounds. Additionally, anaerobic sludge digestion produces a valuable by-product in the form of *methane gas* (the primary constituent of natural gas, which we can burn for heat or convert to electricity). Sludge digestion is carried out in the absence of free oxygen by anaerobic organisms. It is, therefore, anaerobic decomposition. The solid matter in raw sludge is about 70 percent organic and 30 percent inorganic or mineral. Much of the water in wastewater sludge is "bound" water which will not separate from the sludge solids. The facultative and anaerobic organisms break down the complex molecular structure of these solids setting free the "bound" water and obtaining oxygen and food for their growth.

Anaerobic digestion involves many complex biochemical reactions and depends on many interrelated physical and chemical factors. For purposes of simplification, the anaerobic degradation of domestic sludge occurs in two steps. In the first step, acid forming bacteria attack the soluble or dissolved solids, such as the sugars. From these reactions organic acids, at times up to several thousand ppm, and gases, such as carbon dioxide and hydrogen sulfide are formed. This is known as the stage of acid fermentation and proceeds rapidly. It is followed by a period of acid digestion in which the organic acids and nitrogenous compounds are attacked and liquefied at a much slower rate.

Hydrogen Sulfide or H_2S smells like rotten eggs. It is highly dangerous at airborne concentrations greater than 50 ppm. Exposure to H2S above 10 ppm for prolonged periods will cause olefactary saturation – i.e., you will not be able to smell the characteristic rotten egg odor and believe you are not being exposed to an inhalation

In the second stage of digestion, known as the period of intensive digestion, stabilization and gasification, the more resistant nitrogenous materials, such as the proteins, amino-acids and others, are attacked. The pH value must be maintained from 6.8 to 7.4. Large volumes of gases with a 65 or higher percentage of methane are produced. The organisms which convert organic acids to methane and carbon dioxide gases are called methane formers. The solids remaining are relatively stable or only slowly putrescible, can be disposed of without creating objectionable conditions and have value in agriculture.

The whole process of sludge digestion may be likened to a factory production line where one group of workers takes the raw material and conditions it for a second group with different "skills" who convert the material to the end products.

In a healthy, well operating digester, both of the above stages are taking place continuously and at the same time. Fresh wastewater solids are being added at frequent intervals with the stabilized solids being removed for further treatment or disposal at less frequent intervals. The supernatant digester liquor, the product of liquefaction and mechanical separation is removed frequently to make room for the added fresh solids and the gas is, of course, being removed continuously.

While all stages of digestion may be proceeding in a tank at the same time with the acids produced in the first stage being neutralized by the ammonia produced in subsequent stages, best and quickest results are obtained when the over-all pH of 6.8 to 7.4 predominates. The first stage of acid formation should be evident only in starting up digestion units. Once good alkaline digestion is established, the acid stage is not apparent unless the normal digestion becomes upset by overloading, poisonous chemicals or for other reasons. It is critical to the overall process to maintain balanced populations of acid formers and methane formers. The methane formers are more sensitive to environmental conditions and slower growing than the acid forming group of bacteria and control the overall reactions.

The progress of digestion can be measured by the destruction of organic matter (volatile solids), by the volume and composition of gases produced, by the pH, volatile acids, and alkalinity concentration. It is recommended that no on parameter or test be used to predict problems or control digesters. Several of the following parameters must be considered together.

The reduction of organic matter as measured by the volatile solids indicates the completeness of digestion. Raw sludge usually contains from 60 to 70 percent volatile solids while a well digested sludge may have as little as 50 percent. This would represent a volatile solids reduction of about 50 percent. Volatile solids reduction should be measured weekly and trended. Downward trends in volatile solids reduction might mean:

- Temperature too low and/or poor temperature control.
- Digester is overloaded.
- Ineffective mixing of digester contents.
- Grit and/or scum accumulations are excessive.
- Low volatile solids in raw sludge feed.

A well digested sludge should be black in color, have a not unpleasant tarry odor and, when collected in a glass cylinder, should appear granular in structure and show definite channels caused by water rising to the top as the solids settle to the bottom.

For domestic wastewater in a normally operating digestion tank, gas production should be in the vicinity of 12 cu.ft. of gas per day per Lbs of volatile matter destroyed. This would indicate that for a 50 percent reduction of volatile matter, a gas yield of six cu.ft. per Lbs of volatile matter added should be attained. The quantity of gases produced should be relatively constant if the feed rate is constant.

Sharp decreases in total gas production may indicate toxicity in the digester. The gas is usually about 70 percent methane, about 30 percent carbon dioxide and inert gases such as nitrogen. An increasing percentage of carbon dioxide may be an indication that the digestion process is not proceeding properly. In plants with primary and secondary digester, raw sludge is pumped to the primary digester displacing partially digested sludge. The major portion of the digestion with the greatest gas yield is in the primary digester.

A popular figure for sludge from average domestic sewage is an expected gas yield of one cu.ft. per capita per day. Industrial wastes, depending on their character may raise or lower this figure materially.

Volatile acids (mainly acetic acid) are generated by the acid forming bacteria as a result of the initial breakdown of the sludge solids. The volatile acids concentration indicates digestion progress and is probably the best warning sign of trouble. In a well operating digester, the volatile acids concentration should be measured weekly and remain fairly constant. Sudden increases in volatile acids means digester trouble. During periods of digester imbalance, volatile acids should be measured daily.

Bicarbonate Alkalinity indicates the buffering capacity of the sludge, the ability to keep the pH constant, and the ability to neutralize acids. Normally, the bicarbonate alkalinity varies between 1500 and 6000 mg/L (as calcium carbonate). The ratio between the volatile acids and the bicarbonate alkalinity concentrations is an excellent process indicator. Normally if the ratio of volatile acids concentration (mg/Liter) to bicarbonate alkalinity (mg/Liter) < 0.25, the digester is operating properly. A rising volatile acids to bicarbonate alkalinity ratio means possible trouble. Sometimes either decreasing the sludge feed to digester or resting the digester will correct the problem.

Since digestion is accomplished by living organisms, it is desirable to provide an environment in which they are most active and carry on their work in the shortest time. The environmental factors involved are moisture, temperature, availability of proper food supply, mixing and seeding, alkalinity, and pH. To these might be added the absence of chemicals toxic to the organisms. Moisture is always adequate in wastewater sludge.

It has been found that sludge digestion proceeds in almost any range of temperature likely to be encountered, but the time taken to complete digestion varies greatly with the temperature. Also rapid changes in temperature are detrimental. Digester temperature should not vary more than $\pm$ 2° F per day. Pumping excessive quantities of thin sludge can cause significant decreases in digester temperature. Thin, dilute sludge with a high moisture content also waste digester space and reduce solids retention time. The methane forming organisms are extremely sensitive to changes in temperature. At a temperature of 55° F, about 90 percent of the desired digestion is completed in about 55 days. As the temperature increases, the time decreases, so that at 75° F the time is cut to 35 days, at 85° F to 26 days, and at 95° F to 24 days. The theoretical time for sludge digestion at 95° F is one half that at 60° F. Of course, the figures are average, not exact figures for all sludge of varying composition. These digestion times may be materially reduced in digesters provided with efficient mixing of thickened sludge.

The proper amount of food must be provided for the digester organisms. This is in the form of volatile sludge solids from the various wastewater treatment units. The total volume of raw sludge pumped to the digester, the rate at which it is pumped, and the degree to which it is made available to all of the different groups of organisms are vital factors in efficient digester operation. If too much sludge is added to a digester, the first, or acid stage, predominates to such an extent that the environment becomes unfavorable for the organisms responsible for the second stage of digestion, the balance of the whole digestion process is upset, and the digester is said to be overloaded. If this is due to unbalanced plant design whereby the digester capacity is too small in relation to the sludge producing units, the only solution is to provide additional digester capacity. There are, however, other factors which can upset the balance of the digestion process and which are under the control of the operator. In heated digesters, failure to maintain uniform temperatures in the digester within the proper range will upset the digestion process. Adding fresh solids in large volumes at widely separated intervals or removing too much digested sludge at one time will result in temporary overloading. Avoid shock loadings of solids. In unheated digestion tanks similar conditions are to be expected seasonally and during winter months digester organisms are almost dormant, so that with the advent of warm weather there is in the digester an excessive accumulation of almost raw sludge solids. This, together with the normally slower digestion in unheated tanks, necessitates storage capacity twice that needed in heated digesters.

The organisms in a digester are most efficient when food is furnished them in small volumes at frequent intervals. Fresh sludge solids should therefore be pumped to the digester as often as practical, at least twice a day for the smallest plants and more frequently where facilities and operators' attention are available. This, of course, fits in with the proper schedule of removing sludge from settling units before it becomes septic.

In starting a digester unit, quickest results can be obtained by putting in it at the start some digested sludge if this is obtainable from another digester or a nearby plant. In this way all stages of digestion can be started almost simultaneously instead of by successive stages. This seeding supplies an adequate number of organisms of the methane forming type to consume the end products of the first stage and in this way the unit will "ripen" in the shortest time. After normal operation has been established, seeding of the fresh solids as added to the digester by mixing them with the digesting sludge greatly improves the rate of digestion.

POISONS

The presence of various inorganic and organic substances in a digester can cause inhibition or toxicity. Heavy metals, light metal cations, oxygen, sulfides and ammonia are potential toxic materials. The best method of controlling digester toxicity is to prevent the toxic materials from entering the system. A strong industrial waste ordinance which is enforced is the ideal solution. The cause of toxicity should be identified. Temporary relief from toxicity may be achieved by: (1) increasing the mixing the dilute and disperse the toxic material; (2) uniform feeding of sludges rather than slugs; (3) decreasing the feed to the digester; (4) diluting the sludge; (5) pumping in an active seed sludge if the digester is only partially inhibited.

THE ROLE OF MIXING

Mixing plays an important role in digester operation Without well-mixed systems, the processes cannot acceptable levels of efficiency. There are a number of methods or combination of methods whereby proper mixing is attained. These include:

- Stirring by rotating paddles and scum breaker arms.
- Forced circulation of sludge and/or supernatant by pumps or by draft tubes with impeller.
- Discharge of compressed sludge gas from diffusers at the bottom of the digestion tank..

Mixing may be either intermittent or continuous, but however effected it provides all working organisms their proper food requirements and helps maintain uniform temperature. Intermittent mixing allows separation and removal of supernatant from a single stage digester. With continuous mixing the digestion proceeds at a higher rate

Mixing serves several purposes:
The incoming raw solids are intimately mixed with the actively digesting sludge -

- ***Promotes contact between food and organisms.***
- ***Scum formation is reduced.***
- ***The transmission of heat from internal coils or other heating devices to the sludge is improved.***
- ***Distributes alkalinity throughout tank, aiding in pH control.***
- ***Continuously disperses and dilutes inhibitory materials in feed sludge.***

throughout the entire tank, thus reducing the tank capacity needed. Such continuous mixing requires a second digester or storage tank into which digesting sludge may be moved to make room for fresh sludge in the first digester and to make possible separation and removal of supernatant in the secondary digester.

SLUDGE CONDITIONING USING CHEMICALS

Sludge conditioning is a process whereby sludge solids are treated with chemicals or various other means to prepare the sludge for dewatering processes. Chemical conditioning (sludge conditioning) prepares the sludge for better and more economical treatment with vacuum filters or centrifuges. Many chemicals have been used such as sulfuric acid, alum, chlorinated copperas, ferrous sulfate, and ferric chloride with or without lime, and others (refer back to discussions in Chapter 8). The local cost of the various chemicals is usually the determining factor. In recent years the price of ferric chloride has been reduced to a point where it is the one most commonly used. The addition of the chemical to the sludge lowers or raises its pH value to a point where small particles coagulate into larger ones and the water in the sludge solids is given up most readily. There is no one pH value best for all sludge. Different sludge such as primary, various secondary and digested sludge and different sludge of the same type have different optimum pH values which must be determined for each sludge by trial and error. Tanks for dissolving acid salts, such as ferric chloride, are lined with rubber or other acid-proof material. Intimate mixing of sludge and coagulant is essential for proper conditioning.

Feeders are also necessary for applying the chemicals needed for proper chemical conditioning. The most frequently encountered conditioning practice is the use of ferric chloride either alone or in combination with lime. The use of polymers is rapidly gaining widespread acceptance. Although ferric chloride and lime are normally used in combination, it is not unusual for them to be applied individually. Lime alone is a fairly popular conditioner for raw primary sludge and ferric chloride alone has been used for conditioning activated sludges. Lime treatment to a pH of 10.4 or above has the added advantage of providing a significant degree (over 99 percent) of disinfection of the sludge. Organic polymer coagulants, and coagulant aids have been developed in the past 20 years and are rapidly gaining acceptance for sludge conditioning. These polymers are of three basic types:

- **Anionic** (negative charge) -- serve as coagulants aids to inorganic Aluminum and Iron coagulants by increasing the rate of flocculation, size, and toughness of particles.
- **Cationic** (positive charge) -- serve as primary coagulants alone or in combination with inorganic coagulants such as aluminum sulfate.
- **Nonionic** (equal amounts of positively and negatively charged groups in

monomers) -- serve as coagulant aids in a manner similar to that of both anionic and cationic polymers.

The popularity of polymers is primarily due to their ease in handling, small storage space requirements, and their effectiveness. All of the inorganic coagulants are difficult to handle and their corrosive nature can cause maintenance problems in the storing, handling, and feeding systems in addition to the safety hazards inherent in their handling.

SLUDGE CONDITIONING BY THERMAL METHODS

There are two basic processes for thermal treatment of sludge. One, wet air oxidation, is the flameless oxidation of sludge at temperatures of 450 to 550° F and pressures of about 1,200 psig. The other type, heat treatment, is similar but carried out at temperatures of 350 to 400° F and pressures of 150 to 300 psig. Wet air oxidation (WAO) reduces the sludge to an ash and heat treatment improves the dewaterability of the sludge. The lower temperature and pressure heat treatment is more widely used than the oxidation process.

When the organic sludge is heated, heat causes water to escape from the sludge. Thermal treatment systems release water that is bound within the cell structure of the sludge and thereby improves the dewatering and thickening characteristics of the sludge. The oxidation process further reduces the sludge to ash by wet incineration (oxidation). Sludge is ground to a controlled particle size and pumped to a pressure of about 300 psi. Compressed air is added to the sludge (wet air oxidation only), the mixture is brought to a temperature of about 350° F by heat exchange with treated sludge and direct steam injection, and then is processed (cooked) in the reactor at the desired temperature and pressure. The hot treated sludge is cooled by heat exchange with the incoming sludge. The treated sludge is settled from the supernatant before the dewatering step. Gases released at the separation step are passed through a catalytic after burner at 650 to 705° F or deodorized by other means. In some cases these gases have been returned through the diffused air system in the aeration basins for deodorization.

An advantage of thermal treatment is that a more readily dewaterable sludge is produced than with chemical conditioning. Dewatered sludge solids of 30 to 40 percent (as opposed to 15 to 20 percent with chemical conditioning) have been achieved with heat treated sludge at relatively high loading rates on the dewatering equipment (2 to 3 times the rates with chemical conditioning). The process also provides effective disinfection of the sludge. Unfortunately, the heat treatment process ruptures the cell walls of biological organisms, releasing not only the water but some bound organic material. This returns to solution some organic material previously converted to particulate form and creates other fine particulate matter. The breakdown of the biological cells as a result of heat treatment converts these previously particulate cells back to water and fine solids. This aids the dewatering process, but creates a separate problem of treating this highly polluted liquid from the cells. Treatment of this water or liquor requires careful consideration in design of the plant because the organic content of the liquor can be extremely high.

Because the WAO process also aims to reduce sludge volume we will spend more time describing this process under the section dealing with ***Volume Reduction***. The other thermal sludge conditioning method is best-known as sludge pasteurization, and deserves more than just a brief overview.

THE SLUDGE PASTEURIZATION PROCESS

This process is really *sludge disinfection*. Its aim is the destruction or inactivation of pathogenic organisms in the sludge. *Destruction* is defined as the physical disruption or disintegration of a pathogenic organism, while *Inactivation* is defined as the removal of a pathogen's ability to infect.

In the United States procedures to reduce the number of pathogenic organisms are a requirement before sale of sludge or sludge-containing products to the public as a soil conditioner, or before recycling sludge to croplands. Since the final use or disposal of sludge may differ greatly with respect to health concerns, and since a great number of treatment options effecting various degrees of pathogen reduction are available, the system chosen for the reduction of pathogens should be tailored to the specific application. Thermal conditioning of sludge in a closed, pressurized system destroys pathogenic organisms and permits dewatering. The product generally has a good heating value or can be used for land filling or fertilizer base. In this process, sewage sludge is ground and pumped through a heat exchanger and sent with air to a reactor where it is heated to a temperature of 350-400° F. The processed sludge and air are returned through the heat exchanger to recover heat. The conditioned slurry is then discharged to a gravity thickener where the vapors are separated and the solids are concentrated (thickened). The treatment process renders the sludge easily dewaterable without the addition of chemicals.

After thickening, a variety of sludge handling and disposal options are available. For example, the thickened sludge can be applied directly to land. If liquid disposal is not applicable to a specific project, the thickened sludge can be dewatered by centrifugation, vacuum filtration, or filter pressing. The dewatered residue can then be land fill or incinerated. These options are discussed further on.

Thermal sludge conditioning and its effects on the chemical and physical structures of wastewater sludge can be best understood from analyses of typical sewage sludges. Wastewater sludge is a complex mixture of waste solids forming a gelatinous mass that is nearly impossible to dewater without further treatment. The organic fraction of the sludge consists of lipids, proteins, and carbohydrates, all bound by physical-chemical forces in a predominantly water-gel-like structure.

When the sludge is heated under pressure to temperatures above 350° F (176.5° C), the gel-like structure of sludge is destroyed, liberating the bound water. Dewatering by filtration without chemical conditioning is then a simple matter.

There are several characteristics of thermally conditioned sludge which have an important effect on the cost of plants utilizing thermoconditioning. Various factors, such as thickening properties, dewatering properties, heavy metal distribution, heating value, volatile solids solubilization, and others, all have a major impact on the evaluation of various process alternatives and ultimate disposal. Thickening and dewatering properties vary depending on the type of sludge. In general, vacuum filtration rates vary from 2-15 $Lbs/ft^2/hr$, with cake moistures ranging from 50-70 %. Lower values (2-4 $Lbs/ft^2/hr$) are observed for high proportions of waste-activated sludge; the higher values (up to 15 $Lbs/ft^2/hr$) are observed for sludge which are predominantly primary sludge.

Similar results have been obtained for filter pressing. Mixtures of primary and waste-activated sludge of relatively the same proportions which have been thermally conditioned and dewatered at rates from 2,540 Lbs dry solids per ft/hr, with cake moistures ranging from 50-60 percent. Heating values of thermally conditioned sludge cake are typically about 12,500 Btu/Lbs. No marked differences in heating values have been found for different types of sludge. Sludge conditioned with ferric chloride and lime have been found to have heating values in the range of 9,000-10,000 Btu/Lbs. The lower values experienced for chemically conditioned sludge could be due to:

- Selective solubilization of materials of lower heating value in the thermal conditioning process.
- Enclothermic reactions with the conditioning materials.
- Operational differences in the analytical methods used for determining the heating value of the volatile content.

Solubilization of a fraction of the influent-suspended solids can occur as a result of thermal conditioning. In low-pressure, wet-air oxidation, some of the organics present are oxidized as well. Solubilization of the volatile suspended solids produces a supernatant or filtrate of relatively high organic strength.

Ash solubilization and volatile suspended-solids oxidation also decrease the solids loads to downstream solids-handling units.

There are several advantages that thermoconditioning has over chemical conditioning. These include the sterility of the end product and a residue that can be readily thickened. Bacteria are numerous in the human digestive tract; humans excrete up to 10^{13} coliform and 10^{16} of other bacteria in their feces every day. The most important of the pathogenic bacteria are listed in Table 1, together with the diseases they cause which may be present in municipal wastewater treatment sludges. Table 2 lists potential parasites in wastewater sludge.

Sludge stabilization processes are ideally intended to reduce putrescibility, decrease mass, and improve treatment characteristics such as dewaterability. Many stabilization processes also accomplish substantial reductions in pathogen concentration. Sludge digestion is one of the major methods for sludge stabilization. Well-operated digesters can substantially reduce virus and bacteria levels but are less effective against parasitic cysts. The requirement for pasteurization is that all sludge be held above a predetermined temperature for a minimum time period.

Table 1. Pathogenic Human Bacteria Potentially in Wastewater Sludge.

Species	**Disease**
Arizona hinshawii	Arizona infection
Bacillus cereus	*B. cereus* gastroenteritis; food poisoning
Vibrio cholerae	Cholera
Clostridium perfringens	C. perfringens gastroenteritis; food poisoning
Clostridium tetani	Tetanus
Escherichia coli	Enteropathogenic E. coli infection; acute diarrhea
Leptospira sp	Leptospirosis; Swineherd's disease
Mycobacterium tuberculosis	Tuberculosis
Salmonella paratyphi, A, B, C	Paratyphoid fever
Salmonella sendai	Paratyphoid fever
Salmonella sp (over 1,500 serotypes)	Salmonellosis; acute diarrhea
Salmonella typhi	Typhoid fever
Shigella sp	Shigellosis; bacillary dysentery; acute diarrhea
Yersinia enterocolitica	Yersinia gastroenteritis
Yersinia pseudotuberculosis	Mesenteric lymphadenopathy

Table 2. Pathogenic Human and Animal Parasites Found in Wastewater Sludge.

Species	Disease
Protozoa	
Acanthamoeba sp	Amoebic meningoencephalitis
Balantidium coli	Balanticliasis, Balantidial dysentery
Dientamoeba fragilis	Dientamoeba infection
Entamoeba histolytica	Amoebiasis; amoebic dysentery
Giardia lamblia	Giardiasis
Isospora bella	Coccidiosis
Naegleria fowleri	Amoebic meningoencephalitis
Toxoplasma gordii	Toxoplasmosis
Nematodes	
Ancyclostoma dirodenale	Ancylostomiasis; hookworm disease
Ancyclostoma sp	Cutaneous larva migrans
Ascaris lumbricoides	Ascariasis; roundworm disease; Ascaris pneumonia
Enterobius vermicularis	Oxyuriasis; pinworm disease
Necator americanus	Necatoriasis; hookworm disease
Strongyloides stercoralis	Strongyloidiasis; hookworm disease
Toxocara canis	Dog roundworm disease, visceral larva migrans

Species	Disease
Toxocara cati	Cat roundworm disease; visceral larva migrans
Trichusis trichiura	Trichuriasis; whipworm disease
Helminths	
Diphyllobothrium latum	Fish tapeworm disease
Echinococcus granulosis	Hydatid disease
Echinococcus multilocularis	Aleveolar hydatid disease
Hymenolepsis diminuta	Rat tapeworm disease
Tymenolepsis nana	Dwarf tapeworm disease
Taenia saginata	Taeniasis; beef tapeworm disease
Taenia solium	Cysticercosis; pork tapeworm disease

Heat transfer can be accomplished by steam injection or with external or internal heat exchangers. Steam injection is preferred because heat transfer through the sludge slurry is slow and not dependable. Incomplete mixing will either increase heating time, reduce process effectiveness, or both. Overheating or extra detention times are not desirable, however, because trace metal mobilization may be increased, odor problems will be exacerbated, and unneeded energy will be expended. Batch processing is preferable to avoid reinoculations if short circuiting occurs.

The flow scheme for a sludge pasteurization system with a one-stage heat recuperation system is shown in Figure 7. System components include a steam boiler, a preheater, a sludge heater, a high-temperature holding tank, blow-off tanks, and storage basins for the untreated and treated sludge.

Pasteurization is employed extensively in Western Europe. As examples, in Germany and Switzerland it is required before application of sludge to farmlands during the spring-summer growing season. Based on Western European experience, heat pasteurization is a proven technology, requiring skills such as boiler operation and understanding of high-temperature and pressure processes. Pasteurization can be applied to either untreated or digested sludge with little pretreatment. Digester gas, available in many plants, is an ideal fuel and is usually produced in sufficient amounts to disinfect locally produced sludge. The disadvantages of this process include significant odor problems and the need for large sludge storage facilities following the process. The storage facilities are not only a problem because of space requirements, but they offer the opportunity for bacterial pathogens to regrow if the sludge becomes reinoculated.

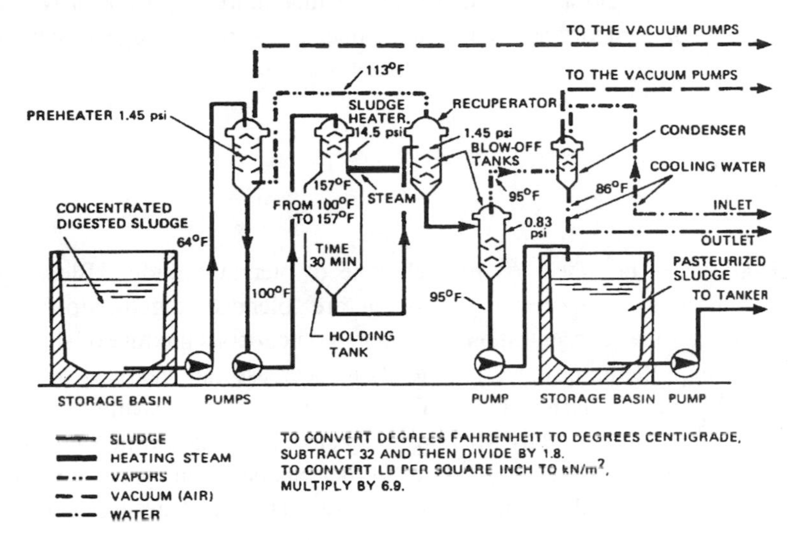

Figure 7. Flow scheme for sludge pasteurization process.

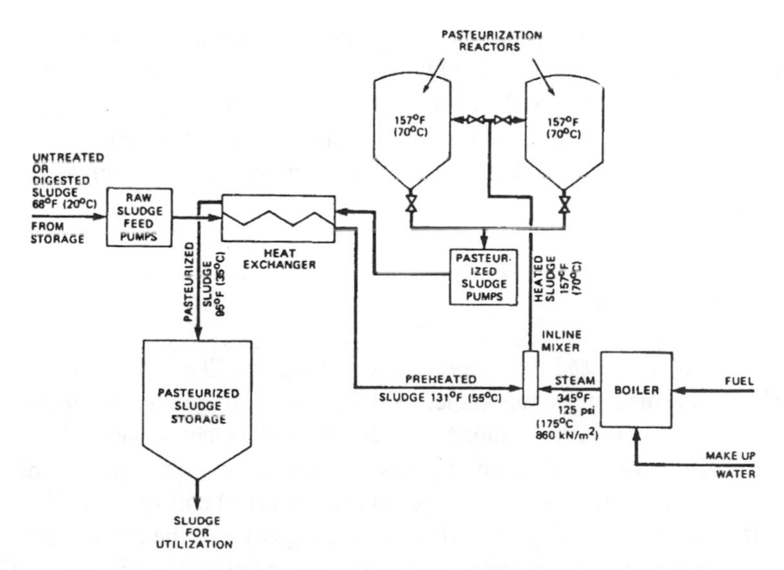

Figure 8. Sludge pasteurization process with heat recovery.

A pasteurization process should be designed to provide uniform minimum temperature of at least 70° C for at least 30 minutes (note - some literature sources argue for higher minium temperatures). Batch processing is necessary in order to prevent short-circuiting and recontamination of the sludge, especially by bacteria. In-line mixing of steam and sludge is normally practiced to ensure uniform heat transfer among the sludge mass. This practice also eliminates the need to mix the sludge while it is held at the pasteurization temperature. Figure 8 provides us a glimpse of what important process components are in a sludge pasteurization process.

BLENDING

Stabilization can be aided by the technique of blending sludge. Blending is a process where two or more types of sludge are "blended" together to facilitate a higher sludge solids concentration and a more homogenous mixture of sludge prior to dewatering. Blending operations tends to decrease the chemical demand for conditioning and dewatering sludge. The blending operation usually takes place in sludge holding tanks normally where primary sludge is mixed with waste activated sludge. The amounts of the sludge to be blended can only be found by experimentation, with the final results being seen at the dewatering operations.

SLUDGE DEWATERING OPERATIONS

Another term for dewatering the sludge is *sludge thickening*. The objective is to concentrate the sludge, and quite frankly - make it as dry as economically possible for post processing and disposal purposes. There are both mechanical and thermal techniques for achieving this. Among the mechanical processes used to dewater sludge are belt fiter presses and drum filters (vacuum technologies), pressure filter presses, and centrifugation.

VACUUM FILTRATION

We have already discussed this important technology in Chapter 5, but a review may be helpful in placing its importance to sludge processing into perspective. The vacuum filter for dewatering sludge is a drum over which is laid the filtering medium consisting of a cloth of cotton, wool, nylon, dynel, fiber glass or plastic, or a stainless steel mesh, or a double layer of stainless steel coil springs. The drum with horizontal axis is set in a tank with about one quarter of the drum submerged in conditioned sludge. Valves and piping are so arranged that, as a portion of the drum rotates slowly in the sludge, a vacuum is applied on the inner side of the filter medium, drawing out water from the sludge and holding the sludge against it. The application of the vacuum is continued as the drum rotates out of the sludge and into the atmosphere. This pulls water away from the sludge, leaving a moist mat or cake on the outer surface. This mat is scraped, blown or lifted away from the drum just before it enters the sludge tank again. The common measure of performance of vacuum filters is the rate in pounds per hour of dry solids filtered per square foot of filter surface. For various sludge this rate may vary from a low of 2.5 for activated sludge to a high of 6 to 11 for the best digested primary sludge. The moisture content in the sludge cake also varies with the type of sludge from 80 to 84 percent, for raw activated sludge to 60 to 68 percent for well digested primary sludge. While operating costs, including conditioning of sludge for vacuum filtration, are usually higher than with sludge beds, filtration has the advantage of requiring much less area, is independent of seasons and weather conditions, and can eliminate the necessity for digestion since raw sludge can be dewatered sufficiently to be incinerated.

Prolongation of the life of the material used as the filter may be effected by proper care. Such care includes washing of the filter material with the spray jets after every period of use, removal of grease and fats with warm soap solution if clogged, treatment with diluted hydrochloric acid for removal of lime encrustations, maintenance of scraper bade in careful adjustment to filter drum to prevent tearing of the filter material.

With regard to chemical use -- diluted ferric chloride solutions (10% to 20%) usually give better results in the conditioning of the sludge. A high calcium lime is preferable or sludge filtration work. One should avoid excessive use of chemicals. The quantities of chemicals used for conditioning can be frequently reduced by careful control of the

Operational Considerations

- *Conditioned sludge should be filtered as quickly as possible after the addition of the chemicals and adequate mixing.*
- *Continuous feeding is preferable to batch conditioning. In raw sludge filtration, fresh sewage solids and sludge filter more readily than stale or septic sludge. The applies to a raw sludge filtration.*
- *Completely digested sludge usually filters more readily than partially digested sludge.*
- *The concentration of sludge to be filtered is critical as sludge with the higher solid content usually filters more readily than that with a lower solid content.*
- *The presence of mineral oils and wastes from dry cleaning establishments makes sludge filtration difficult. Such wastes should, therefore, be kept out of the sewer system and disposed of separately.*
- *After every use the vacuum filter should be cleaned, and all sludge drained from the unit. This sludge and wash water should not be returned to the sludge storage tank but to the raw sewage channel or to a digester.*

mixing and flocculation equipment. The maintenance of a uniform vacuum is necessary for satisfactory operation. Loss or fluctuations in vacuum usually indicate a break in the filter material, poorly conditioned sludge or uneven distribution of the sludge solids in the filter pan.

We should spend just a few minutes talking about the Rotary Drum Precoat Filter. This machine is used to polish solutions having traces of contaminating insolubles, so it is not a dewatering machine per se, but its use is often integrated into the process. To polish the solution the drum deck is precoated with a medium of a known permeability and particle size that retains the fines and produces a clear filtrate. The following materials are used to form the precoat bed: Diatomaceous Earth (or Diatomite) consisting of silicaceous skeletal remains of tiny aquatic unicellular plants; Perlite consisting of glassy crushed and heat-expanded rock from volcanic origin; Cellulose consisting of fibrous light weight and ash less paper like medium; Special ground wood is becoming popular in recent years since it is combustible and reduces the high cost of disposal. There are nowadays manufacturers that grind, wash and classify special timber to permeabilities which can suit a wide range of applications. These materials when related to precoating are wrongly called filter-aids since they do not aid filtration but serve as a filter medium in an analogy to the filter cloth on a conventional drum filter. The Precoat Filter is similar in appearance to a conventional drum filter but its construction is very different. The scraper blade on conventional drum filters is stationary and serves mainly to deflect the cake while it is back-blown at the point of discharge. The scraper on a precoat filter, which is also called "Doctor Blade", moves slowly towards the drum and shaves-off the blinding layer of the contaminants together with a thin layer of the precoating material. This movement exposes continuously a fresh layer of the precoat surface so that when the drum submerges into the tank it is ready to polish the solution. The blade movement mechanism is equipped with a precision drive having an adjustable advance rate of 1 to 10 mm/hr. The selected rate is determined by the penetration of fines into the precoat bed which, in turn, depends on the permeability of the filter aid. Once the entire precoat is consumed the blade retracts at a fast rate so that the filter is ready for a new precoating cycle. The cake discharges on conventional drum filters by blow-back hence a section of the main valve's bridge setting is allocated for this purpose. On precoat filters the entire drum deck is subjected to vacuum therefore there are two design options:

1. A conventional valve that is piped, including its blow-back section, to be open to vacuum during polishing. When the precoat is consumed its blow-back section is turned on to remove the remaining precoat heel over the doctor blade.

2. A valveless configuration in which there is no bridge setting and the sealing between the rotating drum and the stationary outlet is by circumferential "o" rings rather than by a face seal used on conventional valves.

The flow scheme for a conventional precoat filter station typically looks like that shown in Figure 9. The doctor blade discharge configuration for this machine is illustrated in Figure 10.

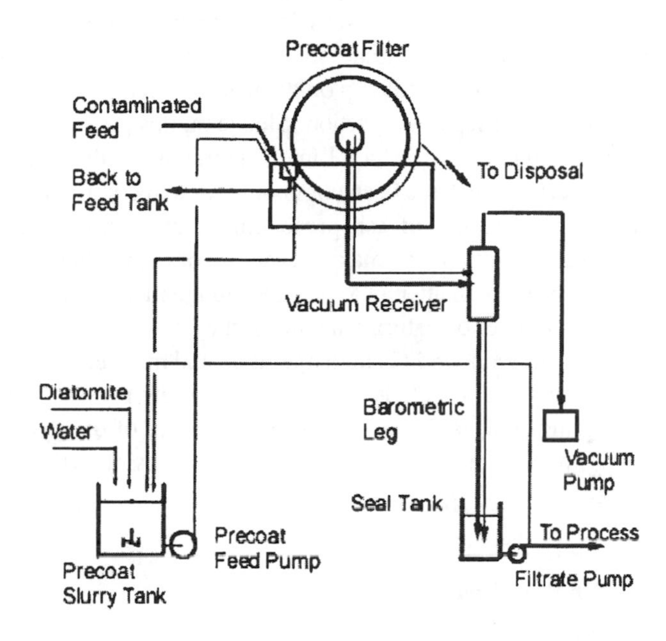

Figure 9. Precoat drum filter flow scheme for polishing operations.

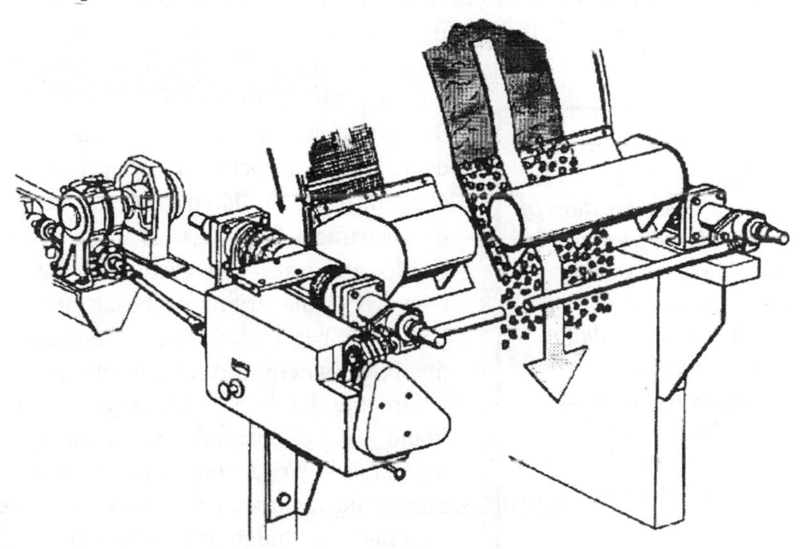

Figure 10. Doctor blade discharge for precoat filter.

PRESSURE FILTRATION

Pressure filtration is a process similar to vacuum filtration where sludge solids are separated from the liquid. Leaf filters probably are the most common type of unit. Refer back to Chapter 5 for a detailed discussion. Like vacuum filtration, a porous media is used in leaf filters to separate solids from the liquid. The solids are captured in the media pores; they build up on the media surface; and they reinforce the media in its solid-liquid separation action. Sludge pumps provide the energy to force the water through the media. Lime, aluminum chloride, aluminum chlorohydrate, and ferric salts have been commonly used to condition sludge prior to pressing. The successful use of ash precoating is also prevalent. Minimum chemical costs are supposed to be the major advantage of press filters over vacuum filters. Leaf filters represent an attempt to dewater sludge in a small space quickly. But, when compared to other dewatering methods, they have major disadvantages, including: (1) batch operation, and (2) high operation and maintenance costs. Some other types of pressure filters include hydraulic and screw presses, which while effective in dewatering sludges, have a major disadvantage of usually requiring a thickened sludge feed. Sludge cakes as high as 75% solids using pressure filtration have been reported.

CENTRIFUGE DEWATERING

Centrifuges are machines that separate solids from the liquid through sedimentation and centrifugal force. In a typical unit sludge is fed through a stationary feed tube along the centerline of the bowl through a hub of the screw conveyor. The screw conveyor is mounted inside the rotating conical bowl. It rotates at a slightly lower speed than the bowl. Sludge leaves the end of the feed tube, is accelerated, passes through the ports in the conveyor shaft, and is distributed to the periphery of the bowl. Solids settle through the liquid pool, are compacted by centrifugal force against the walls of the bowl, and are conveyed by the screw conveyor t the drying or beach area of the bowl. The beach area is an inclined section of the bowl where further dewatering occurs before the solids are discharged. Separated liquid is discharged continuously over adjustable weirs at the opposite end of the bowl. The important process variables are: (1) feed rate, (2) sludge solids characteristics, (3) feed consistency, (4) temperature, and (5)

In centrifuge dewatering centrifugal force is used to accelerate the separation of solid and liquid phases of the liquid sludge stream. The process involves clarification of the sludge and its compaction. Centrifuges separate the sludge into dewatered sludge cakes and clarified liquid, which is called centrate.

chemical additives. Machine variables are: (1) bowl design, (2) bowl speed, (3) pool volume, and (4) conveyor speed. Two factors usually determine the success of failure of centrifugation -- cake dryness and solids recovery. The effect of the various parameters on these two factors are listed below:

To Increase Cake Dryness:	To Increase Solids Recovery:
1. Increase bowl speed	1. Increase bowl speed
2. Decrease pool volume	2. Increase pool volume
3. Decrease conveyor speed	3. Decrease conveyor speed
4. Increase feed rate	4. Decrease feed rate
5. Decrease feed consistency	5. Increase temperature
6. Increase temperature	6. Use flocculents
7. Do no use flocculents	7. Increase feed consistency

Centrifugation has some inherent advantages over vacuum filtration and other processes used to dewater sludge. It is simple, compact, totally enclosed, flexible, can be used without chemical aids, and the costs are moderate. Industry particularly has accepted centrifuges in part due to their low capital cost, simplicity of operation, and effectiveness with difficult-to-dewater sludges. The most effective centrifuges to dewater waste sludges are horizontal, cylindrical -- conical, solid bowl machines. Basket centrifuges dewater sludges effectively but liquid clarification is poor. Disc-type machines do a good job of clarification but their dewatering capabilities leave much to be desired. Centrifuges are being installed in more and more wastewater treatment plants for the following reasons: (1) the capital cost is low in comparison with other mechanical equipment, (2) the operating and maintenance costs are moderate, (3) the unit is totally enclosed so odors are minimized, (4) the unit is simple and will fit in a small space, (5) chemical conditioning of the sludge is often not required, (6) the unit is flexible in that it can handle a wide variety of solids and function as a thickening as well as a dewatering device, (7) little supervision is required, and (8) the centrifuge can dewater some industrial sludges that cannot be handled by vacuum filters.

The disadvantages associated with centrifugation are: (1) without the use of chemicals the solids capture is often very poor, and chemical costs can be substantial; (2) trash must often be removed from the centrifuge feed by screening; (3) cake solids are often lower than those resulting from vacuum filtration; and (4) maintenance costs are high.

The poor quality of the centrate is a major problem with centrifuges. The fine solids in centrate recycled to the head of the treatment plant sometimes resist settling and as a result, their concentrations in the treatment system gradually build up. The centrate from raw sludge dewatering can also cause odor problems when recycled. Flocculents can be used to increase solids captures, often to any degree desired, as well as to materially increase the capacity (solids loading) of the centrifuges. However, the use of chemicals nullifies the major advantage claimed for centrifuges -- moderate operating costs. As noted, three basic types of centrifuges are disc-nozzle, basket, and solid bowl. The latter two types have been used extensively for both dewatering and thickening. The disk-nozzle centrifuge is seldom used for dewatering sludge, but is used more for sludge thickening in the industrial sector. Because the solid bowl design has undergone major improvements throughout the history of its use, this method is used more than any other to dewater sludge. Because of recent improvements in solid bowl centrifuge design, solid concentrations can reach 35%. The solid bowl conveyor centrifuge operates with a continuous feed and discharge rates. It has a solid-walled imperforated bowl, with a horizontal axis of rotation. These centriffiges are enclosed, so they have a limited odor potential compared with other dewatering methods. The laydown area, access area, and centrifuge required space for a large machine (200 m to 700 gpm of sludge feed) is approximately 400 square feet. Compared to other mechanical dewatering machines, this space is significantly smaller. An example of a continuous horizontal solid-bowl centrifuge is illustrated in Figure 11. It consists of a cylindrical rotor with a truncated cone-shaped end and an internal screw conveyor rotating together. The screw conveyor often rotates at a rate of 1 or 2 rpm below the rotor's rate of rotation. The suspension enters the bowl axially through the feed tube to a feed accelerated zone, then passes through a feed port in the conveyor hub into the pond. The suspension is subjected to centrifugal force and thrown against the bowl wall where the solids are separated. The clarified suspension moves toward the broad part of the bowl to be discharged through a port. The solid particles being scraped by the screw conveyor are carried in the opposite direction (to the small end of the bowl) across discharge ports through which they are ejected continuously by centrifugal force. As in any sedimentation centrifuge, the separation takes place in two stages: settling (Figure 11, in the right part of the bowl), and thickening or pressing out of the sediment (left-hand side of the bowl).

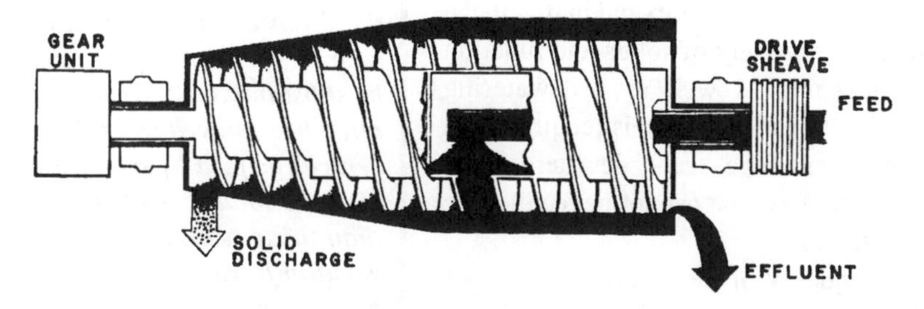

Figure 11. Continuous solid-bowl centrifuge.

Because the radius of the solid discharge port is ususally less than the radius of the liquid overflow at the broader end of the bowl, part of the settled solids is submerged in the pond.

The remainder, closer to the center, is inside the free liquid interface, where they can drain before being discharged. The total length of the "settling" and "pressing out" zones depends on the dimensions of the rotor. Their relative length can be varied by changing the pond level through suitable adjustment of the liquid discharge radius. When the pond depth is lowered, the length of the pressing out zone increases with some sacrifice in the clarification effectiveness. The critical point in the transport of solids to the bowl wall is their transition across the free liquid interface, where the buoyancy effect of the continuous phase is lost. At this point, soft amorphous solids tend to flow back into the pond instead of discharging. This tendency can be overcome by raising the pond level so that its radius is equal to, or less than, that of the solids discharge port. In reality, there are no dry settled solids. The solids form a dam, which prevents the liquid from overflowing. The transfer of solids becomes possible because of the difference between the rotational speed of the screw conveyor and that of the bowl shell. The flights of the screw move through the settled solids and cause the solids to advance. To achieve this motion, it is necessary to have a high circumferential coefficient of friction on the solid particles with respect to the bowl shell and a low coefficient axially with respect to the bowl shell and across the conveyor flights. These criteria may be achieved by constructing the shell with conical grooves or ribs and by polishing the conveyor flights. The conveyor or differential speed is normally in the range of 0.8 % to 5 % of the bowl's rotational speed.

The required differential is achieved by a two-stage planetary gear box. The gear box housing carrying two ring gears is fixed to, and rotates with, the bowl shell. The first stage pinion is located on a shaft that projects outward from the housing. This arrangement provides a signal that is proportional to the torque imposed by the conveyor. If the shaft is held rotational (for example, by a torque overload release device or a shear pin), the relative conveyor speed is equivalent to the bowl rotative speed divided by the gear box ratio. Variable differential speeds can be obtained by driving the pinion shaft with an auxiliary power supply or by allowing it to slip forward against a controlled breaking action. Both arrangements are employed when processing soft solids or when maximum retention times are needed on the pressing out zone. The solids handling capacity of this type centrifuge is established by the diameter of the bowl, the conveyor's pitch and its differential speed. Feed ports should be located as far from the effluent discharge as possible to maximize the effective clarifying length. Note that the feed must be introduced into the pond to minimize disturbance and resuspension of the previously sedimented solids. As a general rule, the preferred feed location is near the intercept of the conical and cylindrical portions of the bowl shell. The angle of the sedimentation section with respect to the axis of rotation is typically in the range of 3 to 15°. A shallow angle provides a longer sedimentation area with a sacrifice in the effective length for clarification. In some designs, a portion of the conveyor flights in the sedimentation area is shrouded (as with a cone) to prevent intermixing of the sedimented solids with the free supernatant liquid in the pond through which they normally would pass. In other designs, the clarified liquid is discharged from the front end via a centrifugal pump or an adjustable skimmer that sometimes is used to control the pond level in the bowl. Some displacement of the adhering virgin liquor can be accomplished by washing the solids retained on the settled layer, particularly if the solids have a high degree of permeability. Washing efficiency ranges up to 90% displacement of virgin liquor on coarse solids. Some configurations enable the settled layer to have two angles; comparatively steep in the wetted portion (10-15°) and shallow in the dry portion (3-5°). A wash is applied at the intersection of these angles, which, in effect, forms a constantly replenished zone of pure liquid through which the solids are conveyed. The longer section of a dry shallow layer provides more time for drainage of the washed solids. In either washing system, the wash liquid that is not carried out with the solids fraction returns to the pond and eventually discharges along with the effluent virgin liquor.

Separation Rates in Tubular-and Solid-Bowl Centrifuges: To evaluate the radial velocity of a particle moving toward a centrifuge wall, the expression for particle settling in a gravitational field is applied:

$$a = u_r^2/r \tag{1}$$

where u_r is the peripheral velocity at a distance, r, from the axis of rotation, and 'a' is centrifugal acceleration. Expressing u_r in terms of the number of rotations, n,

$$u_r = 2\pi rn \tag{2}$$

The centrifugal acceleration is:

$$a = 4\pi^2 rn^2 \tag{3}$$

Depending on the particle diameter and properties of the liquid, the radial motion of particles will be laminar, turbulent or transitional. The motion of large particles at Re > 500 is turbulent. Therefore, their settling velocity in a gravitational field may be expressed as:

$$u = 1.74[d(\gamma_p - \gamma)g/\gamma]^{1/2} \tag{4}$$

Replacing g by centrifugal acceleration, a:

$$u_r = 1.74[d(\gamma_p - \gamma)a/\gamma]^{1/2} \tag{5}$$

where γ_p = specific weight of the solids and γ = specific weight of the liquid. Substituting for "a" into this last expression, we obtain the particle velocity in the radial direction of the wall:

$$u_r = 10.94[d(\gamma_p - \gamma)r/\gamma]^{1/2} \quad (6)$$

Combining the above expressions, we determine the number of times the particle velocity in a centrifuge is greater than that in free particle settling:

$$u_r/u_g = 2\pi n(r/g)^{1/2} \quad (7)$$

For example, at n = 1200 rpm = 20 liter/sec $^{-1}$ and r = 0.5 m, the settling velocity in the centrifuge is almost 28 times greater than that in free settling. Note that the above expressions are applicable only for Re > 500. For small particles, Re < 2, migration toward the wall is laminar. The proper settling velocity expression for the gravitational field is

$$u = d^2(\gamma_p - \gamma)/18\mu \quad (8)$$

Substituting in ρg for γ, we obtain

$$u = d^2(\rho_p - \rho)a/18\mu \quad (9)$$

Replacing 'a' and with some algebraic manipulations we obtain:

$$u_r/u_g = 4\pi^2 n^2(r/g) \quad (10)$$

For the same case of n = 1200 rpm and r = 0.5, we obtain u_r/u_g = 800, whereas for the turbulent regime the ratio was only 28. This example demonstrates that the centrifugal process is more effective in the separation of small particles than of large ones. Note that after the radial velocity u_r is determined, it is necessary to check whether the laminar condition, Re < 2, is fulfilled. For the transition regime, 2 < Re < 500, the sedimentation velocity in the gravity field is:

$$u = [0.153d^{1.14}(\gamma_p - \gamma)^{0.71}]/[(\gamma/g)^{0.29}\mu^{0.43}] \quad (11)$$

The expression for particle radial velocity toward the wall is:

$$u_r = [0.153d^{1.14}(\rho_p - \rho)^{0.71}]/[\gamma^{0.29}\mu^{0.43}] \quad (12)$$

or

$$u_r = [0.136d^{1.14}(\gamma_p - \gamma)^{0.71}(n^2r)^{0.71}]/[\gamma^{0.29}(\mu g)^{0.43}] \quad (13)$$

The ratio of settling velocity in a centrifuge to that of pure gravitational settling is:

$$u_r/u_g = (4\pi^2 n^2 r/g)^{0.71} \qquad (14)$$

This ratio represents an average between similar ratios for the laminar and turbulent regimes. In the most general case, $u_r = f(D, \rho_p, \rho, \mu, r, \omega)$, and hence we may ignore whether the particle displacement is laminar, turbulent or within the transition regime. This enables us to apply the dimensionless Archimedes number (recall the derivation back in Chapter 5):

$$Ar_c = C_D Re^2 = (4/3)d^3\rho(\rho_p - \rho)\omega^2 r/\mu^2 \qquad (15)$$

A plot of the Archimedes number versus Reynolds number is provided in Figure 12.

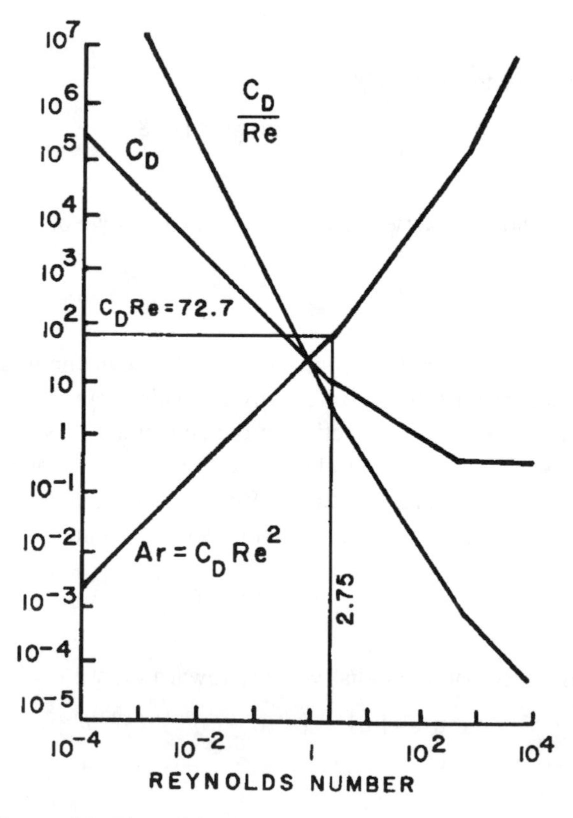

Figure 12. Plot of Archimedes number versus Reynolds number.

Let's consider the following example in order to get a better appreciation for the application of the above expressions. In this problem oil droplets (d_p = 10-4 m, ρ_p = 900 kg/m^3, $\mu = 10^{-3}$ cP) are to be separated in a sedimentation centrifuge. The machine operates at 5000 rpm ($\omega = 2\pi \times 5000/60$). If the distance of a single droplet from the axis of rotation is 0.1 m, determine the droplet's redial settling velocity.

The solution to this problems is as follows:

$$Ar_c = C_D Re^2 = \{4 \times 10^{-12} \times 1000 \times 100(2\pi \times 5000/60)^2 \times 0.1\}/(3 \times 10^{-6})$$

$$= 3650$$

Note that the absolute value of $\rho_p - \rho$ has been assumed. The negative value of this difference indicates that the droplet displacement is centripetal. The value of the Reynolds number corresponding to $Ar_c = 3650$ from Figure 10 is 45. Hence we can determine the radial settling velocity from the definition of the Reynolds number:

$$45 = 10^{-4} \times 1000\ u_r/10^{-3}$$

$$u_r = 0.45 \text{ m/sec.}$$

If the separation were to occur in a gravitational field only, the droplet velocity would be $u_r = 5.45 \times 10^{-4}$ m/sec. For laminar droplet motion, this corresponds to a separation number of 2800. However, in this case the flow is transitional. Particle settling velocity in a centrifuge depends on the particle location. For the laminar regime, the particle velocity is proportional to the centrifuge radius; however, for the turbulent regime, it is related to $\sqrt{r}$. In an actual operation, a particle may blow over different regimes depending on its location. Particles of different sizes and densities may be located at the same point with the same velocity because the larger particle diameter tends to compensate for lower density. Two particles with identical settling velocities in a gravitational field will settle in the same manner in a centrifugal field, provided the regimes of motion remain unchanged. For example, if the gravitational sedimentation is laminar and centrifugal sedimentation is transitional or turbulent, then particles will have different velocities within the centrifuge. Under certain conditions of centrifugal sedimentation (characterized by r and w), the settling velocity for a gravitational field can be applied. This is exactly correct when gravitational settling occurs in the turbulent regime.. However, when a particle passes from one regime to the other, or when it is in the transitional regime, one either must calculate the velocities or extrapolate from experimental results. Extrapolation is valid only in cases in which there are no changes in the flow regime. Consider a particle ($d = 5 \times 10^{-4}$ in) located at a distance of 0.2 m from the axis of a centrifuge operating at 4000 rpm has a velocity of 1 m/sec through water. In this case, $Re = 1 \times 1000 \times 0.0005/0.001 = 500$ and, hence, the regime is turbulent. It is possible to extrapolate to conditions of greater radii and higher rotations. If the centrifuge operates at 5000 rpm and the distance is 0.25 m from axis of rotation, then the settling velocity is:

$$u_r = \{1 \times 5000 \times \sqrt{0.25}\}/\{4000 \times \sqrt{0.20}\}$$

If the centrifuge operates at 4000 rpm and the particle's distance from the axis of rotation is 0.25 m, the settling velocity is only 0.01 m/sec, which corresponds to:

$Re = 1000 \times 10^{-4} \times 0.01/0.001 = 1$

In this latter case the flow is laminar and extrapolation can be only to much smaller rotation velocities and radii. If the particle now settles in the centrifuge operating at 3000 rpm and the distance of the particle from the axis of rotation is 0.20 m, the settling velocity is:

$u_r = 0.01(3000^2 \times 0.20)/(4000^2 \times 0.25) = 0.0045$ m/sec

Note that the above calculations are based on nonhindered sedimentation and, therefore, should be modified for a hindered "fall."

Estimating Capacities of Tubular- and Solid-Bowl Centrifuges: When a rotating centrifuge is filled with suspension, the internal surface of liquid acquires a cylindrical geometry of radius R_1, as shown in Figure 11. The free surface is normal at any point to the resultant force acting on a liquid particle. If the liquid is lighter than the solid particles, the liquid moves toward the axis of rotation while the solids flow toward the bowl walls. The flow of the continuous liquid phase is effectively axial. A simplified model of centrifuge operation is that of a cylinder of fluid rotating about its axis. The flow forms a layer bound outwardly by a cylinder, R_2, and inwardly by a free cylindrical surface, R_1 (Figure13). This surface is, at any point, normal to a resulting force (centrifugal and gravity) acting on the solid particle in the liquid. The gravity force is, in general, negligible compared to the centrifugal force, and the surface of liquid is perpendicular to the direction of centrifugal force. Consider a solid particle located at distance R from the axis of rotation. The particle moves centrifugally with a settling velocity, u, while liquid particles move in the opposite direction centripetally with a velocity u_f

$$u_f = (dV/d\tau)/2\pi R\ell \tag{16}$$

where V = volume (m^3), τ = time (sec), and ℓ = height of the bowl (m).

The resulting velocity will be centrifugal, and the solid particles will be separated, provided that:

$$u_s > (dV/d\tau)/2\pi R\ell \tag{17}$$

The capacity of the centrifuge will be:

$$dV/d\tau = 2\pi R\ell u_s \tag{18}$$

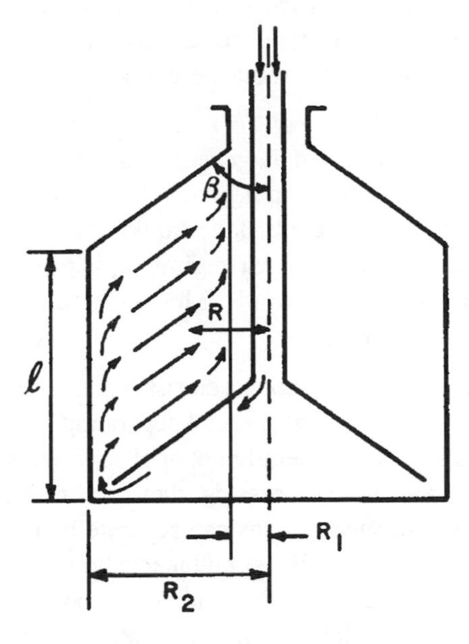

Figure 13. Centrifuge operation.

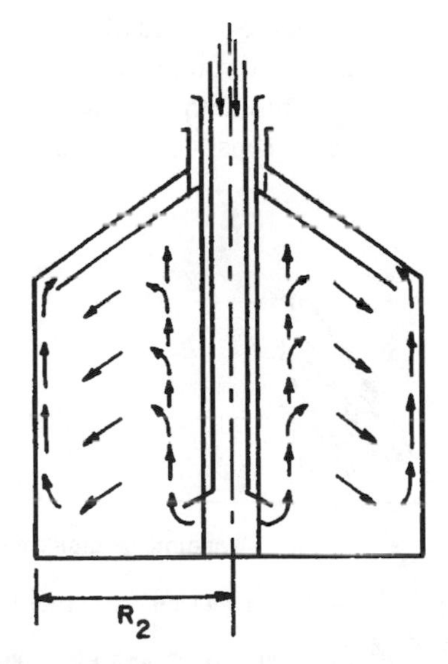

Figure 14. Operation when particle density is less than liquid density.

If the particle's density is lower than that of the liquid, the path of the liquid will be centripetal, as illustrated in Figure 14. Settling will occur when u_s (centripetal) is higher than the radial velocity u_f (centrifugal).

The settling capacity for a given size of particles is a function of R, ℓ and u_s, which itself is proportional to R. In general, for the sedimentation of heavy particles in a suspension it is sufficient that the radial component of u_f be less than u_s at a radius greater than R_2.

Because of turbulence effects, it is generally good practice to limit the settling capacity so that u_s again exceeds u_f near R_1. The same situation occurs when the particles are lighter than the continuous liquid. The relation between u_s and R depends on the regime. In the laminar regime, u_s is proportional to R, whereas in the turbulent regime, it is related to $\sqrt{R}$. Most industrial sedimentation centrifuges operate in the transition regime.

Disk-Bowl Centrifuges - Disk-bowl centrifuges are used widely for separating emulsions, clarifying fine suspensions and separating immiscible liquid mixtures. Although these machines are generally not applied to wastewater applications, and are more usually found in food processing, they can find niche applications in water treatment. More sophisticated designs can separate immiscible liquid mixtures of different specific gravities while simultaneously removing solids. Figure 15 illustrates the physical separation of two liquid components within a stack of disks. The light liquid phase builds up in the inner section, and the heavy phase concentrates in the outer section. The dividing line between the two is referred to as the "separating zone." For the most efficient separation this is located along the line of the rising channels, which are a series of holes in each disk, arranged so that the holes provide vertical channels through the entire disk set. These channels also provide access for the liquid mixture into the spaces between the disks. Centrifugal force causes the two liquids to separate, and the solids move outward to the sediment-holding space.

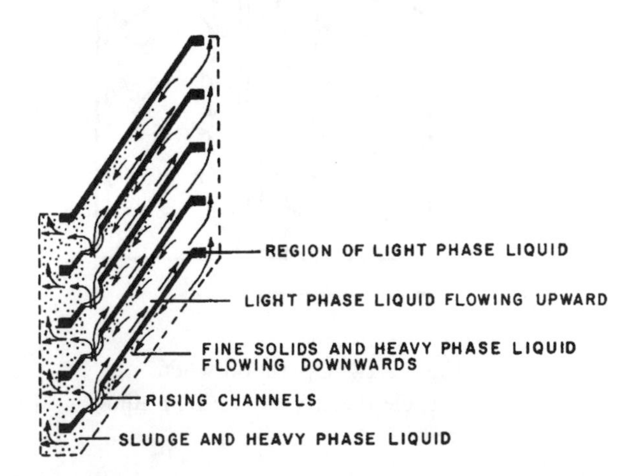

Figure 15. Separation is achieved by use of stack discs.

The position of the separating zone is controlled by adjusting the back pressure of the discharged liquids or by means of exchangeable ring dams. Figure 16 illustrates the main features of a disk-bowl centrifuge, which includes a seal ring (1), a bowl (2) with a bottom (13); a central tube (18), the lower part of which has a fixture (16) for disks; a stack of truncated cone disks (17), frequently flanged at the inside and outer diameters to add strength and rigidity; collectors (3 and 4) for the products of separation; and a feed tank (5) with a tube (6). The bowl is mounted to the tube (14) with a guide in the form of a horizontal pin. This arrangement allows the bowl to rotate along with the shaft. The suspension is supplied from the feed tank (5) through the fixed tube (6), to the central tube (18), which rotates together with the bowl and allows the liquid to descend to the bottom. In the lower part of the bowl, the suspension is subjected to centrifugal force and, thus, directed toward the periphery of the bowl. The distance between adjacent disks is controlled by spacers that usually are radial bars welded to the upper surface of each disk. The suspension may enter the stack at its outside diameter or through a series of vertical channels cut through the disks, as described earlier. The suspension is lifted up through vertical channels formed by the holes in the disks and distributed simultaneously under the action of centrifugal force into the spacings between the discs. These spacings are of tight tolerances and can range from 0.3 to 3 mm.

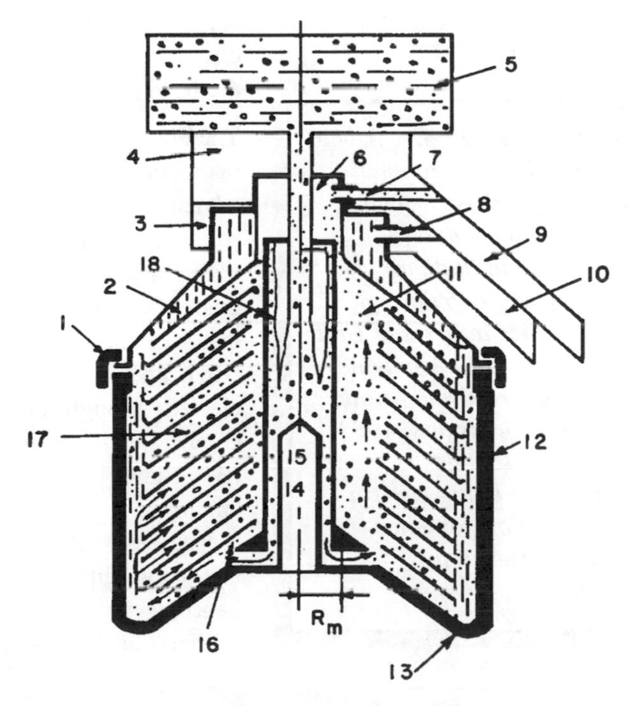

Figure 16. Details of the disc-bowl centrifuge.

Due to a larger diameter, the disk bowl operates at a lower rotational speed than its tubular counterpart. Its effectiveness depends on the shorter path of particle settling. The maximum distance a particle must travel is the thickness of the spacer divided by the cosine of the angle between the disk wall and the axis of rotation. Spacing between disks must be wide enough to accommodate the liquid flow without promoting turbulence and large enough to allow sedimented solids to slide outward to the grit-holding space without interfering with the flow of liquid in the opposite direction.

The disk angle of inclination (usually in the range of 35 to 50°) generally is small to permit the solid particles to slide along the disks and be directed to the solids-holding volume located outside of the stack. Dispersed particles transfer from one layer to the other; therefore, the concentration in the layers and their thickness are variables. The light component from the spacing near central tube (18) falls under the disk; then it flows through the annular gap between tube (18) and the cylindrical end of the dividing disk, where it is ejected through the port (7) into the circular collector (4) and farther via the funnel (9) on being discharged to the receiver. The heavier product is ejected to the bowl wall and raised upward. It enters the space between the outside surface of the dividing disk and the cone cover (2); then passes through the port (8) and is discharged into the collector (3). From there, the product is transferred to the funnel (10).

One variation of the disk-type bowl centrifuge is the nozzle centrifuge, so named because nozzles are arranged on the periphery or on the bottom of the bowl in a circle that is smaller in diameter than the bowl peripheral diameter.

Quite a few years ago, Dr. Azbel and I analyzed the operational requirements for these machines and developed some design formulae. You can find this analysis on pages 646 through 665 in ***Fluid Mechanics and Unit Operations****, David S. Azbel and Nicholas P. Cheremisinoff, Ann Arbor Science Publishers, 1983. There are some sample calculations and sizing criteria that you can follow for some practical exercises in this publication.*

The Figure 17 (A) shows the conceptual operation of such a unit. The design is advantageous because it provides a high solids concentration in the discharge with nozzles of relatively large diameters. As centrifugal force is less in that area than near the periphery of the bowl, the concen- trated solids are ejected through the nozzles under a comparatively low pressure. Figure 17 (B) shows an actual unit being employed as a yeast concentrator. Substances such as yeast and bacteria are very slippery and slide easily; hence, they will not stick or plug up the channels leading to the nozzles.

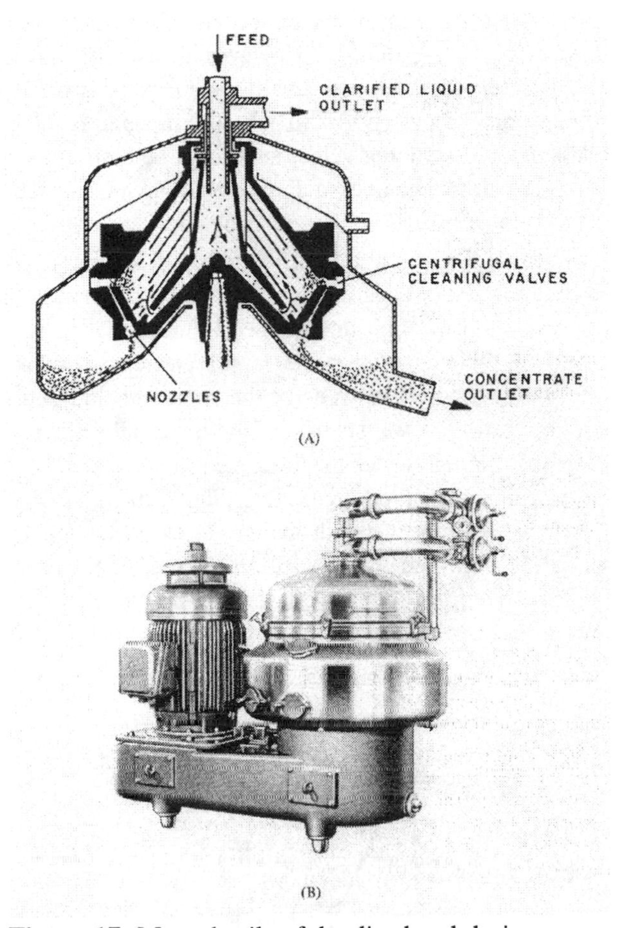

Figure 17. More details of the disc-bowl design.

HYDROCLONES

We can think of these machines as low-energy centrifuges. Hydroclones are employed for the separation of solid particles from medium- to low-viscosity liquids. Like their cyclone counterparts used in gas cleaning applications, hydroclones are simple in design, and the degree of separation can be altered by either varying loading conditions or changing geometric proportions. Unlike other types of solid-liquid separating equipment, they are better suited for classifying than for clarifying because high shearing stresses in a hydroclone promote the suspension of particles which oppose flocculation. However, by properly specifying dimensions and operating conditions, they can be used as thickeners in such a manner that the underflow contains mostly solid particles, while the clear overflow constitutes the largest portion of the liquid.

The fluid vortices and flow patterns characteristic of gas-cyclone operations are equally descriptive of liquid-hydroclones. However, the density differences between particles and liquids are significantly smaller than for gas-solid systems. For example, the density of water is approximately 800 times greater than that of air. This means that high fluid-spinning velocities cannot be employed in hydroclones as excessive pressure drop becomes a limitation. Obviously, the efficiency of hydroclones is low in comparison to gas cyclones. The design features of a hydroclone are illustrated in Figure 18. It consists of an upper short cylindrical section (1) and an elongated conical bottom (2). The suspension is introduced into the cylindrical section (1) through the nozzle (3) tangentially, whence the fluid acquires an intensive rotary motion. The larger particles, under the action of centrifugal force, move toward the walls of the apparatus and concentrate on the outer layers of the rotating flow. Then they move spirally downward along the walls to the nozzle (4), through which the thickened slurry is evacuated. The largest portion of liquid containing small particles (clear liquid) moves in the internal spiral flow upward along the axis of the hydroclone. The cleared liquid is discharged through the nozzle (5) and fixed at the partition (6) and the nozzle (7). The actual flow pattern is more complicated than described because of radial and closed circulating flows. Because of peripheral flow velocities, the liquid column formed at the hydroclone axis has a pressure that is below atmospheric.

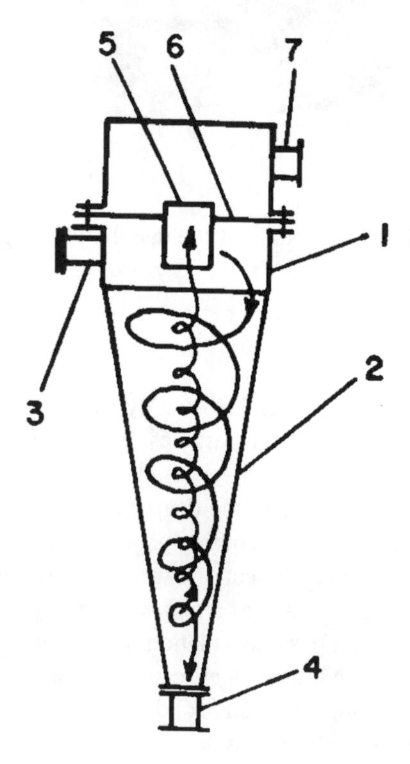

Figure 18. Features of a hydroclone.

The liquid bulk flow limits the upward flow of small particles from the internal side and has a significant influence on the separating effect. Hydroclones are applied successfully for classification, clarification and thickening of suspensions containing particles from 5 to 150 μm in size.

The smaller the hydroclone diameter, the greater the centrifugal forces developed and, consequently, the smaller the size particles that can be separated. The following are typical hydroclone diameters used for various general applications: for classification and degritting process streams, D = 300-350 mm; for thickening of suspensions, D = 100 mm; for clarification (where it is necessary to apply powerful centrifugal fields), D = 10-15 mm. In the last case, multiclones are employed. Figures 19 and 20 provide examples of hydroclones used in a degritting operations. Sand is accumulated in a grit chamber for intermittent blowdown. Such an operation could be used off of a cooling tower installation. Good separation of suspensions is achieved, especially in thickening and clarification, when hydroclones have an elongated shape with the slope of the cone equal to approximately 10-15°. At such a cone shape, the path of solid particles is increased as well as the residence time, which thus increases the separating efficiency.

Figure 19. Hydroclone used in degritting water.

Design methodology consists of determining capacity, approximate sizes of particles settled and horsepower requirements for pumping. The flowrate of a suspension with a density ρ through an inlet nozzle of diameter d_N, at pressure drop ΔP, can be calculated from the following formula:

$$V_{sec} = \mu'(\pi\ d_N^2/4)(2\ \Delta P/\ \rho)^{1/2} \tag{19}$$

where μ' is known as an empirical flowrate constant. The hydroclone diameter id D, and the lower nozzle diameter is $d_{\ell\ell N}$, and so:

$$V_{sec} = C'(Dd_{\ell\ell N})(\Delta P/\ \rho)^{1/2} \tag{20}$$

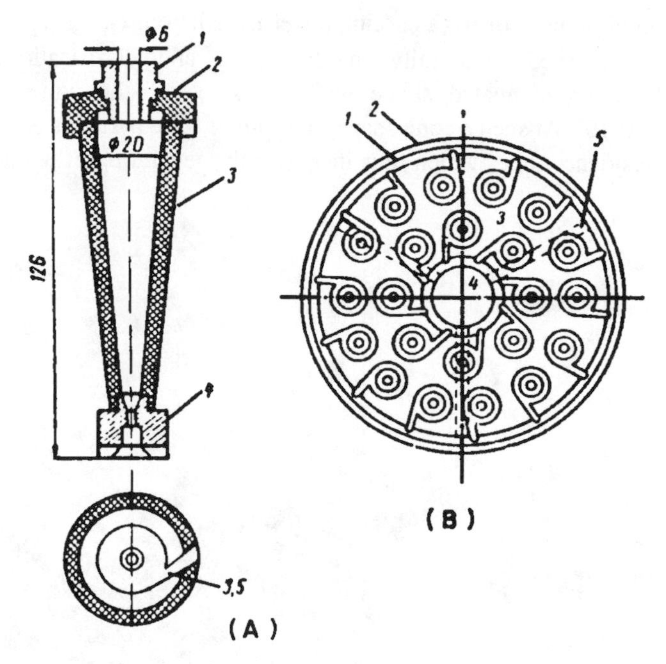

Figure 20. Multiclone details: (A) Front view: 1- bushing; 2 - cover; 3 - body; 4 - endpiece. (B) Top view: 1 - rubber block; 2 - metal body; 3 - multiclones; 4 - central feed pipe; 5 - radial channels.

In the above equation, parameter C′ is:

$$C' = 0.25\mu'(\pi\surd 2)(d_N^2/D\ Dd_{\ell\ell N}) \tag{21}$$

Coefficient C′ is a constant for geometrically similar hydrolclones, and is defined by the following relationship:

$$K = C'/\surd\rho \tag{22}$$

By combining the above expressions, the following formula is obtained for estimating the suspension flowrate:

$$V_{sec} = K\, Dd_{\ell\ell N}(\Delta P)^{1/2} \tag{23}$$

Using SI units, and from some literature reported values for hydroclones with D between 125 and 600 mm, and a cone angle of 38°, a value for coefficient K is 2.8 $\times$ 10^{-4}. The maximum size of particles in the cleared liquid can be estimated from:

$$d = (1.33 \times 10^{-2}\, d_N^{\,2}/\phi_x)(\mu/V_{sec} h \rho_f)^{1/2},\ m \tag{24}$$

where h is the height of the central flow (assumed equal to 1/3 of the cone height), and ϕ_x is a parameter that accounts for the change in fluid velocity. For hydroclones having a cone angle of 38°,

$$\phi_x = -.103D/d_{\ell\ell N} \tag{25}$$

The power required to operate a hydroclone is the horsepower needed for a pump that supplies the capacity V_{sec} with an acceptable head of pressure.

The separating effect is influenced mostly by the ratio $d_N/d_{\ell\ell N}$ of nozzles, which may be assumed to be in the range of 0.37 to 0.40. The diameter of the inlet nozzle, d_N, is assumed to be 0.14 to 0.3 times D, and the diameter of discharge nozzle d_d is in the range of 0.2D to 0.167D. The cone angle for hydroclones used as classifiers is typically 20°; for thickeners it is 10- 15°. The operating arrangements used with hydroclones can be parallel and series operations. The example illustrated in Figure 20 is referred to as a "battery of multiclones".

THICKENERS

We have already discussed this technology back in Chapter 8 and will onlyadd a few more general comments. Thickening is practiced in order to remove as much water as possible before final dewatering of the sludge. It is usually accomplished by floating the solids to the top of the liquid (floatation) or by allowing the solids to settle to the bottom (gravity thickening). Other method of thickening are by centrifuge, gravity belt, and rotary drum thickening, as already described. These processes offer a low-cost means of reducing the volumetric loading of sludge to subsequent steps. In the floatation thickening process air is injected into the sludge under pressure. The resulting air bubbles attach themselves to sludge solids particles and float them to the surface of an open tank. The sludge forms a layer at the top of the tank which is removed by a skimming mechanism. This process increases the solids concentration of activated sludge from 0.5-1 % to 3-6 %.

Gravity thickening has been widely used on primary sludge for many years because of its simplicity and inexpensiveness. In gravity thickening, sludge is concentrated by the gravity- induced settling and compaction of sludge solids. It is essentially a sedimentation process. Sludge flows into a tank that is similar to the circular clarifiers used in primary and secondary sedimentation.

The solids in the sludge settle to the bottom where a scraping mechanism removes them to a hopper. The type of sludge being thickened has a major effect on performance. The best results can be achieved with primary sludge. Purely primary sludge can be thickened from 1-3% to 10% solids. As the proportion of activated (secondary) sludge increases, the thickness of settled solids decreases. There are various designs for sludge thickeners. Figure 21 illustrates a tray thickener.

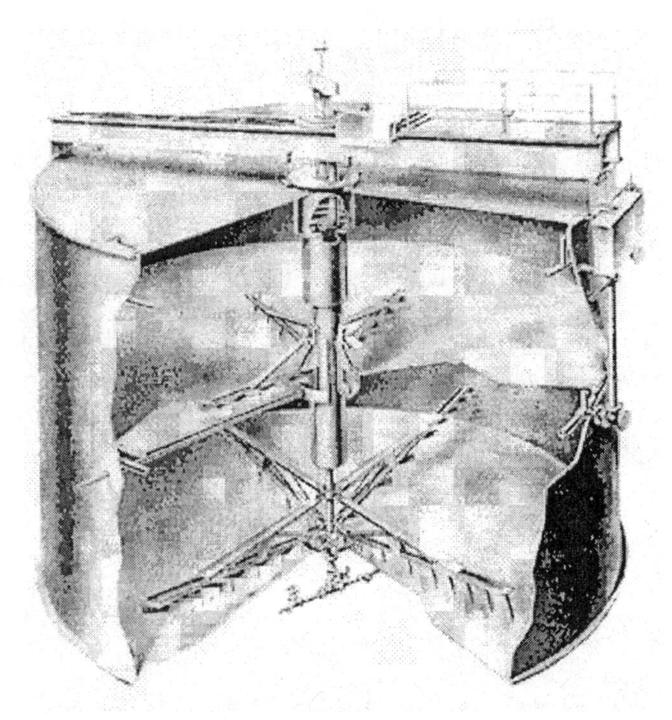

Figure 21. Illustrates a tray thickener.

COMPARING MECHANICAL DEWATERING TECHNOLOGIES

As we see from the above descriptions there are a variety of technologies from which to select from for sludge dewatering operations. Each has its own set of advantages, disadvantages, and limitations in operating ranges. Selection greatly depends on the volumes and nature of the sludge. Table 3 provides a relative comparison between the principle mechanical dewatering techniques.

Table 3. Comparison of the Advantages and Disadvantages of Mechanical Thickening Technologies.

Technology or Method	Advantages	Disadvantages
Gravity	Simple	Potential for obnoxious and harmful odors
	Low operating and maintenance costs	Thickened sludge concentration limited for WAS
	Low operator attention and moderate training requirements	High space requirements for WAS
	Minimal power consumption	
Dissolved Air Flotation	Effective for WAS	Relatively high power consumption
	Can work without conditioning chemicals	Thickening solids concentration limited
	Relatively simple equipment components	Potential for obnoxious and harmful odors
		High space requirements
Centrifugation	Low space requirements	Best suited for continuous operations
	Effective for WAS	Sophisticated maintenance requirements
	Minimum housekeeping and odor problems	Relatively high power consumption
	Highly thickened concentrations available	Relatively high capital cost
Rotating Drum Filter	Low space requirements	Can be polymer dependent

Technology or Method	Advantages	Disadvantages
	Low capital cost	Sensitive to polymer type
	Relatively low power consumption	Housekeeping requirements high
	High solids capture achievable	Potential for obnoxious and harmful odors
		Moderate operator attention and training requirements
Gravity Belt Thickener	Low space requirements	Housekeeping requirements high
	Relatively low power consumption	Can be polymer dependent
	Relatively low capital cost	Moderate operator attention and training requirements
	Can achieve high thickened concentrations and solids capture with minimum power	Potential for obnoxious and harmful odors

DRYING BEDS

This is one of two common methods of dewatering based upon thermaal energy. Drying beds are generally used for dewatering of well digested sludges. Attempts to air dry raw sludge usually result in odor problems. Sludge drying beds consist of perforated or open joint drainage pipe laid within a gravel base. The gravel is covered with a layer of sand. Partitions around and between the drying beds are generally open to the weather but may be covered with ventilated green-house type enclosures where it is necessary to dewater sludge in wet climates. The drying of sludge on sand beds is accomplished by allowing water to drain from the sludge mass through the supporting sand to the drainage piping and natural evaporation to the air. As the sludge dries, cracks develop in the surface allowing evaporation to occur from the lower layers which accelerates the drying process.

There are many design variations used for sludge drying beds, including the layout of the drainage piping, thickness and type of materials in the gravel and sand layers, and construction materials used for the partitions. The major variation is whether or not the beds are covered. Any covering structure must be well ventilated. In the past, some beds were constructed with flat concrete bottoms for drainage without pipes, but this construction has not been very satisfactory. Asphalt concrete (blacktop) has been used in some drying beds.
The only sidestream is the drainage water. This water is normally returned to the raw sewage flow to the plant or to the plant headworks. The drainage water is not normally treated prior to return to the plant. Experience is the best guide in determining the depth of sludge to be applied, however, typical application depth is 8 to 12 inches. The condition and moisture content of the sludge, the sand bed area available, and the need to draw sludge from digesters are factors to consider. It is not advisable to apply fresh sludge on top of dried sludge in a bed.
The best time to remove dried sludge from drying beds depends on a number of factors, such as subsequent treatment by grinding or shredding, the availability of drying bed area for application of current sludge production, labor availability, and, of course, the desired moisture content of the dried sludge. Sludge can be removed by shovel or forks at a moisture content of 60 percent, but if it is allowed to dry to 40 percent moisture, it will weigh only half as much and is still easy to handle. If the sludge gets too dry (10 to 20 percent moisture) it will be dusty and will be difficult to remove because it will crumble as it is removed. Many operators of smaller treatment plants use wheelbarrows to haul sludge from drying beds. Planks are often laid on the bed for a runway so that the wheelbarrow tire does not sink into the sand. Wheelbarrows can be kept close to the worker so that the shoveling distance is not great. Most plants use pick-up trucks or dump trucks to transport the sludge from the drying bed. Dump trucks have the advantage of quick unloading.

Where trucks are used, it is best to install concrete treadways in the sludge drying bed wide enough to carry the dual wheels since the drying bed can be damaged if the trucks are driven directly on the sand. The treadways should be installed so that good access is provided to all parts of the beds. If permanent treadways have not been installed, heavy planks may be placed on the sand. Large plants will normally utilize mechanical equipment for handling the dried sludge. Some communities have encouraged public usage of the dried sludge. In some cases users are allowed to remove the sludge from the beds, but this may not be satisfactory in many cases. Local regulations should be reviewed before attempting to establish a public utilization program.

SLUDGE LAGOONS

This is a technique that relies both on the settling characteristics of sludge and solar evaporation. The considerable labor involved in sludge drying bed operation may be avoided by the use of sludge lagoons. These lagoons are nothing but excavated areas in which digested sludge is allowed to drain and dry over a period of months or even a year or more (refer to Figure 22 for an example). They are usually dug out by bulldozers, or other dirt-moving equipment, with the excavated material used for building up the sides to confine the sludge. Depths may range from two to six feet. Areas vary, and although drainage is desirable, it is not usually provided. Digested sludge is drawn as frequently as needed, with successive drawings on top of the previous ones until the lagoon is filled. A second lagoon may then be operated while the filled one is drying.

After the sludge has dried enough to be moved, a bulldozer, or a tractor with an end-loader, may be used to scoop out the sludge. In some locations it may be pushed from the lagoon by dozers into low ground for fill. Lagoons may be used for regular drying of sludge, re-used after emptying, or allowed to fill and dry, then leveled and developed into lawn.

Figure 22. Sludge lagoon used by a treatment plant in Russia.

They can also be used as emergency storage when the sludge beds are full or when the digester must be emptied for repair. In the latter case it should be treated with some odor control chemicals, such as hydrated or chlorinated lime.

The size of the lagoon depends upon the use to which it will be put. Lagoons may take the place of sludge beds or provide a place for emergency drawings of sludge, but they may be unsightly and even unwanted on a small plant site. However, they are becoming more popular because they are inexpensive to build and operate.

Although lagoons are simple to construct and operate, there can be problems associated with sizing them. These problems largely arise from uncertainty in estimating the solar evaporative capacity.

In semi-arid regions evaporation ponds are a conventional means of disposing of wastewater without contamination of ground or surface waters. Evaporation ponds as defined herein will refer to lined retention facilities. Successful use of evaporation for wastewater disposal requires that evaporation equal or exceed the total water input to the system, including precipitation. The net evaporation may be defined as the difference between the evaporation and precipitation during any time period. Evaporation rates are to a great extent dependent upon the characteristics of the water body. Evaporation from small shallow ponds is usually considered to be quite different than that of large lakes mainly due to differences in the rates of heating and cooling of the water bodies because of size and depth differences. Additionally, in semi-arid regions, hot dry air moving from a land surface over a water body will result in higher evaporation rates for smaller water bodies. The evaporation rate of a solution will decrease as the solids and chemical composition increase. Depending upon its origin, evaporation pond influent may contain contaminates of various amounts and composition. Decreases in evaporation rates compared to fresh water rates can seriously increase the failure potential of ponds designed on fresh water evaporation criteria. Designers of settling ponds and lagoons that rely on evaporation need to know the probability level of their designs being exceeded. Confidence limits for published evaporation normals have not been given, nor have analyses been made of the effects of uncertainty in the estimated normals or of the temporal variation of net evaporation. Definition of the spatial and temporal distribution of parameters such as evaporation and precipitation is difficult in mountainous regions. A concern is that the application of many of the empirical equations, based on climatological data, for estimating evaporation have not been thoroughly tested for high altitude conditions. In particular, the ability of these equations for defining the variability of evaporation basically is unknown. Historically, pan data is the most common means for defining free water evaporation. However, the density of evaporation pan stations is much less than that of weather stations.

Many methods exist for either measuring or estimating evaporative losses from free water surfaces. Evaporation pans provide one of the simplest, inexpensive, and most widely used methods of estimating evaporative losses. Long-term pan records

Pan evaporation is considered an indication of atmospheric evaporative power. Evaporation from a free surface is related to pan evaporation by a coefficient applied to the pan readings. Most evaporation pans in the U. S. are Class A pans made of unpainted galvanized iron or stainless steel 4 feet in diameter and 10 inches deep. The pans are supported on low wooden frames and are filled with 8 inches of water.

are available, providing a potential source of data for developing probabilities of net evaporation. The use of pan data involves the application of a coefficient to measured pan readings to estimate evaporation from a larger water body. Among the most useful methods for estimating evaporation from free water surfaces are the methods which use climatological data. Many of these equations exist, most being based directly upon the a method which was originally intended for open water surfaces, but is now commonly applied to estimates of vegetative, water use.

Monthly evaporation estimates can be made using the Kohler-Nordenson-Fox equation with a pan coefficient of 0.7. The Kohler-Nordenson-Fox equation describes evaporation as the combination of water loss due to radiation heat energy and the aerodynamic removal of water vapor from a saturated surface. The general form for the combination equation is:

$$E = (d / (d + y))\, R_n + (Y / (d+Y))\, E_a \qquad (26)$$

where E is the evaporation in inches per day, d is the slope of the saturation vapor pressure curve at air temperature in inches of mercury per degree F, Y is the psychrometric constant in inches of mercury per degree F, R_n is the net radiation exchange expressed in equivalent inches of water evaporated, and E_a is an empirically derived bulk transfer term of the form :

$$Ea = f(u)\,(e_s - e_d) \qquad (27)$$

Where f(u) is a wind function and $(e_s - e_d)$ is the vapor pressure deficit.

Kohler-Nordenson-Fox evaluated the aerodynamic term using pan data resulting in the form:

$$E_a = (0.37 + 0.0041U_p)(e_s - e_a) \qquad (28)$$

Where e_a is in units of inches of water per day, U_p is the wind speed 2 feet above the ground expressed in miles per day, and e_s and e_a are the saturation vapor pressures at mean air and mean dew-point temperatures, respectively (expressed in inches of mercury). For development of the wind function, an adjustment in the psychrometric constant is generally made to account for the sensible heat conducted through the sides and bottom of the evaporation pan. One may also apply as an approximation the following expression for the psychrometric constant:

$$Y = 0.000367P \quad (29)$$

where P is the atmospheric pressure in inches of mercury. My own experience in designing surface lagoons and evaporation ponds over the years, and substantiated in the literature, has been to apply a pan coefficient of 0.7.

Of concern is that there is very little information often available concerning the effects of common waste waters on evaporation rates. As noted, the evaporation rate of a solution will decrease as the solids and chemical concentrations increase. However, the overall effects on evaporation rates of dissolved constituents as well as color changes and other factors of wastewater are largely unknown.

Evaporation from surface ponds are usually based upon estimates of annual net evaporation. Calculation of annual evaporation rates requires estimates during periods when the surface may be frozen. Most studies related to cold weather evaporation have been concerned with snow rather than ice. In general, the evaporation from a snow pack is usually much less than the amount of melting that occurs. Considering the large percentage of the annual evaporation which occurs during the warmer months and the overall uncertainties involved in estimates of evaporation from water surfaces, the amount of evaporation from frozen ponds during winter can reasonably be neglected in calculating annual evaporation. A more important consideration is the evaporation which occurs during winter from ponds which may remain unfrozen due to the introduction of warm wastewater. In these cases, water temperature will influence the evaporation rates. However, the low value of the saturation vapor pressure of the air above any water body will limit evaporation. Annual estimates of evaporation herein can be made by applying the Kohler-Nordenson-Fox equation throughout the year. Such estimates should provide near maximum possible evaporation estimates. For lined ponds, evaporation will be confined mainly to the water surface area. Evaporation from the soil and vegetation on the banks surrounding the pond should be minimal. However, for ponds which have appreciable seepage to the surrounding area, evaporation from this area will be dependent upon the type and amount of vegetation, as well as the moisture content of the upper soil layers. Methods foe estimating evaporation and/or evapotranspiration in these instances are readily available and you can find some of these studies and estimating procedures by doing a Web search.

If water losses from the surrounding area are a major component of the total evaporative losses of the pond, then soil moisture conditions will be expected to be high. Under non-limiting soil moisture conditions vegetative moisture losses are often defined as "potential" losses. Evaporative losses in this case would not be expected to differ greatly from free water evaporation. The literature recommends in fact that lake evaporation be used as a measure of potential evapotranspiration. Thus, for high soil moisture conditions, evaporation rates calculated for the water surface should be applicable to the surrounding area. The influence upon evaporation of vegetative growth within a pond is uncertain. The literature is inconclusive as to whether vegetation will increase or decrease evaporation compared to an open surface. It appears that the effect may be somewhat dependent upon the size of the water body. Literature studies indicate vegetation will decrease evaporation for extensive surfaces with the effect being less for smaller surface areas. It is very possible, however, that the introduction of vegetation upon the surface of a water body of more limited extent may increase its evaporative water loss, but only while the vegetation remains in a healthy, robust condition. Thus, the effect of the presence of vegetation appears to range from being a water conservation mechanism to that of increasing evaporation. In either case, the potential effects appear to be quite large with reported ratios of vegetative covered to open water evaporation under extreme conditions ranging from 0.38 to 4.5. In most instances, this ratio would be expected to be much closer to unity.

VOLUME REDUCTION

As the title implies, we will now focus our attention to those technologies aimed at reducing the volume of the final form of the sludge. Dewatering or thickening technologies can only bring us so far in concentrating the form of the waste. Ultimately, we must find ways of either disposing of this waste, or in using it. We'll discuss applications later on. Of immediate concern is how we can reduce the volume of so-called "dry" sludge, at solids contents ranging anywhere from 30 to 60 %, even further.

INCINERATION

In all types of incinerators, the gases from combustion must be brought to and kept at a temperature of 1250° F to 1400° F. until they are completely burned. This is essential to prevent odor nuisance from stack discharge. It is also necessary to maintain effective removal of dust, fly ash and soot from the stack discharge. This may be done by a settling chamber, by a centrifugal separator, or by a Cottrell electrical precipitator. The selection depends on the degree of removal efficiency required for the plant location. All types of sludge, primary, secondary, raw or digested sludge, may be dried and burned. Raw primary sludge with about 70 percent volatile solids contains about 7800 Btu per pound of dry solids and when combustion is once started will burn without supplementary fuel, in fact an excess of heat is usually available. Digested sludge may or may not require supplementary fuel, depending on the moisture content of the cake and percent volatile solids or degree of digestion. Raw activated sludge generally requires supplementary fuel for drying and burning. In all cases, supplementary fuel is necessary to start operation and until combustion of the solids has been established.

Incineration of sludge has gained popularity throughout the world, especially at large plants. It has the advantages of economy, freedom of odor, independence of weather and the great reduction in the volume and weight of end product to be disposed of. There is a minimum size of sewage treatment plant below which incineration is not economical. There must be enough sludge to necessitate reasonable use of costly equipment. One of the difficulties in operating an incinerator is variations in tonnage and moisture of sludge handled.

There are two major incinerator technologies used in this process. They are (1) the multiple hearth incinerator, and (2) the fluidized bed incinerator. An incinerator is usually part of a sludge treatment system which includes sludge thickening, macerations, dewatering (such as vacuum filter, centrifuge, or filter press), an incinerator feed system, air pollution control devices, ash handling facilities and the related automatic controls. The operation of the incinerator cannot be isolated from these other system components. Of particular importance is the operation of the thickening and dewatering processes because the moisture content of the sludge is the primary variable affecting the incinerator fuel consumption.

Incineration may be thought of as the complete destruction of materials by heat to their inert constituents. This material that is being destroyed is the waste product (i.e., the sludge). Sewer sludge as sludge cake normally contains from 55 to 85% moisture. It cannot burn until the moisture content has been reduced to no more than 30%. The purpose of incineration is to reduce the sludge cake to its minimum volume, as sterile ash. There are three objectives incineration must accomplish:

3. dry the sludge cake,
4. destroy the volatile content by burning, and
5. produce a sterile residue or ash.

There are four basic types of incinerators used in wastewater treatment plants. They are the multiple hearth incinerator, the fluid bed incinerator , the electric furnace , and the cyclonic furnace. Each system has it's own distinct method of incineration and while one may be more cost efficient, another may have more of an environmental impact.

The basic configuration and features of the multiple hearth incinerator are illustrated in Figure 23. This incinerator is the most prevalent incinerator technology for the disposal of sewage sludge in the U.S. due to it's low ash discharge. Sludge cake enters the furnace at the top. The interior of the furnace is composed of a series of circular refractory hearths, which are stacked one on top of the other. There are typically five to nine hearths in a furnace. A vertical shaft, positioned in the center of the furnace has rabble arms with teeth attached to them in order to move the sludge through the mechanism. Each arm is above a layer of hearth. Teeth on each hearth agitate the sludge, exposing new surfaces of the sludge to the gas flow within the furnace. As sludge falls from one hearth to another, it again has new surfaces exposed to the hot gas. At the top of the incinerator there is an exit for flue gas, an end product of sludge incineration. At the bottom of the furnace there is an exit for the ashes.

The basic configuration and features of the fluid bed incinerator are illustrated in Figure 24. This technology has been around since the early 1960s. In this system, air is introduced at the fluidizing air inlet at pressures of 3.5 to 5 psig. The air passes through openings in the grid supporting the sand and creates fluidization of the sand bed. Sludge cake is introduced into the bed. The fluidizing air flow must be carefully controlled to prevent the sludge from floating on top of the bed. Fluidization provides maximum contact of air with sludge surface for optimum burning. The drying process is practically instantaneous. Moisture flashes into steam upon entering the hot bed. Some advantages of this system are that the sand bed acts as a heat sink so that after shutdown there is minimal heat loss. With this heat containment, the system will allow startup after a weekend shutdown with need for only one or two hours of heating. The sand bed should be at least 1200° F when operating.

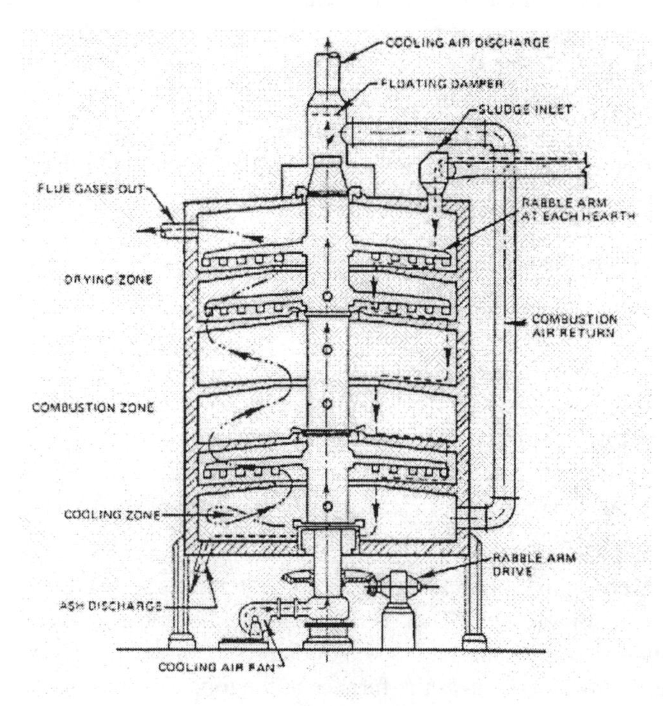

Figure 23. Cross section of multiple hearth furnace.

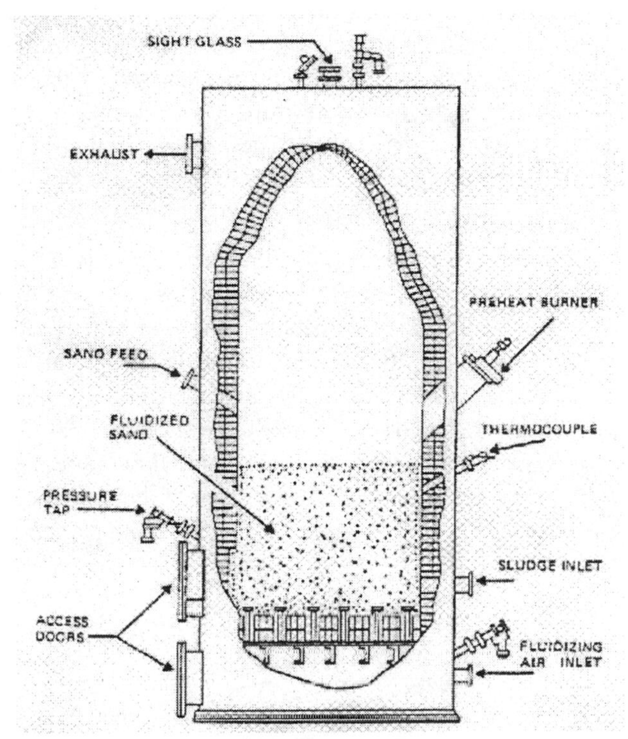

Figure 24. Cross section of a fluid bed incinerator.

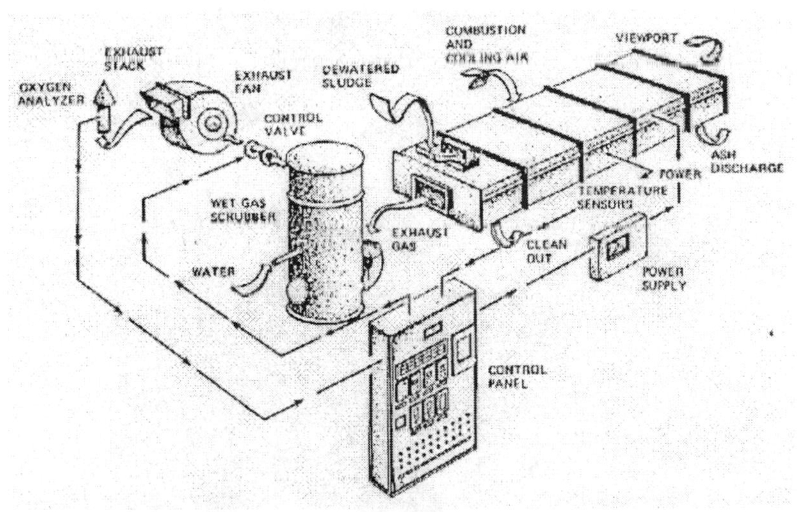

Figure 25. Radiant heat (electric) incineration scheme.

The basic features of the electric furnace are illustrated in Figure 25. The electric furnace is basically a conveyor belt system passing through a long rectangular refractory lined chamber. Heat is provided by electric infrared heating elements within the furnace. Cooling air prevents local hot spots in the immediate vicinity of the heaters and is used as secondary combustion air within the furnace. The conveyer belt is made of continuous woven wire mesh chosen of steel alloy that will withstand the 1300 to 1500° F temperatures. The sludge on the belt is immediately leveled to one inch. The belt speed is designed to provide burnout of the sludge without agitation.

The basic features of the cyclone furnace are illustrated in Figure 26. The cyclonic furnace is a single hearth unit where the hearth moves and the rabble teeth are stationary. Sludge is moved towards the center of the hearth where it's discharged as ash. The furnace is a refractory lined cylindrical shell with a domed top. The air, heated with the immediate introduction of supplemental fuel creates a violent swirling pattern which provides good mixing of air and sludge feed. The air, which later turns into flue gas, swirls up vertically in cyclonic flow through the discharge flue in the center of the doomed roof. One advantage of these furnaces is that they are relatively small and can be placed in operation, at operating temperature within an hour.

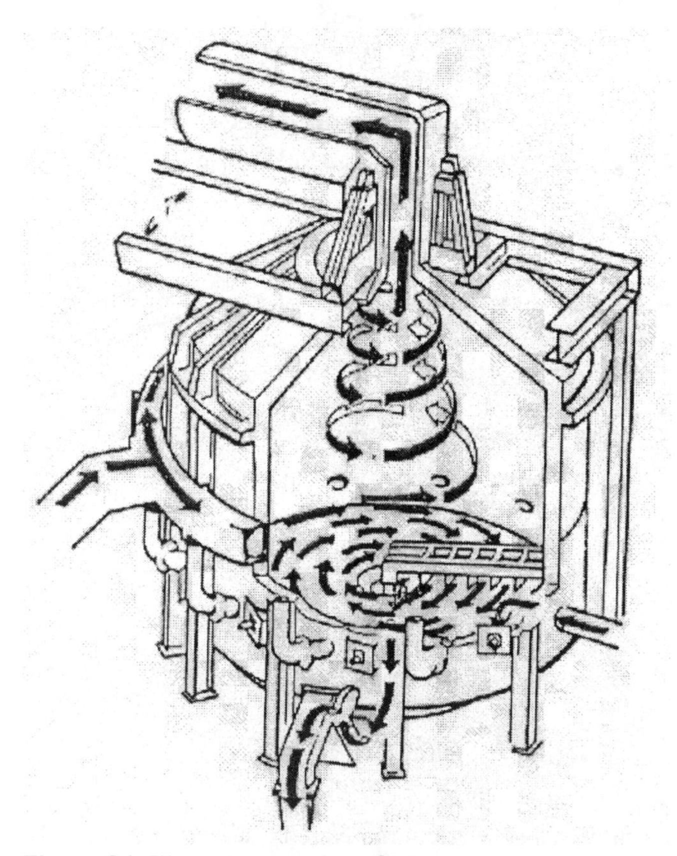

Figure 26. Illustrates a cyclone furnace.

A good question for us to ask at this stage is what does a sludge treatment plant do with the ash that is discharged out of the furnace? As ash falls into a wet sump, turbulence is created by the entrance of water. This turbulence is necessary so that the ash doesn't collect and cake up. This water containing the ash is pumped into a holding pond or lagoon, with a residence time of at least 6 hours. During this time, 95% of the ash will have settled to the bottom and the overflow is taken back to the treatment plant. There has to be a minimum of two lagoons with one being used to hold the ash-water discharge and the other for drying. When dry, the ash is hauled to a landfill or used for concrete. Mixing one part of ash to four parts cement will produce a slow-setting concrete with no loss in strength.

A serious environmental impact that incineration has is on the air. An incinerator's smoke discharge or flue gas should be colorless. Flue gas is an emission mainly made up of nitrogen, carbon dioxide and oxygen. There are traces of chloride and sulfides in the gas and if these levels become to high, they could cause the possibility of corrosion. With respect to the color of the discharge again, if there is a significant amount of particulate matter in the emission, it will be detected by color. The stream can range from a black to white appearance and will have a pale yellow to dark brown trail. The discharge should also have no discernable odor and there should be no detectable noise due to incinerator operation at the property line. Unfortunately colored emissions and odor problems do occur and treatment plants take the proper actions to correct it. Air pollution controls are critical factors that add significant costs onto these technologies. A discussion of these technology options and requirements are quite extensive and beyond the scope of this volume. There are some good references cited at the end of this chapter where you can gain valuable information from.

When dealing with incinerators, fuel is generally the most expensive part of the process from an operational standpoint. There should be a ratio calculated before hand that represents the amount of fuel used for the amount of sludge inputted. If there is a significant change to the amount of fuel consumed, it could mean that there is a problem in the fuel supply system, air flow to the incinerator, or that an extensive furnace cleaning is in order. Minimal cost of operation and equipment maintenance is another economic parameter for sludge incineration. Preventive maintenance is the single most important factor in reduction of operating costs. Semiannual or quarterly appointments must be scheduled to allow time for complete furnace check-out and cleaning (referred to as "turnarounds"). ***Not having your furnace inspected at least semiannually is a federal violation in the U.S.***. The following (Table 2) is a break down of the costs of each incinerator. Essentially costs can be related to one basic parameter, namely - the lower the moisture content is in the sludge, the less expensive the incinerator will be to operate. Also incinerators are bought based on what moisture level of sludge they are going to be effective with. Some incinerators can burn out sludge with 20% moisture levels and some cannot. Table 4 provides some costs for the four basic incinerators plus installation:

Table 4. Estimated Economics for Incineration.

Type Incinerator	Capacity, Lbs/hr	Sludge Moisture Content, %	Installed Cost (U.S. $)
Multiple Hearth Furnace	7000	0	11,000,000
Fluid Bed Incinerator	1000	0	900,000
	2900	20	1,600,000
Electric Furnace	2400	30	1,3000,000
	2400	0	950,000
Cyclonic Incinerator	2000	20	1,000,000

The design cost will be a function of the incinerator cost plus installation which is normally in the range of 4 to7%. This cost should be doubled to include engineering services during project construction. It should be noted that with the electric furnace, the power needed to start up results in a large connected load. In areas of the country where there are high demand charges for electric power, this system can be economically impractical.

SOME FINAL COMMENTS ON INCINERATION

High-temperature processes have been used for the incineration or combustion of municipal wastewater solids since the early 1900s. Popularity of these processes has fluctuated greatly since their adoption from industrial combustion. Thirty years ago, combustion of wastewater solids was both practical and inexpensive. Solids were easily dewatered and the fuel required for combustion was cheap and plentiful. In addition, air-emission standards were virtually nonexistent. Today, wastewater solids are more complex and include sludge from secondary and advanced waste treatment processes. These sludge are more difficult to dewater and thereby increase fuel requirements for combustion. Due to environmental concerns with air quality and the energy crisis, the use of high-temperature processes for combustion of municipal solids is being scrutinized. More efficient solids dewatering processes and advances in combustion technology have renewed an interest in the use of high-temperature processes for specific applications.

High-temperature processes should be considered where available land is scarce, stringent requirements for land disposal exist, destruction of toxic materials is required, or the potential exists for recovery of energy, either with wastewater solids alone or combined with municipal refuse. High-temperature processes have potential advantages over other methods which include:

- Maximum volume reduction. Reduces volume and weight of wet sludge cake by approximately 95 percent, thereby reducing disposal requirements.
- Detoxification. Destroys or reduces toxics that may otherwise create adverse environmental impacts.
- Energy recovery. Potentially recovers energy through the combustion of waste products, thereby reducing the overall expenditure of energy.

Disadvantages of high-temperature processes include:

- Cost. Both capital and operation and maintenance costs, including costs for supplemental fuel, are generally higher than for other disposal alternatives.
- Operating problems. High-temperature operations create high maintenance requirements and can reduce equipment reliability.
- Staffings. Highly skilled and experienced operators are required for high-temperature processes. Municipal salaries and operator status may have to be raised in many locations to attract the proper personnel.
- Environmental impacts. Discharges to atmosphere (particulates and other toxic or noxious emissions), surface waters (scrubbing water), and land (furnace residues) may require extensive treatment to assure protection of the environment.

Combustion is the rapid exothermic oxidation of combustible elements in fuel. Incineration is complete combustion. Classical pyrolysis is the destructive distillation, reduction, or thermal cracking and condensation of organic matter under heat and/or pressure in the absence of oxygen. Partial pyrolysis, or starved-air combustion, is incomplete combustion and occurs when insufficient oxygen is provided to satisfy the combustion requirements. The basic elements of each process are shown on Figure 27. Combustion of wastewater solids, a two-step process, involves drying followed by burning.

A value commonly used in sludge incineration calculations is 10,000 Btu per pound of combustibles. It is important to clearly understand the meaning of combustibles. For combustion processes, solid fuels are analyzed for volatile solids and total combustibles. The difference between the two measurements is the fixed carbon. Volatile solids are determined by heating the fuel in the absence of air. Total combustibles are determined by ignition at 1,336° F (725° C). The difference in weight loss is the fixed carbon. In the volatile-solids determination used in sanitary engineering, sludge is heated in the presence of air at 1,021° F (550° C). This measurement is higher than the volatile-solids measurement for fuels and includes the fixed carbon. Numerically, it is nearly the same as the combustible measurement. If volatile solids are used in the sense of the fuels engineer, it will be followed parenthetically by the designation *fuels usage*. If the term *volatile solids or volatiles* is used without designation, it will indicate sanitary engineering usage and will be used synonymously with *combustibles*. The amount of heat released from a given sludge is a function of the amounts and types of combustible elements present.

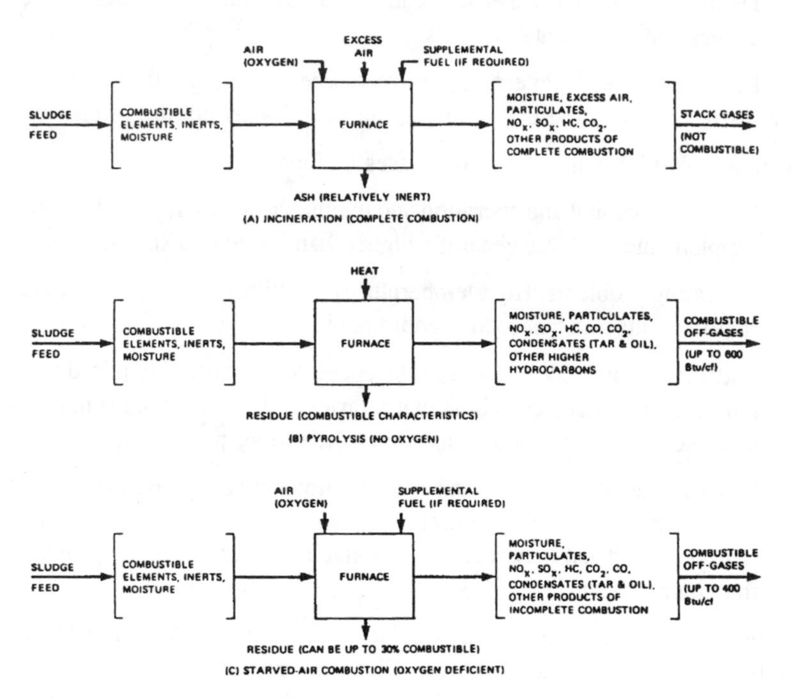

Figure 27. Basic elements of high temperature processes.

The primary combustible elements in sludge and in most available supplemental fuels are fixed carbon, hydrogen, and sulfur. Because free sulfur is rarely present in sewage sludge to any significant extent and because sulfur is being limited in fuels, the contributions of sulfur to the combustion reaction can be neglected in calculations without compromising accuracy. Similarly, the oxidation of metals contributes little to the heat balance and can be ignored. Solids with a high fraction of combustible material (for example, grease and scum) have high fuel values. Those which contain a large fraction of inert materials (for example, grit or chemical precipitates) have low fuel values. Chemical precipitates may also exert appreciable heat demands when undergoing high-temperature decomposition. This further reduces their effective fuel value. Table 5 provides a summary of typical chemical reactions that take place during combustion, along with heating values of the reactions.

Table 5. Chemical Reactions Occurring During Combustion.

Reaction	High Heat Value of Reaction
$C + O_2 \rightarrow CO_2$	-14,100 Btu/Lbs of C
$C + ½O_2 \rightarrow CO$	-4,000 Btu/Lbs of C
$CO + ½O_2 \rightarrow CO_2$	-4,400 Btu/Lbs of CO
$H_2 + ½O_2 \rightarrow H_2O$	-61,100 Btu/Lbs of H_2
$CH_4 + 2O_2 \rightarrow CO_2 + 2H_2O$	-23,900 Btu/Lbs of CH_4
$2H_2S + 3O_2 \rightarrow 2SO_2 + 2H_2O$	-7,100 Btu/Lbs of H_2S
$C + H_2O$ (gas) $\rightarrow CO + H_2$	-4,700 Btu/Lbs of C
Sludge combustibles $\rightarrow CO_2 + H_2O$	-10,000 Btu/Lbs of combustibles

The following are experimental methods from which sludge heating value may be estimated or computed:

- Ultimate analysis-an analysis to determine the amounts of basic feed constituents. These constituents are moisture, oxygen, carbon, hydro- gen, sulfur, nitrogen, and ash. In addition, it is typical to determine chloride and other elements that may contribute to air emissions or ash- disposal problems. Once the ultimate analysis has been completed, Dulong's formula can be used to estimate the heating value of the sludge, Dulong's formula is:

$$\text{Btu/Lbs} = 14{,}544C + 62{,}208(H_2 - O_2/8) + 4{,}050S \qquad (30)$$

where C, H_2, O_2, and S represent the weight fraction of each element determined by ultimate analysis. This formula does not take into account endothermic chemical reactions that occur with chemically conditioned or physical-chemical sludge. The ultimate analysis is used principally for developing the material balance, from which a heat balance can be made.

- Proximate analysis -- a relatively low-cost analysis in which moisture content, volatile combustible matter, fixed carbon, and ash are determined. The fuel value of the sludge is calculated as the weighted average of the fuel values of its individual components.
- Calorimetry - this is a direct method in which heating value is determined experimentally with a bomb calorimeter. Approximately 1 gram of material is burned in a sealed, submerged container. The heat of

combustion is determined by noting the temperature rise of the water bath. Several samples must be taken and then composited to obtain a representative 1-gram sample. Several tests should be run, and the results must be interpreted by an experienced analyst. New bomb calorimeters can use samples up to 25 grams and this type of unit should be used where possible.

The preceding tests give approximate fuel values for sludge and allow the designer to proceed with calculations which simulate operations of an incinerator. If a unique sludge will be processed, or unusual operating conditions will be used, pilot testing is advised. Many manufacturers have test furnaces especially suited for pilot testing.

WET AIR OXIDATION

When the organic sludge is heated, heat causes water to escape from the sludge. Thermal treatment systems release water that is bound within the cell structure of the sludge and thereby improves the dewatering and thickening characteristics of the sludge. The oxidation process further reduces the sludge to ash by wet incineration (oxidation). Sludge is ground to a controlled particle size and pumped to a pressure of about 300 psi. Compressed air is added to the sludge (wet air oxidation only), the mixture is brought to a temperature of about 350° F by heat exchange with treated sludge and direct steam injection, and then is processed (cooked) in the reactor at the desired temperature and pressure. The hot treated sludge is cooled by heat exchange with the incoming sludge. The treated sludge is settled from the supernatant before the dewatering step. Gases released at the separation step are passed through a catalytic after-burner at 650 to 705° F or deodorized by other means. In some cases these gases have been returned through the diffused air system in the aeration basins for deodorization. The same basic processes is used for wet air oxidation of sludge by operating at higher temperatures (450 to 640° F) and higher pressures (1200 to 1600 psig). The wet air oxidation (WAO) process is based on the fact that any substance capable of burning can be oxidized in the presence of water at temperatures between 250° F and 700° F. Wet air oxidation does not require preliminary dewatering or drying as required by conventional air combustion processes. However, the oxidized ash must be separated from the water by vacuum filtration, centrifugation, or some other solids separation technique.

Wet-air oxidation (also called liquid-phase thermal oxidation) is not a new technology; it has been around for over forty years and has already demonstrated its great potential in wastewater treatment facilities. Despite this, there are some very important issues that remain to be addressed before a wet oxidation process can be scaled-up: the kinetics of oxidation of many important hazardous compounds are as yet unavailable, to mention only one among them. However, the kinetic models that predict solely the disappearance rate of mother compounds usually reported in the open literature are not enough; what is needed is a model capable of predicting complete conversion of all organic species present in a wastewater. Such models have to rely on the use of lumped parameters such as total organic carbon (TOC), chemical oxygen demand (COD), and biochemical oxygen demand (BOD). To point out a reaction engineering problem associated with the designing of a well established subcritical wet oxidation reactor, one can assume the TOC reduction to be linearly dependent on both reactant, organic compounds and oxygen. With a large excess of oxygen, the oxidation can be considered as a pseudo-first order reaction with the Hatta number defined as:

$$Ha = (D_{O2}k^*C_{TOC})^{1/2}/k_L \qquad (31)$$

If $Ha^2 >> 1$, the entire oxidation reaction occurs within the liquid film and when $Ha^2 << 1$, then most of the reaction occurs beyond the film, i.e. in the bulk liquid phase. In the latter case, the bulk liquid volume controls the rate with no benefit from an increase of the interfacial area. However, the Hatta number is a criterion which tells us whether the oxidation would occur in the bulk liquid-phase, therefore necessitating a large volume of liquid, or completely in the boundary layer, which calls for contacting devices that provide for a large interfacial area. In order to employ the principle of the Hatta number, in a system of complex reactions such as oxidation of organics in wastewater, a multidimensional space has to be reduced by lumping the species with similar reactivity. For kinetic analysis and design purposes, the species originally present in a wastewater or produced during the course of oxidation are conveniently divided at least into three lumps: (i) *original compounds and relatively unstable intermediates,* (ii) *high molecular mass organic acids,* and (iii) *low molecular mass organic acids.* As it has been shown for catalytic cracking, the lumped oxidation kinetics in many cases also obey the power-law form, with the exponent for a continuous-stirred tank reactor (CSTR) being lower than that for a plug-flow reactor (PFR) or a batch reactor (BR). It has been demonstrated that the kinetic behavior of a reactive mixture of organics in a batch system is governed by the most refractory lump, i.e. the lump of low molecular mass acids, while this is not the case with CSTR. Consequently, the lumped kinetics developed from BR data cannot be used for predicting TOC conversions in CSTR.

The Catalytic Wet Air Oxidation (CWAO) process is capable of converting all organic contaminants ultimately to carbon dioxide and water, and can also remove oxidizable inorganic components such as cyanides and ammonia. The process uses air as the oxidant, which is mixed with the effluent and passed over a catalyst at elevated temperatures and pressures. If complete COD removal is not required, the air rate, temperature and pressure can be reduced, therefore reducing the operating cost. CWAO is particularly cost-effective for effluents that are highly concentrated (chemical oxygen demands of 10,000 to over 100,000 mg/Liter) or which contain components that are not readily biodegradable or are toxic to biological treatment systems. CWAO process plants also offer the advantage that they can be highly automated for unattended operation, have relatively small plant footprints, and are able to deal with variable effluent flow rates and compositions. The process is not cost-effective compared with other advanced oxidation processes or biological processes for lightly contaminated effluents (COD less than about 5,000 mg/Liter).

Wet Oxidation is the oxidation of soluble or suspended oxidizable components in an aqueous environment using oxygen (air) as the oxidizing agent. When air is used as the source of oxygen the process is referred to as wet air oxidation (WAO). The oxidation reactions occur at elevated temperatures and pressures.

The CWAO process is a development of the wet air oxidation (WAO) process. Organic and some inorganic contaminants are oxidized in the liquid phase by contacting the liquid with high pressure air at temperatures which are typically between 120° C and 310° C.

In the CWAO process the liquid phase and high pressure air are passed co-currently over a stationary bed catalyst. The operating pressure is maintained well above the saturation pressure of water at the reaction temperatures (usually about 15-60 bar) so that the reaction takes place in the liquid phase. This enables the oxidation processes to proceed at lower temperatures than those required for incineration. Residence times are from 30 minutes to 90 minutes, and the chemical oxygen demand removal may typically be about 75% to 99%. The effect of the catalyst is to provide a higher degree of COD removal than is obtained by WAO at comparable conditions (over 99% removal can be achieved), or to reduce the residence time. Organic compounds may be converted to carbon dioxide and water at the higher temperatures; nitrogen and sulphur heteroatoms are converted to to molecular nitrogen and sulphates. The process becomes autogenic at COD levels of about 10,000 mg/l, at which the system will require external energy only at start-up. A simplified process diagram of the wet air oxidation process is shown in Figure 28. Typical wet oxidation applications have a feed flow rate of 1 to 45 ml/hr (5 to 100 gpm) per train, with a Chemical Oxygen Demand (COD) between 10,000 and 100,000 mg/Liter. Wet air oxidation can involve any or all of the following reactions:

$\text{Organics} + O_2 \rightarrow CO_2 + H_2O + RCOOH$

$\text{Sulfur Species} + O_2 \rightarrow SO_4^{-2}$

$\text{Organic Cl} + O_2 \rightarrow Cl^{-1} + CO_2 + RCOOH$

$\text{Organic N} + O_2 \rightarrow NH4^{+1} + CO_2 + RCOOH$

$\text{Phosphorus} + O_2 \rightarrow PO_4^{-3}$

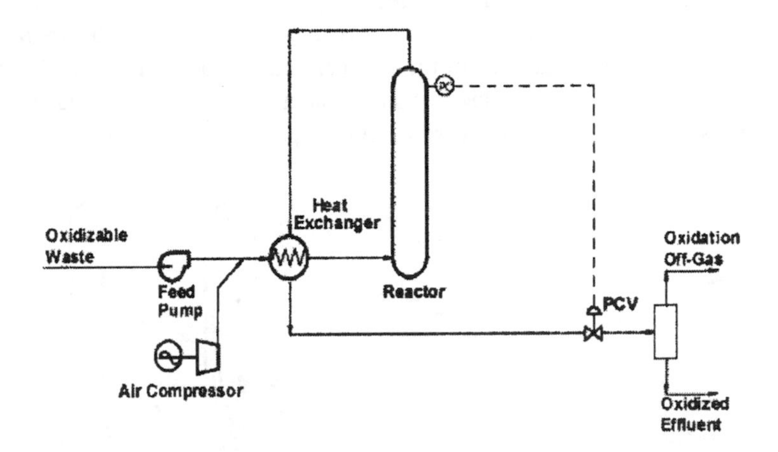

Figure 28. Process scheme for Wet Air Oxidation (WAO).

Note that RCOOH denotes short chain organic acids such as acetic acid which make up the major fraction of residual oxidation intermediates in a typical wet oxidation effluent. Properties of wet oxidation liquid effluent include: negligible NO_x and SO_2, negligible particulate matter, and some VOCs, depending on the waste. Wet oxidation is a mature technology with a long history of development and commercialization. Wet oxidation is applicable to numerous types of waste and is used commercially for the treatment of high strength industrial wastewater, ethylene and refinery spent caustic sludge. There are two other processes that we should mention that are used in conjunction with the WAO process. The first of these is thermal sludge conditioning/low pressure oxidation (LPO). Thermal sludge conditioning is used for the conditioning of biological sludge for dewatering. Thermal conditioning is accomplished using temperatues of 175 to 200° C (350 to 400° F). The low temperature allows for low operating pressures. Thermal conditioning is most commonly used for municipal wastewater treatment sludge. It has also been applied to industrial sludge processing. The technology is applicable to any organic sludge which is difficult to dewater or that contains pathogenic components. The LPO process heats sludge to a point where the biosolids break apart, releasing much of the water trapped within the cell structures, allowing filter presses, vacuum filters , belt presses and other dewatering technologies to perform their jobs more effectively. This process along with dewatering achieves a 90 to 95 % sludge volume reduction, while at the same time destroys any pathogens in the sludge. A schematic of the process is illustrated in Figure 29. The second process used in conjunction with WAO is wet air regeneration. This is a liquid phase reaction in water using dissolved oxygen to oxidize sorbed contaminants and biosolids in a spent carbon slurry, while simultaneously regenerating the powdered activated carbon. The regeneration is conducted at moderate temperatures of 400 to 500° F and at pressures from 700 to 1000 psig. The process converts organic contaminants to CO_2, water and biodegradable short chain organic acids; sorbed inorganic constituents such as heavy metals can be converted to stable, non-leaching forms that can be separated from the regenerated carbon. The technology can be more cost and energy efficient than incineration, and the regeneration is accomplished in a slurry without NO_x, SO_x or particulate air emissions.

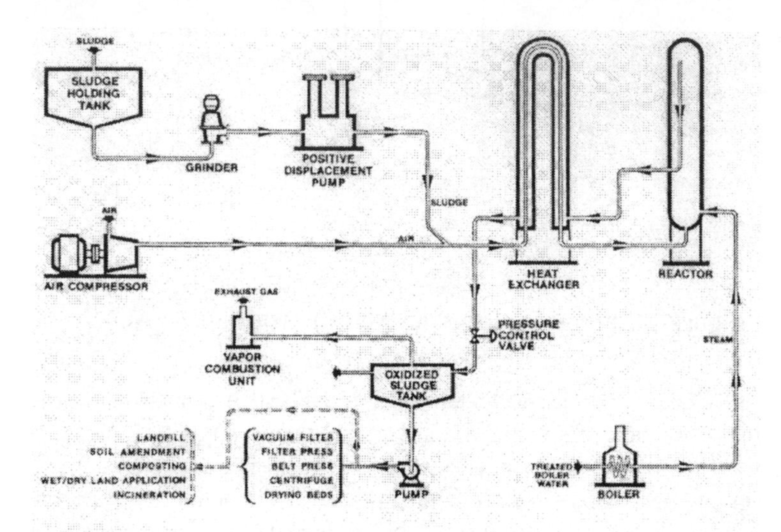

Figure 29. Schematic of low pressure oxidation process.

Here Are Some Important Terms for You to Remember

BOD: *Biological Oxygen Demand*
COD: *Chemical Oxygen Demand*
Hydrothermal Oxidation Process: *Processes which involves oxidation/reduction in an aqueous matrix at elevated temperatures. Examples of hydrothermal oxidation processes include:*

Wet Oxidation (WO): *The oxidation of oxidizable substances in water using the oxygen in air, pure or enriched oxygen, hydrogen peroxide, nitric acid or some other oxidizing agent as the source of the oxidant. The oxidation process is conducted at subcritical temperatures (<374°C).*

Wet Air Oxidation (WAO): *Wet oxidation using air as the oxygen source.*

Supercritical Water Oxidation (SCWO): *Wet oxidation occurring in supercritical water at temperatures greater than 374°C (705°F) and pressures greater than 221 bar (3204 psig).*

Some More Important Terms for You to Remember

Hydrothermal Treatment Process: *Processes which involve the use of hydrolysis or that use oxidation/reduction in an aqueous matrix at elevated temperatures.*

SCWO: *Supercritical Water Oxidation*
TKN: *Total Kjeldahl Nitrogen*
TOC: *Total Organic Carbon*
TSS: *Total Suspended Solids*
WAO: *Wet Air Oxidation*

APPLICATION OF HYDROLYSIS

In addition to the WAO process, hydrolysis, a technology similar to wet oxidation, can be applied for the treatment of wastewaters when oxygen is not a necessary reactant. In hydrolysis, certain constituents of wastewaters and sludges can react directly with water at elevated temperatures and pressures to yield a treated effluent which is detoxified or meets the desired treatment objective. Wastewaters and sludges which contain cyanide, phosphorus or other hydrolyzable constituents, can potentially be treated by hydrolysis without the addition of an oxidizing agent. Cyanide can react with water to yield formate ion and ammonia. Phosphorus can react with water to yield phosphate ion. A variety of other wastewater constituents can also be treated by hydrolysis to yield environmentally friendly products. When used as a wastewater treatment process, hydrolysis is usually employed as a pretreatment step and is followed by a polishing step, e.g., biological treatment. Common hydrolysis reactions include:

$$CN^- + 2H_2O \rightarrow NH_3 + HCO_2^-$$

$$P4 + 3CaO + 3H_2O \rightarrow PH_3 + 3CaHPO_2$$

WHAT FINALLY HAPPENS TO SLUDGE AFTER VOLUME REDUCTION

The sidebar discussion provides us with a summary of the overall scheme of wastewater treatment covered over the last several hundred pages. At the end of the day, what we are left with is *ultimate sludge*. If we choose incineration, we still have a solid waste left to deal with, ash. If we choose another route to sludge volume reduction, we still have a solid waste residue to deal with. There is no ultimate destruction of sludge, only ultimate sludge that we are left with. The final engineering solution we need to devise is how to ultimately handle this waste. It simply boils down to whether we select a so-called pollution prevention related technology or a final disposal option for the solid waste. In the remaining sections we will explore the options available to us.

AN OVERVIEW OF THE OPTIONS

The solids that result from wastewater treatment may contain concentrated levels of contaminants that were originally contained in the wastewater. A great deal of concern must be directed to the proper disposal of these solids to protect environmental considerations. Failure to do this may result in a mere shifting of the original pollutants in the waste stream to the final disposal site where they may again become free to contaminate the environment and possibly place the public at risk. A more reasonable approach to ultimate solids disposal is to view the sludge as a resource that can be recycled or reused. That concept embodies the spirit of pollution prevention.

SUMMARIZING SOLID-LIQUID SEPARATIONS TECHNOLOGY

Suspension
A
Pretreatment
A
Solids Concentration
A
Solids Stabilization
Liquid Polishing
A
Solids Separation
Clean Water
A
Post Treatment
Ultimate Solids
A - Addition of Additives

As already noted, all the sludge produced at a treatment plant (whether it be sewage or industrial in origin) must be disposed of ultimately. Treatment processes such as have been described may reduce its volume or so change its character as to facilitate its disposal, but still leave a residue which in most cases must be removed from the plant site. Like the liquid effluent from the treatment plant, there are two broad methods for the disposal of sludge :

- disposal in water, and
- disposal on land.

This applies regardless of whether or not the sludge is treated to facilitate or permit the selected method of disposal. Before discussing the specific options in detail, let's first take a look at the big picture for each option.

Disposal in water is one option to consider. This is an economical but not common method because it is contingent on the availability of bodies of water adequate to permit it. At some seacoast cities, sludge either raw or digested is pumped to barges and carried to sea (the context of these discussions is strictly sewage) to be dumped in deep water far enough off shore to provide huge dilution factors and prevent any ill effects along shore. In the past few years there has been an increased problem of pollutional loads, well above safe standards, affecting the south-shore beaches on Long Island, facilitating the closing of the beaches to the public. Some of these pollutional loads have been attributed to sludge deposits coming to shore form off-shore sludge barging operations. Where barged to sea, the value of some treatment such as thickening or digestion, depends on the relative cost of the treatment and savings in cost by barging smaller volumes, or the value of gas produced by digestion. Overall, this is an environmentally unfriendly option, and the bottom line is that it is no different that straight landfill, and in fact can be more environmentally damaging. It is plain and simple, an end-of-pipe solution with trad-offs.

Under land disposal the following methods may be included:

- Burial
- Fill
- Application as fertilizer or soil conditioner

Burial is used principally for raw sludge, where, unless covered by earth, serious odor nuisances are created. The sludge is run into trenches two to three feet wide and about two feet deep. The raw sludge in the trenches should be covered by at least 12 inches of earth. Where large areas of land are available, burial of raw sludge is probably the most economical method of sludge disposal as it eliminates the costs of all sludge treatment processes. It is, however, rarely used and even then as a temporary makeshift because of the land area required. The sludge in the trenches may remain moist and malodorous for years so that an area once used cannot be reused for the same purpose or for any other purpose for a long period of time.

The option of using sludge for **Fill** is confined almost entirely to digested sludge which can be exposed to the atmosphere without creating serious or widespread odor nuisances. The sludge should be well digested without any appreciable amount of raw or undigested mixed with it. Either wet or partially dewatered sludge, such as obtained from drying beds or vacuum filters can be used to fill low areas. Where wet sludge is used the area becomes a sludge lagoon. When used as a method of disposal, the lagoon area is used only until filled, and then abandoned. When used as a method of treatment, the sludge after some drying, is removed for final disposal and the lagoon reused. Lagoons used for disposal are usually fairly deep. Sludge is added in successive layers until the lagoon is completely filled. Final disposal of digested sludge by lagoons is economical as it eliminates all dewatering treatments. It is applicable, however, only where low waste areas are available on the plant site or within reasonable piping distance. They are frequently used to supplement inadequate drying bed facilities. Dewatered digested sludge from drying beds and vacuum filters can be disposed of by filling low areas at the plant site or hauled to similar areas elsewhere without creating nuisances. The ash from incinerators is usually disposed of by using it for fill. Where fill area is available close to the incinerator, the ash can be made into a slurry with water when removed from the ash hopper and pumped to the point of disposal. If the fill area is remote, the ash should be sufficiently wet to suppress the dust and transported by truck or railroad cars to the point of disposal. It should be clear to you that the above options for sludge sill are temporary solutions, and they still have environmental trade-offs. In the end, they too represent environmentally unfriendly solutions and are end-of-pipe disposal technologies that add costs to treatment.

Well what about pollution prevention type technologies? The two we will explore in some detail are **Soil Conditioning or Fertilizer** and **Composting**.

Sewage sludge contains many elements essential to plant life, such as nitrogen, phosphorous, potassium, and in addition, at least traces of minor nutrients which are considered more or less indispensable for plant growth, such as boron, calcium, copper, iron, magnesium, manganese, sulfur, and zinc. In fact, sometimes these trace elements are found in concentrations, perhaps from industrial wastes, which may be detrimental. The sludge humus, besides furnishing plant food, benefits the soil by increasing the water holding capacity and improving the tilth, thus making possible the working of heavy soils into satisfactory seed beds. It also reduces soil erosion. Soils vary in their requirements for fertilizer, but it appears that the elements essential for plant growth may be divided into two groups: those which come from the air and water freely and those which are found in the soil or have to be added at certain intervals. In the first group are hydrogen, oxygen and carbon. In the second group are nitrogen, phosphorous and potassium and several miscellaneous elements usually found in sufficient quantities in the average soil, such as calcium, magnesium, sulfur, iron, manganese, and others. The major fertilizing elements are nitrogen, phosphorous and potassium, and the amount of each required depends on the soil, climatic conditions and crop. Nitrogen is required by all plants, particularly where leaf development is required. Thus, it is of great value in fertilizing grass, radishes, lettuce, spinach, and celery. It stimulates growth of leaf and stem. Phosphorous is essential in many phases of plant growth. It hastens ripening, encourages root growth and increases resistance to disease. Potassium is an important factor in vigorous growth. It develops the woody parts of stems and pulps of fruits. It increases resistance to disease, but delays ripening and is needed in the formation of chlorophyll. Dried or dewatered sewage sludge makes an excellent soil conditioner and a good, though incomplete fertilizer, unless fortified with nitrogen, phosphorous and potassium. Head dried, raw activated sludge is the best sludge product, both chemically and hygienically, although some odor may result from its use. Heat dried, digested sludge contains much less nitrogen and is more valuable for its soil conditioning and building qualities than for its fertilizer content. For some crops it is deleterious. It is practically odorless when well digested.

Sludge cake from vacuum filters, because of its pasty nature, cannot be readily spread on land as a fertilizer or soil conditioner. It must be further air-dried. At some plants the sludge cake is stockpiled on the plant site over winter. Freezing, thawing and air drying result in a material which breaks up readily. Digested sludge has been said to be somewhat comparable to farm manure in its content of fertilizer constituents, their relative availability and the physical nature of the material.

➪ Raw primary sludge, unless composted, is unsatisfactory as a soil conditioner because of its effect on the soil and on growing plants, and because of the health hazards involved.

➪ Raw activated sludge, after heat drying, is established as a superior sludge product. Such sludge retains most of its organic solids and it contains more nitrogen than other sludge.

➪ Digested sludge from all sewage treatment processes are materials of moderate but definite value as a source of slowly available nitrogen and some phosphorous. They are comparable with farm yard manure except for a deficiency of potash. Their principal value is the humus content resulting in increased moisture-holding capacity of the soil and a change in soil structure which results in a greater friability.

Before sludge digestion was so widely adopted, the application of raw sludge to fields was sometimes detrimental because the grease content was difficult for the soil to absorb and caused it to become impervious. In digested sludge, however, fat has been reduced and become so finely divided that it does not adversely affect the porosity of the soil. The continued use of digested sludge tends to lower the pH value of soil and it is recommended that either lime or ground limestone be applied occasionally. In some tests it has been found that activated sludge used as an organic carrier for added inorganic forms of nitrogen, has given better results for crops with a short growing season than activated sludge alone. The inorganic nitrogen is quickly available while that from the organic portion is available more slowly and lasts over a period of time.

There is a potential hazard of transmission of parasitic infections with air-dried digestion sludge as a result of handling the sludge or from sludge contaminated vegetables eaten raw. Spreading of digested sludge in the fall and allowing it to freeze in cold climates in the winter is believed helpful in killing these organisms. Heat-dried sludge is considered safe for use under all conditions because of the destructive action of heat upon bacteria.

Let's talk about composting. A good compost could contain up to 2 percent nitrogen, about 1 percent phosphoric acid, and many trace elements. Its most valuable features, however, are not its nutrient content, but its moisture retaining and humus forming properties. Many types of microorganisms are involved in converting the complex organic compounds such as carbohydrates and proteins into simpler materials, but the bacteria, actinomycetes, and fungi, predominate. These organisms function in a composting environment that is optimized by copying the natural decomposition process of nature where, with an adequate air supply, the organic solids are biochemically degraded to stable humus and minerals. Compost is generally considered as a material to be used in conjunction with fertilizer, rather than as a replacement for fertilizer unless it is fortified with additional chemical nutrients. Compost benefits the soil by replenishing the humus, improving the soil structure, and providing useful nutrients and minerals. It is particularly useful on old, depleted soils and soils that are drought-sensitive. In horticulture applications, compost has been useful on heavy soils as well as sandy and peat soil. It has been commonly applied to parks and gardens because it increases the soil water absorbing capacity and improves the soil structure.

Composting is the process of aerobic thermophilic decomposition of organic wastes to a relatively stable humus. Decomposition results from the biological activity of microorganisms which exist in the waste.

All composting processes attempt to create a suitable environment for thermophilic facultative aerobic microorganisms. If the environmental conditions for biological decomposition are appropriate, a wide variety of organic wastes can be composted. The most important criteria for successful composting are: (1) complete mixing of organic solids, (2) nearly uniform particle size, (3) adequate aeration, (4) proper moisture content, (5) proper temperature and pH, and (6) proper carbon-nitrogen ratio in the raw solids. The smaller the particles, the more rapidly they will decompose; size is controlled by grinding. Air is necessary for aerobic organisms to function in a fast, odor-free manner. Aeration is enhanced by blending wastes to form a porous solids structure in the composting materials. Some composting systems use blowers while others aerate by frequent turning of compost placed in windows and bins. The solids to be composted must not, of course, contain high concentrations of materials toxic to the decomposing microorganisms. A proper moisture content is the most important composting criteria. Microorganisms need moisture to function but too much moisture can cause the process to become anaerobic and develop the characteristic odor and slow decomposition rate associated with anaerobic processes.

Composting mixtures should have a pH near 7 (neutral) for optimum efficiency. The temperatures vary a great deal but those in the thermophilic range (greater than 110°F) produce a more rapid rate of decomposition than those in the lower mesophilic range. Higher temperatures also cause a more efficient destruction of pathogenic organisms and weed seeds. An essential requirement of the composting process is control of the ratio of carbon to nitrogen in the raw materials. Microorganisms need both carbon and nitrogen, but they must be available in the proper amounts of decomposition will be prolonged. The time required to complete composting varies, depending on the climate, materials composted, the degree of mechanization, whether the process is enclosed, and the desired moisture content of the final product. Composting detention times from a couple of weeks to several months have been reported.

Many types of wet solids have been successfully used in composting operations. These include sewage sludge, cannery solids, pharmaceutical sludge, and meat packing wastes. Sewage sludge has been frequently used as an additive when composting dry refuse and garbage. It enhances the composting operation because: (1) it serves as a seeding material to encourage biological action, (2) it helps to control the moisture content in the composting mixture, (3) it enhances the value of the compost by contributing nitrogen and other nutrients, and (4)it can be used to control the important carbon/nitrogen ratio. Normally, blending sewage sludge with other compost raw materials required prior dewatering of the sludge. If the dewatering step is omitted, the moisture content of the mixture is too high and odors develop. Reducing sludge moisture from 90 to 70 percent by vacuum filtration or centrifugation allows good aerobic composting with garbage at a blended moisture content of 53 percent. In favorable climates, the composting of digested sludge with sawdust, straw, and wood shavings has been successful.

POLLUTION PREVENTION OPTIONS

The balance of our discussions focus on the pollution prevention technologies for sludge management and use. When you surf the Internet or look at some of the technical and trade journal references, you will more commonly see these subjects referred to as *sludge and biosolids resuse*.

As we have seen, sewage sludge has many characteristics that are good for soils and plants, if applied properly. Research has shown that the organic matter in sludge can improve the physical properties of soil. Reused sludge is also considered biosolids, which is a slightly more attractive name -- *don't you think?* Used as a soil additive, sludge improves the bulking density, aggregation, porosity and water retention of the soil. When added properly, sludge enhances soil quality and makes it better for vegetation. Vegetation also benefit from the nitrogen, phosphorus and potassium in sludge. When applied to soils at recommended volumes and rates, sludge can supply most of the nitrogen and phosphorus needed for good plant growth, as well as magnesium and many other essential trace elements like zinc, copper and nickel.

There are alternative systems to the marketability of biosolids from wastewater treatment plants. In fact, there are more than a dozen systems encompassing Class A pathogen-reduction technologies, but among these the most promising and widely used are *alkaline stabilization*, *thermal drying*, and *composting*.

We only briefly mentioned alkaline stabilization, but in reality this is a variation of sludge pasteurization. The basic process uses elevated pH and temperature to produce a stabilized, disinfected product. The two alkaline stabilization systems most common in the U.S. are a lime pasteurization system and a cement kiln dust pasteurization system. The lime pasteurization product has a wet-cake consistency, while the kiln dust pasteurization has a moist solid like consistency. Both products can be transported to agricultural areas for ultimate use. Literature studies show that the kiln dust product can capture a marketable value of $6.60/Mg ($6.00/ton) to offset hauling costs, while the lime product does not appear to be able to capture financial credits for product revenues at this point in time. The reasons for this are not entirely clear.

In contrast, composting processes utilize a mixture of solids and yard waste under controlled environmental conditions to produce a disinfected, humus-like product. Three common composting systems are a horizontal agitated reactor, a horizontal nonagitated reactor, and an aerated static pile system (nonproprietary). Compost can be marketed as a soil conditioner in competition with such products as peat, soil, and mulch. Although a large potential market exists, significant effort is required to penetrate this market. Yard waste revenue of $6.50/m^3 ($5/yd^3) and product revenue of $2.00/m^3 ($1.50/yd^3) appear to be reasonable market values based on various studies reported on the Web.

The lime stabilization system has advantages of low capital costs, process reliability, flexibility, and operability. The main disadvantage attributed to this system are questionable product marketability because of the uncertain availability of suitable agricultural land in some parts of the country where the product could be locally marketed. The steam drying alternative has the advantages of small facility land requirements, good public acceptance, and favorable product marketability. The disadvantages of this system included relatively high capital costs, reduced expansion flexibility, and complex operational requirements. These advantages and disadvantages apply to all of the thermal drying alternatives.

Land application is the largest beneficial use for sewage sludge. Since municipal sludges are a by by-product of the foods we eat, they contain important nutrients such as nitrogen, phosphorus and potassium. Proper land application provides a way to recycle these nutrients and return them to the soil safely. Sludge can also be processed into heat dried pellets that are marketed as fertilizers and soil conditioners. The pelletization process also reduces disease causing organisms. Golf courses, parks, cemeteries, nurseries and municipal landscaping projects provide markets for such pelletized sludge products.

Composting is another way to recycle nutrients and organic matter in sludge. The benefits from using sludge composts include increased water and nutrient holding capacity and increase aeration and drainage of soils. Composted sludge's also provide the soil with low levels of plant nutrients. Sludge compost is currently being produced and marketed by municipalities around the United States. More and more cities are turning to composting as a method to beneficially manage sludge's.

Philadelphia and Washington D.C. both market sludge compost for use as a mine spoils cover, landscaping material, soil amendment for public lands, and potting material. Sludge composts can also be used along roadsides to establish vegetation and reduce erosion, uses which require only a single or infrequent permit application.

There are concerns that land application of sludge will result in an increase of pathogenic bacteria, viruses, parasites, chemicals and metals in drinking water reservoirs, aquifers, and the food chain. This raises additional concerns of cumulative effects of metals in cropped soils. Research shows that if metals such as zinc, copper, lead, nickel, mercury, and cadmium are allowed to build up in soils due to many applications of sludges over the years, they could be released at levels harmful to crops, animals, and humans. While some of these metals are necessary micronutrients, at higher levels they may be harmful to plants, particularly those grown on acid soils (soils with a low pH). Cadmium, a suspected carcinogen, and mercury cause even greater concern because of their toxic effects on animals and humans. Likewise, synthetic organic compounds such as dioxins and PCBs, if present, cause concern about ecological and human health impacts. The degree of risk depends directly on the initial sludge quality, the way the sludges are processed and how the amended soil is managed during and after land application. Current state and federal legislation requires sludge treatment processes to reduce pathogens prior to land application. Furthermore, state and federal standards mandate specific limits for metals contained in sludge. Since metal concentrations depend mainly on the type and amount of industrial waste that flows into the wastewater treatment system, strongly enforced pretreatment and source control programs could effectively reduce the metals content of sludge.

Providing proper employee training and applying the best management practices will yield the best sludge use program. The fate of sludge components is also influenced by factors such as climate (rainfall and temperature), soil management (irrigation, drainage, liming, fertilization, and addition of amendments), and composition of the sludge. In the past, the success of land application has been hurt by the mismanagement of important factors such as soil pH. For example, the uptake of many metals, such as cadmium, is related to soil pH. If pH drops below a certain level, heavy metals will be released, increasing the chances of leaching and plant uptake. In addition, nutrient contamination of surface waters through nonpoint source pollution needs to be carefully monitored. While not a concern for human health and the environment, odors associated with poorly managed sludge application can be a serious concern to those living near application sites. Prompt incorporation of sludges and sludge products into the soil and avoidance of stockpiling can help to prevent odor problems. It is essential for sludge management programs to have knowledgeable staff available to teach people how to apply and monitor the sludges and the treated area correctly.

In general, researchers agree that the effects of organic compounds, certain pesticides and metals are not dangerous when managed properly at regulated levels. However, they caution that additional study of organic compounds and long-term fate of materials is needed before unlimited application of sludge can occur safely on all lands.

Sludge landfill can be defined as the planned disposal of wastewater solids including sludge , grit, and ash at a designated site where it is buried and monitored. The sludge is delivered to the landfill by trucks that pick up the sludge from the wastewater treatment plants. There are several different types of landfilling, these are all listed below under disposal methods, but the most frequent method used is dewatering then burial. This method is done by the plant dewatering the sludge then trucks pick up the sludge which is approximately 80% moisture and 20% solids. The trucks then dump the sludge into the landfill, where tractors bury the sludge using one of two special burial techniques. These techniques utilize space most efficiently and develop a grade for drainage of precipitation.

A typical biosolids application has the potential to supplement the soil with:

- **135 kg/ha (120 lbs/acre) nitrogen**
- **250 kg/ha (223 lbs/acre) total phosphorous**
- **250 kg/ha (223 lbs/acre) organic nitrogen**
- **30 kg/ha (27 lbs/acre) total potash**
- **4000 kg/ha (3600 lbs/acre) organic matter**
- **and other nutrients such as magnesium, zinc and copper**

Many municipalities and state regulatory agencies do not want sludge to be landfilled. Most states require special permission to do so. Landfills must be monitored regularly with monitoring wells and a few other environmental safety measures. The municipality are the state determine where and how the sludge will be disposed of. Once they are designated to be a part of land use, the sludge is either landfilled or if it is usable or the right grade, which is usually grade A, the sludge is used for composting. The essential difference between land application and landfill is that land application leads to treatment or assimilation, while landfill leads to containment and only for an unspecified time.

A landfill has two major drawbacks, these drawbacks are leachate and the gases of decomposition. These impacts can be some what monitored and minimized by the specifications listed under design. Siting and design of landfill operations to avoid disturbing water quality should be based on geological and hydrological considerations. The disposal options we have available to us are:

- Dumped in sand and gravel within open pits previously dug by bulldozer, pits then filled to control odor and other problems.
- Dumped at a site and leveled.
- Dumped on top of fill and mixed with refuse during compaction.
- Dumped into pit.
- Dewatered by the treatment plant, moved to landfill , dumped , and immediately buried.
- Only air-dried digested sludge accepted.
- City landfill disposal of sludge unregulated.

The objectives of a properly designed landfill are to:

- ***Protect groundwater quality***
- ***Protect air quality and conserve energy by installing a landfill gas recovery system***
- ***Minimize impact upon adjacent surface waters and wetlands***
- ***Utilize landfill space efficiently and extend site life***
- ***Provide maximum use of land after completion.***

The most important factor of a landfill is to build it properly so that the environment is not disturbed in any fashion. There are several components to the design of a environmentally friendly landfill. These components are that the landfill should be placed on a compacted low permeable medium, preferably a clay layer. This layer is then covered by a impermeable membrane which is then covered by a granular substance to act as a secondary drainage system. Layers upon layers are built up, while each layer is separated by a granular membrane. This is done over and over again until the entire landfill is full. Then they cap off the landfill to prevent excess amounts of surface water from entering. The design of the landfill layers and the mound are:

- an above grade containment mound, sloped to support the weight of the waste and cover
- a liner system across the base to retard entry of water and subsequent percolation of leachate
- a leachate collection and removal system that is drained freely by gravity, with drainage above ground

- a cover system consisting of a layer with gas collection equipment, a composite liner, a drainage liner, and a permanent vegetative cover
- a monitoring system

Investigating a site

Investigating a site is to first look at the type of soil and its bearing capacity. This should be done by digging boreholes at several designated locations over the entire landfill design site. There are several parameters which should be evaluated on the soil and they are :
A) unit weight of the soil
B) moisture content
C) void ratios
D) angle of internal friction
E) cohesion
F) transmissibility
G) solution holding capacity

To determine the cost analysis of landfilling sludge you must evaluate the steps preceding it. After the sewage treatment plant has treated the sludge they send it to a dewatering site. This site reduces the sludge to 20% solid and 80% water. The actual cost of operating a dewatering facility is depends upon size and technology. This cost is not accessible, but after the dewatered sludge leaves the plant it is hauled by truck to the landfill site, which costs the sewage treatment plant approximately $91 a wet ton. Take into consideration that biosolids reduce the need for commercial fertilizers and can reduce fertilizer cost by over $100.00 an acre. The two contaminants of environmental concern from refuse disposal are gas and leachate. The leachate is generated because of the water the penetrates the landfill and the gas is due to the decomposing of the organic matter. Gas production from the organic matter begins before it is actually landfilled. the principal gases that are generated from the decomposing matter is carbon dioxide and methane. Carbon dioxide is important in the surrounding areas water quality, because it is soluble in water, unlike the other gases that can be produced within the landfill which are insoluble. When carbon dioxide is dissolved in water it lowers the pH , which creates a corrosive environment. It also creates an increase in water hardness.

Usually the effects of carbon dioxide are at a maximum during the first few months of decomposition and could continue on for a few years. As time goes on carbon dioxide values decrease and pose a lesser problem as the years go by. Leachate production within a landfill depends on the amount of water that enters the landfill. Leachate results when the amount of water entering exceeds the amount of water that can be retained by the waste. This is a major reason why site investigation and soil characteristics are so very essential in landfill design. The primary causes of excess water intrusion are due to a raise in groundwater elevation. Another consideration that should be evaluated is the topography and the climate of the area, because these two factors can cause a dramatic impact on the landfill if they are not assessed properly. The best approach to leachate management is to prevent or limit its production from the beginning. This is why proper design and elaborate research of an area are so very essential to a landfill and its operation.

BIOSOLIDS REGULATIONS

The U.S. EPA has developed comprehensive federal biosolids use and disposal regulations , which are organized in five parts. These parts are general provisions, land application, surface disposal, pathogens and vector attraction reduction, and incineration. Parts of the regulations which address standards for land application, surface disposal, and incineration practices consist of general requirements, pollutant limits, operational requirements, management practices, frequency of monitoring, recordkeeping, and reporting requirements for biosolids processing facilities to abide by. Regulatory considerations play a key role in determining how to efficiently use sludge. The EPA has adopted a sludge management policy intended to encourage the beneficial use of sludge while protecting public health and the environment. The EPA's recent revisions of the Clean Water Act part 503 regulations promote the beneficial use of clean sludge that contain low levels of pollutants, sludge of "exceptional quality". The revisions follow many years of sludge used in field studies that analyzed the effect of toxic elements in land-applied sludge and sludge composts. The regulations and technical support documents are interpreted by some scientists to reinforce the safety of using sludge on both agricultural and non-agricultural lands while ensuring the protection of soils, water quality, the food chain, and human health. However, scientific uncertainties remain particularly with respect to the long-term and ecological safety of sludge application. New York State has adopted an integrated waste management policy that involves a hierarchy of solid waste management methods intended to reduce dependency on landfills for waste disposal. The hierarchy is incorporated into New York's Environmental Conservation Law in order of preference: Reduction, recycle/reuse, incineration, and landfilling.This policy is the cornerstone of the state's solid waste management program. Several components of the program assist local governments in managing their wastes safely and efficiently. One component is the New York State Department of Environmental Conservation's Part 360 regulations, requiring state facility siting and operation and a comprehensive recycling analysis. These regulations contain specific guidelines for land application and other sludge management options that must be considered by a municipality or purveyor during its planning process. These regulations are currently under revision.

The part 503 rule applies to biosolids generated from the treatment of domestic wastewater and includes domestic septage. Compliance with the part 503 standards is required within 12 months of publication of the regulation. However, if new pollution control facilities need to be constructed to achieve compliance, then compliance is required by February 19, 1995. Compliance with monitoring, recordkeeping, and reporting provisions is required provisions is required by July 15, 1993. For the most part, the rule is "self implementing," which means that citizen suits or EPA can enforce the regulation even before permits are issued. The standards will be incorporated into National Pollution Discharge Elimination System (NPDES) permits issued by EPA or permits issued by states with approved biosolids management programs. EPA will work closely with the states to encourage their adoption of approved biosolids management programs that can carry out delegated programs. Land Applied Biosolids must Meet Quality Requirements. Land application includes all forms of applying bulk or bagged biosolids to land for beneficial uses at argonomic rates (rates designed to provide the amount of nitrogen needed by the vegetation while minimizing the amount that passes below the root zone). These include application to agricultural land; pasture and range land; nonagricultural land such as forests; public contract sites such as parks and golf courses; disturbed land such as mine spoils, construction sites and gravel pits; and home lawns and gardens. Selling or giving away biosolids products is addressed under land application of domestic septage (liquid or solid material removed from a septic tank). The person who prepares biosolids for land application or applies biosolids to the land must obtain and provide the necessary information needed to comply with the rule. For example, the person who prepares bulk biosolids that are land applied must provide the person who applies it to land with all information necessary to comply with the rule, including the total nitrogen concentration of the biosolids. The rule establishes two levels of biosolids quality with respect to heavy metal concentrations- pollutant ceiling concentrations and pollutant concentrations ("high quality" biosolids); two levels of quality with respect to pathogen densities- class A and class B; and two types of approaches for meeting vector attraction reduction- biosolids processing or the use of physical barriers. (Vector attraction reduction reduces the potential for spreading infectious disease agents by vectors, that is, flies, rodents, and birds.)

To qualify for land application, biosolids must meet the pollutant ceiling concentrations, class B requirements for pathogens, and vector attraction reduction requirements. Bulk biosolids applied to lawns and home gardens must meet the pollutant concentration limits, class A pathogen reduction requirement, and vector attraction reduction using biosolids using biosolids processing. Bulk biosolids applied to agricultural and non-agricultural land must meet at a minimum the pollutant ceiling concentrations and cumulative pollutant loading, at least class B pathogen reduction requirements, and one of the vector attraction reduction requirements.

Management practices that apply to land applied biosolids (other than "exceptional quality" biosolids products) include:

- no application to flooded, frozen, or snow covered ground;
- no application at rates above argonomic rates (reclamation projects may be excepted)
- no application if threatened endangered species are adversely affected;

- labeling of biosolids that are sold or given away;
- a required 10m buffer from U.S. waters.

If the biosolids are of "exceptional quality"- that is, they meet the pollutant concentration limits, class A pathogen reduction requirements, and a vector attraction processing option- they are usually exempt. However, when biosolids meeting class B pathogen reduction requirements are applied to the land, additional site restrictions are required. Table 6 provides a summary of the land application pollution limits for biosolids as they currently stand.

Table 6. Land Application Pollutant Limits[(a)].

Pollutant	Ceiling Concentration Limits,[(b)] mg/kg	Cummulative Pollutant Loading Rates, kg/ha	High Quality Pollutant Concentration Limits,[(c)] mag/kg	Annual Pollutant Loading Rates, kg/ha/yr
Arsenic	75	41	41	2.0
Cadmium	85	39	39	1.9
Chromium	3000	3000	1200	150
Copper	4300	1500	1500	75
Lead	840	300	300	15
Mercury	57	17	17	0.85
Molybdenum	75	18	18	0.90
Nickel	420	420	420	21
Selenium	100	0	36	5.0
Zinc	7500	2800	2800	140

(a) - EPA Part 503 standards. All weights are on a dry-weight basis.

(b) - Absolute values

(c) - Monthly averages

SOME CASE STUDIES FOR YOU TO PONDER

Studies have shown that the biosolids application reduce runoff from treated rangeland. Results were attributed to increase increased ground surface roughness and water adsorption by the dry biosolids. Over time, the biosolids should decompose and have a less direct effect on surface runoff. However, the increase in vegetative cover in response to the fertilizer effect of the biosolids should further improve the surface hydrology of the treated rangeland. Potential surface water contamination by constituents in biosolids does not appear to be a problem. Nitrate-nitrogen, copper, and cadmium concentrations in the runoff water, both during natural and simulated rainfall, were below limits for groundwater and for livestock and wildlife watering areas. Potential surface water pollution by biosolids contaminants was not a problem with a one time application of 22.5 to 45 dry Mg/ha (10 to 20 dry ton/ac). The following are some case studies that you can read through to get an idea of typical applications.

IMPORTING BIOSOLIDS FROM NEW YORK
(Case Study1)

Biosolids from New York, N.Y., have been imported into southeastern Colorado for application to dry land wheat, grassland, and reclamation sites. The Colorado Department of Health and the EPA have issued land application permits for about 7700 ha (19,000 ac). Annual precipitation in the area averages 36 cm. Biosolids from this project were used to reclaim a 104 ha (258 ac) sand blowout on private land near Granada, Colorado. The sandy dunes began as a 12 ha (30 ac) area subject to wind erosion. Like the many other blowouts in the Arkansas River valley, the dunes resisted all previous efforts at reclamation. Dewatered biosolids were mixed with wood chips and seeded with an improved range grass mixture. Approximately 4500 dry tons of the biosolid mixture were applied at rates of 20 and 40 dry tons/ac. This site has shown rapid germination and establishment of grasses. Biosolids were also applied to dry land wheat and range grass in the area. The inability to provide sufficient quantities of biosolids to satisfy the demands of local farmers and ranchers has been a problem.

GUNNISON, COLO.
(Case Study 2)

This southern Colorado town uses about 32.5 ha (80 ac) of grass on a private ranch for application of biosolids slurry from its oxidation ditch. Application is approximately 56 dry Mg/yr (62 dry tons/yr) through a sprinkler gun. Other dry land parcels on ranches have also been used.

Case Study 2 Continued

Average annual precipitation in Gunnison is 28 cm (11 in.); the typical yearly application rate is 5 dry Mg/ha (2.3 dry ton/ac). Grazing on the biosolids-treated land has increased from once to twice per year. Gunnison is also experimenting with applying grass seed with biosolids to increase production.

APPLYING BIOSOLIDS REDUCES SOIL EROSION (Case Study 3)

Scientists from the U.S. Department of Agriculture Rocky Mountain Forest and Range Experiment Station conducted a 5-year study of the effects of biosolids application to degraded semiarid rangeland. Dewatered, anaerobically digested biosolids from Albuquerque, N. Mex., were surface applied to a grass land site in the Rio Puerco Watershed resource area of New Mexico. The Rio Puerco basin is a degraded watershed were livestock have grazed extensively. The project site consisted of a broom snakeweed and blue grama plant community located on a moderately deep, medium textured soil. Mean annual precipitation during the 5-year study was 25 cm. Biosolids were surface applied at one time application rates of 0, 22.5, 25, and 90 Mg/ha (0, 10,20,40 dry ton/acre). Each application rate was replicated four times; plots were in a randomized block design, with a total of 16 plots. Each plot was 3m x 20m (10ft x 65 ft) in size. To minimize disturbance of the existing grassland vegetation, plots were not disked or tilled after biosolids were added. Plant nutrients in the soil, including total nitrogen, available phosphorus, and potassium, increased linearly with increasing biosolids application during the first year of study. Organic matter in soil below the biosolids layer did not increase until after the fifth year because of a lag between increased nutrient availability and subsequent plant productivity and microbial activity belowground. Although soil pH decreases from 7.8 to 7.4 because of biosolids, only DTPA-extractable copper and cadmium increased in concentrations above desirable levels. This occurred after the fifth year and only in plots receiving applications greater than 45 dry Mg/ha (20 dry ton/ac). Blue grama grass, a desirable species for grazing livestock and wildlife, increased during the first and second years, with yield from 1.5 to 2.7 times greater in the treated plots than in the control plots in the treated plots than in the control plots. Throughout the study, blue grama production remained higher in plots receiving 45 to 90 dry Mg/ha (20 and 40 dry ton/ac) of biosolids than in control plots.

The application of biosolids also increases the nutritional value of blue grama. Tissue levels of nitrogen, phosphorus, potassium, and crude protein increased to recommended tissue concentrations with biosolids treatments. Trace metals in blue grama grass did not increase during the study, thereby eliminating concerns that toxic amounts of these elements could be transferred to grazing animals.

Case Study 3 Continued

Results indicated that a one time biosolids treatment ranging from 22.5 to 45 dry Mg/ha (10 to 20 dry ton/ac) yielded positive vegetative response without harm to the environment. An unexpected benefit from the biosolids treatments was a decrease in broom snakeweed, a toxic, non palatable competitive range plant. Following the addition of biosolids, the number of broom snakeweed plants in the biosolids-treated plots decreased over the course of the study.

U.S. AIR FORCE ACADEMY
(Case Study 4)

The U.S. Air Force Academy near Colorado Springs, Colo., applies 27 to 36 dry Mg/yr (30 to 40 dry tons/yr) of anarobically digested biosolids to 40.5 ha (100 ac) of native vegitation on the academy property. The program began in 1988 with the permitting of a 405 ha (1000 ac) site. The area, which recieves 39 cm/yr (15 in./yr) of precipitation, consists of highly erodible solids. Through the application of biosolids, the academy has been able to improve vegitative cover, reduce soil erosion, and improve wildlife habitat.

FINAL COMMENTS AND EVALUATING ECONOMICS

A great deal of our discussions have focused on municipal treatment applications, particularly in this chapter. However, most if not all of the principles throughout the book are readily applicable to industrial water treatment applications. Try to approach each water treatment assignment from a first principles standpoint, and then develop design-specific cases with as much information on the chemistry, physical and thermodynamic properties of the wastewater stream and sludge to be handled. In all assignments, be sensitive to the cost issues. Engineering projects are not complete unless we have evaluated the project economics. Some cost factors for different technologies have been included in our discussions, but no real effort has been made for detailed comparisons between technologies. This really has to be performed on a case specific basis. What we can do before closing this volume is review some of the generalized project cost estimating parameters that are applicable to assessing the investments that may be needed in upgrading and/or installing wastewater treatment facilities and various solid-liquid separation equipment.

PROJECT COST ESTIMATING PRINCIPLES

Approaches to project cost estimating have changed in recent years, to the point where smart companies and engineers apply principles of Total Cost Accounting. The term *total-cost accounting* (TCA) has come to be more commonly known as *life-cycle costing* (LCC). LCC is a method aimed at analyzing the costs and benefits associated with a piece of equipment, plant or a practice over the entire time of intended use. The idea actually originated in the federal government and was first applied in procuring weapons systems. Experience showed that the up-front purchase price was a poor measure of the total cost. Instead, costs such as those associated with maintainability, reliability, disposal and salvage value, as well as employee training and education, had to be given equal weight in making financial decisions. By the same token, justifying the investment into a piece of requires that all benefits and costs be clearly defined in the most concrete terms possible, and projected over the life of each technology option. The following are important definitions and calculations you should apply when developing cost estimates for equipment and processes that you are considering in a wastewater treatment project.

PRESENT WORTH OR PRESENT VALUE

The importance of *present worth*, also known as *present value*, lies in the fact that time is money. The preference between a dollar now and a dollar one year from now is driven by the fact that a dollar in-hand can earn interest. Present value can be expressed by a simple formula:

$$\mathbf{P = F/(1 + \mathit{i})^n} \qquad (32)$$

where P is present worth or present value, F is future value, i is the interest or discount rate, and n is the number of periods. As a simple example, if we have or hold \$1,000, in one year at 6 percent interest compounded annually, the \$1,000 would have a computed present value of:

$$P = \$1{,}000/(1 + 0.06)^1 = \$943.40$$

Because our money can "work" at 6% interest, there is no difference between \$943.40 now and \$1,000 in one year because they both have the same value now. Economically, there is an additional factor at work in present value, and that factor is *pure time preference,* or *impatience*. However, this issue is generally ignored in business accounting, because the firm has no such emotions, and opportunities can be measured in terms of financial return.

But going back to our \$1,000, if the money was received in three years, the present value would be:

$$P= \$1{,}000/(1 + 0.06)^3 = \$839.62$$

In considering either multiple payments or cash into and out of a company, the

present values are additive. For example, at 6 percent interest, the present value of receiving both $1,000 in one year and $1,000 in three years would be $943.40 + $839.62 = $1,783.06. Similarly, if one was to receive $1,000 in one year, and pay $1,000 in 3 years the present value would be $943.40 - $839.62 = $103.78.
It is common practice to compare investment options based on the present-value equation shown above. We may also apply one or all of the following four factors when comparing investment options: Payback period; Internal rate of return; Benefit-to-cost ratio; and Present value of net benefit.

WHAT PAYBACK PERIOD IS

The payback period of an investment is essentially a measure of how long it takes to break even on the cost of that investment. In other words, how many weeks, months, or years does it take to earn the investment capital that was laid out for a project or a piece of equipment?
Obviously, those projects with the fastest returns are highly attractive. The technique for determining payback period again lies within present value; however, instead of solving the present-value equation for the present value, the cost and benefit cash flows are kept separate over time.
First, the project's anticipated benefit and cost are tabulated for each year of the project's lifetime. Then, these values are converted to present values by using the present-value equation, with the firm's discount rate plugged in as the discount factor. Finally, the cumulative total of the benefits (at present value) and the cumulative total of the costs (at present value) are compared on a year-by-year basis. At the point in time when the cumulative present value of the benefits starts to exceed the cumulative present value of the costs, the project has reached the payback period. Ranking your equipment and technology options for an intended application then becomes a matter of selecting those project options with the shortest payback period. So for example, if we compare a rotary drum filter versus a centrifuge for a dewatering application, an overall payback period for each option can be determined based on capital investment for each equipment, operating, maintenance, power costs, and other factors.
Although this approach is straightforward, there are dangers in selecting technology or equipment options based upon a minimum payback-time standard. For example, because the equipment's use generally extends far into the future, discounting makes its payoff period very long. Because the payback period analysis stops when the benefits and costs are equal, the projects with the quickest positive cash flow will dominate. Hence, for a project, with a high discount rate, the long-term costs and benefits may be so far into the future that they do not even enter into the analysis. In essence, the importance of life-cycle costing is lost in using the minimum payback-time standard, because it only considers costs and benefits to the point where they balance, instead of considering them over the entire life of the project or piece of solids-liquid separations equipment.

THE INTERNAL RATE OF RETURN

You are likely more familiar with the term *return on investment*, or ROI, than internal rate of return. The ROI is defined as the interest rate that would result in a return on the invested capital equivalent to the project's return. For illustration, if we had a water treatment project with a ROI of 30 percent, that's financially equivalent to investing resources in the right stock and having its price go up 30 percent. As before, this method is based in the net present value of benefits and costs; however, it does not use a predetermined discount rate. Instead, the present-value equation is solved for the discount rate i. The discount rate that satisfies the zero benefit is the rate of return on the investment, and project selection is based on the highest rate. From a simple calculation standpoint, the present value equation is solved for i after setting the net present value equal to zero, and plugging in the future value, obtained by subtracting the future costs from the future benefits over the lifetime of the project. This approach is frequently used in business; however, the net benefits and costs must be determined for each time period, and brought back to present value separately. Computationally, this could mean dealing with a large number of simultaneous equations, which can complicate the analysis.

THE BENEFIT-TO-COST RATIO

The benefit to cost (B/C) ratio is a benchmark that is determined by taking the total present value of all of the financial benefits of a water treatment project and dividing it by the total present value of all the costs of the project. If the ratio is greater than unity, then the benefits outweigh the costs, and we may conclude that the project is economically worthwhile. The present values of the benefits and costs are kept separate, and expressed in one of two ways. First, as already explained, there is the pure B/C ratio, which implies that if the ratio is greater than unity, the benefits outweigh the costs and the project is viable. Second, there is the net B/C ratio, which is the net benefit (benefits minus costs) divided by the costs. In this latter case, the decision criteria is that the benefits must outweigh the costs, which means that the net ratio must be greater than zero (if the benefits exactly equaled the costs, the net B/C ratio would be zero). In both cases, the highest B/C ratios are considered as the best projects. The B/C ratio can be misleading. For example, if the present value of a filter press' benefits were $10,000 and costs were $6,000, the B/C ratio would be $10,000/$6,000 or 1.67. But what if, upon further reassessment of the project, we find that some of the costs are not 'true' costs, but instead simply offsets to benefits? In this case, the ratio could be changed considerably. For argument sake, let's say that $4,500 of the $6,000 total cost is for lower energy costs over say a thermal dewatering technology, and that $7,000 of the $10,000 in benefits is due to power savings; one could then use them to offset each other. Mathematically then, both the numerator and denominator of the ratio could be reduced by $4,500 with the following effect:

(\$10,000 - \$4,500) / (\$6,000 - \$4,500) = 3.7

Without changing the project, the recalculated B/C ratio would make the project seem to be considerably more attractive.

PRESENT VALUE OF NET BENEFITS

The present value of net benefits (PVNB) shows the worth of a project in terms of a present-value sum. The PVNB is determined by calculating the present value of all benefits; doing the same for all costs; and then subtracting the two totals. The result is an amount of money that would represent the tangible value of undertaking the project. This comparison evaluates all benefits and costs at their current or present values. If the net benefit (the benefits minus costs) is greater than zero, the project is worth undertaking; if the net is less than zero, the project should be abandoned on a financial basis. This technique is firmly grounded in microeconomic theory and is ideal for total-cost analysis (TCA).
Even though it requires a preselected discount rate, which can greatly discount long-term benefits, it assures that all benefits and costs over the entire life of the project are included in the analysis. Once you know the present value of all options with positive net values, the actual ranking of equipment and technology options using this method is straightforward; those with the highest PVNBs are funded first. There are no hard and fast rules as to which factors one may apply in performing life-cycle costing or total-cost analysis; however, conceptually, the PVNB method is preferred. There are, however, many small-scale equipment projects where the benefits are so well defined and obvious that a comparative financial factor as simple as a ROI or the payback period will suffice.

ESTABLISHING BASELINE COSTS

To properly determine the cost of any engineering project, we first need to establish a baseline for comparative purposes. If nothing else, a baseline defines for management the option of maintaining the status quo. If we are faced with meeting a legal discharge limit, then obviously we need to do something other than status quo to remain in business. But here is where we can develop some interesting and very detailed justifications for one water treatment technology or piece of equipment over another. Changes in material consumption, utility demands, staff time, etc., for options being considered can be measured as either more or less expensive than a certain baseline. The baseline may be arbitrary for comparison. For example, we could choose as a baseline a traditional industry average. By starting with those technologies that comprise a conventional treatment plant, we have a baseline for the cost analysis. But to determine true cost benefits to your specific application, you really need to compare technologies and costs. This is an optimization exercise, and one which is well worthwhile in establishing the right level of investment for a water treatment plant or operation.

I highly recommend you follow the methodology of McHugh (McHugh, R.T., *The Economics of Waste Minimization*, McGraw-Hill Book Publishers, 1990). McHugh defines four tiers of potential costs, which the author applies to pollution prevention, but the principles and methodology are universal:

- *Tier 0:* Usual or normal costs, such as direct labor, raw materials, energy, equipment, etc.
- *Tier 1:* Hidden costs, such as monitoring expenses, reporting and record keeping, permitting requirements, environmental impact statements, legal, etc.
- *Tier 2:* Future liability costs, such as remedial actions, personal injury under the OSHA regulations, property damage, etc.
- *Tier 3:* Less tangible costs, such as consumer response and confidence, employee relations, corporate image, etc.

Tier 0 and Tier 1 costs are direct and indirect costs. They include the engineering, materials, labor, construction, contingency, etc., as well as waste-collection and transportation services, raw-material consumption (increase or decrease), and production costs. Tier 2 and Tier 3 represent intangible costs. They are much more difficult to define, and include potential corrective actions under the Clean Water Act (CWA); possible more stringent discharge limitations in the future ; and benefits of improved safety and work environments. Although these intangible costs often cannot be accurately predicted, they can be very important and should not be ignored when assessing an equipment or technology option.

A present-value analysis that contains such uncertain factors generally requires a little ingenuity in assessing the full merits of an engineering project.

When analyzing the financial impact of projects, it is often useful to further categorize costs as either *procurement costs* or *operations costs*. This distinction better enables the projection of costs over time, because procurement costs are short-term, and refer to all costs required to bring a new piece of equipment or a new procedure on-line. Conversely, operations costs are long-term, and represent all costs of operating the equipment or performing the procedure in the post-procurement phase.

Tier 2 and 3 costs are difficult to quantify or predict. While Tier 2 costs include potential liabilities, such as changing water discharge limits, Tier 3 costs are even harder to predict -- for example, a typical Tier 3 cost could be associated with public acceptance or rejection of a particular technology. In many cases, there is a probability that can be connected with a particular event. This enters into the calculation of *expected value*. The expected value of an event is the probability of an event occurring, multiplied by the cost or benefit of the event. Once all expected values are determined, they are totaled and brought back to present value as done with any other benefit or expense. Hence, the expected value measures the central tendency, or the average value that an outcome would have. For example, there are a number of games at county fairs that involve betting on numbers or colors, much like roulette. If the required bet is $1, and the prize is worth $5, and there are 10 selections, the expected value of the game can be computed as:

(Benefit of Success) x (Probability of Success) - (Cost of Failure) x (Probability of Failure), or
($5) x (0.1) - ($l) x (0.9) = -$0.40

On the average, the player will lose -- meaning the game operator will win -- 40 cents on every dollar wagered. For tier 2 and 3 expenses, the analysis is the same. For example, there is a great deal of data available from Occupational Safety and Health Administration (OSHA) studies regarding employee injury in the workplace. If one technology poses a higher risk to occupational exposure than another, then the probability of injury and a cost can be found, and the benefit of the project can be computed.
The concept of expected value is not complicated, though the calculations can be cumbersome. For example, even though each individual's chance of injury may be small, the number of employees, their individual opportunity costs, the various probabilities for each task, etc., could mean a large number of calculations. However, if one considers the effect of the sum of these small costs, or the large potential costs of having to replace a technology or consider significant upgrades within 5 years, then the expected value computations can be quite important in the financial analysis.

REVENUES, EXPENSES, AND CASH FLOW

Because it is the goal of any business to make a profit, the costs-and-benefits cash flows for each option can be related to the basic profit equation:

REVENUES - EXPENSES = PROFITS

The most important aspect of this is that profits can be increased by either an increase in revenues or a decrease in expenses. Water treatment operations are by and large end-of-pipe treatment technologies, and hence from the standpoint industry applications that must treat water, the investments required increase expenditures and decrease profit. Municipal facilities view their roles differently, because their end-product is clean water which is saleable, plus they may have add-on revenues when biosolids are developed and sold into local markets. There are different categories of revenues and expenses, and it is important to distinguish between them.
Obviously, revenue is money coming into the company; from the sale of goods or services, from rental fees, from interest income, etc. The profit equation shows that an increase in revenue leads to a direct increase in profit, and vice versa if all other revenues and expenses are held constant. Note that we are going to assume that the condition of other expenses/revenues are held constant in the discussions below.

Revenue impacts must be closely examined. For example, companies often can cut wastewater treatment costs if water use (and, in turn, the resulting wastewater flow) is limited to nonpeak times at the wastewater treatment facility. However, this limitation on water use could hamper production. Consequently, even though the company's actions to regulate water use could reduce wastewater charges, revenue could also be decreased, unless alternative methods could be found to maintain total production. Conversely, a change in a production procedure as a result of a technology change could increase revenue. For example, moving from liquid to dry paint stripping can not only reduce water consumption, but also affect production output. Because clean-up time from dry paint-stripping operations (such as bead blasting) is generally much shorter than from using a hazardous, liquid based stripper, it could mean not only the elimination of the liquid waste stream (this is a pollution prevention example), but also less employee time spent in the cleanup operation. In this case, production is enhanced and revenues are increased by the practice. Another potential revenue effect is the generation of marketable by-products such as biosolids. Such opportunities bring new, incremental revenues to the overall operation of the plant. The point to remember is that the project has the potential to either increase or decrease revenues and profits - and that's the reason for doing a financial analysis.

Expenses are monies that leave the company to cover the costs of operations, maintenance, insurance, etc. There are several major cost categories:

- Insurance expenses
- Depreciation expenses
- Interest expenses
- Labor expenses
- Training expenses
- Auditing and demo expenses
- Floor-space expenses

Each of these should be carefully considered in your analysis.

Insurance Expenses. Depending upon the project, insurance expenses could either increase or decrease. Insurance premiums can be increased depending on the technology option chosen for a plant design.

Depreciation Expenses. By purchasing capital equipment with a limited life the entire cost is not charged against the current year. Instead, depreciation expense calculations spread the equipment's procurement costs (including delivery charges, installation, start-up expenses, etc.) over a period of time by taking a percentage of the cost each year over the life of the equipment. For example, if the expect life of a piece of equipment is 10 years, each year the enterprise would charge an accounting expense of 10 percent of the procurement cost of the equipment. This is known as the *method of straight-line depreciation*. Although there are other methods available, all investment projects under consideration at any given time should use a single depreciation method to accurately compare alternative projects' expense and revenue effects. Because straight-line depreciation is easy to compute, it is the preferred method. Note that even though a company must use a different depreciation system for tax purposes (e.g., the Accelerated Cost Recovery System, or ACRS), it is acceptable to use other methods for bookkeeping and analysis. In any event, any capital equipment must be expensed through depreciation.

Interest Expense. Investment in equipment implies that one of two things must occur: Either a company must pay for the project out of its own cash, or it must finance the cost by borrowing money from a bank, by issuing bonds, or by some other means. When a firm pays for a project out of its own cash reserves, the action is sometimes called an *opportunity cost*. If you must borrow the cash, there is an interest charge associated with using someone else's money. It is important to recognize that interest is a true expense and must be treated, like insurance expense, as an offset to the project's benefits. The magnitude of the expense will vary with bank lending rates, the interest rate offered on the corporate notes issued, etc. In any case, there will be an expense. The reason companies account for equipment purchases as a cost is this: If cash is used for the purpose of pollution control, it is unavailable to use for other opportunities or investments. Revenues that *could* have been generated by the cash (for example, interest from a certificate of deposit at a bank) are treated as an expense and thus reduce the value of the project. But again - we may not have a choice if the project is driven by legal requirments such as the CWA.

Although the reasoning seems sound, opportunity costs are not really expenses. Though it is true that the cash will be unavailable for other investments, opportunity cost should be thought of as a comparison criteria and not an expense. The opportunity forgone by using the cash is considered when the project competes for funds and is expressed by one of the financial analysis factors discussed earlier (net value of present worth, pay back period, etc.). It is this competition for company funds that encompasses opportunity cost, so opportunity cost should not be accounted directly against the project's benefits.

Many companies apply a minimum rate of return, or *hurdle rate*, to express the opportunity-cost competition between investments. For example, if a firm can draw 10 percent interest on cash in the bank, then 10 percent would be a valid choice for the hurdle rate as it represents the company's cash opportunity cost. Then, in analyzing investment options under a return-on-investment criteria, not only would the highest returns be selected, but any project that pays the firm a return of less than the 10 percent hurdle rate would not be considered.

Labor Expenses. In the majority of situations, projects will cause a company's labor requirements to change. This change could be a positive effect that increases available productive time, or there could be a decrease in employees' production time depending upon the practice. When computing labor expenses, the Tier 1 costs could be significant. Labor expense calculations can be simplistic or comprehensive. The most direct and basic approach is to multiply the wage rate by the hours of labor. More comprehensive calculations include the associated costs of payroll taxes, administration, and benefits. Many companies routinely track these costs and establish an internal 'burdened' labor rate to use in financial analysis.

Training Expenses. Your project may also involve the purchase of equipment that requires additional operator training. In computing the total training costs, the enterprise must consider as an expense both the direct costs and the staff time spent in training. Remember that some of the technologies discussed require more extensive worker training than others. In addition, any other costs for refresher training, or for training for new employees, that is above the level currently needed must be included in the analysis. Computing direct costs is simply a matter of adding the costs of tuition, travel, per diem, etc., for the employees. Similarly, to compute the labor costs, simply multiply the employees' wage rates by the number of hours spent away from the job in training.

Auditing and Demo Expenses. Labor and other expenses associated with defining the engineering project are often overlooked. Although these tend to be small for low-investment projects, some contemplated operations may require pilot testing, or sending personnel off-site to work with vendors in their shops. This can happen when dealing with exotic sludge or unique waste waters that require treatment. Pilot or plant trials can incur significant up-front costs from production down times, personnel, monitoring equipment, and laboratory measurements, as well as engineering design time and consultant-time charges. Some enterprises may prefer to absorb these costs as part of their R&D budget -- for organizations these expenses simply are a part of the baseline cost of operations.

Floor Space Expenses. As with any costs, the floor-space costs must be based on the value of alternative uses. Unfortunately, computing floor-space opportunity cost is not always straightforward, as it is in the case of training costs. In instances where little square footage is required, there may be no other use for the floor space, which implies a zero cost. Alternatively, as the square footage required increases, calculating floor-space costs becomes more straightforward. For example, if a new building is needed to house the water treatment equipment, it's easy to compute a cost. The average-square-foot cost for a new or used warehouse (or administrative or production space) that would be charged to procure the space on the local market is the average market worth of a square foot of floor space. Unless there is a specific alternative proposal for the floor space, this market analysis should work as a proxy.

Though cash flow does not have a direct effect on a company's revenues or expenses, the concept must be considered. If the project involves procurement costs, they often must be paid upon delivery of the equipment - yet cash recovery could take many months or even years. Three things about any project can affect a firm's available cash. First, cash is used at the time of purchase. Second, it takes time to realize financial returns from the project, through either enhanced revenues or decreased expenses. Finally, depreciation expense is calculated at a much slower rate than the cash was spent. As a result of the investment, a company could find itself cash-poor. Even though cash flow does not directly affect revenues and expenses, it may be necessary to consider when analyzing your project.

INCOME TAXES

Though most companies use only revenue and expense figures when comparing investment projects, income-tax effects can enter into each calculation if either revenues or expenses change from the baseline values. More expenses mean lower profits and less taxes, and vice versa. If an company needs to know the effect of income taxes on profit, the computations are simple and can be done during or after the analysis. As with expenses and revenues, you do not need to compute the total tax liability for each option. Instead, you only need to look at the options' effect on revenues and/or expenses, and the difference in tax liability resulting from deviations from the baseline. The profit equation reflects gross or pretax profits. Income tax is based on the gross profit figure from this equation and cannot be computed until you know what effect the options will have on revenues and/or expenses.

FINAL, FINAL COMMENTS

We have introduced some concepts above as they relate to *total-cost accounting* and the *total-cost assessment*. Total-cost accounting is applied in management accounting to represent the allocation of all direct and indirect costs to specific products, the lives of products, or to operations as we have considered in this volume. It should be thought of as a long-term, comprehensive analysis of the entire range of costs and savings associated with the investment. Life-cycle cost assessment represents a methodical process of evaluating the life-cycle costs of a product, product line, process, system, or facility - starting with raw-material acquisition such as the chemical additives used in water conditioning, and going all the way to disposal of the sludge - by identifying the environmental consequences and assigning monetary value. A more detailed discussion of this subject is beyond our scope, but you will find some good references to refer to below.
The references provided below are organized by general subject category. I have looked at all of these and relied on quite a few in y own consulting practice. The last section provides you with some challenging exercises that you can work through with your colleagues.

RECOMMENDED RESOURCES FOR THE READER

The following are good references to obtain information on sludge treatment technologies and applications:

1. Brunner, Calvin. *Design of Sewage Sludge Incineration Systems*. Noyes Data Corporation: 1980.

2. Michael Ray Overcash and Dhiraj Ray. *Design of Land Treatment Systems For Industrial Wastes*. Michigan : Ann ArborScience , 1979.
3. P. Aarne Vesilind. *Treatment and Disposal Of Wastewater Sludges*. Michigan : Ann Arbor Science , 1979.
4. Stanley E. Manahan. *Enviromental Chemistry*. Florida : CRC Press, 1994.
5. JFWEF and ASCE. *Design of Municipal Wastewater Treatment Plants, Volume II*. Book Press, Inc.: Brattleboro, Vermont, 1991.
6. Davis, M. L. and Cornwell, D. A. *Introduction to Environmental Engineering*. McGraw-Hill, Inc.: New York, 1991.
7. Craig Cogger. *Recycling Municipal Wastewater Sludge in Washington.*Washington State University, November 1991.
8. DEC Devision of Solid Waste.*Municiple Sewage Sludge Management Practices in New York State*, April 1989.
9. Chaney and J.A. Ryan.*The Future of Residuals Management After 1991*.AWWA/WPCF Joint Residuals Management Conference, Water Pollution Control Federation, Arlington, Va.,August 1991.
10. U.S. Environmental Protection Agency, 1993.40 CFR Parts 257,403 and 503. Standards for use or disposal of sewage sludge. page 3, Federal Registry 58.9248-9415.
11. Cheremisinoff, N. P. and P. N. Cheremisinoff, *Water Treatment and Waste Recovery: Advanced Technologies and Application*, Prentice hall Publishers, New Jersey, 1993.

The following are good references to obtain pollution prevention information on, many of which cover water management and treatment practices. With the exception of reference 12, they can all be obtained through the U.S. EPA at minimal to no cost:

12. Cheremisinoff, Nicholas P. and Avrom Bendavid-Val, *Green Profits: The Manager's Handbook for ISO 14001 and Pollution Prevention*, Butterworth-Heinemann and Pollution Engineering Magazine, MA, 2001,
13. ERIC: DB54 Cleaning Up Polluted Runoff with the Clean Water State Revolving Fund, March 1998 832/F-98-001 NSCEP:
14. Enforcement Requirements: Case Studies [Fact Sheet] 832/F-93-007 NSCEP: 832/F-93-007.
15. Environmental Pollution Control Alternatives: Centralized Waste Treatment Alternatives for the Electroplating Industry, June 1981 625/5-81-017.
16. Environmental Pollution Control Alternatives: Municipal Wastewater, 1976 625/5-76-012 ERIC: W437; NTIS: PB95-156709.
17. Environmental Pollution Control Alternatives: Municipal Wastewater, November 1979 625/5-79-012 ERIC: W438; NTIS: PB95-156691.
18. Environmental Pollution Control Alternatives: Sludge Handling, Dewatering, and Disposal Alternatives for the Metal Finishing Industry, October 1982

625/5-82-018 NSCEP: 832/F-93-007; ERIC: W439; NTIS: PB95-157004.
19. Facility Pollution Prevention Guide, May 1992 600/R-92-088 NSCEP: 600/R-92-088; ERIC: W600; NTIS: PB92-213206.
20. Guides to Pollution Prevention: Non-Agricultural Pesticide, July 1993 625/R-93-009 NSCEP: 625/R-93-009; ERIC: W316; NTIS: PB94-114634.
21. Guides to Pollution Prevention: The Commercial Printing Industry, August 1980 625/7-90-008 NSCEP: 625/7-90-008; ERIC: WA06; NTIS: PB91-110023.
22. Guides to Pollution Prevention: The Fabricated Metal Products Industry, July 1990 625/7-90-006 NSCEP: 625/7-90-006; ERIC: WA07; NTIS: PB91-110015.
23. Guides to Pollution Prevention: Wood Preserving Industry, November 1993 625/R-94-014 ERIC: WA08; NTIS: PB94-136298.
24. Pollution Prevention Information Exchange System (PIES): User Guide Version 2.1, November 1992 600/R-92-213 NSCEP: 600/R-92-213; ERIC: W390.
25. Pollution Prevention Opportunity Checklists: Case Studies, September 1993 832/F-93-006 NSCEP: 832/F-93-006; ERIC: W543.
26. Waste Minimization Opportunity Assessment Manual, July 1988 625/7-88-003 ERIC: W423; NTIS: PB92-216985.
27. Water-Related GISs (Geographic Information Systems) Along the United States-Mexico Border, July 1993 832/B-93-004 NSCEP: 832/B-93-004; ERIC: W358; NTIS: PB94-114857.

QUESTIONS FOR THINKING AND DISCUSSING

1. We have an emulsion of oil in water that we need to separate. The oil droplets have a mean diameter of 10^{-4} m, and the specific gravity 0f the oil is 0.91. Applying a sedimentation centrifuge to effect the separation at a spedd of 5,000 rpm, and assuming that the distance of a droplet to the axis of rotation is 0.1 m, determine the droplet's radial settling velocity.

2. Determine the settling velocity of a particle ($d = 4 \times 10^{-4}$ m and $\rho_p = 900$ kg/m^3) through water in a sedimentation centrifuge operating at 4,000 rpm. The particle velocity is a function of distance from the axis of rotation, as shown by the following data:

Distance, m	Settling Velocity, m/sec	Reynolds Number
0.01	0.0155	0.62
0.03	0.0465	1.90

Distance, m	Settling Velocity, m/sec	Reynolds Number
0.05	0.070	2.8
0.10	0.1125	4.5
0.20	0.185	7.4
0.30	0.245	9.8
1.00	0.437	17.5

3. A solid-bowl centrifuge has the following dimensions: R_2 = 0.30 m, R_3 = 0.32 m, ℓ = 0.30 m. It is designd to operate at 5,000 rpm, separating particles from a suspension where the particle specific gravity is 7.8. Determine the required horsepower needed to set the centrifuge into operation.

4. A hydroclone will be used to separate out grit from cooling water that is recycled to plant process heat exchangers. The unit's diameter, D_c, is 32 inches. The waverage temperature of the suspension is 88° F and the specific gravity of the solids is 2.1. The volumetric flowrate of the susepnsion is 300 gpm and the solids concentration of the influent suspensnion is 7.8 % (weight basis). The average particle size is 300 μm. (a) Determine the overall separation efficiency of the hydroclone; (b) Determine the minium size horsepower requirements for the pump (*you will need to make some assumptions for head*); (c) If the process requirements demand that the return water only contain 1 weight % solids, will additional units (i.e., multiclones) be needed? If so, size these additional units.

5. For the above problem, develop a design basis for a settling chamber as an alternative.

6. Take the results for questions 4 and 5 and do a comparative cost analysis. First go the the Web and find suitable equipment suppliers that will provide the equipment in the size ranges you have calculated. Obtain some vendor quotes (rough ones will do). Then perform the fowllowing analysis: (a) What are the comparative costs between the two oprions for energy use?; (b) What are the comparative costs between the two options in terms of maintanance and labor costs?; (c) Can you combine both equipment options into a single process, and if so, can you justify this and how? Assume in the above that the reduction in solids concentration must meet the 1 % weirht criteria described in question 4.

7. A clarifying settler has the following characteristics: 750 mm bowl diameter; 600 mm bowl depth; 95 mm liquid layer thickness. The specific gravity of the susepnsion is 1.5, and that of the solids is 1.9. The particle cut size is 60 μm and the viscosity of the susepension is 15 cP. (a) Determine the capacity of the centrifuge in untis of gpm. (b) Determine the horesepoer requirments needed.

8. We wish to separate titanium dioxide particles from a water suspension. The method chosen is centrifugation. The unit is a continuous solid-bowl type with a bowl diamter of 400 mm, a length to width ratio of 3.0, and the unit operates at 2,000 rpm. The feed contains 18 % (weight basis) solids and is fed to the unit at 2,500 Liters/hr at a temperature of 95° F. The average particle size is 65 μm. (a) Determine the amount of solids recovered per hour; (b) Determine the solids concentration in the centrate; (c) Determine the horsepower requirments for the centrifuge; (d) Size a graviy settler to remove an additional 15 % of the solids.

9. The investment for a sludge dewatering and pasteurization process for a small municipal treatment facility is 4.5 million dollars. It is estimated that the operation can generate about 18 tons per year of a sludge suitable as a composting material that will support a local market. This offteake would represent about 10 % of the total market demand and resale values for the treated sludge range from $6.35 to $ 6.80 per ton. A market survey suggests that consumption will grow at a modest rate of 3.5 % per year over a five year projection. Labor and energy costs for the operation are estimated to be $ 165,000 per year. Determine whether this investment is practical and worthwhile.The current practice at the facility is to haul untreated wates off-site to a municipal landfill. Costs for transportation and disposal are typically 28 dollars per ton, and there is concern that these costs could escalate by 15 % over the next 5 years. In performing the analysis, consider several project investment paramters (e.g., payback period, ROI, B/C ratio, others).

10. For question 9, the municipal landfill has had public relations problems with the community. There has been concern over both odor issues and possible groundwater contamination. Taking these concerns into consideration, can you develop addional arguments that make the investment more finacially attractive?

11. We have an aeration basin that currently operates at 3.2 mg/Liter DO. Compare this operation where the DO concentration is 1.3 mg./Liter. The temperature of the basin is 18.0° C and 200 kW of aeration power is used. The average electricity cost is 8 cents per kWhr. Determine: (a) the current average electricity consumption for aeration; (b) the daily electricity costs for the operation; (c) what you could save on a daily basis and per year by lowering the DO concentration; (d) determine the yearly savings on a percentage basis.

12. When dealing with water treatment applications you cannot avoid pipe flow calculations. We have a pipeline in which the throughput capacity of 500 Liter/sec. The flow is split into two pipelines and the inside diamter of the pipe is 350 mm. The length of the pipeline is 55 m. The entry loss is 0.70 and the exit loss is 1.00. There are two 45° bends and two 90° bends in the lines. (a) Determine the flow per pipe; (b) determine the line velocity; (c) determine the resulting hydraulic loss in meters.

13. Holly's (Holly, Michigan) original Wastewater Treatment Plant, WWTP, was built a trickling filter plant built in 1957 and had a design flow of 500,000 gallons per day. As the community grew it became necessary to construct a new plant. The

majority of the plant was constructed in 1980 at a cost of approximately 6.3 million dollars. Seventy-five percent was funded by the Federal Government and five percent was funded by the State Government. The type of treatment used in the Holly Wastewater Treatment Plant is advanced treatment. Since the WWTP has a large impact on the receiving stream, the Shiawassee River, effluent and discharge limits are very stringent. For the past 10 years averages for BOD are 3.9 mg/Liter and suspended solids 4.2 mg/Liter. Plant was designed for an average flow of 1.5 million gallons per day (MGD). Presently average flow is 1.0 MGD. Sewage enters the plant via two thirty inch sanitary sewers. preliminary treatment consist of bar screen, aerated grit removal, and two 60,000 gallon primary clarifiers. The heart of the treatment system are rotating biological contactors. The RBC System consists of 3 rows of discs, with 4 discs per row with a total surface area of 1,500,000 ft^2. Ferrous chloride is used at the head end of the treatment process (aerated grit tank) to aid in the removal of suspended solids and phosphorus. After the RBC's , wastewater enters into two final clarifiers. The sludge that is pumped out is much lighter in solids content (<1%). The sludge from the secondary clarifiers is then pumped back to the head end of the primary clarifiers. This helps to get the full use of the primary chemical added and thickens the sludge for better treatment and storage capacity in the digester. The sewage from the secondary clarifiers then flows into the filter feed wet well. Secondary effluent is pumped through four mixed media pressure sand filters. Filtration of secondary effluent is considered advanced or tertiary treatment and makes it possible to achieve excellent water quality. Effluent quality from the pressure filters averages below 2 mg/Liter for BOD and suspended solids during the summer months. Sludge stabilization process consists of an anaerobic digester. The digester provides anaerobic fermentation of the sludge in the enclosed tank. When operating, the destruction of the organisms produces methane gas. The gas is then used to heat the contents of the digester and other plant buildings. When the sludge is digested, it is transferred over the sludge storage tank. This simple tank holds 320,000 gallons of digested sludge and uses gravity to thicken the sludge. When the heavier sludge settles to the tank bottom, the remaining water or supernatant may be drawn off through a series of valves to the equalization basin. The treated biosolids is 8 - 9.5% solids is finally removed and injected 8 inches into farmland as a fertilizer supplement. Develop a detailed process flow sheet for the WWTP. Then develop a cost breakdown for each major component. Next, try to develop a qualitative energy audit, listing those operations in order of their highest energy consumption first. You can obtain more information on this plant's design by going to the following Web site: http://www.w-ww.com..

14. The following information has been extracted from the design basis for an actual wastewater treatment plant:

Loadings

Population, 1,500

Flow, mgd

Average Annual, 0.30

Maximum Day, 0.30

Peak Hour, 0.91

BOD and SS, PPD

Average Annual, 424

Maximum Month, 694

Maximum Day, 868

Headworks

Bar Screen

Type, MANUALLY CLEANED

Number, 1

Size, inches, 16

Bar Spacing, inches, 1

Comminutor

Type, IN-LINE

Number, 1

Size, inches, 12

Motor, hp, 6

Flow Measurement

Type, PARSHALL FLUME
Number, 1
Throat Width, inches. 6

Aeration
Number of basins, 1
Volume, mgal, 0.67

Theoretical hydraulic residence in time hours:
Average Annual, 53
Maximum Month, 32

Design Waste Sludge Production, ppd, 420
Design Mixed Liquor Concentration, mg/l, 3,000
Design Sludge Mean Cell Residence Time, Days, 40

Design Temperature, °C
Low Month, 10
Average Month, 15

Aerators

Type, HIGH SPEED MECHANICAL FLOATING SURFACE
Number, 2
Size, hp (ea), 25
Design Maximum Oxygen Transfer, ppd, 1,500

Secondary Sedimentation

Number of tanks, 1
Diameter, ft, 35

Overflow rate, gpd/sf
Average Annual, 312

Peak Hour, 946

Solids Loading Rate, ppd/sf
Average Annual (50% return), 12
Peak Hour (100% return), 47

Return Sludge Pumps
Number, 2
Design Maximum Capacity, gpm, 630

Infiltration Basins
Number, 8
Surface Area, 0.44

Net Hydraulic Loading rate, in/wk
Average Annual, 21.9
Maximum Month, 36.6

Sludge Disposal System
Type, LIQUID SLUDGE LAND APPLICATION
Land Area, acres, 13
Sludge Loading, Tons per acre per year, 5.3

The wastewater enters the plant through the headworks where it passes through a bar screen, comminutor and Parshal flume. Following the headworks, the wastewater enters the aeration basin where floating surface aerator aerate and mix the sewage. Biological growth in an aeration basin is carried with the effluent to secondary sedimentation. Here the growth settles to the bottom of the tank. It is raked to the center of the tank by rotating arms and flows to the return activated sludge pump station. The clarifier effluent flows by gravity to the infiltration basins where it seeps into the ground. The sludge from the return activated sludge pump station is pumped back into the aeration basin on the anoxic zone side. The return sludge seeds the incoming waste water and increases the BOD removal capacity. Excess sludge is removed from the clarifier of the aeration basin by the waste activated sludge pump. The excess sludge is disposed by spraying on an adjacent forest land with a permanent spray irrigation system. Based on the above information, do the following exercises:

(a) Develop a list of any terms that you are not familiar with and not covered in detail in this volume. Obtain the definitions and an understanding of those terms as they apply to this design case.

(b) Develop a detailed process flowsheet for the plant. Show flow rates and mass flows for major process streams on your system diagram.

(c) Develop an inventory list of the chemicals needed for water conditioning in this plant.

(d) Develop an estimate for the horsepower requirements needed for the return sludge pumps.

(e) Develop a plot plan layout for the plant based on the information given above. Roughly determine the amount of plot area needed for this plant.

(f) Work with your design team to develop a estimate for the cost of installing such a plant. Include in your estimates engineering, site preparation, start-up, and training costs.

(g) Based on the cost estimate you develop, discuss with your team options for financing such an investment.

15. A large settling lagoon (approximately 0.5 ac in area and 75 ft deep) is used to separate a solid waste product whose particle density is roughly 1,700 kg/m^3. The density of the dilute slurry is roughly 1,300 kg/m^3 and its viscosity is 3.2 cp. The particles are spherical in nature with a 50 wt% size of 210 μm. If the lagoon is filled to 90% capacity with a solids concentration of 40%, how long will it take to achieve an 85% separation of sludge from the slurry? First analyze this problem by ignoring any evaporation losses. Next, analyze the problem considering evaporation losses. Assume that pan evaporation data from a local weather station show a yearly average of 53 in./yr. (*Note - a standard evaporation pan is about 2 ft in diameter and 36 in. deep*).

16. The lagoon described in the above question operates in the summer months at a mean temperature of 65° F. The mean ambient air temperature between the

months of June and September is about 75° F. Assuming an average wind velocity of 5 m/s, determine the following: (a) estimated losses due to evaporation; and (b) the concentration of the dilute slurry at then end of four months.

GLOSSARY

A

Abiotic factors: Non living; moisture, soil, nutrients, fire, wind, temperature, climate.

Absolute filtration rating (largest particle passed): The diameter of the largest hard spherical particle that will pass through a filter under specified test conditions. This is an indication of the largest opening in the filter cloth.

Absorption: The taking in or soaking up of one substance into the body of another by molecular or chemical action (as tree roots absorb dissolved nutrients in the soil).

Absorption Field: A system of properly sized and constructed narrow trenches partially filled with a bed of washed gravel or crushed stone into which perforated or open joint pipe is placed. The discharge from the septic tank is distributed through these pipes into trenches and surrounding soil. While seepage pits normally require less land area to install, they should be used only where absorption fields are not suitable and well-water supplies are not endangered.

Acetogenic bacterium: Prokaryotic organism that uses carbonate as a terminal electron acceptor and produces acetic acid as a waste product.

Acetylene-block assay: Estimates denitrification by determining release of nitrous oxide (N_2O) from acetylene-treated soil.

Acetylene-reduction assay: Estimates nitrogenase activity by measuring the rate of acetylene reduced to ethylene.

***N*-Acetylglucosamine** and ***N*-Acetylmuramic acid**: Sugar derivatives in the peptidoglycan layer of bacterial cell walls.

Acid: A substance that dissolves in water with the formation of hydrogen ions, contains hydrogen which may be replaced by metals to form salt, and/or is corrosive.

Acidity: The capacity of water or wastewater to neutralize bases. Acidity is expressed in milligrams per liter of equivalent calcium carbonate. Acidity is not the same as Ph.

Acidophile: Organism that grows best under acid conditions (down to a pH of 1).

Acid soil: Soil with a pH value < 6.6.

Actinomycete: Nontaxonomic term applied to a group of high G + C base composition, Gram-positive bacteria that have a superficial resemblance to fungi. Includes many but not all organisms belonging to the order Actinomycetales.

Activated sludge: Sludge particles produced in raw or settled wastewater (primary effluent) by the growth of organisms (including zoogleal bacteria) in aeration tanks in the presence of dissolved oxygen. The term "activated" comes from the fact that the particles are teeming with fungi, bacteria, and protozoa. Activated sludge is different from primary sludge in that the sludge particles contain many living organisms which can feed on the incoming wastewater.

Activation energy: Amount of energy required to bring all molecules in one mole of a substance to their reactive state at a given temperature.

Active site: Region of an enzyme where substrates bind.

Adenosine triphosphate (ATP): Common energy-donating molecule in biochemical reactions. Also an important compound in transfer of phosphate groups.

Adsorption: The gathering of a gas, liquid, or dissolved substance on the surface or interface zone of another substance.

ADP: Adenosine diphosphate. See ATP.

Aeration: The process of adding air to water. In wastewater treatment, air is added to freshen wastewater and to keep solids in suspension.

Aeration tank: The tank where raw or settled wastewater is mixed with return sludge and aerated. This is the same as an aeration bay, aerator, or reactor.

Aerobe: An organism that requires free oxygen for growth.

Aerobic: (i) Having molecular oxygen as a part of the environment. (ii) Growing only in the presence of molecular oxygen, as in aerobic organisms. (iii) Occurring only in the presence of molecular oxygen, as in certain chemical or biochemical processes such as aerobic respiration.

Aerotolerant anaerobes: Microbes that grow under both aerobic and anaerobic conditions, but do not shift from one mode of metabolism to another as conditions change. They obtain energy exclusively by fermentation.

Air flow/air permeability: Measure of the amount of air that flows through a filter - a variable of the degree of contamination, differential pressure, total porosity, and filter area. Expressed in either cubic feet/minute/square foot or liters/minute/square centimeter at a given pressure.

Agar: Complex polysaccharide derived from certain marine algae that is a gelling agent for solid or semisolid microbiological media. Agar consists of about 70% agarose and 30% agaropectin. Agar can be melted at temperature above 100° C; gelling temperature is 40-50° C.

Agarose: Nonsulfated linear polymer consisting of alternating residues of D-galactose and 3,6-anhydro-L-galactose. Agarose is extracted from seaweed, and agarose gels are often used as the resolving medium in electrophoresis.

Akinete: Thick-walled resting cell of cyanobacteria and algae.

Alkaline substance: Chemical compounds in which the basic hydroxide (OH-) ion is united with a metallic ion, such as sodium hydroxide (NaOH) or potassium hydroxide (KOH). These substances impart alkalinity to water and are employed for neutralization of acids. Lime is the most commonly used alkaline material in wastewater treatment.

Alga (plural, algae): Phototrophic eukaryotic microorganism. Algae could be unicellular or multicellular. Blue-green algae are not true algae; they belong to a group of bacteria called cyanobacteria.

Aliphatic: Organic compound in which the main carbon structure is a straight chain.

Alkaline soil: Soil having a pH value >7.3.

Alkalophile: Organism that grows best under alkaline conditions (up to a pH of 10.5).

Alkane: Straight chain or branched organic structure that lacks double bonds.

Alkene: Straight chain or branched organic structure that contains at least one double bond.

Allochthonous flora: Organisms that are not indigenous to the soil but that enter soil by precipitation, diseased tissues, manure, and sewage. They may persist for some time but do not contribute in a significant way to ecologically significant transformations or interactions.

Allosteric site: Site on the enzyme other than the active site to which a nonsubstate compound binds. This may result in a conformational change at the active site so that the normal substrate cannot bind to it.

Alum: Astringent crystalline double sulfate of an alkali. $K_2SO_4AL_2\ (SO_4)_3\ 24H_2O$. Used in the processing of pickles and as a flocking agent. Excess aluminum in the environment can be hazardous.

Amensalism (antagonism): Production of a substance by one organism that is inhibitory to one or more other organisms. The terms antibiosis and allelopathy also describe cases of chemical inhibition.

Ambient temperature: Temperature of the surroundings.

Amino group: An --NH_2 group attached to a carbon skeleton as in the amines and amino acids.

Ammonia oxidation: Test drawn during manufacturing process to evaluate the ammonia oxidation rate for the nitrifiers.

Ammonification: Liberation of ammonium (ammonia) from organic nitrogenous compounds by the action of microorganisms.

Amoeba (plural, amoebae): Protozoa that can alter their cell shape, usually by the extrusion of one or more pseudopodia.

Anabolism: Metabolic processes involved in the synthesis of cell constituents from simpler molecules. An anabolic process usually requires energy.

Anaerobe: An organism that lives and reproduces in the absence of dissolved oxygen, instead deriving oxygen from the breakdown of complex substances.

Anaerobic: (i) Absence of molecular oxygen. (ii) Growing in the absence of molecular oxygen, such as anaerobic bacteria. (iii) Occurring in the absence of molecular oxygen, as a biochemical process.

Anaerobic respiration: Metabolic process whereby electrons are transferred from an organic, or in some cases, inorganic compounds to an inorganic acceptor molecule other than oxygen. The most common acceptors are nitrate, sulfate, and carbonate.

Anamorph: Asexual stage of fungal reproduction in which cells are formed by the process of mitosis.

Anhydrous: Very dry. No water or dampness is present.

Anion: A negatively charged ion in an electrolyte solution, attracted to the anode under the influence of a difference in electrical potential. Chloride is an anion.

Anion exchange capacity: Sum total of exchangeable anions that a soil can adsorb. Expressed as centimoles of negative charge per kilogram of soil.

Anoxic: Literally "without oxygen." An adjective describing a microbial habitat devoid of oxygen.

Anoxygenic photosynthesis: Type of photosynthesis in green and purple bacteria in which oxygen is not produced.

Antagonist: Biological agent that reduces the number or disease-producing activities of a pathogen.

Antheridium: Male gametangium found in the phylum Oomycota (Kingdom Stramenopila) and phylum Ascomycota (Kingdom Fungi).

Anthropogenic: Derived from human activities.

Antibiosis: Inhibition or lysis of an organism mediated by metabolic products of the antagonist; these products include lytic agents, enzymes, volatile compounds, and other toxic substances.

Antibiotic: Organic substance produced by one species of organism that in low concentrations will kill or inhibit growth of certain other organisms.

Antibody: Protein that is produced by animals in response to the presence of an antigen and that can combine specifically with that antigen.

Antigen: Substance that can incite the production of a specific antibody and that can combine with that antibody.

Antiseptic: Agent that kills or inhibits microbial growth but is not harmful to human tissue.

Antistatic: Material that minimizes static charge generation, provides "controlled" static charge dissipation, or both.

API separator: A facility developed by the Committee on Disposal or Refinery Wastes of the American Petroleum Institute for separation of oil from wastewater in a gravity differential and equipped with means for recovering the separated oil and removing sludge.

Aromatic: Organic compounds which contain a benzene ring, or a ring with similar chemical characteristics.
Arthropod: Invertebrate with jointed body and limbs (includes insects, arachnids, and crustaceans).

Ascoma (plural, ascomata): Fungal fruiting body that contains ascospores; also termed an ascocarp.

Ascospore: Spores resulting from karyogamy and meiosis that are formed within an ascus. Sexual spore of the Ascomycota.

Ascus (plural, asci): Saclike cell of the sexual state formed by fungi in the phylum Ascomycota containing ascospores.

Aseptic: Free from living germs of disease, fermentation or putrefaction.

Aseptic technique: Manipulating sterile instruments or culture media in such a way as to maintain sterility.

Assimilate: To take in, similar to eating food.

Assimilatory nitrate reduction: Conversion of nitrate to reduced forms of nitrogen, generally ammonium, for the synthesis of amino acids and proteins.

Associative dinitrogen fixation: Close interaction between a free-living diazotrophic organism and a higher plant that results in an enhanced rate of dinitrogen fixation.
Associative symbiosis: Close but relatively casual interaction between two dissimilar organisms or biological systems. The association may be mutually beneficial but is not required for accomplishment of a particular function.

Attached growth processes: Wastewater treatment processes in which the microorganisms and bacteria treating the wastes are attached to the media in the reactor. The wastes being treated flow over the media. Trickling filters, bio-towers, and RBCs are attached growth reactors. These reactors can be used for removal of BOD, nitrification, and denitrification.
Attenuation: Reduction of the signal power of field strength as a function of distance through a material. Also refers to shielding effectiveness.
ATP: Adenosine triphosphate. Chemical energy generated by substrate oxidations is conserved by formation of high-energy compounds such as adenosine diphosphate (ADP) and adenosine triphosphate (ATP) or compounds containing the thioester bond.
Autoclave: Vessel for heating materials under high steam pressure. Used for sterilization and other applications.
Autolysis: Spontaneous lysis.
Autoradiography: Detecting radioactivity in a sample, such as a cell or gel, by placing it in contact with a photographic film.

Autotroph: Organism which uses carbon dioxide as the sole carbon source.

Autotrophic nitrification: Oxidation of ammonium to nitrate through the combined action of two chemoautotrophic organisms, one forming nitrite from ammonium and the other oxidizing nitrite to nitrate.

Autotrophy: A unique form of metabolism found only in bacteria. Inorganic compounds (e.g., NH3, NO2-, S2, and Fe2+) are oxidized directly (without using sunlight) to yield energy. This metabolic mode also requires energy for CO2 reduction, like photosynthesis, but no lipid-mediated processes are involved. This metabolic mode has also been called chemotrophy, chemoautotrophy, or chemolithotrophy.

AWT: Advanced Waste Treatment - any process of water renovation that upgrades treated wastewater to meet reuse requirements.

Axenic: Literally "without strangers." A system in which all biological populations are defined, such as a pure culture.

B

Bacillus- Bacterium with an elongated, rod shape.

Bacteria: Living organisms, microscopic in size, which usually consist of a single cell. Most bacteria use organic matter for their food and produce waste products as a result of their life processes.

Bacteriochlorophyll- Light-absorbing pigment found in green sulfur and purple sulfur bacteria.

Bacteriocin- Agent produced by certain bacteria that inhibits or kills closely related isolates and species.

Bacteriophage- Virus that infects bacteria, often with destruction or lysis of the host cell.

Bacterial Photosynthesis: A light-dependent, anaerobic mode of metabolism. Carbon dioxide is reduced to glucose, which is used for both biosynthesis and energy production. Depending on the hydrogen source used to reduce CO_2, both photolithotrophic and photoorganotrophic reactions exist in bacteria.

Bacteroid- Altered form of cells of certain bacteria. Refers particularly to the swollen, irregular vacuolated cells of rhizobia in nodules of legumes.

Base: A substance which dissociates (separates) in aqueous solution to yield hydroxyl ions, or one containing hydroxyl ions (OH-) which reacts with an acid to form a salt or which may react with metal to form a precipitate.

Base composition- Proportion of the total bases consisting of guanine plus cytosine or thymine plus adenine base pairs. Usually expressed as a guanine + cytosine (G + C) value, e.g. 60% G+C.

Basidioma (plural, basidiomata)- Fruiting body that produces basidia; also termed a basidiocarp.

Basidiospore- Spore resulting from karyogamy and meiosis that are formed on a basidium that usually is formed on a basidium. Sexual spore of the Basidiomycota.

Basidium (plural, basidia)- Clublike cell of the sexual state formed by fungi in the phylum Basidiomycota.

Batch process: A treatment process in which a tank or reactor is filled, the wastewater (or solution) is treated or a chemical solution is prepared and the tank is emptied. The tank may then be filled and the process repeated. Batch processes are also used to cleanse, stabilize or condition chemical solutions for use in industrial manufacturing and treatment processes.

Bench scale analysis: Also known as: "bench test". A method of studying different ways of treating wastewater and solids on a small scale in a laboratory. Alken-Murray offers several such test kits including: Alken Clear-Flo® Bench Test 1 and Alken PCB Bench Test.

Benzene: An aromatic hydrocarbon which is a colorless, volatile, flammable liquid. Benzene is obtained chiefly from coal tar and is used as a solvent for resins and fats in dye manufacture.

Binary fission: During binary fission, a single cell divides transversely to form two new cells called daughter cells. Both daughter cells contain an exact copy of th geneticinformation contained in the parent cell.

Biocatalysis: Chemical reactions mediated by biological systems (microbial communities, whole organisms or cells, cell-free extracts, or purified enzymes aka catalytic proteins).

Binary fission- Division of one cell into two cells by the formation of a septum. It is the most common form of cell division in bacteria.

Binomial nomenclature- System of having two names, genus and specific epithet, for each organism.

Bioaccumulation-Accumulation of a chemical substance in living tissue.

Biochemical oxygen demand (BOD)- Amount of dissolved oxygen consumed in five days by biological processes breaking down organic matter.

Biodegradable- Substance capable of being decomposed by biological processes.

Biofilm: A slime layer which naturally develops when bacteria attach to an inert support that is made of a material such as stone, metal, or wood. There are also non-filamentous bacteria that will produce an extracellular polysaccharide that acts as a natural glue to immobilize the cells. In nature, nonfilament-forming microorganisms will stick to the biofilm surface, locating within an area of the biofilm that provides an optimal growth environment (i.e., pH, dissolved oxygen, nutrients). Since nutrients tend to concentrate on solid surfaces, a microorganism saves energy through cell adhesion to a solid surface rather than by growing unattached and obtaining nutrients randomly from the medium. *Pseudomonas* and *Nitrosomonas* strains are especially well known for their ability to form a strong biofilm.

Bioflocculation: The clumping together of fine, dispersed organic particles by the action of certain bacteria and algae.

Biogeochemistry- Study of microbially mediated chemical transformations of geochemical interest, such as nitrogen or sulfur cycling.
Biomagnification- Increase in the concentration of a chemical substance as it is progresses to higher trophic levels of a food chain.

Biomass: A mass or clump of living organisms feeding on the wastes in wastewater, dead organisms and other debris.
Bioremediation- Use of microorganisms to remove or detoxify toxic or unwanted chemicals from an environment.
Biosolid- The resides of wastewater treatment. Formerly called sewage sludge.
Biosphere- Zone incorporating all forms of life on earth. The biosphere extends from deep in sediment below the ocean to several thousand meters elevation in high mountains.
Biotrophic- Nutritional relationship between two organisms in which one or both must associate with the other to obtain nutrients and grow.

Biostimulation: Any process that increases the rates of biological degradation, usually by the addition of nutrients,oxygen, or other electron donors and acceptors so as to increase the number of indigenous microorganisms available for degradation of contaminants.
Biosynthesis- Production of needed cellular constituents from other, usually simpler, molecules.
Biotechnology- Use of living organisms to carry out defined physiochemical processes having industrial or other practical application.

Biotic potential: All the factors that contribute to a species increasing its number. Reproduction, migration, adaptation etc.
BOD: Biochemical Oxygen Demand - the rate at which microorganisms use the oxygen in water or wastewater while stabilizing decomposable organic matter under aerobic conditions. In decomposition, organic matter serves as food for the bacteria and energy results from this oxidation.
BOD test: A procedure that measures the rate of oxygen use under controlled conditions of time and temperature. Standard test conditions include dark incubation at 20 C for a specified time (usually 5 days).
Bolting cloth (silk): Screens woven of twisted multifilament natural silk.
Bolting grade (wire cloth): Weaves that are uniformly woven of stainless steel to provide high strength and the largest possible pore openings.

Bio-Tower: An attached culture system. A tower filled with a media similar to rachet or plastic rings in which air and water are forced up a counterflow movement in the tower.
Blinding: The clogging of the filtering medium of a microscreen or a vacuum filter when the holes or spaces in the media become sealed off due to a buildup of grease or the material being filtered.
Brown rot fungus: Fungus that attacks cellulose and hemicellulose in wood, leaving dark-colored lignin and phenolic materials behind.

Bubble point test: A test to determine the maximum pore size opening of a filter.
Buffer: A solution or liquid whose chemical makeup neutralizes acids or bases without a great change in pH.
Bulk density, soil - Mass of dry soil per unit bulk volume (combined volume of soil solids and pore space).
Bulked yarn: A yarn that has been geometrically changed to give it the appearance of having greater volume than a conventional yarn of the same linear density.
Bulking sludge: Clouds of billowing sludge that occur throughout secondary clarifiers and sludge thickeners when sludge becomes too light and will not settle properly. In the activated sludge process, bulking is usually caused by filamentous bacteria.

C

Cake: The solids discharged from a dewatering apparatus.

Calendering: A process by which fabric or wire is passed through a pair of heavy rolls to reduce thickness, to flatten the intersections of the threads/wires and to control air permeability. Rolls are heated when calendering synthetic materials.
Carbonized threads: Nylon or polyester therads that have been treated to include varrying degrees of carbon.

Cation exchange capacity: The ability of a soil or other solid to exchange cations (positive ions such as calcium) with a liquid.
Cess Pools: This system is similar to a septic tank. in performance. Sewage water usually seeps through the open bottom and portholes in the sides of the walls. These can also clog up with overuse and the introduction of detergents and other material which slow up the bacterial action.
CFU: Viable micro-organisms (bacteria, yeasts & mould) capable of growth under the prescribed conditions (medium, atmosphere, time and temperature) develop into visible colonies (colony forming units) which are counted. The term colony forming unit (CFU) is used because a colony may result from a single micro-organism or from a clump / cluster of micro-organisms.
Chemoautotroph: An organism that obtains its energy from the oxidation of chemical compounds and uses only organic compounds as a source of carbon. Example: nitrifiers.
Chemotroph: An organism that obtains its energy from the oxidation of chemical compounds.

Chemical precipitation: Precipitation induced by addition of chemicals; the process of softening water by the addition of lime and soda ash as the precipitants.
Chloramines: Compounds formed by the reaction of hypochlorous acid (or aqueous chlorine) with ammonia.
Chlorination: The application of chlorine to water or wastewater, generally for the

purpose of disinfection, but frequently for accomplishing other biological or chemical results.

Clarification: A process in which suspended material is removed from a wastewater. This may be accomplished by sedimentation, with or without chemicals, or filtration.

Clarifier: Settling tank, sedimentation basin. A tank or basin in which wastewater is held for a period of time, during which the heavier solids settle to the bottom and the lighter material will float to the water surface.

Coagulants: Chemicals which cause very fine particles to clump (floc) together into larger particles. This makes it easier to separate the solids from the water by settling, skimming, draining, or filtering.

Coliform bacteria: Non-pathogenic microbes found in fecal matter that indicate the presence of water pollution; are thereby a guide to the suitability for potable use.

Colloids: Very small, finely divided solids (particles that do not dissolve) that remain dispersed in a liquid for a long time due to their small size and electrical charge.

Combined available chlorine: The concentration of chlorine which is combined with ammonia (NH3) as chloramine or as other chloro derivatives, yet is still available to oxidize organic matter.

Combined sewer: A sewer designed to carry both sanitary wastewaters and storm or surface-water runoff.

Ciliates: A class of protozoans distinguished by short hairs on all or part of their bodies.

COD: Chemical oxygen demand - the amount of oxygen in mg/l required to oxidize both organic and oxidizable inorganic compounds.

Commensalism: When two organisms coexist, one organism benefits, the other is not affected.

Comminution: Shredding. A mechanical treatment process which cuts large pieces of waste into smaller pieces so that they won't plug pipes or damage equipment.

Contact stabilization: Contact stabilization is a modification of the conventional activated sludge process. In contact stabilization, two aeration tanks are used. One tank is for separate reaeration of the return sludge for at least four hours before it is permitted to flow into the other aeration tank to be mixed with the primary effluent requiring treatment.

Conventional treatment: The preliminary treatment, sedimentation, flotation, trickling filter, rotating biological contactor, activated sludge and chlorination of wastewater.

Conversion: Changing from one substance to another. As food matter is changed to cell growth or to carbon dioxide.

CRT: Cell residence time - the amount of time in days that an average "bug" remains in the process. Also termed "sludge age".

D

DAF: Dissolved air flotation - one of many designs for waste treatment.
Decitex (dtex): The mass in grams of 10,000 meters of fiber or yarn. A direct yarn numbering system used to define size of fiber or yarn. The higher the number, the coarser (larger) the yarn.
Declining growth: A growth phase in which the availability of food begins to limit cell growth.
Degradation: A growth phase in which the availability of food begins to limit cell growth.
Deionized water: Water that goes through an ion exchange process in which all positive and negative ions are removed.
Denier: The mass in grams of 9000 meters of fiber or yarn. A direct numbering system used to define size of fiber or yarn. The higher the number, the coarser (larger) the yarn.

Denitrification: An anaerobic biological reduction of nitrate nitrogen to nitrogen gas, the removal of total nitrogen from a system, and/or an anaerobic process that occurs when nitrite ions are reduced to nitrogen gas and bubbles are formed as a result of this process. The bubbles attach to the biological floc in the activated sludge process and float the floc to the surface of the secondary clarifiers. This condition is often the cause of rising sludge observed in secondary clarifiers or gravity thickeners. (See Nitrification).
Depth filter: A filter medium consisting of randomly distributed particles or fibers resulting in openings with a non-uniform and tortuous path.
Detritus: Dead plant and animal matter, usually consumed by bacteria, but some remains.
Dew Point: The temperature to which air with a given quantity of water vapor must be cooled to cause condensation of the vapor in the air.
D/I unit: Deionizing unit, frequntly used to maintain water quality in aquariums. Advantages: does not waste water like the R/O unit, is designed to be hooked up to either a faucet or household piping system, the anion & cation resins can be regenerated (with another expensive unit) indefinitely, and these systems allow a larger water flow (up to 2,000 gallons a day), than an R/O system, but cost dramatically more too.

Diatomaceous earth: A fine, siliceous (made of silica) "earth" composed mainly of the skeletal remains of diatoms (single cell microscopic algae with rigid internal structure consisting mainly of silica). Tests prove that DE leaches unacceptable amounts of silicate into the water for fish health. If used as a filter substance, a silicone removing resin should be employed afterwards.

Differential pressure: The difference in pressure between two points of a system, such as between two sides of an orifice.

Diffused Air Aeration: A diffused air activated sludge plant takes air, compresses it, and then discharges the air below the water surface of the aerator through some type of air diffusion device.
Digester: A tank in which sludge is placed to allow decomposition by microorganisms. Digestion may occur under anaerobic (most common) or aerobic conditions.
Disinfection: The process designed to kill most microorganisms in wastewater, including essentially all pathogenic (disease-causing) bacteria. There are several ways to disinfect, with chlorine being the most frequently used in water and wastewater treatment plants.
Distribution box: Serves to distribute the flow from the septic tank evenly to the absorption field or seepage pits. It is important that each trench or pit receive an equal amount of flow. This prevents overloading of one part of the system.
Dissolved solids: Chemical substances either organic or inorganic that are dissolved in a waste stream and constitute the residue when a sample is evaporated to dryness.
Distributor: The rotating mechanism that distributes the wastewater evenly over the surface of a trickling filter or other process unit.
DO: Dissolved Oxygen - a measure of the oxygen dissolved in water expressed in milligrams per liter.
DOUR: Dissolved Oxygen Uptake Ratio.

Downstream side: The side of a product stream that has already passed through a given filter system; portion located after the filtration unit.
Dual chamber test method: Measures near field shielding effectiveness by indicating the signal attenuation caused by passage through test material.
Dyeing: The process of adding color to textiles in either fiber, yarn or fabric form.

E

Ecology: The study of all aspects of how organisms interact with each other and/or their environment.
Ecosystem: Groupings of various organisms interacting with each other and their environment.
E-coli: Escherichia coli - one of the non-pathogenic coliform organisms used to indicate the presence of pathogenic bacteria in water.
Effective area: The total area of the porous medium exposed to flow in a filter element.
Efficiency: The ability, expressed as a percent, of a filter to remove specified artificial contaminant at a given contaminant concentration under specified test conditions.

Effluent: Wastewater or other liquid - raw (untreated), partially or completely treated - flowing from a reservoir, basin, treatment process, or treatment plant.

E-Field (Electric field): The dominant component of a high impedance electromagnetic field produced by a near field source such as a short diapole, or the electric component of a far field plane wave. Expressed in V/m..

EGL: Energy grade line - a line that represents the elevation of energy head in feet of water flowing in a pipe, conduit, or channel.

Electrolytic process: A process that causes the decomposition of a chemical compound by the use of electricity.

Electromagnetic Interference (EMI): Electromagnetic energy that causes interference in the operation of electronic equipment. Can be conducted, coupled or radiated. Can be natural or man-made.

Electromagnetic Capability (EMC): The capability of electronic equipment of systems to be operated in the intended operational electromagnetic environment at designed levels of efficiency.

Emulsion: A liquid mixture of two or more liquid substances not normally dissolved in one another, one liquid held in suspension in the other.

Endogenous respiration: A reduced level of respiration (breathing) in which organisms break down compounds within their own cells to produce the oxygen they need.

Endotoxin: A toxin produced by bacteria. The toxin is present in the environment only after death of the bacteria.

Enteric: Of intestinal origin, especially applied to wastes or bacteria.

Enzyme: Organic substances (proteins) produced by living organisms and act as catalysts to speed up chemical changes.

Environmental resistance: All biotic and abiotic factors combining to limit explosion.

Equalizing basin: A holding basin in which variations in flow and composition of liquid are averaged. Such basins are used to provide a flow of reasonably uniform volume and composition to a treatment unit. Also called a balancing reservoir.

Estuaries: Bodies of water which are located at the lower end of a river and are subject to tidal fluctuations.

Eurythermal: Bodies of water which are located at the lower end of a river and are subject to tidal fluctuations.

Extractables: Substances that can be leached from a filter during the filtration process or under other specified conditions.

F

Facultative anaerobe: A bacterium capable of growing under aerobic conditions or anaerobic conditions in the presence of an inorganic ion ie. SO_4, NO_3.

Facultative pond: The most common type of pond in current use. The upper portion (supernatant) is aerobic, while the bottom layer is anaerobic. Algae supply most of the oxygen to the supernatant.

Faraday cage: A spherical cage made of conductive material. Static fields and discharges do not pass through it. Electromagnetic energy passing through the skin or shield is attenuated to varying degrees.

Feed: The material entering a filter processing unit for treatment.

Fermentation: A type of heterotrophic metabolism in which an organic compound rather than oxygen is the terminal electron (or hydrogen) acceptor. Less energy is generated from this incomplete form of glucose oxidation than is generated by respiration, but the process supports anaerobic growth.

Filamentous organisms: Organisms that grow in a thread or filamentous form. Common types are *Thiothrix, Actinomycetes, and Cyanobacteria* (aka blue-green algae). This is a common cause of sludge bulking in the activated sludge process. Variously known as "pond scum", "blue-green algae", or "moss", when it appears in a pond/lake, and confused with algae because it looks a lot like algae. Cyanobacteria forms a symbiotic relationship with some varieties of algae, making the combination very difficult to combat in lakes and ponds. Filamentous organisms and Actinomycetes will naturally stick to solid surfaces. Common types of Cyanobacteria are: *Oscillatoria, Anabaena, and Synechococcus.* Other filament formers include: *Spirogyra, Cladophora, Rhizoclonium, Mougeotia, Zygnema and Hydrodictyon. Nocardia* is another filament former, which causes foaming and interferes with flocculation in a waste treatment plant.

Filter aid: A chemical (usually a polymer) added to water to help remove fine colloidal suspended solids.

Filter life: Measure of the duration of a filter's useful service. This is based on the amount of standard contaminant required to cause differential pressure to increase to an unacceptable level–typically 2-4 times the initial differential pressure, a 50-80% drop in initial flow, or a downstream measure of unacceptable particulate.

Filter media: A porous material for separating suspended particulate matter from fluid.

Filter medium: The permeable portion of a filtration system that provides the liquid-solid separation, such as screens, papers, non-wovens, granular beds and other porous media.

Filtrate: The discharge liquor in filtration.

Filtration: A process of separating particulate matter from a fluid by passing it through a permeable material.

Floating matter: Matter which passes through a 2000 micron sieve and separates by flotation for an hour.

Floc: Clumps of bacteria and particulate impurities or coagulants that have come together and formed a cluster. Found in aeration tanks and secondary clarifiers.

Flocculation: The process of forming floc particles when a chemical coagulant or

flocculent such as alum or ferric chloride is added to the wastewater.
Flow rate: Measure of the amount of fluid passing through the filter. This is always a variable of filter area, porosity, contamination and differential pressure.
F: Food - represents BOD in the F/M ratio. Expressed in pounds.
FOG: Fats, Oils and Greases. A measure of the non-petroleum based fats in waste treatment.
F/M: A ratio of the amount of food to the amount of organisms. Used to control an activated sludge process.
Flow equalization system: A device or tank designed to hold back or store a portion of peak flows for release during low-flow periods.
Food chain: Very simple pathway of nutrient flow. Ex. Carnivore > herbivore > plant.

Frazier test: Measures the amount of air transmitted through a filter under selected differential pressures. Historically used for textile products.
Frequency: Number of complete cycles of current per second, expressed in Hertz (Hz). Megahertz (MHz) is 106 Hz.

G

Gasification: The conversion of soluble and suspended materials into gas during anaerobic decomposition. In clarifiers the resulting gas bubbles can become attached to the settled sludge and cause large clumps of sludge to rise and float on the water surface. In anaerobic sludge digesters, this gas is collected for fuel or disposed of using a waste gas burner.
Generation time: The time required for a given population to double in size. This time can be as short as 20 minutes or as long as a week.
Glyoxylate cycle: A modification of the Krebs cycle, which occurs in some bacteria. Acetyl coenzyme A is generated directly from oxidation of fatty acids or other lipid compounds.
GMPs: Good Manufacturing Practices. Food and Drug Administration regulations governing the manufacture of drugs and medical devices (Ref. Code of Federal Regulations 21CFR).
Gram positive: Bacterial cells which retain the crystal violet stain during a staining procedure. Most strains of bacilli are gram positive.
Gram negative: Bacteria cells which lose the crystal violet during the decolorizing step and are then colored by the counterstain. *Pseudomonas* and *Thiobacillus* are examples of gram negative strains.
Grit: The heavy material present in wastewater, such as sand coffee grounds, eggshells, gravel and cinders.

H

Halophilic or Halotolerant: Bacteria which thrive in a highly salt environment, up to 25% NaCl.
Headworks: The facilities where wastewater enters a wastewater treatment plant. The headworks may consist of bar screens, comminutors, a wet well and pumps.
Heterotroph: A microorganism which uses organic matter for energy and growth.
HRT: Hours of Retention Time.
House Sewer: The pipeline connecting the house and drain and the septic tank.

Humus: The dark organic material in soils, produced by the decomposition of soils. The matter that remains after the bulk of detritus has beenconsumed (leaves, roots). Humus mixes with top layers of soil (rock particles), supplies some of the nutrients needed by plants -increases acidity of soil; inorganic nutrients more soluble under acidic conditions, become more available, EX. wheat grows best at pH 5.5-7.0. Humus modifies soil texture, creates loose, crumbly texture, that allows water to soak in and nutrients retained; permits air to be incorporated into soil.
Hydraulic loading: Hydraulic loading refers to the flows (MGD or m3/day) to a treatment plant or treatment process.
Hydrogen sulfide gas: Hydrogen sulfide is a gas with a rotten egg odor. This gas is produced under anaerobic conditions. Hydrogen sulfide is particularly dangerous because it dulls your sense of smell so that you don't notice it after you have been around it for a while and because the odor is not noticeable in high concentrations. The gas is very poisonous to your respiratory system, explosive, flammable, and colorless.
Hydrolysis: The process in which carbohydrates and starches are simplified into organic soluble organics, usually by facultative anaerobes.
Hydrophilic: Having an affinity for water and aqueous solutions.
Hydrophobic: Cannot be wetted by aqueous and other high surface tension fluids.

Hygroscopic: Absorbing or attracting moisture from the air.

I

Incineration: The conversion of dewatered wastewater solids by combustion (burning) to ash, carbon dioxide, and water vapor.
Infiltration: The seepage of groundwater into a sewer system, including service connections. Seepage frequently occurs through defective or cracked pipes, pipe joints, connections or manhole walls.
Influent: The liquid - raw (untreated) or partially treated - flowing into a reservoir, basin, treatment process or treatment plant.
Inoculate: To introduce a seed culture into a system.

Inorganic waste: Waste material such as sand, salt, iron, calcium, and other mineral materials which are only slightly affected by the action of organisms. Inorganic wastes are chemical substances of mineral origin; whereas organic wastes are chemical substances usually of animal or plant origin.
Interface: The common boundary layer between two substances such as between water and a solid (metal) or between water and a gas (air) or between a liquid (water) and another liquid (oil).
Intraspecies: Within same species; Elk vs. Elk.
Interspecies: Between two different species, such as tomato and weeds.

Ionization: The process of adding electrons to, or removing electrons from, atoms or molecules, thereby creating ions. High temperatures, electrical discharges, and nuclear radiation can cause ionization.

K

Kick net: 500 micron white mesh net is designed to meet the requirements of groups performing USEPA Rapid Bioassessment Protocols for Benthic Invertebrates. (Benthic = bottom dwelling).
Knit fabric: A fabric structure made by interlooping yarns.
Krebs Cycle: The oxidative process in respiration by which pyruvate (via acetyl coenzyme A) is completely decarboxylated to CO_2. The pathway yields 15 moles of ATP (150,000 calories).

L

Liquefaction: The conversion of large solid particles of sludge into very fine particles which either dissolve or remain suspended in wastewater.
Loaded (plugged): A filter element that has collected a sufficient quantity of insoluble contaminants such that it can no longer pass rated flow without excessive differential pressure.
Log growth: A growth phase in which cell production is maximum.
Lysing: A disintegration or breakdown of cells which releases organic matter.

M

MacConkey Streak: Laboratory test for the presence of gram negative bacteria.

We use this test to detect contamination of *Bacillus* products such as CF 1000, 1002, 4002 and som of the Enz-Odor® products.
Macronutrient: An element required in large proportion by plants and other life forms for survival and growth. Macronutrients include Nitrogen (N), Potassium (K), and Phosphorous (P).
Market grade (wire cloth): Screens that meet general industrial specifications and have a low percentage of open area.
Masking agent: Substance used to cover up or disguise unpleasant odors. Liquid masking agents are dripped into wastewater, sprayed into the air, or evaporated (using heat) with the unpleasant fumes or odors and then discharged into the air by blowers to make an undesirable odor less noticeable.
M: Microorganisms - small organisms which require a microscope to be seen. M represents the SS in the mixed liquor and is part of the F/M ratio.
MCRT: Mean Cell Retention Time - days. An expression of the average time that a microorganism will spend in an activated sludge process.
Mean filtration rating: Derived from Bubble Point test method. This data should be used as a guide only to compare overall retention capabilities between fabrics and should not be considered a guarantee of the particle size that the fabric will retain.
Mechanical aeration: The use of machinery to mix air and water so that oxygen can be absorbed into the water. Some examples are paddle wheels, mixers, rotating brushes to agitate the surface of an aeration tank; pumps to create fountains; and pumps to discharge water down a series of steps forming falls or cascades.
Media: The material in the trickling filter on which slime accumulates and organisms grow. As settled wastewater trickles over the media, organisms in the slime remove certain types of wastes thereby partially treating the wastewater. Also the material in a rotating biological contactor (RBC) or in a gravity or pressure filter.

MEK: Methyl ethyl ketone.
Membrane: A thin polymeric film with pores.
Mercaptans: Compounds containing sulfur which have an extremely offensive skunk-like odor. Also sometimes described as smelling like garlic or onions.
Mesh count: The number of threads in a linear inch of fabric/wire cloth.
Mesh opening: See **Pore size**.

Mesophilic bacteria: A group of bacteria that grow and thrive in a moderate temperature range between 68 F (20 C) and 113 F (45 C).
Metabolism: All of the processes or chemical changes in an organism or a single cell by which food is built up (anabolism) into living protoplasm and by which protoplasm is broken down (catabolism) into simpler compounds with the exchange of energy.
Metalized screens: Screens that have been metalized with nickel. These screens will bleed off static charges, promote EMC and reflect electromagnetic energy.
MGD: Million gallons daily - refers to the flow through a waste treatment plant.

Mg/L: Milligrams per liter = ppm (parts per million) - expresses a measure of the concentration by weight of a substance per unit volume.
Micron: A unit of length. One millionth of a meter or one thousandth of a millimeter. One micron equals 0.00004 of an inch.

Micronutrient: An element required by plants and bacteria, in proportionately smaller amounts, for survival and growth. Micronutrients include: Iron (Fe), Managanese (MN), Zinc (Zn), Boron (B), and Molybdenum (Mo).
MIK: Methyl Isobutyl Ketone.
Mill grade (wire cloth): Stainless steel screens that have moderate open area and good strength.
Molecule: The smallest division of a compound that still retains or exhibits all the properties of the substance.
Monoculture: Aquaculture in which one species is grown.
Motile: Motile organisms exhibit or are capable of movement.
ML: Mixed Liquor - the combination of raw influent and returned activated sludge. (no, not mixed drinks for human consumption)
MLSS: Mixed Liquor Suspended Solids - the volume of suspended solids (see SS) in the mixed liquor (see ML) of an aeration tank.
MLVSS: Mixed Liquor Volatile Suspended Solids - the volume of organic solids that can evaporate at relatively low temperatures (550 C) from the mixed liquor of an aeration tank. This volatile portion is used as a measure or indication of microorganisms present. Volatile substances can also be partially removed by air stripping.
MPN index: Most Probable Number of coliform-group organisms per unit volume of sample water. Expressed as a density or population of organisms per 100 mL of sample water.
MSDS: Material Safety Data Sheet - a document that provides pertinent information and a profile of a particular hazardous substance or mixture. An MSDS is normally developed by the manufacturer or formulator of the hazardous substance or mixture. The MSDS is required to be made available to employees and operators whenever there is the likelihood of the hazardous substance or mixture being introduced into the workplace. Some manufacturers prepare MSDS for products that are not considered to be hazardous to show that the product or substance is not hazardous.

Mutualism: Two species living together in a relationship in which both benefit from the association.

N

NPDES Permit: **N**ational **P**ollutant **D**ischarge **E**limination **S**ystem permit is the regulatory agency document issued by either a federal or state agency which is

designated to control all discharges of pollutants from point sources into U.S. waterways. NPDES permits regulate discharges into navigable waters from all point sources of pollution, including industries, municipal wastewater treatment plants, sanitary landfills, large agricultural feed lots and return irrigation flows.
Nitrification: An aerobic process in which bacteria change the ammonia and organic nitrogen in wastewater into oxidized nitrogen (usually nitrate). The second-stage BOD is sometimes referred to as the "nitrification stage" (first-stage BOD is called the "carbonaceous stage").
Nitrifying bacteria: Bacteria that change the ammonia and organic nitrogen in wastewater into oxidized nitrogen (usually nitrate).
Nitrogen fixation: Conversion of atmospheric nitrogen into organic nitrogen compounds available to green plants; a process that can be carried out only by certain strains of soil bacteria.
Non-woven: A porous web or sheet produced by mechanically, chemically or thermally bonding together polymers, fibers or filaments.
Nucleic acid: An organic acid consisting of joined nuceleotide complexes; the principal tyes are deoxyribonucleic acid (DNA) and ribonucleic acid (RNA).
Nutrients: Substances which are required to support living plants and organisms. Major nutrients are carbon, hydrogen, oxygen, sulfur, nitrogen and phosphorus. Nitrogen and phosphorus are difficult to remove from wastewater by conventional treatment processes because they are water soluble and tend to recycle.

O

Obligate aerobe: Bacteria which require the presense of oxygen, such as Pseudomonas flourescens. A few strains of this species are capable of utilizing nitrate to allow anaerobic respiration.
Oil Retention Boom: A floating baffle used to contain and prevent the spread of floating oil on a water surface.
Open area: The proportion of total screen area that is open space. Expressed as a percent.
Organic matter: All of the degradable organics. Living material containing carbon compounds. Used as food by microorganisms.
Organic nitrogen: The nitrogen combined in organic molecules such as proteins, amines, and amino acids.
ORP: Oxidation reduction potential - the degree of completion of a chemical reaction by detecting the ratio of ions in the reduced form to those in the oxidized form as a variation in electrical potential measured by an ORP electrode assembly.
OSHA: The Williams-Steiger Occupational Safety and Health Act of 1970 (OSHA) is a law designed to protect the health and safety of industrial workers and treatment plant operators. It regulates the design, construction, operation and maintenance of industrial plants and wastewater treatment plants. The Act does not apply directly

to municipalities, EXCEPT in those states that have approved plans and have asserted jurisdiction under Section 18 of the OSHA Act. Wastewater treatment plants have come under stricter regulation in all phases of activity as a result of OSHA standards, which also refers to the federal and state agencies which administer OSHA.

Organic waste: Waste material which comes mainly from animal or plant sources. Organic waste generally can be consumed by bacteria and other small organisms. Inorganic wastes are chemical substances of mineral origin.

Organism: Any form of animal or plant life.

Oxidation: Combining elemental compounds with oxygen to form a new compound. A part of the metabolic reaction.

Oxidizing bacteria: Any substance such as oxygen (O_2) and chlorine (Cl_2), that can accept electrons. When oxygen or chlorine is added to wastewater, organic substances are oxidized. These oxidized organic substances are more stable and less likely to give off odors or to contain disease bacteria.

Ozonation: The application of ozone to water, wastewater, or air, generally for the purposes of disinfection or odor control.

P

Parisitism: One organism living on or in another to obtain nourishment, without provviding any benefit to the host organism..

Particle: A relatively small subdivision of matter ranging in diameter from a few angstroms (as with gas molecules) to a few millimeters (as with large raindrops). The particle can have various shapes and dimensions.

Particulate: Free suspended solids.

Pathogenic organisms: Bacteria, viruses or cysts which cause disease (typhoid, cholera, dysentery) in a host (such as a person). There are many types of bacteria (non-pathogenic) which do NOT cause disease. Many beneficial bacteria are found in wastewater treatment processes actively cleaning up organic wastes.

PAH: Polycyclic Aromatic Hydrocarbons. (rarely used as abbreviation for polyaluminum hydroxide)

PCB: Polychlorinated biphenyls. Aka polychloro-biphenyls. Difficult to remediate chemical used in old-style transformers. Concentrated PCBs used to be referred to as "1268".

Percolation: The movement or flow of water through soil or rocks.

Peristaltic pump: A type of positive displacement pump.

Permeability: Ability of a membrane or other material to permit a substance to pass through it.

pH: pH is an expression of the intensity of the basic or acidic condition of a liquid.

Mathematically, pH is the logarithm (base 10) of the reciprocal of the hydrogen ion concentration. The pH may range from 0 to 14, where 0 is most acidic, 14 most basic, and 7 is neutral. Natural waters usually have a pH between 6.5 and 8.5.

Phenol: An organic compound that is an alcohol derivative of benzene.

Phototroph: A microorganism which gains energy from sunlight (radiant energy).

PIB: Product Information Bulletin. General information on a product.

Pin Floc: Excessive solids carryover. May occur from time to time as small suspended sludge particles in the supernatant. There are two kinds: grey -ashlike, inert, has low BOD - indicates old sludge; and brown, but a portion neither settles nor rises, has high BOD - indicates young sludge.

Plain weave: Another name for square weaves. See **weave patterns**.

Plane wave: An electromagnetic wave with electric and magnetic components perpendicular to, and in phase with, each other.

ppm: Parts Per Million - the unit commonly used to designate the concentration of a substance in a wastewater in terms of weight ie. one pound per million pounds, etc. ppm is synonymous with the more commonly used term mg/L (milligrams per liter).

Pollution: The impairment (reduction) of water quality by agriculture, domestic or industrial wastes (including thermal and radioactive wastes) to such a degree as to hinder any beneficial use of the water or render it offensive to the senses of sight, taste, or smell or when sufficient amounts of waste creates or poses a potential threat to human health or the environment.

Polyculture: Fish farming in which 2 or more compatible or symbiotic species of fish are grown together. Also known as Multiculture.

Polymer: A chemical formed by the union of many monomers (a molecule of low molecular weight). Polymers are used with other chemical coagulants to aid in binding small suspended particles to form larger chemical flocs for easier removal from water. All polyelectrolytes are polymers, but not all polymers are polyelectrolytes.

Pore size: The distance between two adjacent warp or weft threads, measured in the projected plane. Only applies to fabrics above 10 microns.

Potable water: Water that does not contain objectionable pollution, contamination, minerals, or infective agents and is considered satisfactory for drinking.

POWT: Publicly Owned Treatment Works, as opposed to an industrially owned facility or pipe system.

Predation: One species benefits at the expense of another.

Preliminary treatment: The removal of metal, rocks, rags, sand, eggshells, and similar materials which may hinder the operation of a treatment plant. Preliminary treatment is accomplished by using equipment such as racks, bar screens, comminutors, and grit removal systems.

Pretreatment facility: Industrial wastewater treatment plant consisting of one or more treatment devices designed to remove sufficient pollutants from wastewaters to allow an industry to comply with effluent limits established by the US EPA General and Categorical Pretreatment Regulations or locally derived prohibited discharge requirements and local effluent limits. Compliance with effluent limits allows for a legal discharge to a POTW.

PRD: Plain Reverse Dutch weave. See **weave patterns**.

Primary treatment: A wastewater treatment process that takes place in a rectangular or circular tank and allows those substances in wastewater that readily settle or float to be separated from the water being treated.

Procaryotic organism: Microorganisms which do NOT have an organized nucleus surrounded by a nuclear membrane. Bacteria and blue-green algae fit in this category.

Protozoa: A group of motile microscopic animals (usually single-celled and aerobic) that sometimes cluster into colonies and often consume bacteria as an energy source.

Psychrophilic bacteria: Bacteria whose optimum temperature range is between 0 and 20° C (32 to 68° F).

Putrefaction: Biological decomposition of organic matter with the production of ill-smelling products associated with anaerobic conditions.

Pyrogenic: A fever-producing substance. The presence of these substances is determined by the Limulus Amebocyte Lysate (LAL) test and measured in EU/ml (endotoxin units per milliliter).

R

Rack: Evenly spaced parallel metal bars or rods located in the influent channel to remove rags, rocks, and cans from wastewater.

Radio frequency interference (RFI): EMI in electronic equipment caused by radio frequencies, ranging typically from 10 kHz (104 Hz) to 1000 MHz (109 Hz or 1 GHz).

RAS: Return activated sludge - settled activated sludge that is collected in the secondary clarifier and returned to the aeration basin to mix with incoming raw settled wastewater.

RASVSS: Return Activated Sludge Volatile Suspended Solids.

RBC: Rotating biological contactor - an attached culture wastewater treatment system..

Reagent: A pure chemical substance that is used to make new products or is used in chemical tests to measure, detect, or examine other substances.

Recycle: The use of water or wastewater within (internally) a facility before it is discharged to a treatment system.

REDOX: Biological reductions/oxidations. These reactions usually require enzymes to mediate the electron transfer. The sediment in the bottom of a lake, sludge in a sewerage works or septic tank will have a very low redox potential and will likely be devoid of any oxygen. This sludge or waste water will have a very high concentration of reductive anaerobic bacteria, indeed the bulk of the organic matter may in fact be bacteria. As the concentration of oxygen increases the oxidation potential of the water will increase. A low redox potential or small amount of oxygen is toxic to anaerobic bacteria, therefore as the concentration of oxygen and redox potential increases the bacterial population changes from reductive anaerobic bacteria to oxidative aerobic bacteria. Measurement of REDOX potential is also referred to as ORP.

Reducing agent: Any substance, such as the base metal (iron) or the sulfide ion that will readily donate (give up) electrons. The opposite of an oxidizing agent.
Refractory materials: Material difficult to remove entirely from wastewater such as nutrients, color, taste, and odor-producing substances and some toxic materials.
Residual shrinkage: The amount of shrinkage remaining in a fabric after it has undergone all fabric weaving, washing and heat setting steps.
Respiration: The energy producing process of breathing, by which an organism supplies its cells with oxygen and relieves itself of carbon dioxide. A type of heterotrophic metabolism that uses oxygen in which 38 moles of ATP are derived from the oxidation of 1 mole of glucose, yielding 380,000 cal. (An additional 308,000 cal is lost as heat.)
Rhizosphere: Soil surrounding plant roots.
Retentate: Substance retained in the upstream side of a filter.
RF (radio frequency) welding: Utilizes specific bands of radio frequency waves which are directed through specially constructed tooling to form localized melting/joining of certain dielectric thermoplastic materials. Can be used to form hermetic seals. Also known as high frequency or dielectric welding.
R/O unit: Reverse Osmosis Unit for water purification in small aquariums and miniature yard-ponds, utilizes a membrane under pressure to filter dissolved solids and pollutants from the water. Two different filter membranes can be used: the CTA (cellulose triacetate) membrane is less expensive, but only works with chlorinated water and removes 50-70% of nitrates, and the TFC membrane, which is more expensive, removes 95% of nitrates, but is ruined by chlorine. R/O wastes water and a system that cleans 100 gallons a day will cost from $400 to $600 with membrane replacement adding to the cost. A unit that handles 140 gallons a day will cost above $700.00.
RR: Respiration rate - the weight of oxygen utilized by the total weight of MLSS in a given time.
Runoff: Water running down slopes rather than sinking in (again,result of poor humus content) Ex. erosion due to deforestation

S

Saprophytic: Bacteria that breakdown bodies of dead plants and animals (non-living organic material), returning organic materials to the food chain. Saprophytic bacteria are usually non-pathogenic, too. Most Alken Clear-Flo® products are saprophytic.

SAR: Sodium Adsorption Ratio - this ratio expresses the relative activity of sodium ions in the exchange reactions with the soil.

SCFM: Cubic feet of air per minute at standard conditions of temperature, pressure and humidity (0 , 14.7 psi and 50% relative humidity).

Secondary Treatment: A wastewater treatment process used to convert dissolved or suspended materials into a form more readily separated from the water being treated. Usually the process follows primary treatment by sedimentation. The process commonly is a type of biological treatment process followed by secondary clarifiers that allow the solids to settle out from the water being treated.

Sedimentation: The process of subsidence and deposition of suspended matter from a wastewater by gravity.

Seeding: Introduction of microorganisms (such as ALKEN CLEAR-FLO® 1000 series for aquaculture, 4000 series for grease, and 7000 series for industrial and municipal wastewater) into a biological oxidation unit to minimize the time required to build a biological sludge. Also referred to as inoculation with cultured organisms.

Seine net: A net designed to collect aquatic organisms inhabiting natural waters from the shoreline to 3' depths is called a seine net. Most often a plankton seine.

Selvage: A loom finished edge that prevents cloth unravelling.

Septic: A condition produced by anaerobic bacteria. If severe, the wastewater turns black, gives off foul odors, contains little or no dissolved oxygen and creates a high oxygen demand.

Septicity: Septicity is the condition in which organic matter decomposes to form foul-smelling products associated with the absence of free oxygen. If severe, the wastewater turns black, gives off foul-odors, contains little or no dissolved oxygen and creates a heavy oxygen demand.

Septic Tank: Untreated liquid household wastes (sewage) will quickly clog your absorption field if not properly treated. The septic tank is a holding tank in which this treatment can take place. When sewage enters the septic tank, the heavy solids settle to the bottom of the tank; the lighter solids, fats and greases partially decompose and rise to the surface and form a layer of scum. The solids that have settled to the bottom are attacked by bacteria and form sludge.

Settleable solids: Those solids in suspension which will pass through a 2000 micron sieve and settle in one hour under the influence of gravity.

Sewage: The used water and water-carried solids from homes that flow in sewers to a wastewater treatment plant. The preferred term is wastewater.

Shock load: The arrival at a plant of a waste which is toxic to organisms in sufficient quantity or strength to cause operating problems. Possible problems include odors and sloughing off of the growth or slime on a trickling-filter media. Organic or hydraulic overloads also can cause a shock load.
Sieve: A screen with apertures of uniform size used for sizing granular materials.
Sloughings: Trickling-filter slimes that have been washed off the filter media. They are generally quite high in BOD and will lower effluent quality unless removed.
Sludge: The settleable solids separated from liquids during processing; the deposits of foreign materials on the bottoms of streams or other bodies of water.
Sludge age: A measure of the length of time a particle of suspended solids has been retained in the activated sludge process.
Slugs: Intermittent releases or discharges of industrial wastes.
Soluble: Matter or compounds capable of dissolving into a solution.
Soluble BOD: Soluble BOD is the BOD of water that has been filtered in the standard suspended solids test.
Solution: A liquid mixture of dissolved substances, displaying no phase separation.
Specific gravity: Weight of a particle, substance or chemical solution in relation to an equal volume of water.
Spec. Sheet: Specification Sheet. Detailed information of a product including, tests, color, odor, specific gravity, bacterial strains, other major ingredients, etc.

Surface media: Captures particles on the upstream surface with efficiencies in excess of depth media, sometimes close to 100% with minimal or no off-loading. Commonly rated according to the smallest particle the media can repeatedly capture. Examples of surface media include ceramic media, microporous membranes, synthetic woven screening media and in certain cases, wire cloth. The media characteristically has a narrow pore size distribution.
Surface resistivity (W/o): Expressed in ohms/square. It is numerically equal to the resistance between two electrodes forming opposite sides of a square on the surface of a material. The size of the square is irrelevant. For conductive materials, surface resistivity is the ratio of the volume resistivity to the fabric thickness (r/t).

T

Tangential crossflow filtration: Process where the feed stream "sweeps" the membrane surface and the particulate debris is expelled, thus extending filter life. The filtrate flows through the membrane. Most commonly used in the separation of high-and-low-molecular weight matter such as in ultrapure reverse osmosis, ultrafiltration, and submicron microfiltration processes.

Taxonomy: The classification, nomenclature, and laboratory identification of organisms (Do not confuse with taxidermy - stuffing dead animals)
TDS: Total Dissolved Solids is commonly estimated from the electrical conductivity

of the water. Pure water is a poor conductor of electricity. Impurities dissolved in the water cause an increase in the ability of the water to conduct electricity. Conductivity, usually expressed in units of microsimens, formerly micromhos or in mg/l, thus becomes an indirect measure of the level of impurities in the water.
TOC: Total organic carbon - a measure of the amount of organic carbon in water.
Thermophilic bacteria: Hot temperature bacteria. a group of bacteria that grow and thrive in temperatures above 113° F (45° C), such as bacillus licheniformis. The optimum temperature range for these bacteria in anaerobic decomposition is 120° F (49° C) to 135° F (57° C).
Thread diameter: The cross-sectional measurement of an individual fabric thread/yarn or wire.
Throughput: The amount of solution which will pass through a filter prior to clogging.
Toxic: A substance which is poisonous to a living organism.
Toxicity: The relative degree of being poisonous or toxic. A condition which may exist in wastes and will inhibit or destroy the growth or function of certain organisms.
Transpiration: The process by which water vapor is released to the atmosphere by living plants, a process similar to people sweating.

Trickling filter: An attached culture wastewater treatment system. A large tank generally filled with rock or rings (see Bio-Tower). Wastewater is sprayed over the top of the media, providing the opportunity for the formation of slimes or biomass to remove wastes from the wastewater, through revolving arms which have spray nozzles. Water is pumped from the bottom of a trickle filter to a secondary clarifier.
TSS: Total suspended solids.
Turbidity: The amount of suspended matter in wastewater, obtained by measuring its light scattering ability.

U

Unicellular: Single celled organism, such as bacteria.

Upset: An upset digester does not decompose organic matter properly. The digester is characterized by low gas production, high volatile acid/alkalinity relationship, and poor liquid-solids separation. A digester in an upset condition is sometimes called a "sour" or "stuck" digester.

Ultrasonic (processes): Process which utilizes specially designed tooling usually vibrating at 15-80 KHz. Processes are designed to cause localized heating of thermoplastic materials which, in turn, will provide some type of welded or fused joint. Benefits are elimination of fillers and minimized heat stress on surrounding materials.

Upstream side: The feed side of the filter.

Uronic acid: Class of acidic compounds of the general formula $HOOC(CHOH)_nCHO$ that contain both carboxylic and aldehydic groups, are oxidation products of sugars, and occur in many polysaccharides; especially in the hemicelluloses.

Useful life: Determined when contamination causes an adverse flow rate, low efficiency or high differential pressure.

V

Vadose zone: Unsaturated zone of soil above the groundwater, extending from the bottom of the capillary fringe all the way to the soil surface.

Vector: (i) Plasmid or virus used in genetic engineering to insert genes into a cell. (ii) Agent, usually an insect or other animal, able to carry pathogens from one host to another.

Vegetative: Actually growing state.

Vegetative cell: Growing or feeding form of a microbial cell, as opposed to a resting form such as a spore.

Vesicles: Spherical structures, formed intracellularly, by some arbuscular mycorrhizal fungi.

Viable: Alive; able to reproduce.

Viable but nonculturable: Organisms that are alive but cannot be cultured on laboratory media.

Viable count: Measurement of the concentration of live cells in a microbial population.

Vibrio: (i) Curved, rod-shaped bacterial cell. (ii) Bacterium of the genus *Vibrio*.

Virion: Virus particle; the virus nucleic acid surrounded by protein coat and in some cases other material.

Virulence: Degree of pathogenicity of a parasite.

Virus: Any of a large group of submicroscopic infective agents that typically contain a protein coat surrounding a nucleic acid core and are capable of growth only in a living cell.

Volatile: A volatile substance is one that is capable of being evaporated or changed to a vapor at a relatively low temperature. Volatile substances also can be partially removed by air stripping.

Volume resistivity: Or specific resistivity of a material, expressed in W/cm. Resistance to electrical current flow through the bulk of an object.

VS/L: Measure of volatile solids, usually expressed as g VS/L/day-grams volatile solids per liter per day.

W

WAS: Waste activated sludge, mg/L. The excess growth of microorganisms which must be removed from the process to keep the biological system in balance.
Wastewater: The used water and solids from a community that flow to a treatment plant. Storm water, surface water, and groundwater infiltration also may be included in the wastewater that enters a wastewater treatment plant. The term "sewage" usually refers to household wastes, but this word is being replaced by the term "wastewater".
Water content: Water contained in a material expressed as the mass of water per unit mass of oven-dry material.
Water-retention curve: Graph showing soil-water content as a function of increasingly negative soil water potential.
Weathering: All physical and chemical changes produced in rock by atmospheric agents.

Weir: A wall or plate placed in an open channel and used to measure the flow of water.

White rot fungus: Fungus that attacks lignin, along with cellulose, and hemicellulose, leading to a marked lightening of the infected wood.
Wild type: Strain of microorganism isolated from nature. The usual or native form of a gene or organism.
Winogradsky column: Glass column with an anaerobic lower zone and an aerobic upper zone, which allows growth of microorganisms under conditions similar to those found in nutrient-rich water and sediment.
Woronin body: Spherical structure associated with the simple pore in the septa separating hyphal compartments of fungi in the phylum Ascomycota.

X

Xenobiotic: Compound foreign to biological systems. Often refers to human-made compounds that are resistant or recalcitrant to biodegradation and decomposition.

Xerophile: Organism adapted to grow at low water potential, i.e., very dry habitats.

Y

Yeast: Fungus whose thallus consists of single cells that multiply by budding or fission.

Z

Zoogleal film: A complex population of organisms that form a "slime growth" on a trickling-filter media and break down the organic matter in wastewater.

Zoogleal mass: Jelly-like masses of bacteria found in both the trickling filter and activated sludge processes.

Zoospore: An asexual spore formed by some fungi that usually can move in an aqueous environment via one or more flagella.

Zygospore: Thick-walled resting spore resulting from fusion of two gametangia of fungi in the phylum Zygomycota.

Zygote: In eukaryotes, the single diploid cell resulting from the union (fusion) of two haploid gametes.

Zymogenous flora: Refers to microorganisms, often transient or alien, that respond rapidly by enzyme production and growth when simple organic substrates become available. Also called *copiotrophs*.

INDEX